V International Scientific and Technical Conference "Energy Systems" (ICES 2020)

IOP Conference Series: Materials Science and Engineering
Volume 1089

Belgorod, Russia
19-20 November 2020

ISBN: 978-1-7138-4590-4
ISSN: 1757-8981

Printed from e-media with permission by:

Curran Associates, Inc.
57 Morehouse Lane
Red Hook, NY 12571

Some format issues inherent in the e-media version may also appear in this print version.

Institute of Physics
Dirac House, Temple Back
Bristol BS1 6BE UK

Phone: 44 1 17 929 7481
Fax: 44 1 17 920 0979

techtracking@iop.org

Additional copies of this publication are available from:

Curran Associates, Inc.
57 Morehouse Lane
Red Hook, NY 12571 USA
Phone: 845-758-0400
Fax: 845-758-2633
Email: curran@proceedings.com
Web: www.proceedings.com

TABLE OF CONTENTS

IOP Conf. Series: Materials Science and Engineering 1089 (2021) 011001 doi:10.1088/1757-899X/1089/1/011001

Preface

Information about the conference
The V International Scientific and Technical Conference “Energy systems” (ICES - 2020) was hosted by the federal state budget institution of higher education “Belgorod State Technological University named after V. G. Shukhov” (BGTU named after V.G. Shukhov) on November 19-20, 2020.

The conference suggested the solution to fundamental scientific problems in the field of energy. The main point is a systematic approach. The studied systems are complex and require the management of large systems and integration of knowledge. Also, one of the objectives of the conference was the wide involvement of young scientists – students, undergraduates, and graduate students.

Conference website: http://ensyscon.ru/.

The working languages of the conference were English and Russian. 78 reports were selected for the conference. As a result of the conference 33 Russian-language reports were prepared in the main section of the conference, 56 reports were in English. This volume contains reports in English. Reports in Russian have been published in the journal "Energy Systems" (http://j-es.ru/).

Due to the COVID-19 epidemic the organizing committee decided to hold NMCI 2020 virtually/online in ZOOM.

Fundamental and practical issues in the scientific fields of heat technology, energy, and resource-saving in industrial production have been considered.

The conference demonstrated further approval of innovative approaches which are based on the fundamental research. The identified problems and new challenges that help to strengthen interdisciplinary ties in the energy industry have been shown. It was demonstrated that the domestic industry is in ever-increasing need of modern scientific and technical developments, as well as strengthening of the interaction between universities and industry.

The young scientists have demonstrated a great interest in this conference, which were coauthors of the work of their senior colleagues, and the sole authors of the reports.

Belgorod, Russia, 2020
Prof. Pavel Trubaev
E-mail trubaev@gmail.com

IOP Conf. Series: Materials Science and Engineering 1089 (2021) 011001 doi:10.1088/1757-899X/1089/1/011001

Conference sections

The main problems of the research were determined by the selected sections based on the analysis of the received applications:

Topic 1. Heat and power engineering (10 reports)

The fundamental and practical issues in the scientific fields of heat and mass transfer processes, heat generation, heat conversion to other types of energy, providing consumers with energy resources were considered.

Speaker
Alexey Biryukov, DSc

Speaker
Michael Averbukh, DSc

Topic 2. Electric power and electrical engineering (12 reports)

Fundamental and practical issues in the scientific fields of production, distribution, transformation, and use of electric energy, development, and optimization of electronic components were considered.

Topic 3. Energy-saving and energy efficiency, power ecology (13 reports)

Fundamental and practical questions in scientific areas of energy resources cost reduction, increases in the energy efficiency of power and heat technical systems, the energy efficiency of buildings, buildings, constructions, impacts of power on the environment, elimination of emissions and pollution, waste recycling for energy were considered.

Speaker
Pavel Trubaev, DSc

Speaker
Evgeniy Gasho, DSc

Topic 4. Alternative and renewable energy sources (12 reports)

Fundamental and practical issues in the scientific fields of production, distribution, transformation, and use of electrical energy, development, and optimization of electronic components were considered.

Topic 5. Energy and resource-saving in technologies (12 reports)

Fundamental and practical issues in the scientific fields of heat technology, energy, and resource-saving in industrial production were considered.

Speaker
Nikolay Mozgovoy, DSc

Program Committee of the Conference

- **Pavel Trubaev**, Chairman,
  Dr. of Tech. Sciences (DSc), Professor of the Department of Energy engineering of heat technology BSTU named after V. G. Shukhov (Belgorod, Russia)
- **Alexandr Belousov**, Vice-Chairman
  Cand. of Tech. Sciences (PhD), Head of the Faculty of Energy, information technology and control systems BSTU named after V. G. Shukhov (Belgorod, Russia)
- **Alexander Bazhanov**
  Cand. of Tech. Sciences (PhD), Associate Professor (Reader) of the Department of Technique cybernetics BSTU named after V. G. Shukhov (Belgorod, Russia)
- **Alexey Birukov**
  Dr. of Tech. Sciences (DSc), Professor, Head of the Department of Industrial Thermal Physics Donetsk National Technical University (Donetsk, Ukraine)
- **Evgeniy Gasho**
  Dr. of Tech. Sciences (DSc), Professor of the Department of Industrial heat and energy systems Moscow Power Engineering Institute (Moscow, Russia), Expert Analytical Center Under the Government of the Russian Federation (Moscow, Russia)
- **Marina Kozhukhova**
  PhD, Adjunct Professor University of Wisconsin-Milwaukee (Milwaukee, USA)
- **Natalya Kozhukhova**
  Cand. of Tech. Sciences (PhD), Associate Professor (Reader) of the Department of Materials science and materials technology BSTU named after V. G. Shukhov (Belgorod, Russia)
- **Konstantin Logachev**
  Dr. of Tech. Sciences (DSc), Professor of the Department of Heat and gas supply and ventilation BSTU named after V. G. Shukhov (Belgorod, Russia)
- **Nikolay Mozgovoy**
  Dr. of Tech. Sciences (DSc), Professor, Head of the Department of Industrial ecology and Safety Voronezh State Technical University (Voronezh, Russia)
- **Vadim Radchenko**
  Cand. of Econom. Sciences (PhD), Vice-chairman of the Belgorod City Council,
  CEO of the Center of Energy Technologies Ltd (Belgorod, Russia)
- **Victoria Nelyubova**
  Cand. of Tech. Sciences (Phd), Associate Professor (Reader) of the Department of Materials science and materials technology BSTU named after V. G. Shukhov (Belgorod, Russia)
- **Leonid Sivachenko**
  Dr. of Tech. Sciences (DSc), Professor of Department of Transport and Technological Machines Belarusian-Russian University (Mogilev, Belarus)
- **Valeria Strokova**
  Dr. of Tech. Sciences (DSc), Professor, Head of the Department of Materials science and materials technology BSTU named after V. G. Shukhov (Belgorod, Russia)
- **Zlata Udovichenko**
  Cand. of tech. Sciences (PhDS), Associate Professor (Reader) of the Department of Heat and gas supply and ventilation Donbass Academy of Architecture and Construction (Makeyevka, Ukraine)

Organizing Committee of the Conference

- **Sergey Glagolev**, Chairman,
  Dr. of Econ. Sciences, Rector of BSTU named after V. G. Shukhov
- **Pavel Trubaev**, Deputy Chairman,
  Dr. of Tech. Sciences, Professor of BSTU named after V. G. Shukhov
- **Alexandr Belousov**, Cand. of tech. Sciences (Phd),
  Director of the Institute of Energy, Information Technology and Control Systems of BSTU named after V. G. Shukhov
- **Natalya Kornilova**,
  Senior teacher of Department of of electricity and automation of BSTU named after V. G. Shukhov

IOP Conf. Series: Materials Science and Engineering 1089 (2021) 011002 doi:10.1088/1757-899X/1089/1/011002

Peer review declaration

All papers published in this volume of *IOP Conference Series: Materials Science and Engineering* have been peer reviewed through processes administered by the Editors. Reviews were conducted by expert referees to the professional and scientific standards expected of a proceedings journal published by IOP Publishing.

- **Type of peer review:** Double-blind review
- **Conference submission management system:** Via mail to EnergoBGTU@mail.ru
- **Number of submissions received:** 82
- **Number of submissions sent for review:** 82
- **Number of submissions accepted:** 56
- **Acceptance Rate:** 68%
- **Average number of reviews per paper:** 2
- **Total number of reviewers involved:** 24
- **Any additional info on review process:**

  Detailed information of the used peer-review process:

  The peer-review process includes technical expertise of the submitted papers, external review and paper's peer review by the program committee.

  1. Entry technical expertise.

  The submitted reports are evaluated by the Chairman of the Program Committee and manager of the organizing committee for compliance with themes of the conference and features of scientific pub-lication.

  Some papers are checked for originality by way of random fragments searching, usually abstracts and conclusions, in google.com. The papers which are on iThenticate conference hosts are unable to check because of the high cost and the small number of files.

  The allowed to proceed papers are being distributed to sections.

  2. External review.

  After the technical expertise, the papers are sent to the reviewers which are chosen according to the scientific contacts of the conference hosts. The reviewers are also chosen among the authors of the corresponding section from the previous conferences or based upon analysis of the paper's literature list.

  3. Final scientific expertise.

  The Chairman of the Program Committee or appointed by him members of the Program Committee conduct the paper's peer review considering the review. Result of the peer review:

  a) accept the paper;

  b) make improvements in the paper according to the stated by the reviewer and Program Committee members remarks;

  c) decline the paper.

- **Contact person for queries:** Prof. Pavel Trubaev, trubaev@gmail.com

IOP Conf. Series: Materials Science and Engineering 1089 (2021) 012001 doi:10.1088/1757-899X/1089/1/012001

Analysis of the dual-fuel combined cycle power plants with a parallel operating scheme efficiency

A S Voronkova and O V Borush

Department of Thermal Power Plants, Novosibirsk State Technical University, K. Marks St., 20, Novosibirsk, 630073, Russia

E-mail: voronkova.nastya@mail.ru

Abstract. The article addresses the problems of local power plants operating on fossil fuels such as gas and coal. As a solution to the problem, it is proposed to consider dual-fuel combined cycle power plants with a parallel operating scheme. This type of combined cycle technology is good variant for integration into coal predominating districts. Therefore, the efficiency of dual-fuel combined cycle power plants operating at subcritical (p_0 = 13 MPa, t_0 = 540 °C) and supercritical (p_0 = 24 MPa, t_0 = 545 °C) initial steam parameters are discussed in the article. The thermal model and a calculation methodology for the installation under consideration are proposed. The results of the installation's technical and economic indicators analysis are presented. Additionally article sets out the assessment of capital costs, Net Present Value and the possibility of using the technology under study within the framework of the considering region.

1. Introduction

Nowadays a significant part of electricity (~ 65%) in Russia is generated at thermal power plants operating on fossil fuels (gas, coal). The main problem of such stations is the low efficiency of electric power generation, which is ~ 28 – 34 % [1]. The reason for such a low electricity generation efficiency is the equipment's depreciation (due to the fact that projected service life of the equipment has been repeatedly extended) [2], the obsolescence of the equipment (most of the power plants are operated at subcritical steam parameters) [3], as well as a low percentage of implementation units using advanced technologies such as combined cycle gas turbine units (CCGT) [4]. Thus, increasing the efficiency of electric power generation is one of the key directions in the development of energy not only in Russia but also in the world [5].

The trend of recent years shows that combined-cycle power units that use natural gas as the main fuel are being implemented in many regions of the country. [6] However, for regions where coal predominates in the territorial fuel balance, for example, Siberian Federal District, Ural Federal District, Far Eastern Federal District (Figure 1), the CCGT using is not even considered. [7] One of the variant for integration coal into combined cycle technologies can be dual-fuel CCGT [8].

Thus, it is relevant to study dual-fuel combined-cycle power plants with a parallel scheme of work [9] with the designing of the scheme and calculation methodology for the technology under consideration. In addition it is also interesting to analyze the plant's operation with different initial parameters of steam [10] and to estimate capital costs.

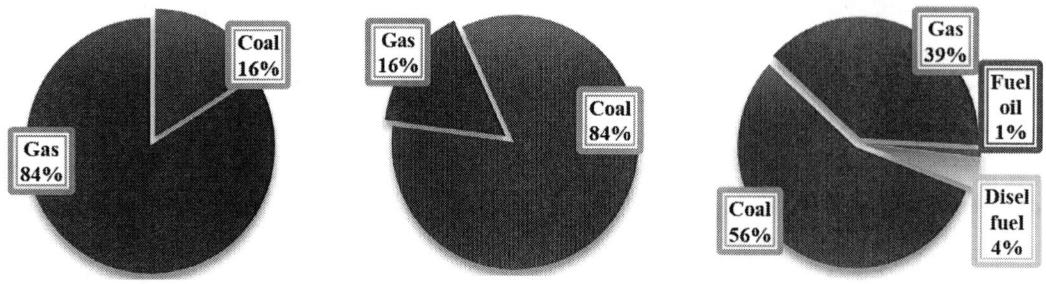

a) fuel balance of the Ural Federal District b) fuel balance of the Siberian Federal District c) fuel balance of the Far Eastern Federal District

Figure 1. Fuel balance of the coal predominate Districts.

2. Materials and Methods

Let us consider the operation principle and calculation methodology of the technology under consideration [11].

2.1. Operation principle and thermal model

The analytical thermal model of the installation is shown in Figure 2.

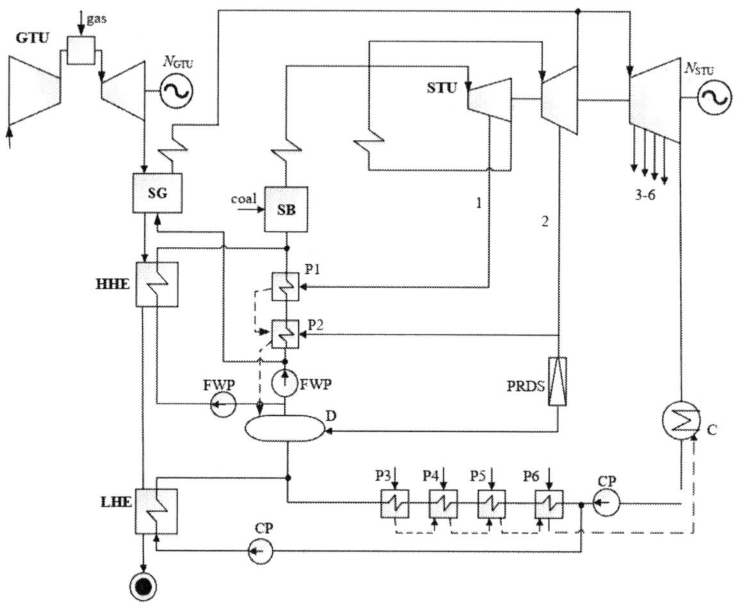

Figure 2. Example of CCPP analytical thermal model: GTU – gas turbine unit; STU – steam turbine unit; SG – steam generator of waste-heat recovery unit; HHE, LHE – heat ex-changers of high and low pressure; SB – steam boiler; P1, P2 – high-pressure preheaters; P3, P4, P5, P6 – low-pressure preheaters; D – deaerator; C – condenser; CP, FWP – condenser and feed water pump; PRDS – pressure-reducing desuperheating station; NGTU, NSTU – electrical output of gas and steam turbine unit.

Gas is supplied to the combustion chamber (CH) of the gas turbine unit (GTU), due to its combustion, electric energy is generated in a part of the gas turbine unit. The gases exhausted in the GTU are sent to the waste heat boiler (WHB) that consists of steam-generator block, where a part of the steam which is subsequently combined with the main steam flow of the steam turbine unit is generated, and two heat exchangers (high and low pressure) for heating the feed water. The steam turbine unit (STU) consists of a steam turbine with an electric generator, a regeneration system, and a steam boiler fueled by coal. Thus, the plant burns two types of fuel: gas and coal. The parallelism of work involves dividing the feed water flow, which is heated and generated into steam in the part of the STU and GTU and after this it is combined in the part of the steam turbine unit.

2.2. Calculation methodology

Combined cycle power plants calculation methodology consists of steps are described below.

2.2.1. Gas turbine unit calculation.

Determination of parameters at the main points of the Brayton cycle, calculation of specific work(l, kJ/kg), determination of efficiency (η_{GTU}) and equivalent fuel consumption of a GTU(b_{GTU}, KgCE./kilowatt-hour):

$$l = (T_0 - T_k) - (T_1 - T_*)$$

$$\eta_{GTU} = \frac{l}{(T_0 - T_1)\eta_{EM}}$$

$$b_{GTU} = \frac{0.123}{\eta_{GTU}}$$

where:

T_0 – temperature of gases after combustion chamber, K;

T_k – temperature after GTU, K;

T_1 – compressed air temperature behind the compressor, K;

T_* – ambient temperature, K;

η_{EM} – electromechanical losses.

2.2.2. Steam turbine unit calculation.

Determination of parameters at characteristic points of the steam expansion process, calculation of steam flow condition per turbine (G_0, Kg/s), efficiency (η_{STU}), equivalent fuel consumption (b_{STU}, KgCE./kilowatt-hour):

$$G_0 = \frac{k_G N_{STU} 10^3}{H_0 \eta_{oi}}$$

$$\eta_{STU} = \eta_C \eta_{SB} \eta_{oi} \eta_{EM}$$

$$b_{STU} = \frac{0.123}{\eta_{STU}}$$

where:

k_G – correction index for steam flow condition per turbine;

N_{STU} – electrical output of steam turbine unit, kW;
H_0 – available heat drop per turbine, kJ/kg;
η_{oi} – internal relative efficiency of the turbine;
η_{C} – Carnot cycle efficiency;
η_{SB} – stem boiler efficiency.

2.2.3. Waste heat boiler and steam boiler calculation.

Determination of the amount of heat transferred to each heat-delivery surface (Q_i, kW), and it's area (F_i , m^2):

$$Q_i = c_p G_i \Delta t$$

$$F_i = \frac{10^3 Q_i}{k \Delta t}$$

where:
c_p – heat capacity at constant pressure, kJ/kg;
G_i – steam/gas flow, Kg/s;
Δt – temperature difference, K;
k – heat-transfer coefficient, W/(m^{2}K).

2.2.4. Layout and calculation of the thermal circuit in the autonomy mode and in the CCGT mode.

Determination of the installation efficiency (η_{CCPP}):

$$\eta_{\mathrm{CCPP}} = \frac{N_{\mathrm{GTU}}\eta_{\mathrm{BOP}}^{\mathrm{GTU}} + N_{\mathrm{STU}}\eta_{\mathrm{BOP}}^{\mathrm{STU}}}{\left(N_{\mathrm{GTU}}\frac{0.123}{3.6\eta_{\mathrm{GTU}}} + N_{\mathrm{STU}}\frac{0.123}{3.6\eta_{\mathrm{STU}}}\right)29.3 - Q_{\mathrm{WHB}}}$$

where:
N_{GTU}, N_{STU} – electrical output of gas and steam turbine unit consequently, kW;
$\eta_{\mathrm{BOP}}^{\mathrm{GTU}}$, $\eta_{\mathrm{BOP}}^{\mathrm{STU}}$ – balance of plant needs efficiency;
Q_{WHB} – waste heat boiler supplied heat, kW.

2.3. Capital costs and the payback period

Together with technical and economic indicators another important factor in assessing the effectiveness of the application of a particular technology is capital costs and the payback period.

Within the framework of the study, the assessment of the power unit under consideration cost was made on the basis of recommendations for assessing specific capital costs and their distribution over equipment. [12]

It is proposed to evaluate the efficiency of commissioning a new power unit using net present value (NPV) [13]:

$$NPV = \sum_{t=0}^{N} \frac{CF^t}{(1+i)^t} - IC$$

where:

t – the period of time for which it is necessary to calculate the net present value;
N – the number of periods (months, quarters, years) for which the estimated project needs to be calculated;
CF^t – expected cash flow (net) for a specified time period;
i – the estimated discount rate for the estimated investment option;
IC – the amount of initial investment.

3. Results

According to proposed methodology calculations were performed to analyze the technical and economic efficiency of dual-fuel CCPP with a parallel operation scheme. Two options for applying the technology under consideration were considered: for subcritical initial steam parameters (p_0 = 13 MPa, t_0 = 540 °C) and for supercritical (p_0 = 24 MPa, t_0 = 545 °C).

Based on the calculation results given in table 1 it can be seen that an increase of steam's initial parameters makes it possible to increase the efficiency of electric power generation by an average of 2 %. In passing from autonomy mode to a CCPP mode the efficiency increases by 22 %.

Table 1. Technical and economic indicators of the installation.

Quantity	UM	13 MPa, 540 °C	24 MPa, 545 °C
Heat flow supplied to the gas turbine	MW	211.34	211.34
Steam flow condition per turbine	Kg/s	163.92	163.92
Heat supplied during the operation of a STU as part of a CCPP	MW	362.18	351.99
STU net efficiency in autonomy mode	%	39.68	40.59
CCPP net efficiency	%	48.65	49.53

In addition, it should be noted that the equivalent fuel consumption has decreased (Figure 3). With an increase of steam's initial parameters, the equivalent fuel consumption is reduced by an average of 3 %, and the operation of the STU as part of the CCPP makes it possible to reduce the equivalent fuel consumption by 21 %.

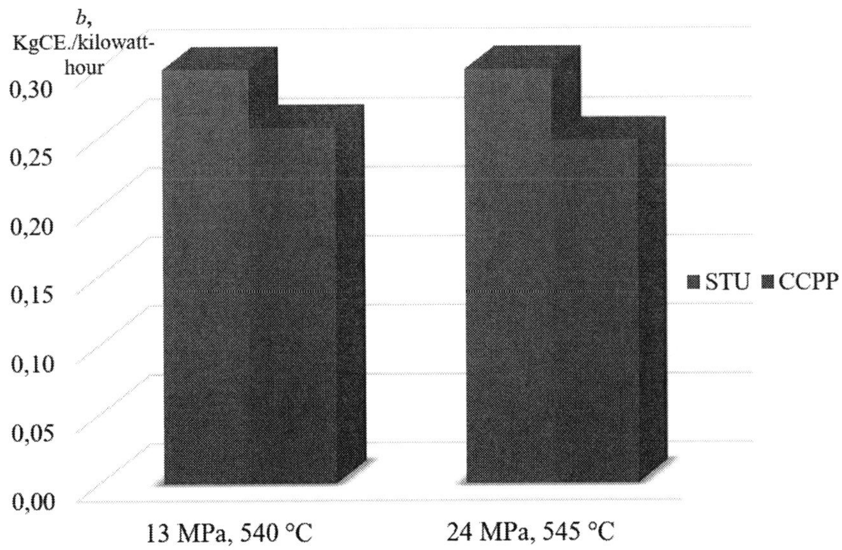

Figure 3. Equivalent fuel consumption's changing.

Table 2 shows the results of calculating capital investments and costs in the construction of a power unit "from scratch". As is seen from the results, with an increasing of initial steam parameters, the total capital costs increase. This is primarily due to the capital costs increasing for the steam boiler: since an increasing of initial parameters leads to steel intensity increasing and more heat-proof steels using.

A graphical interpretation of the calculating results of the net discounted income for unit operating on the subcritical and supercritical parameters are shown at Figure 4 respectively.

Table 2. The result of the calculation of capital investments and costs.

Quantity	UM	13 MPa, 540 °C	24 MPa, 545 °C
	Capital investments		
Steam boiler	RUB million	7 124.05	8 824.68
Steam turbine	RUB million	2 966.04	2 892.96
Gas turbine	RUB million	2 955.57	2 955.57
Waste heat boiler	RUB million	202.00	202.00
Electrical part	RUB million	2 046.24	1 980.72
APCS	RUB million	1 713.60	1 668.24
Steam-water tract	RUB million	3 356.64	3 984.12
Installation and commissioning	RUB million	1 837.08	1 779.12
Construction works	RUB million	4 427.64	4 205.88
Project cost	RUB million	1 333.08	1 401.12
Total capital investments	RUB million	27 961.94	29 894.41
	Costs		
Fuel costs	RUB million	1 188.75	1 174.65
Fixed costs	RUB million	3 705.02	3 960.10
Total costs	RUB million	4 893.77	5 134.76
Cost of electricity	RUB	2.67	2.80

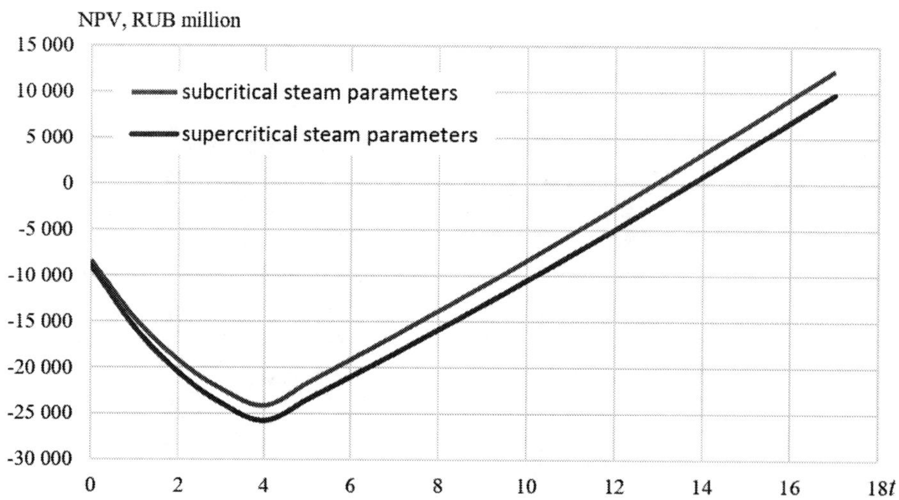

Figure 4. NPV for CCPP

4. Conclusion

Thus, the results of the calculations shows that using of a dual-fuel CCPP with a parallel operation scheme makes it possible to increase the efficiency of electric power generation by an average of 22 % compared to operation in the coal-fired STU mode. In addition, the transition to the CCPP mode allows reducing the equivalent fuel consumption by 21%. The achievement of such results is primarily due to the division of the feed water flow into two parallel flows by integrating a gas turbine. [14] It is also should be noted the improvement of the power unit's efficiency indicators in passing to supercritical steam parameters by an average of 2 %. It is worth noting that the increasing of the electric power generation efficiency obtained in this work is confirmed by studies performed earlier [15]. Capital costs for the construction of the considered power unit "from scratch" amount to 29 billion rubles with a discounted payback period – 13 years from the beginning of construction.

References

[1] Xu L, Yuan J and Wang J 2017 Dynamic modeling and solution algorithm of the evaporation system for the ultra-supercritical power plant *Int. J. of Heat and Mass Transfer* **105** 589-96

[2] Mills A, Wiser R and Seel J 2017 *Power Plant Retirements: Trends and Possible Drivers* (Berkeley: Berkeley Lab)

[3] Beér J M 2007 High efficiency electric power generation: The environmental role *Progress in Energy and Combustion Science* **33** 107-34

[4] Dolotovskii I and Larin E 2017 Stabilization of gas turbine unit power *J. of Physics: Conf. Ser.* **891** 012245

[5] Hoeven M et al 2012 *Technology Roadmap: High-efficiency low-emissions coal-fired power generation* (Paris: IEA)

[6] Shchinnikov P A and Borush O V 2018 Results of national generation reform and predictive information about power engineering market *J. of Physics: Conf. Ser.* **1111** 012007

[7] Tańczuk M, Olszewski E and Goluzda A 2018 Technical and economic issues of reconfiguration of a coal-fired district heating source *MATEC Web Conf.* **174** 04015

[8] Bartnik R 2013 *The Modernization Potential of Gas Turbines in the Coal-Fired Power Industry* (London: Springer)

[9] Bartnik R, Skomudek W, Buryn Z, Hnydiuk-Stefan A and Otawa A 2018 Methodology and mathematical model with the continuous time for the selection of the optimal power of the gas turbine set for the dual fuel gas-steam combined cycle in a parallel system *Applied Thermal Engineering* **141** 1161-72

[10] Cetin B and Ozen E 2018 Comparative energy and exergy analysis of a power plant with super-critical and sub-critical *J. of Thermal Engineering* **4** 2423-31

[11] Voronkova A S, Borush O V, Shchinnikov P A and Zueva A A 2019 Dvuxtoplivny`e PGU s parallel`noj sxemoj na supersverxkriticheskie parametry [Dual-fuel combined cycle power plants with a parallel operating scheme at ultra supercritical initial steam parameters] *Novoe v Rossijskoj E`lektroe`nergetike* **5** 6-11 [In Russia]

[12] Ashinyancz S A 2013 *Ocenka stoimosti stroitel`stva i e`kspluatacii ugol`ny`x TE`S: obzor* [Estimating the cost of construction and operation of coal-fired power plants: an overview] (Moscow: JSC «VTI») [In Russia]

[13] Kryzia D, Kopacz M and Kryzia K 2020 The valuation of the operational flexibility of the energy investment project based on a gas-fired power plant *Energies* **13(7)** 1567

[14] Carapellucci R and Giordano L 2014 Energy, economic and environmental assessments for gas-turbine integration into an existing coal-fired power plant *Energy Procedia* **45** 1175-84

[15] Sanpasertparnich T and Aroonwilas A 2009 Simulation and optimization of coal-fired power plants *Energy Procedia* **1** 3851-8

Analysis of various mathematical models of turbulence when calculating the gas dynamics in a classifier with coaxially arranged pipes

V E Zinurov[1], A V Dmitriev[1], G R Badretdinova[1], A R Galimova[1] and O S Dmitrieva[2]

[1]Department of Theoretical Basis of Thermotechnics, Kazan State Power Engineering University, Krasnoselskaya St., 51, Kazan, 420066, Russia
[2]Department of Food Production Equipment, Kazan National Research Technological University, Karl Marx St., 68, Kazan, 420015, Russia

E-mail: vadd_93@mail.ru

Abstract. This scientific article deals with the problem of choosing the mathematical model for calculating the gas dynamics in a classifier with coaxially arranged pipes. Various mathematical models were analyzed in ANSYS Fluent software package. Laminar, Spalart-Allmaras, *k-ω*, RSM and Transition SST were considered as mathematical models. The influence of number of cells in the calculation grid on the pressure loss change in the classifier for various mathematical models was also analyzed. The results of study showed that the pressure loss in the classifier with coaxially arranged pipes is not more than 395 Pa at the inlet gas flow rate of 10 m/s. Among the considered models i.e. laminar, Spalart-Allmaras, *k-ω*, RSM and Transition SST, the most adequate one is Transition SST model. When changing the number of cells in the calculation grid from 1241551 to 3159388 for this model, the deviation of pressure loss in the classifier was 0.43%. The article includes an equation for the dependence of pressure loss in the classifier on the inlet rate. The distribution of gas flow rate over various sections of rectangular gaps is shown.

1. Introduction

Nowadays, most of industrial enterprises pay great attention to energy and resource saving issues. At the same time, an important problem is to increase the efficiency of production processes. As a rule, this problem is solved by updating and upgrading the devices, used in production processes. One of the important requirements for innovative devices is relatively low energy costs, which allow them to be included in a new developed concept of enterprises, including such criteria as high efficiency, resource and energy saving. Therefore, when developing new designs of devices to be used in production processes, an important task is to meet the 3 above mentioned criteria.

The authors of this scientific article were tasked with developing and creating an energy-efficient classifier for the fractionation of bulk material, based on silica gel, of up to 30 microns in size from a gas flow in the production line of the Salavat catalyst plant. This bulk material is obtained by crushing it in the mills. It should be noted that the volume fraction of collected material larger than 30 microns should be at least 98% of total fraction of collected material. Most of classic devices cannot be used to solve this problem, since they are not designed for the fractionation of finely dispersed powdery materials [1-5].

At the moment, there are various types of powder classifiers: centrifugal, dynamic, gravitational, sieve, etc. [6, 7]. The centrifugal classifier separates the finely dispersed bulk material of fractional composition within the range from 5 to 80 microns by means of centrifugal force and air flow. The range of obtained particles directly depends on the rotor rotation speed and a change in the amount of air flow. The centrifugal classifiers are divided into static ones, when the air flow is created by the movement of blades, and dynamic ones, when the movement of air flow is provided by a rotating rotor. The advantages of centrifugal classifier are the ability to control the range of material separation fineness and provide the greatest accuracy for the particle classification [8]. A significant disadvantage of this type of classifier is that large powder particles fall into the fine fraction, and small particles fall into the large separation product, therefore, there is a decrease in the efficiency of particle separation and a small efficiency of device [9]. The gravitational classifiers carry out the material separation by means of interaction of resistance and gravity forces. The separation of raw material occurs as a result of different rates of sedimentation of solid particles in a liquid medium under the influence of gravitational forces. The large particles settle down faster, and the small particles settle down slower [10-12]. The main advantage of gravity method is the simplicity of device design due to the absence of moving parts in it. The disadvantage of gravity classifier is that it is impossible to classify the finely dispersed particles. At the industrial enterprises, the most common method is the sieve method of classifying the powders, which is directly carried out by the material spilling through a set of different sieves, ensuring separation as per the required parameters by fractional composition. This type of device has simple design and is suitable for classifying the particles larger than 40 microns. Despite the positive aspects of this method, it also has certain disadvantages, such as low efficiency of sieving classifiers for the finely dispersed powders and significantly low operational life of sieves due to rapid clogging caused by sticking of particles to their surface [13].

2. Materials and Methods

In order to solve this problem, the authors developed the design of classifier with coaxially arranged pipes [14], shown in figure 1. The classifier with coaxially arranged pipes is a simple structure, consisting of cylindrical body *6*, inside of which there is an internal cylindrical pipe; in the lower part of this pipe there is an even number of rectangular holes *3* with a certain step, and the bottom part of pipe has a cone-shaped hole *4*. In the middle part of inner cylindrical pipe, there is a screen with coaxially arranged pipes *2*. There is an inlet nozzle in the upper central part of device *1*. The device also has an outlet nozzle *5*. The bunker is located in the lower part of device *7*. When assembling the device, the inner cylindrical pipe is placed in the cylindrical body through the lower hole and inserted into the central hole in its upper part, forming an inlet nozzle *1*. The strength of structure is provided by fixing the inner cylindrical pipe to the cylindrical body *6* by means of screen with coaxially arranged pipes *2* and by welding the outer part of inner cylindrical pipe, forming an inlet nozzle *1*, to the device body *6* (figure 1).

The process of bulk material fractionation in a classifier with coaxially arranged pipes can be described as follows: the gas flow with silica gel-based particles, dispersed in it, is fed into the device through the inlet nozzle *1*, then the flow moves to the rectangular holes *3*, after that some part of flow passes through them and enters the inner area of classifier, which is located between the inner wall of device body and the outer wall of inner cylindrical pipe. It should be noted that the other part of flow carries out further movement along the inner cylindrical pipe and comes through the narrowing cone-shaped hole of classifier *4*; then the flow diverges in an axisymmetric direction and moves to the upper part of device. The largest particles, due to gravitational and inertial forces, are knocked out of flow structure and settle down in the bunker *7*.

An important design feature of classifier is that when the flow moves through each rectangular hole *3*, the flow is further divided into two jets in equal parts, which move in different directions relative to each other, reaching the inner wall of cylindrical body *6*, while the gas jets turn around and move in a vortex to the upper part of device. With the rotational movement, the gas jets form vortices with high values of centrifugal forces. Each vortex has two areas of contact with neighboring vortices that increases the flow movement and centrifugal forces. The gas dynamics of described process, simulated in ANSYS Fluent software package, is shown in figure 2.

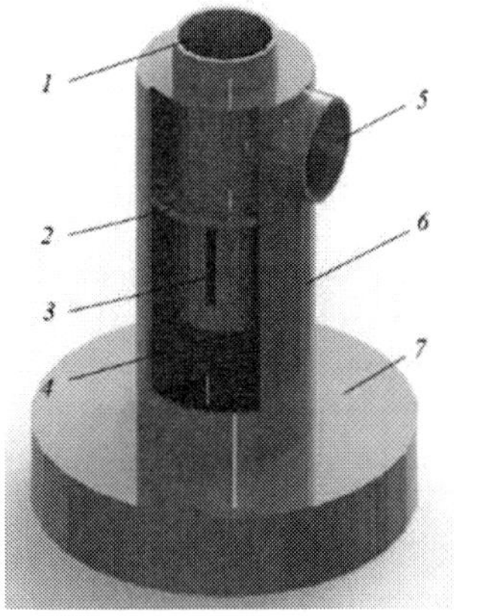

Figure 1. The classifier with coaxially arranged pipes: 1 – inlet nozzle, 2 – screen with coaxially arranged pipes, 3 – rectangular holes, 4 – additional hole, 5 – outlet nozzle, 6 – device bode (case), 7 – bunker.

During the rotational movement due to inertial and centrifugal forces, the particles of silica gel-based material are knocked out of the flow, fall into the "dead zones" between the vortices and settle down in the bunker *7*. Then, the gas flow passes through the screen with coaxially arranged pipes *2* and exits the device through the outlet nozzle *5*. It should be noted that when gas moves through a screen with coaxially arranged pipes, some part of silica gel-based particles cut into the screen wall and they are knocked out of the flow structure that contributes to additional collection. Thus, the collection of silica gel-based material particles of up to 30 microns in size in a classifier with coaxially arranged pipes is carried out mainly by means of centrifugal, inertial and gravitational forces. The most effective areas where the particles of silica gel-based material fall out of the structured flow are the following: 1) transition from the narrowing cone-shaped hole to the inner part of classifier, 2) area of vortices formation, 3) contact area of gas flow and a screen with coaxially arranged pipes.

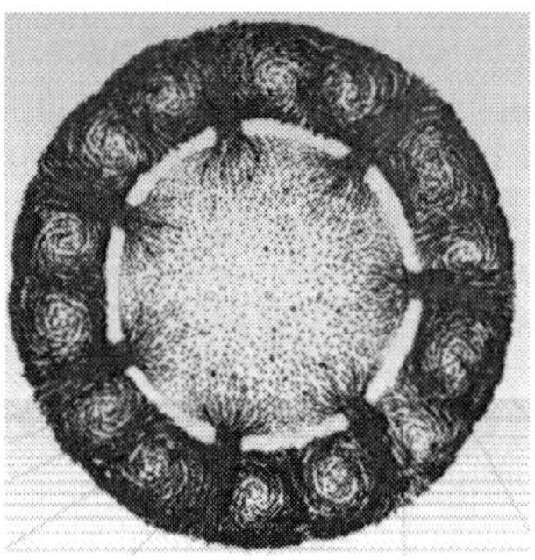

Figure 2. The gas dynamics in a classifier with coaxially arranged pipes; vortex effect (top view of cross section).

It should be noted that the construction of classifier with coaxially arranged pipes was performed with "engineering intuition". At the moment, it is necessary to create an engineering method for calculating this device. The most physically effective and cost-effective way to obtain the significant calculated dependencies between technological and design parameters is the numerical simulation of gas-dynamic processes in the proposed device. The simulation of these processes is associated with a large number of difficulties; in particular, in order to speed up the calculations and obtain reliable data, it is necessary to choose the most suitable turbulence model for this device and to determine the most optimal number of cells in the calculation grid. Therefore, the purpose of this study is to analyze various mathematical models of turbulence when calculating the gas dynamics in a classifier with coaxially arranged pipes.

ANSYS Fluent software package was chosen as a program for analyzing various mathematical models of turbulence. This software package is designed for modeling complex flows of liquids and gases with a wide range of changes in thermal properties by providing various modeling parameters and using the multigrid methods with improved convergence. The program uses the finite element method, which is a grid-based approach. Depending on the chosen turbulence model, the differential equations with partial derivatives are derived (the Navier-Stokes equation) [15-17].

$$\frac{\partial \vec{v}}{\partial t} = -(\vec{v} \cdot \nabla)\vec{v} + \nu \Delta \vec{v} - \frac{1}{\rho}\nabla p + \vec{f}, \quad (1)$$

where:
∇ – nabla;
Δ – vector Laplace operator;
t – time period, c;
ν – kinematic viscosity coefficient, m^2/s;
ρ – density, kg/m^3;
p – pressure, Pa;
$\vec{v}$ – velocity vector field;
$\vec{f}$ – vector field of mass forces.

The Navier-Stokes equation is supplemented with the continuity equation:

$$\frac{\partial \rho}{\partial t} + \nabla \cdot (\rho \vec{v}) = 0. \quad (2)$$

These equations are also supplemented with the specified boundary conditions. During preparation for the numerical simulation, a three-dimensional geometry of classifier with coaxially arranged pipes was created in ANSYS Fluent software package, shown in figure 1. The following turbulence models were chosen for the study: Spalart-Allmaras, SST k-ω, transition SST, and RSM. The laminar model (Laminar) was also considered.

The standard SST k-w model is based on the equations of transferring the turbulence kinetic energy k and the specific dissipation rate ω. These 2 parameters can be determined by equations (3) and (4). It should be noted that the main differences between SST k-ω and k-ω models are the following: 1) sequential change of internal flow in the boundary layer to the high-Reynolds k-ε model for the outer area of boundary layer in SST k-ω model is done; 2) turbulent viscosity formula is modified in order to take into account the effects of the main turbulent shear stress transfer.

$$\frac{\partial}{\partial t}(\rho k) + \frac{\partial}{\partial x_i}(\rho k v_i) = \frac{\partial}{\partial x_j}\left(\Gamma_k \frac{\partial k}{\partial x_j}\right) + G_k - Y_k + S_k, \quad (3)$$

$$\frac{\partial}{\partial t}(\rho w) + \frac{\partial}{\partial x_i}(\rho w u_i) = \frac{\partial}{\partial x_j}\left(\Gamma_w \frac{\partial w}{\partial x_j}\right) + G_w - Y_w + D_w + S_w, \quad (4)$$

where:
G_k – generation of turbulence kinetic energy by averaged flow gradients;
G_w – generation as per w parameter;
Γ_k, Γ_w – effective diffusion k and w respectively;
Y_k, Y_w – dissipation of k and w under the influence of turbulence, respectively;
D_w – lateral diffusion; S_k, S_w – source terms to be defined depending on the task.

The advantage of k-ω model is a high accuracy in describing the wall turbulence and sensitivity to boundary conditions in the external flow.

Transition SST model, which is an improved standard SST model, was considered as well. This model provides a more accurate description of turbulence by introducing additional transfer equations, one for intermittency, and the other one for the velocity head, calculated for the thickness of momentum loss.

Spalart-Allmares model is a relatively simple one-parameter turbulence model, with one additional turbulent viscosity transfer equation for calculating the kinematic coefficient of vortex viscosity:

$$\frac{\partial v_t}{\partial t} + u_i \frac{\partial v_t}{\partial x_i} = P_{v_t} + D_{v_t} + \varepsilon_{v_t} . \quad (5)$$

The left part of equation accounts for convection, the right part – for energy generation, diffusion, and dissipation. It should be noted that the use of this model allows to describe the entire flow area, including the wall layers. The advantages of this model are its stability and a good degree of convergence. The disadvantages include inaccuracy of calculation for shear and separated flows, and damping of turbulence.

RSM model (Reynolds Stress Model – full model of the Reynolds stress). This model is based on one of the above described models, but additionally to two differential equations of these models it is supposed to solve 6 more differential equations which describe the transfer of each of the 6 Reynolds stresses. When the model is supplemented with additional differential equations, the calculation time increases and the convergence of calculation processes as a whole becomes more complicated, which is one of its disadvantages. However, the use of the Reynolds stress model provides a deeper understanding of turbulence process.

When performing the numerical simulation, the following constant values were taken: the gas flow rate at the inlet nozzle was assumed to be 10 m/s, and pressure at the outlet of classifier was assumed to be equal to the atmospheric pressure – 101325 Pa. The geometric dimensions of classifier with coaxially arranged pipes were taken as follows: diameter of inlet and outlet nozzle – 60 mm; diameter of classifier body – 100 mm; height of classifier – 290 mm; height of inner cylindrical pipe – 192 mm; diameter of coaxially arranged pipes in the screen – 12 mm; diameter of hole in the lower conical part of inner cylindrical pipe – 16 mm; height and width of rectangular holes – 60 mm; depth and diameter of bunker – 50 and 210 mm, respectively.

In the course of study, a three-dimensional model of classifier with coaxially arranged pipes was divided into 1241551, 2080907 and 3159388 cells (figure 3).

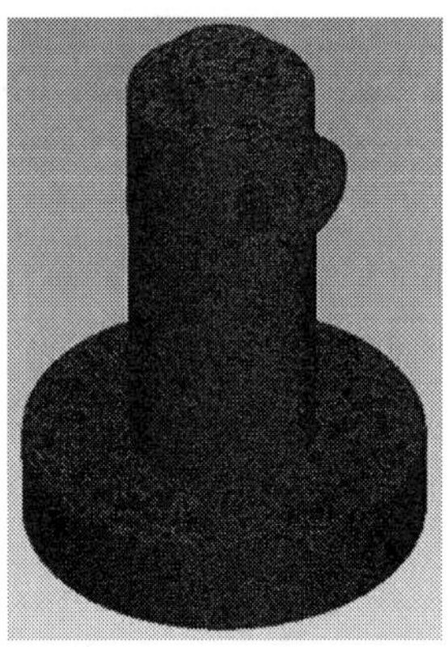

Figure 3. Grid generation. Number of cells $N = 1241551$.

When generating the dependence diagrams, the pressure loss Δp in the classifier with coaxially arranged pipes was calculated by using the following equation:

$$\Delta p = p_1 - p_5 , \quad (6)$$

where p_1 – pressure at the inlet nozzle *1*, Pa; p_5 – pressure at the classifier outlet nozzle *5*, Pa.

3. Results and discussion

The results showed that the pressure loss in the classifier with coaxially arranged pipes is not more than 395 Pa at the inlet gas flow rate W=10 m/s. The pressure loss in the classifier was on average equal to 383, 385, 386, 390, 389 Pa when using laminar, Spalart-Allmaras, k-ω, RSM and Transition SST as the calculation models, respectively, with the number of cells from 1241551 to 3159388. The difference between the calculation models: laminar, Spalart-Allmaras, k-ω, RSM and Transition SST, when determining the pressure loss, was not more than 6%. It should be noted that with the number of cells 1241551, 2080907 and 3159388, the difference between the maximum and minimum values of pressure loss in the classifier for different calculation models was on average equal to 2.92, 4.54 and 2.48%, respectively. The use of laminar, Spalart-Allmaras, k-ω, RSM and Transition SST models, when increasing number of cells from 1241551 to 3159388, showed that the deviation values for the pressure loss in the classifier are not more than 4.44, 3.04, 1.05, 1.22, and 0.43%, respectively (figure 4).

The obtained results show that for the numerical calculations of gas dynamics in a classifier with coaxially arranged pipes, it is possible to use any of models, specified above, with the number of cells at least 1241551. However, the most suitable turbulence model for calculating the developed classifier is Transition SST, since there is the smallest change in the pressure loss in the device when the number of cells in the calculated geometry changes (figure 4).

Based on the chosen Transition SST turbulence model for a classifier with coaxially arranged pipes, a calculation was conducted to determine the pressure loss change in this device when the inlet gas flow rate changes from 5 to 10 m/s. In the course of numerical simulation, it was found that the pressure loss in the classifier is on average equal to 234 Pa at the inlet rate of 5–10 m/s (figure 5).

The following power function equation, describing an impact of inlet gas flow rate on the pressure loss in the classifier, was obtained

$$\Delta p = 3.89W^{1.99}. \quad (7)$$

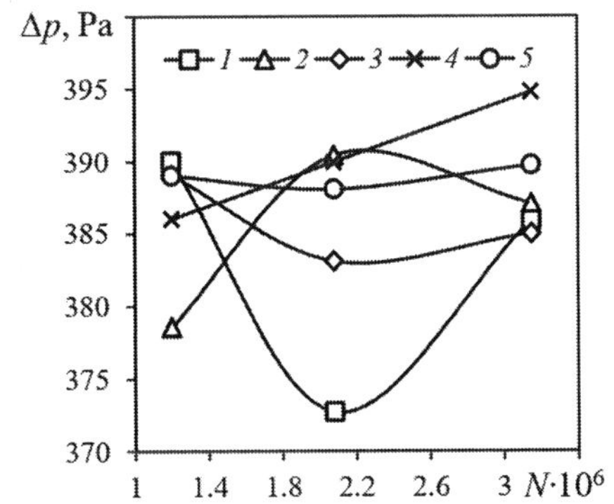

Figure 4. The dependence of pressure loss change in a classifier with coaxially arranged pipes on the number of cells for different calculation models: *1* – Laminar; *2* – Spalart-Allmaras; *3* – *k-w*; *4* – RSM; *5* – Transition SST. The inlet rate of gas flow is W= 10 m/s.

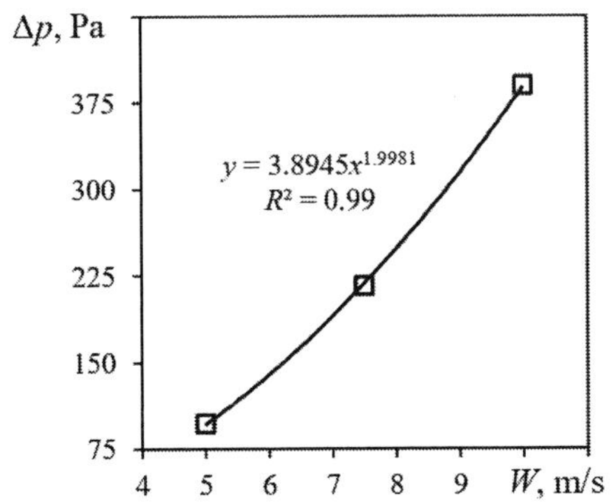

Figure 5. The dependence of pressure loss change in a classifier with coaxially arranged pipes on the inlet rate of gas flow.

Also, when using Transition SST turbulence model, the change of gas flow rate in rectangular gaps was studied, depending on the measurement point. In order to determine the gas flow rate at various points of rectangular gap, it was divided into 7 sections with a step of 10 mm. In the first and the last sections, the gas rate was equal to zero, since these were the extreme points in the upper and lower parts of gap that were directly in contact with the inner cylindrical pipe. The gas rate in the intermediate sections was on average equal to 12.4 m/s. At the same time, the maximum gas rate was reached in the upper sections, which was up to 14.8 m/s. The minimum gas flow rate of 11 m/s was determined in the central part of rectangular gaps. The gas flow rate increased again to 12.7 m/s as it moved away from the central part of gap to the lower parts under the study (figure 6).

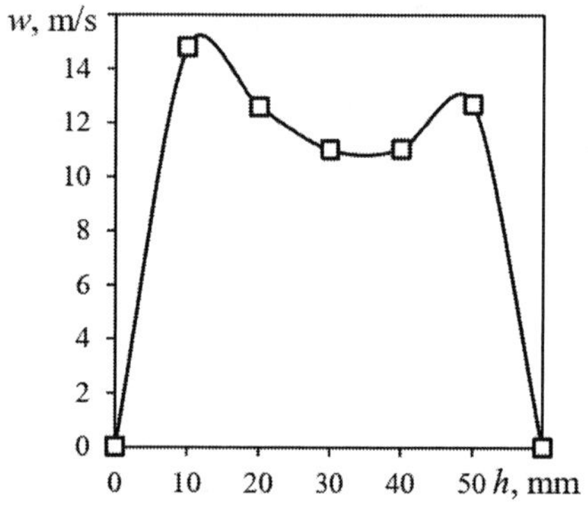

Figure 6. The dependence of gas flow rate change from a rectangular gap on the location of area where the rate was determined.

Such distribution of gas flow rates can be explained as follows: when gas moves through the inner cylindrical pipe, most part of flow is directed immediately to the narrowing areas, i.e. to the upper sections of rectangular gaps, as a result of which the maximum rate values are reached there. The rest part of gas flow is distributed to the central and lower parts of rectangular gaps and to the area, located after the hole in the cone-shaped part of inner cylindrical pipe.

When gas passes through the lower hole in the cone-shaped part of pipe, it is directed to the upper part of device, while skirting the rectangular gaps from the outside, and, as a result of this process, the flow structure changes in the lower and central parts of rectangular gaps, which leads to a change in the values of gas flow rate within them. In particular, the upward flow picks up and pushes gas, leaving the lower sections of rectangular gaps. However, as it moves towards the central part of gaps, it becomes an additional resistance for the outgoing flow from the gap, and, as a result of which the gas rate in the central parts of gaps decreases (figure 6).

Thus, the study showed that when developing the devices, the use of numerical simulation can significantly simplify this process and reduce economic costs.

4. Conclusion

In the course of study, it was found that the most correctly chosen turbulence model, combined with the optimal number of cells in the calculation geometry, contributes to the acceleration of numerical simulation of gas-dynamic processes in the developed device. Among the considered models i.e. laminar, Spalart-Allmaras, *k-ω*, RSM and Transition SST, the most adequate one is Transition SST model. When changing the number of cells in the calculation grid from 1241551 to 3159388 for this model, the deviation of pressure loss in the classifier was 0.43%.

It was found that with a relatively large number of cells in the calculation grid (this paper includes the study of change in the number of cells from 1241551 to 3159388), it is possible to use any model i.e. laminar, Spalart-Allmaras, *k-w*, RSM or Transition SST, since the deviation for the pressure loss calculation is not more than 6%, which is an acceptable value for the numerical simulation (modeling). When using Transition SST turbulence model, an equation was obtained for the dependence of pressure loss in a classifier with coaxially arranged pipes on the gas flow rate, which shows that the pressure loss in the device increases with an increase of rate. The gas dynamics in rectangular gaps was also analyzed, and showed that in the central sections of gaps, the gas rate is minimal, reaching 11 m/s, and in the upper and lower sections of rectangular gaps, the rate increases up to 14.8 m/s, due to the device design features.

The advantages of developed classifier with coaxially arranged pipes are the following: design simplicity, ease of operation and high efficiency.

5. Acknowledgments

The study was supported by the grant of the President of the Russian Federation No. MK-616.2020.8.

References

[1] Koeninger B, Spoetter C, Romeis S, Weber A P and Wirth K-E 2019 Classifier performance during dynamic fine grinding in fluidized bed opposed jet mills *Adv. Powder Technol.* **30(8)** 1678-86

[2] Sun Z, Sun G, Liu J and Yang X 2017 CFD simulation and optimization of the flow field in horizontal turbo air classifiers *Adv. Powder Technol.* **28(6)** 1474-85

[3] Wu S, Liu J and Yu Y 2017 Design of a new double layer spreading plate for a turbo air classifier *Powder Technol.* **312** 277-86

[4] Sun Z, Sun G, Peng P, Liu O and Yu X 2019 A new static cyclonic classifier: Flow characteristics, performance evaluation and industrial applications *Chem. Eng. Res. Des.* **145** 141-9

[5] Sun Z, Liu Q and Yu X 2019 Experimental and CFD study on a cyclonic classifier with new flow pattern *Adv. Powder Technol.* **30** 10 2276-84

[6] Petit H A, Paulo C I, Cabrera O A and Irassar E F 2020 Modelling and optimization of an inclined plane classifier using CFD-DPM and the Taguchi method *Applied Mathematical Modelling.* **77(1)** 617-34

[7] Toprak N A and Benzer A H 2019 Effects of grinding aids on model parameters of a cement ball mill and an air classifier *Powder Technol.* **344** 706-18

[8] Sun Z, Sun G, Yang X, Yuan Y, Wang Q and Liu J 2017 Effects of fine particle outlet on performance and flow field of a centrifugal air classifier *Chem. Eng. Res. Des.* **117** 139-48

[9] Altun O, Toprak A, Benzer H and Darilmaz O 2016 Multi component modelling of an air classifier *Miner. Eng.* **93** 50-6

[10] Johansson R and Evertsson 2012 An empirical study of a gravitational air classifier *Miner. Eng.* **31** 10-6

[11] Li Y, Zhang X, Shen X, Li H and Gao Ke 2015 Experiment and simulation study on optimized structure of a gravitational air classifier *Int. J. Miner. Process.* **141** 44-50

[12] Zinurov V E, Dmitrieva O S and Popkova O S 2020 Collecting finely-dispersed particles from the gas flow in a centrifugial separator with coaxially arranged pipes *Matec Web Conf.* **315** 03003

[13] Bosma J C and Hoffmann A C 2003 On the capacity of continuous powder classification in a gas-fluidized bed with horizontal sieve-like baffles *Powder Technol.* **134(1-2)** 1-15

[14] Zinurov V E, Dmitriev A V, Ruzanova M A and Dmitrieva O S 2020 Classification of bulk material from the gas flow in a device with coaxially arranged pipes *E3S Web Conf.* **193** 01056

[15] Yan C and McDonald J G 2020 Hyperbolic turbulence models for moment closures *J. Comput. Methods Phys.* **422** 109753

[16] Avila V C, Tessaro I C and Cardozo N S M 2019 The role of turbulence models in the modeling of membrane bioreactors *Chem. Eng. Process.* **144** 107639

[17] Jofre L, Dodd M S, Grau J and Torres R 2020 Near-interface flow modeling in large-eddy simulation of two-phase turbulence *Int. J. Multiphase Flow* **132** 103406

Enhancing thermal performance of vertical fin array heat sink using mist assisted evaporative cooling

A H Abed[1], H A Hussein[1] and V M Pakhaluev[2]

[1]Department of Electromechanical Engineering, University of Technology, Baghdad, Iraq
[2]Department of Nuclear Power Plants and Renewable Energy Sources, Ural Federal University named after the first President of Russia B. N. Yeltsin
19 Mira St., Yekaterinburg 620002, Russia

E-mail: akraaam82@yahoo.com

Abstract. In this paper, natural convection heat transfer of parallel-plate rectangular fins heat sink positioned at a certain angle on the vertical substrate was experimentally investigated. Experiments were done with an aim to examine and evaluate the enhancement of heat transfer by deploying fine water mist into ambient air under a range of input power. To empower this investigation, a special purpose experimental set-up was constructed and developed so as to perform experiments on both air and water mist cooling scheme. The water mist concentration influences on the heat transfer enhancement was examined for (m_w=90 – 430 mL/h), while the surface temperature varies from 29 to 110 °C. Experimental results revealed that the heat transfer enhancement increase and reached its maximum value under m_w= 430 mL/h. The average heat transfer enhancement increases by about 1.28%, 1.81%, 2.37%, and 2.83% when compared against traditional air cooling under (m_w=90 – 430 mL/h). In addition, influence of water mist on the fin spacing, fins number, and dissipation area was also examined.

1. Introduction

Due to the development of electronic industry and emergence of modern specialized devices that need to processing data at high speed, reliable and efficient cooling techniques were required. The central processing unit of electronic devices assembled with millions of microscopic transistors to be able to process more data speedily. As a result of passes an electrical signal, the temperature and amount of heat generation in the central processing unit will be increased, making it less reliable and increase the chance of breakdown of data processing in a few years. Dissipation heat it's a very important case to ensure dependable continuous operation for electronic devices and that can be done by several ways using heat sink [1], heat pipes [2], jet impingement [3] and piezoelectrically-driven droplets [4], etc. Altun et al. [5] have experimentally investigated the impact of sinusoidal wavy fins on the natural convection heat transfer augmentation with different amplitudes. They found an appreciable augmentation in the heat transfer with wavy fins than among the rectangular fins. Also, they found that increasing wave amplitude led to a negative effect on the heat transfer augmentation due to the air motion blockage. Numerical and experimental investigation of natural convection heat transfer mechanism was performed by Zhang et al. [6] with a vertical W-type finned heat sink. In comparison with parallel-plate finned heat sink, they pointed out that the average temperature decrease by about 2.9 °C while the maximum temperature decrease by about 4.6 °C. In addition, they found that the heat dissipation area reduced by 10%. Another experimental investigation of rectangular fin array heat sink positioned on the horizontal substrate was performed by Taji et al. [7] under a natural and assisting mode of mixed convection. They showed that the average heat transfer coefficient of mixed convection has increased by 69.46% at low energy input from 0.3 to 0.65 W. Güvenç et al. [8] examined the impacts of fin spacing and fin height on the heat transfer coefficient. They concluded that the essential geometric parameter is fin spacing, and it should be selected as characteristic dimension. Using water mist technique as a cooling medium has many exceptional features and has been effectively implemented in many industrial and engineering applications [9-12]. Compared with air cooling, the essential characteristics of water mist can be illustrated as follows: absorbs a great deal of energy in the form of latent heat during the water mist evaporation process by direct contact of the water droplets with the heated surface, increases the humidity, and specific heat of the mixture [13]. For example, a number of numerical and experimental of previous studies found in the literature will be presented, but not limited. Numerical simulation and parametric analyses [14] conducted to investigate the effect of water mist concentration and water droplet diameter on the fin array heat sink heat transfer augmentation under forced flow. They reported that the heat transfer rate obtained with mist flow increases dramatically with increased concentration of water mist and highlighted the possibility to reduce thermal resistance of heat sink up to 97%. Bahadur et al. [15] suggested a new cooling scheme in which mist injected directly into the air cooling, then the evaporated mist can be considered and re-circulated back to their inlets. The results illustrated that the proposed mist cooling scheme provides a significant promise for giving high-flux heat removal methods for telecommunications center. In the present work, the experimental investigation was performed taking into account both natural convection and radiation to analyze and understand the water mist influences on the thermal performance and heat dissipation area of a parallel-plate rectangular fins heat sink. The influence of the different working conditions such as water mist concentration and input power on heat transfer characteristics was examined.

2. Experimental setup and instrumentation

The main goal of the experimental study is to examine and evaluate the water mist influences on the natural convection heat transfer from a finned heat sink. To empower this investigation, the experimental set-up was constructed and developed so as to perform experiments on both air and water mist cooling scheme. The experimental setup has primarily consisted of a parallel-plate finned heat sink, electrical heater, insulated cover, supported frame, mist generator, and data acquisition system. The schematic layout of experimental set-up and geometry of the finned heat sink is given in figure 1. The heat sink comprising rectangular fins was made of the aluminum alloy due to high-emissivity and high thermal conductivity and positioned at a certain angle on the vertical substrate. A.C. silicon band drum heater whose dimension are 150mm-high and 100mm-width was used as a heat source to heat the heat sink. The maximum power and working temperature of heater are 50W and 160 °C respectively. The input power has been calculated by multiplying the measured voltage and current using a digital multimeter type APPA 79 with (±0.06%) in accuracy. A voltage regulator type (EQ-VT250) was used to modify and control heater input power so as to achieve the required surface temperature. To avoid any thermal resistance between heater and heat sink, a thin layer of high thermal conductivity paste was applied. The insulated cover was made of a wooden box that has dimensions of 260, 180, and 30mm for high, width, and depth respectively. In order to minimize the heat loss from the rear side of heater and maintain the insulation of the four lateral surfaces of heat sink, the insulated cover was loaded with 30mm thickness fibreglass wool insulated material. The surface temperatures of base plate are gauged using calibrated thermocouples (type-K) that are implanted in drilled holes at 6 different locations as presented in figure 1. The ambient temperature and temperature over the heat sink for two cases (air and water mist) were also measured. Simultaneously, the humidity over the heat sink was measured using humidity meter type (GM1362) with accuracy (±4.5%). All thermocouples are connected to (MSD200) type digital data logger with accuracy (±1.0%). Micro-sized water mists (2–10μm) were deployed in the air using an ultrasonic atomizer consisting of a piezoelectric transducer working with high-frequency oscillation of 1.7MHz. This high-frequency oscillation led to create a natural floating water mist without any mechanical agent or heating sources.

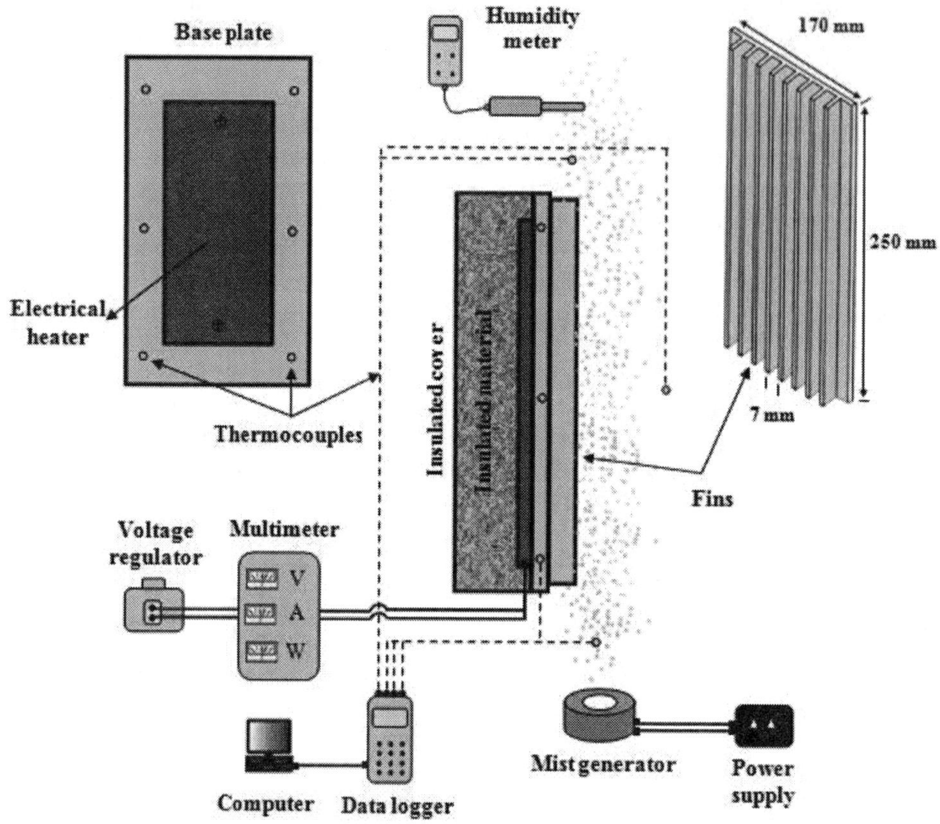

Figure 1. Schematic layout of experimental set-up and geometry dimensions of the finned heat sink.

3. Data processing and uncertainty analysis

The experiments were done for a finned heat sink with and without water mist for various input power. Initially, the results for the finned heat sink without water mist was obtained and regarded as reference data for the other experiments. For experiments without water mist, the heat transfer coefficient is estimated from the following equations [16].

$$h = \frac{Q_{conv.}}{A \cdot (T_s - T_\infty)}; \tag{1}$$

where:
$T_s - T_\infty$ – temperature difference,
$Q_{conv.}$ – heat transfer rate by convection which was evaluated as:

$$Q_{conv.} = Q_P - Q_{rad.}; \tag{2}$$

where:
Q_p – total input power,
$Q_{rad.}$ – heat transfer rate by radiation which was evaluated as:

$$Q_{rad.} = A \cdot \varepsilon \cdot \sigma \cdot (T_s^4 - T_\infty^4); \tag{3}$$

Rayleigh number (Ra) can be defined as:

$$Ra = Gr \cdot Pr = \frac{\Delta T \cdot \rho^2 \cdot g \cdot \beta \cdot L^3}{\mu^2} \cdot \frac{\mu \cdot c_p}{k}; \tag{4}$$

Herein, the thermal expansion coefficient can be obtained as:

$$\beta = \frac{1}{T_f}; \tag{5}$$

where T_f - represent the film temperature at which all thermo-physical properties of air were estimated, it is can be defined as:

$$T_f = \frac{T_s + T_\infty}{2}; \tag{6}$$

For experiments with water mist, the heat transfer rate is estimated from the following equations [17].

$$h_m \cdot A \cdot (T_s - T_\infty) + k_c \cdot h_{fg} \cdot (x_s - x_{sat.}) = Q; \tag{7}$$

where:
h_m – heat transfer coefficient by water mist,
A – total surface area,
k_c – mass transfer coefficient,
h_{fg} – the latent heat of evaporation,
x_s – moisture content of saturated air by surface temperature,
x_{sat} – moisture content of saturated air by ambient temperature.

When water is exposed to the adequately ultrasonic field, separation of water begins to form water mist and they are expelled from water interface into the ambient air. For obtaining the average diameter of water mist, Lang [18] formulated an equation depending on the ultrasonic frequency as follow:

$$d_p = C \cdot \left(\frac{8\pi\sigma}{\rho f^2} \right)^{0.333}; \tag{8}$$

where C – constant fraction of the capillary wavelength equal to 0.34, and f - the ultrasonic mist generator frequency. The Weber number (We) can be defined as:

$$We = \frac{m_w^2 \cdot d_p}{2A^2 \rho \sigma}; \tag{9}$$

For a vertical heat sink, the optimum fin spacing is obtained as follow [19]:

$$S_{opt.} = 2.714 \cdot \frac{L}{Ra^{0.25}}; \tag{10}$$

The fins number of the optimum fin spacing case was calculated as follow:

$$n = \frac{W}{S + t}; \tag{11}$$

The uncertainty of experimental data was calculated with an aim to evaluate the experimental facilities reliability. Uncertainties of the heat transfer coefficient and Rayleigh number can be estimated as follow [20]:

$$\left(\frac{\Delta h}{h}\right)^2=\frac{1}{\alpha}\left[\left(\frac{\partial}{\partial Q}(h)\Delta Q\right)^2+\left(\frac{\partial}{\partial T_s}(h)\Delta T_s\right)^2+\left(\frac{\partial}{\partial T_\infty}(h)\Delta T_\infty\right)^2\right] \tag{12}$$

$$\left(\frac{\Delta Ra}{Ra}\right)^2=\frac{1}{Ra}\left[\left(\frac{\partial}{\partial \Delta T}(Ra)\Delta T\right)^2+\left(\frac{\partial}{\partial \rho}(Ra)\Delta \rho\right)^2+\left(\frac{\partial}{\partial \mu}(Ra)\Delta \mu\right)^2+\left(\frac{\partial}{\partial L}(Ra)\Delta L\right)^2\right] \tag{13}$$

The maximum uncertainties of heat transfer coefficient and Rayleigh number were found to be 4.1%, and 5.3%, respectively.

4. Results and discussion

Herein, the experimental results of heat transfer characteristics, thermal performance, fin spacing, and the overall area of heat sink associated with the uses of water mist are presented. Before start experiments, a trial analysis were performed to evaluate the reliability of the experimental setup and experimental procedures employed. The present results were validated and compared with the previous experimental and numerical data of Harahap F et al. [21] and Senol Baskaya et al. [22] as depicted in figure 2. As observed, the present heat transfer coefficients are in good agreement with previous studies which illustrates that the experimental setup and experimental procedures in the present work are satisfactory and reliable.

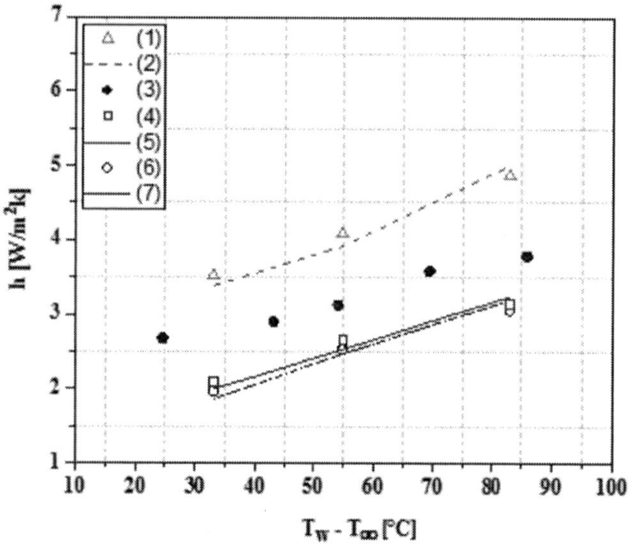

Figure 2. Validation of the heat transfer coefficient: present work compared with experimental results by Harahap F et al. [21] and CFD results by Baskaya S et al. [22]:
(1) Harahap F et al. (127·38·6.4);
(2) Baskaya S et al. (127·38·6.4);
(3) Present work (250·20·7);
(4) Harahap F et al. (254·6.3·6.4);
(5) Baskaya S et al. (254·6.3·6.4);
(6) Harahap F et al. (254·25·8);
(7) Baskaya S et al. (254·25·8).

Figures 3, 4 gives the variations of average surface temperature (base plate temperature) and heat transfer coefficient with input power with and without water mist. It can be seen from figure 3 that the surface temperature increases as input power increase for all cases. For air cooling case, cold air enters the space between fins generally from the heat sink bottom side and leaves from the top side after exchanging heat with the heat sink fins. During this time, the air is significantly heated and resulting in increasing its temperature along the heat sink length. In the water mist case, the influence of water evaporation was clearly observed and it can be noticed that the average surface temperature decreases as water mist concentration increases. The average surface temperature decreases by about 19.1%, 34%, 43.5%, and 48.2% in comparison with traditional air cooling under a range of water mist concentration (m_w=90 – 430 mL/h). Depending on equation (7) the heat transfer rate by water mist is equal to the sum of heat dissipated by water mist evaporation for wetted regions and normally heat transfer by convection. In view of this, the heat transfer coefficient increases by about 27.5%, 79.1%, 135.4%, and 281% in comparison with traditional air cooling under (m_w=90 – 430 mL/h) as shown in figure 4.

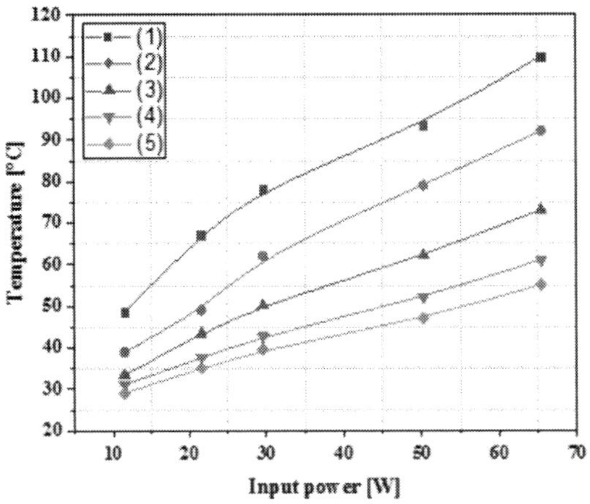

Figure 3. Variation of the average surface temperature: (1) only air, (2) M_w= 90 mL/h, (3) M_w= 180 mL/h, (4) M_w= 340 mL/h, (5) M_w= 430 mL/h for all cases of the input

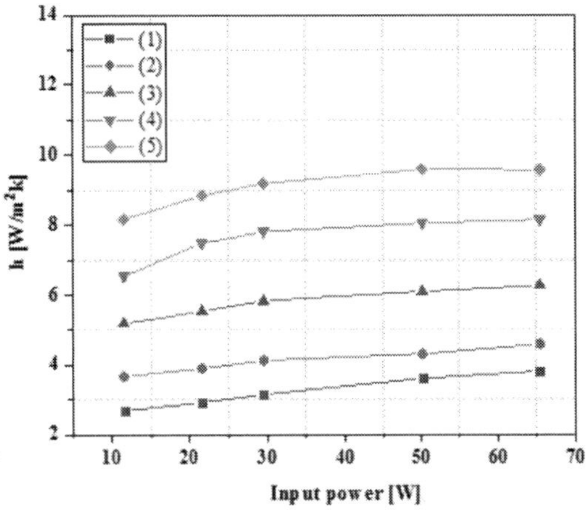

Figure 4. Variation of the average heat transfer coefficient: (1) only air, (2) m_w= 90 mL/h, (3) m_w= 180 mL/h, (4) m_w= 340 mL/h, (5) m_w= 430 mL/h for all cases of the input

At the same time as heat transfer measurement, the relative humidity over the heat sink for both cases was also measured with a 60 sec time interval as presented in figure 5. Initially, the air humidity was equal to the ambient, and then as the temperature of the heat sink became warmer and thus air temperature due the convection flow, the air humidity began to decrease gradually. For the water mist case, an opposite pattern was found, the relative humidity measured drastically increased with time and reached up to 100% after 300 sec.

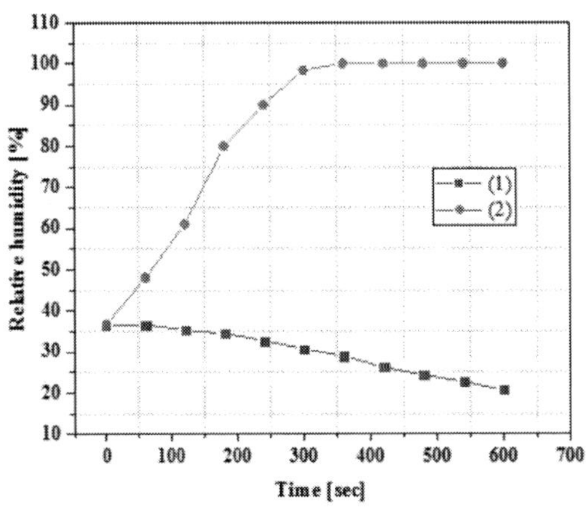

Figure 5. Relative humidity over the heat sink in the time function: (1) only air, (2) with water mist.

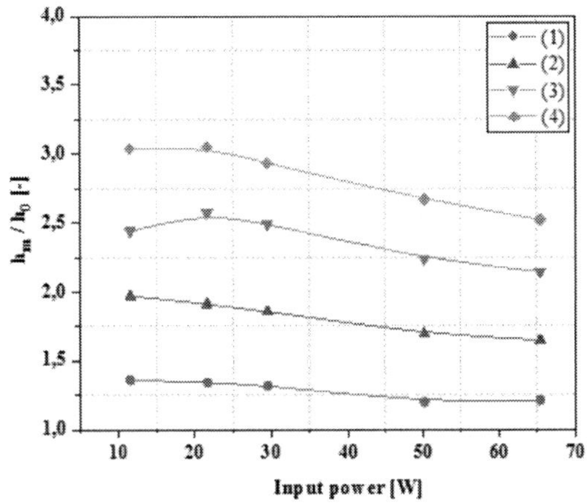

Figure 6. Variations of heat transfer enhancement: (1) mw= 90 mL/h, (2) mw= 180 mL/h, (3) mw= 340 mL/h, (4) mw= 430 mL/h for all cases of the input power.

The ratio of heat transfer coefficient for the finned heat sink with water mist to that when water mist was not deployed at an equal input power can indeed be characterized as the enhancement rate in heat transfer. Figure 6 gives variations of heat transfer enhancement as a function of input power under (m_w=90 – 430 mL/h). It can be viewed that the heat transfer enhancement for all water mist concentration was greater than unity and that refers to the important role of water mist on heat transfer process. As the water mist increases, the heat transfer enhancement increase and reached its maximum value under mw= 430 mL/h. The average heat transfer enhancement increases by about 1.28%, 1.81%, 2.37%, and 2.83% in comparison with traditional air cooling. The approximation of the experimental data by the least-squares method made it possible to obtain an empirical correlation between heat transfer enhancement and the regime parameters of Rayleigh and Weber numbers as follow:

$$h_m / h_0 = 1 + (Gr \cdot \Pr)^{0,19} We^{0,35}; \quad (14)$$

With respect to fin spacing, fins number, and dissipation area, one of the important features of water mist in parallel with heat transfer enhancement it's the reduction of total heat dissipation area. As mentioned above, using water mist as a working medium of heat sink cooling led to decrease the surface temperature by about 43.5%, which contributes to the reduction of the dimensionless Rayleigh number and hence increase the fin spacing and decrease the fin numbers according to equations (10). It is experimentally confirmed that the fin spacing increased as water mist concentration increases, while the fins number was decreased as water mist concentration increases, which ultimately led to a decrease in the dissipation area as shown in figure 7.

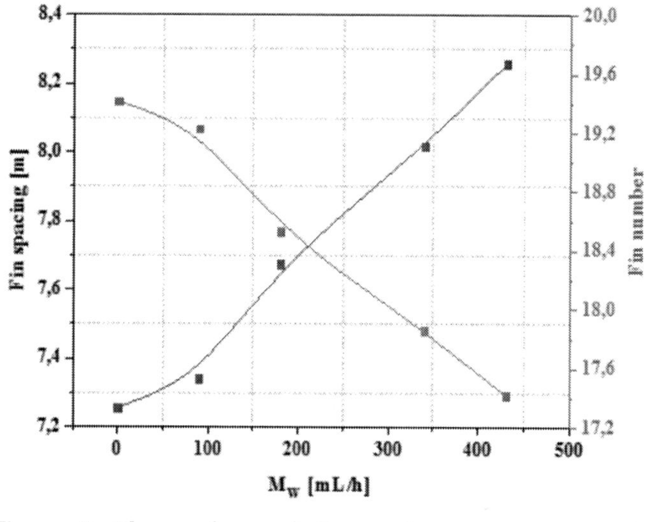

Figure 7. Fin spacing and fin number versus water mist for P=65.44 W.

5. Conclusions

The water mist influences on the natural convection heat transfer of a finned heat sink is experimentally investigated. Thermal performances of the finned heat sink with and without water mist were compared. Based on the obtained experimental results, it can summarize the following:

- Water mist decreased the surface temperature substantially over the use of air alone. The average surface temperature decreases by about 19.1%, 34%, 43.5%, and 48.2% in comparison with traditional air cooling under (m_w=90 – 430 mL/h).

- The heat transfer coefficient increases by about 27.5%, 79.1%, 135.4%, and 281% in comparison with traditional air cooling.
- The heat transfer enhancement increase and reached its maximum value under mw= 430 mL/h. The average heat transfer enhancement increases by about 1.28%, 1.81%, 2.37%, and 2.83% in comparison with traditional air cooling.
- The fin spacing increased as water mist concentration increases, while the fins number was decreased as water mist concentration increases.
- Empirical heat transfer enhancement correlation based on the experimental data obtained from the related experiment was developed and introduced for practical use.

References

[1] Yeh R-H and Chang M 1995 *International Communications in Heat and Mass Transfer* **22 (3)** 445-60
[2] Taheri A, Ghasemian Moghadam M, Mohammadi M, Passandideh-Fard M and Sardarabadi M 2020 *Energy Conversion and Management* **206** 112485
[3] Mira-Hernández C, Clark M D, Weibel J A and Garimella S V 2018 *Int. J. of Heat and Mass Transfer* **124** 782-93
[4] Yu Y, Simon T W, Zhang M, Yeom T, North M T and Cui T 2014 *Int. J. of Heat and Mass Transfer* **68** 184-93
[5] Altun A H and Ziylan O 2019 *Int. J. of Heat and Mass Transfer* **139** 425-31
[6] Kai Z, Li M-J, Wang F-L and He Y-L 2020 *Int. J. of Heat and Mass Transfer* **152** 119315
[7] Taji S G, Parishwad G V and Sane N K 2014 *Int. J. of Heat and Mass Transfer* **72** 250-9
[8] Yildiz Ş and Yüncü H 2004 *Heat and Mass Transfer* **40** 239-51
[9] Abed A H, Shcheklein S E and Pakhaluev V M 2019 *IOP Conf. Ser.: Mater. Sci. Eng.* **481(1)** 012033
[10] Khangembam C, Singh D, Handique J and Singh K 2020 *Int. J. of Heat and Mass Transfer* **150** 119368
[11] Yu F W and Chan K T 2011 *Applied Thermal Engineering* **31(4)** 537-44
[12] Lee S L, Yang Z H and Hsyua Y 1994 *Journal of Heat Transfer* **116(1)** 167-72
[13] Abed A H, Shcheklein S E and Pakhaluev V M 2019 *IOP Conf. Ser.: Mater. Sci. Eng.* **552(1)** 012001
[14] Kumari N, Vaibhav B, Marc H, Salamon T, Kolodner P, Lyons A and Garimella S V 2010 *Int. J. of Heat and Mass Transfer* **53(15-16)** 3346-56
[15] Bahadur V, Hodec M, Lyons A, Krishnan S and Garimella S V 2008 *11th Intersociety Conference on Thermal and Thermomechanical Phenomena in Electronic Systems* (Orlando: IEEE) pp 1191-8
[16] Yalcin H G, Baskaya S and Sivrioglu M 2008 *Int. Communications in Heat and Mass Transfer* **35(3)** 299-311
[17] Barrow H, and Pope C W 2007 *Applied energy* **84(4)** 404-12
[18] Kudo T, Sekiguchi K, Sankoda K, Namiki N and Nii S. 2017 *Ultrasonics sonochemistry* **37** 16-22
[19] Cengel Y 2003 *Heat and Mass Transfer a Practical Approach* (Columbus: Mc-Graw Hill Education)
[20] Moffat R 1988 *Describing the uncertainties in experimental results experimental thermal and fluid science* (New York: Elsevier Science Publishing Company Inc)
[21] Harahap F and McManus Jr H N 1967 *J. of Heat Transfer* **89(1)** 32–8
[22] Baskaya S, Sivioglu M and Ozek M 2000 *Int. J. of Thermal Sciences* **39(8)** 797-805

IOP Conf. Series: Materials Science and Engineering 1089 (2021) 012004 doi:10.1088/1757-899X/1089/1/012004

Mathematical model of the axial compressor blade in system assessing technical condition of the industrial gas turbine

V L Blinov and I S Zubkov

Department of Turbines and Engines, Ural Federal University named after the first President of Russia B. N. Yeltsin,, 19 Mira St., Ekaterinburg, 620002, Russia

E-mail: lamqtada@gmail.com

Abstract. The paper presents the key stages of developing a geometric model of the axial compressor blade. The model is based on Bezier curves, which are used to design the parts of airfoil and allow making local changes to the airfoil and blade geometry. This model is designed to assess the technical condition of the axial compressor being a part of an industrial gas turbine or gas-turbine units, depending on the defects of axial compressor blade row (in the gas-flow channel) and the turbine as a whole. As a part of the study the algorithm, which describes the process of the blade formation, were presented. Although the study presents the main equations of the model based on the key geometric parameters of the airfoil, provides examples of airfoil, blade and stage modelling using the presented method, and shows the results of model verification using numerical simulation methods.

1. Introduction

Today, gas turbine units (GTU) are one of the most widely used machines for generating and converting energy. They cover almost the entire power range required: they are applied in power industry (as part of a combined cycle power unit), transport and other industries (oil and gas, chemical, etc.). In addition, the demand for gas turbines tends to increase across the world, which makes it important to maintain their technical condition and efficiency at a high level.

These parameters, however, largely depend on the technical condition of certain GTU components [1-3]. For instance, the axial compressor substantially contributes to the overall condition and performance of a GTU, since the power consumed by the former significantly exceeds half of the power generated by the turbine. The life cycle of both axial compressor and gas turbine unit will be determined in this case by the life cycle of the elements the most subjected to load – the blade row [4]. According to the statistics, blade damages are the second largest cause of accidents and shutdowns, while various geometric deviations of certain axial compressor blades occurring during its operation have a huge impact on the GTU performance. Making blade leading edge thinner may decrease the amount of losses (when operating at the rated speed) or increase them (when operating outside the optimal range of angles of attack), as well as decrease the operational stability margin of the axial compressor. Some more examples are presented in table 1. Almost all blade defects have such an inconsistent impact [5, 6], so their assessment and analysis, taking into account the operating mode of the stage, is a relevant research area [7-9].

For this purpose, a mathematical description is needed to model a wide range of airfoils, as well as to make various changes to the shape of the blade in all its regions and sections to simulate defects. The coordinates of airfoil points are then needed to be exported to CAE systems to numerically analyze the impact of these deviations on the performance of the airfoil, blade, axial compressor and GTU as a whole.

Table 1. Defects' impact to the axial compressor and GTU performance/

№	Geometry change description	Impact description	Possible impact to[a]	
			Axial Compressor	GTU
1	Leading edge thinning	1. Leading edge loss level decrease; 2. Gas-dynamic stability margin decrease.	N_c ↓[b] η_{ad} ↑ π~ G_{air}~[c]	N_e ↑ η_e ↑
2	Trailing edge thickening	1. Vibration reliability increase 2. Stages mutual operation deterioration due to trailing edge vortices	N_c ↑ η_{ad} ↓ π~ G_{air}~	N_e ↓ η_e ↓
3	Bending angle increase	1. Suction side flow separation	N_c ↑ η_{ad} ↓ π ↓ G_{air} ↓	N_e ↓ η_e ↓
4	Radial clearance increase	1. Secondary loss level increase due to overflows through the radial clearance 2. Probability of rotor grazing the stator decrease	N_c ↑ η_{ad} ↓ π ↓ G_{air}↓	N_e ↓ η_e ↓
5	Axial clearance decrease	1. Stages mutual operation deterioration due to trailing edge vortices	N_c ↑ η_{ad} ↓ π~ G_{air}~	N_e ↓ η_e ↓
6	Airfoil thickness increase	1. Friction loss level increase 2. Blade passage configuration deterioration	N_c ↑ η_{ad} ↓ π ↓ G_{air} ↓	N_e ↓ η_e ↓

[a] N_c – axial compressor power consumption; η_{ad} – axial compressor adiabatic efficiency; π – pressure ratio; G_{air} – Mass flow rate; N_e – GTU efficient power; η_e – GTU efficient efficiency.
[b] «↑» – increase; «↓» – decrease; «~» d – doesn't change *or* difficult to assess the impact.
[c] Presented impact is observed for nominal operating mode.

2. Materials and Methods

In view of the above, a geometric model of the GTU axial compressor blade was developed, which allows taking into account changes in certain geometric parameters using the data on real post-operational and post-repair geometric deviations of certain axial compressor blades. The developed airfoil mathematical description is based on the use of key geometric parameters (pitch angle (β_y), chord (b), radii of the leading and trailing edge (R_{in}; R_{out}) and their edge angles (φ_1; φ_2), blade angles at leading and trailing edge (β_{1b}; β_{2b})

to find the position of the control points of Bezier curves (figure 1). This approach ensures highly precise design of the initial airfoil and allows flexibly changing its geometry [6, 10, 11]. Moreover, it has previously proven itself in solving airfoil optimization problems [10, 12, 13]. The developed mathematical description uses various Bezier curves – second-order Bezier curves for edge modeling, and third-order Bezier curves for the suction and pressure sides of the blade. Fourth-order Bezier curves were also used to describe the center of the mass line of the blade airfoil to set the tangent offset (figure 4-b) [14].

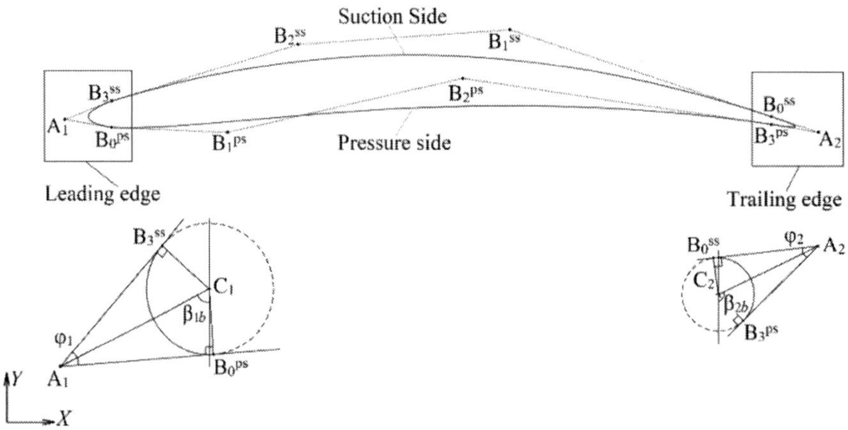

Figure 1. Airfoil description.

3. Results

According to figure 2, the initial stage of modelling is to set the initial airfoil data.

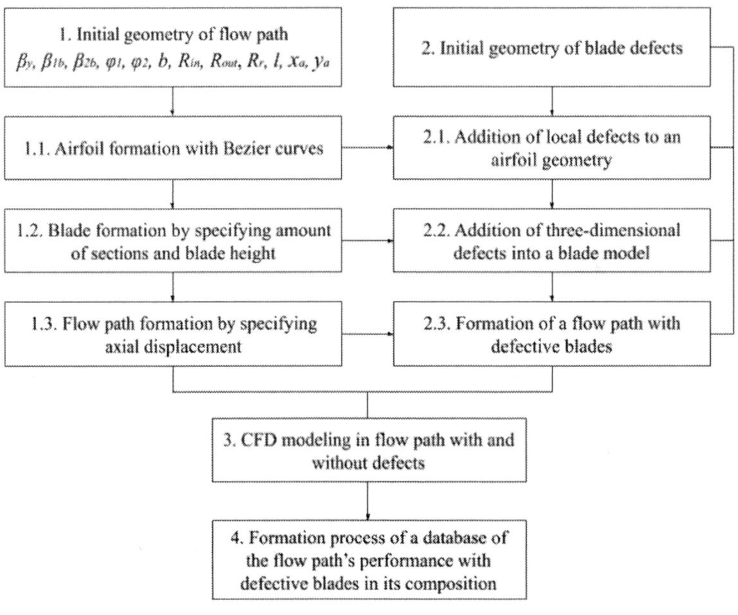

Figure 2. Axial compressor gas-flow channel outline.

The starting point for modeling process is the circular arc center of the leading edge – its vertical and horizontal coordinates are assumed to be equal to the spherical radius of the leading edge R_{in}, which help clearly set the position of the leading edge and use it as a basis for airfoil design. The present paper assumes that the smallest distance between the coordinates of the circular arc centers of the leading and trailing edges is equal to the chord minus the radii of the airfoil edges. The next stage is to find the position of the control points of Bezier curves. The coordinates of A1 (1) are intersection points of tangents to the leading edge:

$$\begin{cases} X_{A1} = R_{in} - \dfrac{R_{in}}{sin\left(\frac{\varphi_1}{2}\right)} * \sin\beta_{1b} \\ Y_{A1} = R_{in} - \dfrac{R_{in}}{sin\left(\frac{\varphi_1}{2}\right)} * cos\beta_{1b} \end{cases}. \quad (1)$$

The coordinates of control points B_0^{ps} (2) and B_3^{ss} (3):

$$\begin{cases} X_{B_0^{ps}} = X_{A1} + R_{in} * ctg\left(\frac{\varphi_1}{2}\right) * \sin\left(\beta_{1b} + \frac{\varphi_1}{2}\right) \\ Y_{B_0^{ps}} = Y_{A1} + R_{in} * ctg\left(\frac{\varphi_1}{2}\right) * \cos\left(\beta_{1b} + \frac{\varphi_1}{2}\right) \end{cases}; \quad (2)$$

$$\begin{cases} X_{B_3^{ss}} = X_{A1} + R_{in} * ctg\left(\frac{\varphi_1}{2}\right) * \sin\left(\beta_{1b} - \frac{\varphi_1}{2}\right) \\ Y_{B_3^{ss}} = Y_{A1} + R_{in} * ctg\left(\frac{\varphi_1}{2}\right) * \cos\left(\beta_{1b} - \frac{\varphi_1}{2}\right) \end{cases}. \quad (3)$$

The coordinates of A2 (4) are intersection points of tangents to the leading edge:

$$\begin{cases} X_{A2} = \left[R_{in} + (b - R_{in} + R_{out}) * \sin\beta_y\right] + \dfrac{R_{out}}{sin\left(\frac{\varphi_2}{2}\right)} * \sin\beta_{2b} \\ Y_{A2} = \left[R_{in} + (b - R_{in} + R_{out}) * \cos\beta_y\right] + \dfrac{R_{out}}{sin\left(\frac{\varphi_2}{2}\right)} * cos\beta_{2b} \end{cases}. \quad (4)$$

The coordinates of control points B_3^{ps} (5) and B_0^{ss} (6):

$$\begin{cases} X_{B_3^{ps}} = X_{A2} - R_{out} * ctg\left(\frac{\varphi_2}{2}\right) * \sin\left(\beta_{2л} + \frac{\varphi_2}{2}\right) \\ Y_{B_3^{ps}} = Y_{A1} + R_{out} * ctg\left(\frac{\varphi_2}{2}\right) * \cos\left(\beta_{1л} + \frac{\varphi_2}{2}\right) \end{cases}; \quad (5)$$

$$\begin{cases} X_{B_0^{ss}} = X_{A1} - R_{out} * ctg\left(\frac{\varphi_2}{2}\right) * \sin\left(\beta_{1л} - \frac{\varphi}{2}\right) \\ Y_{B_0^{ss}} = Y_{A1} + R_{out} * ctg\left(\frac{\varphi_2}{2}\right) * \cos\left(\beta_{1л} - \frac{\varphi}{2}\right) \end{cases}. \quad (6)$$

The coordinates of control points B_2^{ss} (7) and B_1^{ss} (8):

$$\begin{cases} X_{B_2^{ss}} = X_{B_3^{ss}} + \dfrac{2 * \left(X_{1^{ss}} - X_{B_2^{ss}}\right)}{3} \\ Y_{B_2^{ss}} = Y_{B_3^{ss}} + \dfrac{2 * \left(Y_{1^{ss}} - Y_{B_2^{ss}}\right)}{3} \end{cases}; \quad (7)$$

$$\begin{cases} X_{B_1^{ss}} = X_{1^{ss}} + \dfrac{\left(X_{B_0^{ss}} - X_{1^{ss}}\right)}{3} \\ Y_{B_1^{ss}} = Y_{1^{ss}} + \dfrac{\left(Y_{B_0^{ss}} - Y_{1^{ss}}\right)}{3} \end{cases}. \quad (8)$$

Where X_{1ss} and Y_{1ss} (9) are the coordinates of the auxiliary point, calculated as follows:

$$\begin{cases} X_{1^{ps}} = \dfrac{\left(\left[Y_{B_3^{ps}} - tg\left(90° + \beta_{2л} + \frac{\varphi_2}{2}\right) * X_{B_3^{ps}}\right] - \left[Y_{B_0^{ps}} - tg\left(90° - \beta_{1л} - \frac{\varphi_1}{2}\right) * X_{B_0^{ps}}\right]\right)}{\left[tg\left(90° - \beta_{1л} - \frac{\varphi_1}{2}\right) - tg\left(90° + \beta_{2л} + \frac{\varphi_2}{2}\right)\right]} \\ Y_{1^{ps}} = \left[tg\left(90° + \beta_{2л} + \dfrac{\varphi_2}{2}\right) * X_{1^{ps}}\right] + \left[Y_{B_3^{ps}} - tg\left(90° + \beta_{2л} + \dfrac{\varphi_2}{2}\right) * X_{B_3^{ps}}\right] \end{cases}. \quad (9)$$

The coordinates of control points B_2^{ps} (10) and B_1^{ps} (11):

$$\begin{cases} X_{B_1^{ps}} = X_{B_0^{ps}} + \dfrac{2 * \left(X_{1^{ps}} - X_{B_0^{ps}}\right)}{3} \\ Y_{B_1^{ps}} = Y_{B_0^{ps}} + \dfrac{2 * \left(Y_{1^{ps}} - Y_{B_0^{ps}}\right)}{3} \end{cases}; \quad (10)$$

$$\begin{cases} X_{B_2^{ps}} = X_{1^{ps}} + \dfrac{\left(X_{B_3^{ps}} - X_{1^{ps}}\right)}{3} \\ Y_{B_2^{ps}} = Y_{1^{ps}} + \dfrac{\left(Y_{B_3^{ps}} - Y_{1^{ps}}\right)}{3} \end{cases}. \quad (11)$$

Where X_{1ps} and Y_{1ps} (12) are the coordinates of the auxiliary point:

$$\begin{cases} X_{1^{ps}} = \dfrac{\left(\left[Y_{B_3^{ps}} - tg\left(90° + \beta_{2л} + \frac{\varphi_2}{2}\right) * X_{B_3^{ps}}\right] - \left[Y_{B_0^{ps}} - tg\left(90° - \beta_{1л} - \frac{\varphi_1}{2}\right) * X_{B_0^{ps}}\right]\right)}{\left[tg\left(90° - \beta_{1л} - \frac{\varphi_1}{2}\right) - tg\left(90° + \beta_{2л} + \frac{\varphi_2}{2}\right)\right]} \\ Y_{1^{ps}} = \left[tg\left(90° + \beta_{2л} + \dfrac{\varphi_2}{2}\right) * X_{1^{ps}}\right] + \left[Y_{B_3^{ps}} - tg\left(90° + \beta_{2л} + \dfrac{\varphi_2}{2}\right) * X_{B_3^{ps}}\right] \end{cases}. \quad (12)$$

Other geometric parameters used in the proposed description are the radius of the root section (R_r) and the height of the blade (l). By setting them, a model of the blade is formed from a certain number of airfoils. In this case, the vertical coordinate Z (13) of the airfoil points is defined as follows:

$$Z = R_r + nl; \quad (13)$$

where $n \in [0; 1]$ is a parameter that determines the position of the section (for $n = 0$ the airfoil will be located at the root of the blade, and for $n = 1$ – in its periphery). The proposed description allows for specifying 5 or more blade sections, which ensures high accuracy of its design.

The description provides for setting the airfoil axial displacement (along the x and y axes) in order to form the flow by offsetting all airfoils which constitute one blade, as well as to ensure that the blade offset can be adjusted (by offsetting certain airfoils). The center of the mass line of the blades was also described with the help of a fourth-order Bezier curve. This algorithm is analyzed in detail in [14].

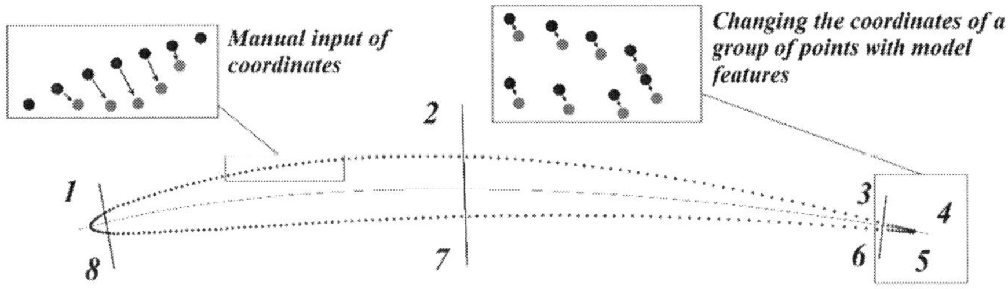

Figure 3. Introducing airfoil defects

The developed description provides for two ways of changing the airfoil point coordinates in order to introduce defects. Firstly, if the geometry of a certain airfoil (a set of coordinates) and the defect parameters (for example, the depth and location of the nick) are known, the coordinates can be manually changed in a certain part of the airfoil – in one of the 8 zones according to figure 3. Secondly, it is possible to change the coordinates of several points simultaneously by changing the position of the corresponding Bezier curve control point, which can be useful, for example, in the case of bended edges. Defects for a blade 3D model can be introduced likewise at the stage of remodeling a separate airfoil, or when working with the 3D model.

4. Discussion

The derived mathematical description allows flexibly changing the airfoil shape. The possibility to make a large number of sections (for example, 10 or more), in turn, helps create the most precise model of a blade. The coordinates of airfoil points can be exported to various CAD and CAE systems for creating blade and flow 3D models, as well as for performing numerical studies. Figure 4-a shows how airfoil forming a single blade are displayed in the developed description. Figure 4-b shows a 3D model of the blade and a scheme for offsetting. Figure 4-c shows an example of a typical axial compressor stage model.

To verify the geometry using the developed description, numerical studies of the NACA-65 airfoils were conducted [15]. Figure 5-a shows an example of a NACA 65-(27)10 airfoil model. Figure 5-b shows a comparison of its design and experimental performance. Figure 5-c shows the distribution of the pressure coefficient S along the airfoil for the angle of attack $i = 33°$. The problem was solved in a quasi-three-dimensional formulation: using the selected airfoil, a blade was formed and located at the maximum possible distance from the axis of rotation. The number of blades, as well as the boundary conditions (total pressure and temperature, flow direction at the inlet, flow rate at the outlet) were selected according to the conditions presented in the report [15].

According to the results obtained, the model can be considered accurate: the design performance almost completely coincides with the experimental one. The increase in the performance difference stems from the failures on the suction side of the blade in the region of 0.6-0.8b occurring during the calculations. However, no failures were observed during the experiment [15]. The maximum deviation was 7.8% (absolute). This result can be associated primarily with the modeling assumptions.

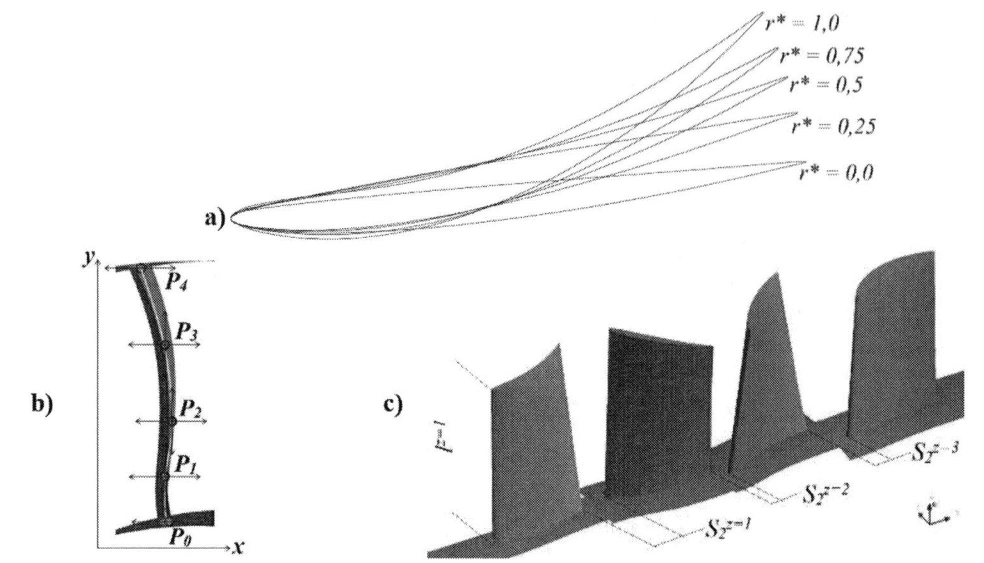

Figure 4. Blade formation (a), blade model and scheme for tangent offsetting (b), flow model (c).

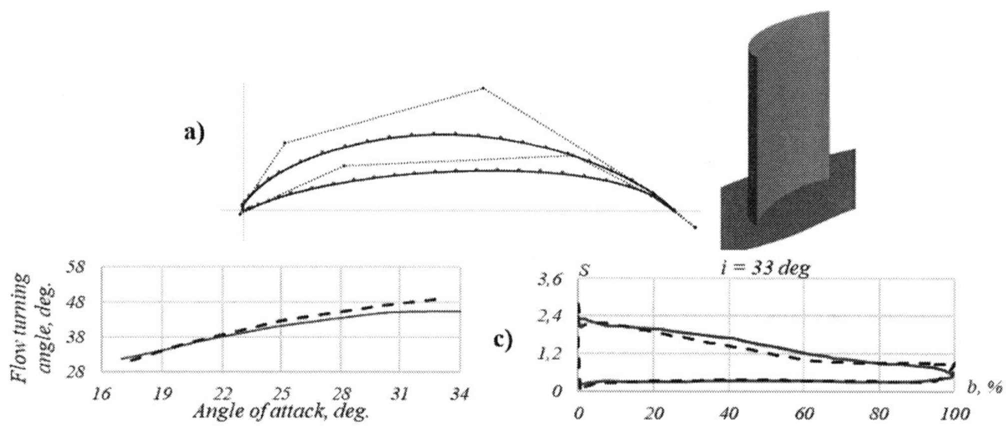

Figure 5. The verification results.

The authors aim to emphasize that the verification of the computational model parameters is the most important task of numerical studies, but it is not a priority in the current work. Demonstrating the capabilities of the developed approach is of greater importance.

5. Conclusion

The present paper presents the key results of developing a mathematical description of the airfoil using Bezier curves. The coordinates of the control points are calculated based on the key geometric parameters of the airfoil. The paper provides equations for calculating the coordinates of Bezier curve points and describes the algorithm of blade and flow modeling. Finally, this work shows the results of the verification of the developed mathematical description of the blade geometry using numerical modeling of the cascade flow.

In general, the model meets the specified requirements, providing for flexible airfoil and blade modeling, as well as making local changes in their geometry. It is designed to become the main tool for the numerical analysis of the defect influence on the characteristics of the axial compressor stage, gas-flow channel or GTU as a whole. This model can be used during the blade row repair or subsequently, with the sufficient data accumulated, to predict the development of various defects of the blades based on the degradation characteristics of the axial compressor and GTU while in operation).

References

[1] Blinov V L, Komarov O V and Zaslavskiy E A 2020 Estimation of the driven gas turbine unit technical performance using the standard measuring systems *E3S Web Conf.* **178** 01044

[2] Błachnio J, Spychała J, Pawlak W, and Zasada D 2015 The attempt to assess the technical condition of A gas turbine blade when information on its operating condition is limited *J. of KONBiN* **3(1)** 75-86

[3] Tejedor A T, Singh R and Pilidis P 2013 Advanced gas turbine asset and performance management *Modern Gas Turbines Systems* (Cambridge: Woodhead Publishing) pp 515-64

[4] Koller U, Monig R, Kusters B and Schreiber H A 1999 Development of Advanced Compressor Airfoils for Heavy-Duty Gas Turbines: Part I — Design and Optimization *ASME Int. Gas Turbine and Aeroengine Congress and Exhibition*(Indianapolis: ASME) pp 406-415

[5] Kurz R and Brun K 2016 Gas Turbine Performance *Asia Turbomachinery & Pump Symposia* (Singapure) pp 131-146

[6] Revzin B S 2000 *Osevyye kompressory gazoturbinnykh gazoperekachivayushchikh agregatov: Uchebnoye posobiye* [Axial Compressors of the Gas Turbine Gas-Pump Units: Classbook] (Ekaterinburg: USTU)

[7] Tabakoff W, Hamed A and Shanov V 1998 Blade Deterioration in a gas turbine engine. *Int. J. of Rotating Machinery* **4(4)** 233-41

[8] Al-Hamdan Q Z and Ebaid M S Y 2006 Modeling and Simulation of a Gas Turbine Engine for Power Generation *J. of Eng. for Gas Turbines and Power* **128(2)** 302-11

[9] Ogbonaya E A, Johnson K T, Ubabuike U H and Orji C U 2010 Component model-based condition monitoring of a gas turbine *ARPN J. of Eng. and Applied Sciences* **5(3)** 40-9

[10] Komarov O V, Blinov V L , Serkov S A and Sedunin V A 2016 Optimization approach and some results for 2D compressor airfoil *Int. J. of Gas Turbine, Propulsion and Power Systems* **8(3)** 39-46

[11] Jaiswal S 2017 Shape parametrization of airfoil shapes using Bezier curves *Innovative Design and Development Practices in Aerospace and Automotive Engineering* (Singapure: Springer) pp 79-85

[12] Blinov V L, Sedunin V A and Komarov O V 2019 Multi-criteria optimisation of subsonic axial compressor blading *IOP Conf. Ser.: Mater. Sci. and Eng.* **43** 012133

[13] Alm-Eldien A M, Abdel Gawad A F, Hafaz G and Abd El Kreim M 2013 Aerodynamic Shape-Optimization of Axial-Compressor Blades *Eleventh Int. Conf. of Fluid Dynamics* (Alexandria) *pp* 1-11

[14] Serkov S A, Blinov V L and Sedunin V A 2016 Vliyaniye ravnomernosti parametrov po vysote lopatok vysokonagruzhennogo osevogo kompressora na yego integral'nyye kharakteristiki i zapas gazodinamicheskoy ustoychivosti [The influence of lean dlades of highly loaded axial compressor at its integral characteristics and the stock of gas-dynamic stability] *Proceedings of the II Scientific and Technical Conf. of Young Scientists of the Ural Power Eng. Institute* (*Ekaterinburg*: URFU) pp 87-92

[15] Emery J C, Herrig L J and Erwin J R 1958 *Systematic two-dimensional cascade test of NACA 65-series compressor blades at low speeds: NACA Report 1368* (Washington: NACA)

On the possibility of obtaining ultra-supercritical steam parameters at Nuclear Power Plants with fast neutron reactors using non-nuclear steam superheating

V S Kostarev, I A Shirmanov, A A Anikin and S E Shcheklein

Department of Nuclear Power Plants and Renewable Energy Sources, Ural Federal University named after the first President of Russia B. N. Yeltsin, Mira St., 19, Ekaterinburg, 620078, Russia

E-mail: slavakostarev@yandex.ru

Abstract. Nuclear and thermal power plants constitute the largest part of Russia's power grid (about 67%). Today the world thermal power industry has already begun the transition to ultra-supercritical steam parameters, which makes it possible to increase the efficiency and reduce fuel consumption, and, accordingly, the discharge of harmful substances into the environment. Nuclear power plants (NPPs) need to increase their performance to keep up with other electric energy producers in the market. One of the ways to increase the energy efficiency of nuclear power plants with fast reactors is to improve the thermodynamic cycle. The paper presents computer-simulated steam-cycle arrangements for NPPs with a BN-1200 reactor using non-nuclear superheating of steam, as well as steam compression to obtain super-supercritical steam parameters, and an assessment of the efficiency of using these cycle arrangements.

1. Introduction

One of the most important indicators of the efficiency of nuclear power plants is the share of useful heat used by consumers from consumed part (released during a chain reaction of nuclear fuel fission). Since the overwhelming majority of nuclear power plants operating in the world use thermal neutron reactors, this predetermines relatively low initial parameters of water vapor for steam turbine plants (since the choice of pressure in the primary circuit of light water thermal reactors of high power (15.7 MPa) is associated with a limitation temperature, equal to 350 °C for cladding of fuel elements made of zirconium alloys) and leads to low values of thermal efficiency (30-35%) [1].

Fast neutron reactors operating in Russia (BN-600, BN-800 and the project of BN-1200) currently use liquid sodium, the temperature of which at the outlet from the core is about 550 ° C (the boiling point of sodium is about 878 °C). This makes it possible to generate superheated steam of high parameters (P=13 MPa, t=505 °C) in steam generators, which significantly increases the thermodynamic efficiency of nuclear power plants. Due to this, the thermal efficiency of nuclear power plants with fast reactors is about 10% higher than that of thermal reactors [2].

The fast neutron reactor system with a sodium coolant and a closed nuclear fuel cycle (NFC) is included in the list of proposed nuclear power systems of the fourth generation, which are characterized by increased indicators in the field of ensuring sustainable development, safety and reliability, as well as competitiveness.

This paper presents the results of modeling various variants of thermal schemes of the BN-1200 reactor (standard cycle, standard cycle with initial fired steam superheating, cycle with compression and one-stage intermediate, cycle with compression and two-stage intermediate fired steam superheating) using

CAD United Cycle. The aim of the work is to determine the most efficient variant of modification of the cycle arrangement of a nuclear power plant with BN-1200 reactor (by using compression and non-nuclear fired steam superheating to instruct ultra-supercritical steam parameters) from a thermodynamic point of view.

2. Materials and Methods

2.1. Research Subject

Today several methods are known to increase the energy efficiency of nuclear power plants by using both nuclear and non-nuclear steam superheating.

The Increasing of the efficiency of the NPP cycle by using nuclear steam superheating has been successfully used in the AMB-100 and AMB-200 channel type reactors, however, it has not received further development, mainly due to the need to use high-temperature steels for the reactor core, reducing the efficiency of using uranium fuel [3].

Non-nuclear steam superheating was used at the Indian Point NPP (USA), however, due to the thermodynamic inefficiency of using superheated low-pressure steam, it did not bring the expected results.

However, in heat power engineering, due to the development of power engineering technologies and the creation of high-temperature steam turbines and compressors, it became possible to widely use thermodynamic cycles at ultra-supercritical steam parameters (P=30 MPa, t=650 °C), which makes it possible to consider the possibility of achieving super-supercritical steam parameters both at NPPs with thermal and fast neutron reactors [4, 5].

Previously, the authors carried out calculations to determine the most thermodynamically effective modification of the cycle arrangement of a nuclear power plant with a VVER-1200 reactor (the article is being published in the international scientific journal «Alternative Energy and Ecology»). In the case of light water reactors (LWR), it also becomes possible to lower the parameters of the primary and secondary circuits, with subsequent compression and fired steam superheating. Lowering the initial steam parameters at the LWR will increase the depth of nuclear fuel burnup and reduce the thickness of the reactor vessel [6].

Modeling of cycle arrangements carried out by the authors showed that in the case of a VVER-1200 reactor, the cycle arrangement with compression (up to 30 MPa) and single intermediate fired steam superheating (up to 650 °C) has the highest thermodynamic efficiency, while the steam leaves the steam generator with the following parameters: t=250 °C, P=3.98 MPa).

With these parameters, according to the results of calculations, the electric power of the installation increased by 50% and amounted to 1749.09 MW with the capacity of the standard cycle arrangement being 1200 MW; The net efficiency increased by 20% and amounted to 0.429 while the efficiency of the standard cycle arrangement was 0.355. Also, due to a decrease in the parameters of the primary circuit to P=8 MPa and t=275 °C, the nuclear fuel burn-up of the reactor increased by 3% and amounted to 44.70 MW·day/kgU with a value of the standard scheme of 43.44 MW·day/kgU.

Since the use of compression and fired steam superheating to achieve ultra-supercritical parameters has proven its effectiveness in modeling cycle arrangements of light-water reactors [7], it is obvious that it is necessary to analyze the possibility of their application to fast neutron reactors [10-11].

When developing, designing and creating thermal power plants and nuclear power plants, knowledge of the quantitative and qualitative laws inherent in the systems under consideration is necessary [8-16]. This information can be obtained using mathematical modeling methods.

To solve the problems of modeling cycle arrangements of nuclear power plants, the authors used the United Cycle CAD software package, designed to solve problems and determine the best structure and composition of the equipment of a heat-and-power facility and calculate stationary operating modes.

This package has the following features:

- a highly detailed representation of the cycle arrangement being simulated;
- an advanced graphical design and visualization environment;
- models of machinery elements adjusted and refined over the years;

- highly accurate heat and mass balance sheet;
- a multi-level testing system at each step of operating mode simulation, calculation, and analysis.

2.2. Research Objective

Prior to simulation, the following initial parameters were set: the live steam temperature was taken equal to 505 °C, and the pressure to 16.30 MPa. Steam flow rate was set to 4796.10 t/h.

Several potential cycle arrangements were then designed in the United Cycle CAD system to be calculated further. The following arrangements were considered:

- BN-1200 cycle arrangement with standard parameters;
- BN-1200 cycle arrangement with standard parameters and initial fired steam superheating (to 650°C);
- arrangement with compression (to 30 MPa) and single intermediate fired steam superheating (to 650°C);
- arrangement with compression (to 30 MPa) and two intermediate fired steam superheatings (to 650°C).

Initially, cycle arrangements were calculated with no regard for regenerative heating. Later, however, net efficiency was determined for all arrangements, and regenerative heating was additionally calculated for the standard arrangement and for those with lower initial parameters and highest efficiency to determine the final net efficiency including regeneration.

3. Results

Based on the results of the calculations, an analysis of the effectiveness of using various options for cycle arrangements was carried out. Figure 1 and Figure 2 shows electric output and efficiency values obtained as a result of calculations.

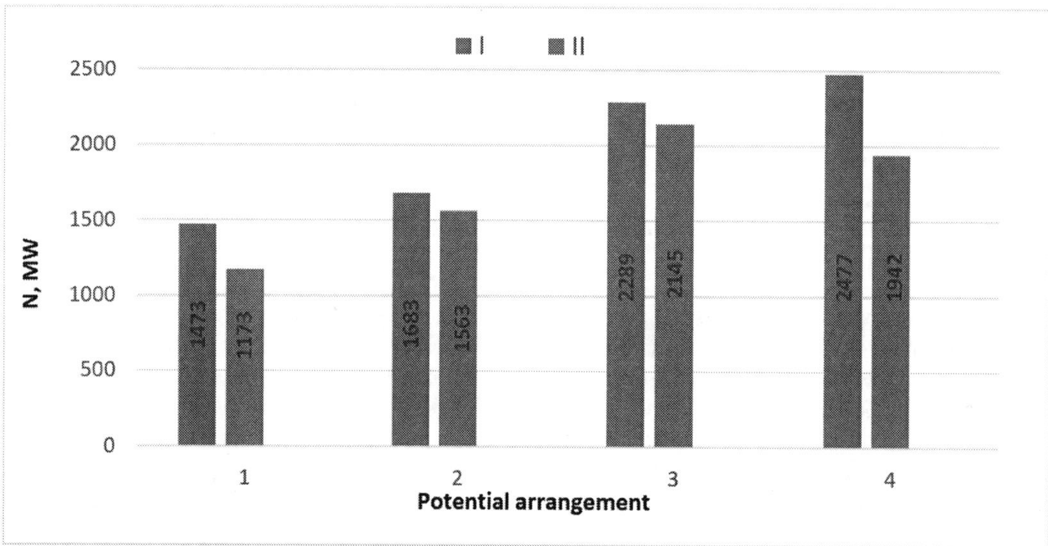

Figure 1. Net electric output: I – cycle arrangements excluding the regenerative feed water heating system: II – cycle arrangements including regenerative feed water heating system. 1 – BN-1200 cycle arrangement with standard parameters; 2 – BN-1200 cycle arrangement with standard parameters and initial fired steam superheating; 3 – arrangement with compression and single intermediate fired steam superheating; 4 – arrangement with compression and two intermediate fired steam superheatings.

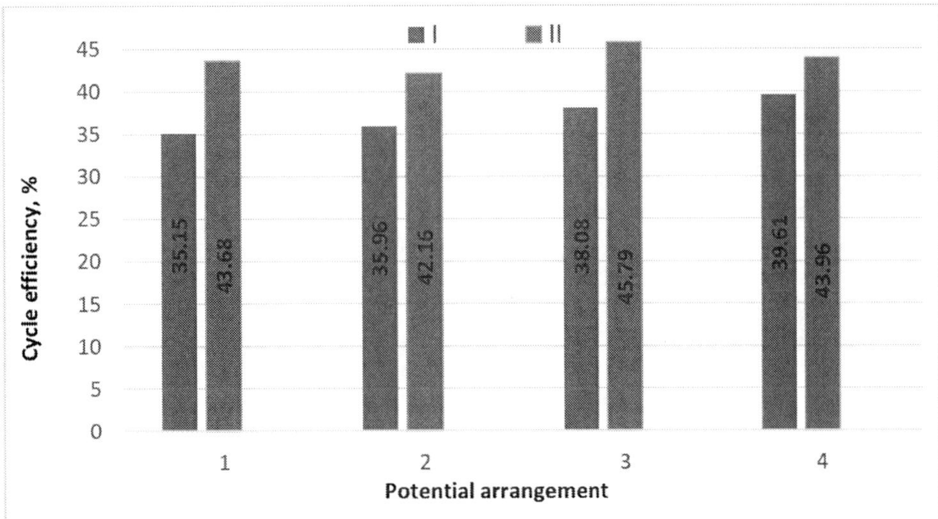

Figure 2. Net cycle efficiency. Numeric designations of the arrangements are similar to those in Figure 1.

Also, to estimate the share of nuclear power in the total output of the station, nuclear power utilization factor, ξ, was additionally calculated as:

$$\xi = \frac{Q_{SG}}{Q_{SG}+\sum Q_b}, \tag{1}$$

where Q_{SG} is the steam generator thermal output, ΣQb is the total thermal output of the steam boilers in the arrangement. The results are shown in Figure 3.

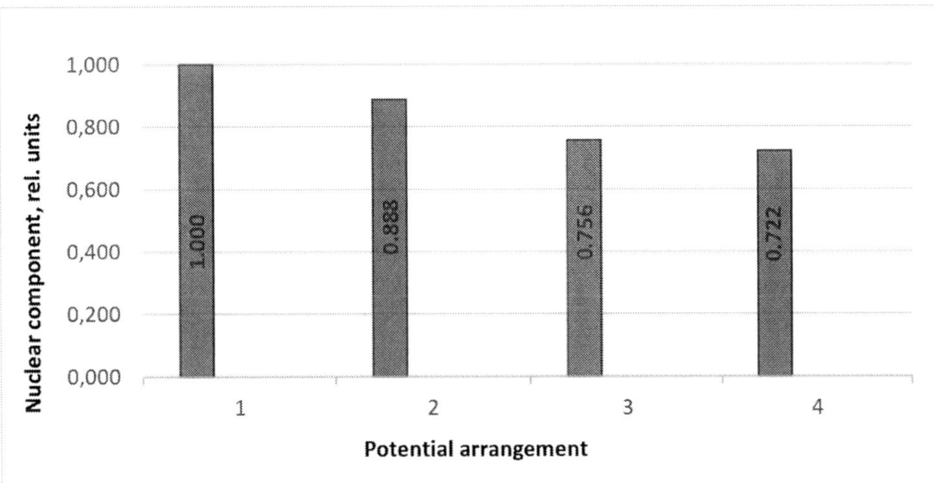

Figure 3. Nuclear power utilization factor. Numeric designations of the arrangements are similar to those in Figure 1.

From the analysis of cycle arrangements excluding regeneration, it can be seen that circuit 4 has the highest efficiency and the highest electric power (thermal circuit with compression and two intermediate fire superheats of steam). At the same time, the contribution of the nuclear component to the total power of the installation for this scheme is the smallest of all the calculated options, however, despite this fact, it remains quite high.

The process of steam expansion in a turbine for all variants of the cycle arrangements, except for the standard one, takes place in the region of deep ultra-supercriticality, which has a positive effect on the thermodynamic efficiency of the cycle. The transition to moist steam is carried out only in the last stages of the low-pressure cylinder, which is due to the operating conditions of the condenser, where superheated steam should not enter under normal conditions.

For further analysis, in the United Cycle CAD system, the authors modeled and calculated variations for cycle arrangements 1-4, taking into account the inclusion of a regenerative feed water heating (RFWH) in the turbine system (Figure 1 and Figure 2).

The values of steam parameters obtained by calculating the standard BN-1200 cycle arrangement in United Cycle software were found to match the design values (the difference did not exceed 5%), which suggests that calculations made with the software are reliable.

4. Discussion

Parameters of extraction and parameters before the condenser for each variant of the cycle arrangement, with the exception of the standard one, were selected based on the need to obtain the maximum efficiency as a result of calculations. According to the results of calculations, the highest efficiency (45.79%) and the highest power (2145.36 MW) is obtained by using the cycle arrangement with arrangement with compression and single intermediate fired steam superheating. At the same time, in order to increase the thermodynamic efficiency, it was decided to lower the steam pressure at the end of the expansion process in the LPC to from 0.052 to 0.031 kgf/cm^2, to reduce the final steam moisture from 9.73 to 6.2%, and also to exclude HPH-6 from the RFWH system (the feed water after passing HPH-5 is sent directly to the steam generator). This version of the cycle arrangement is shown in Figure 4.

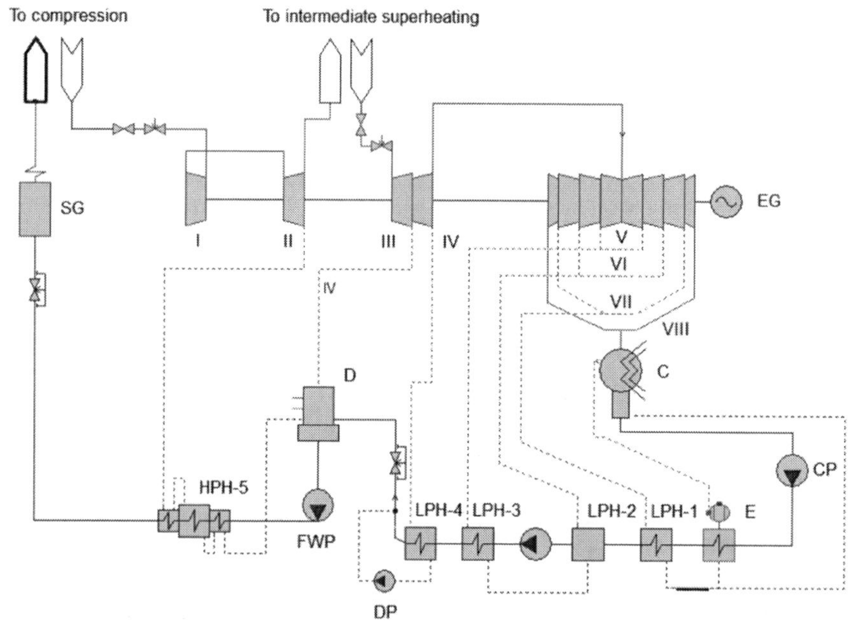

Figure 4. Cycle arrangement with compression and single-stage fossil-fired superheating, created in United Cycle software: I, II, III, IV, V, VI, VII, VIII – the turbine compartments; D – a deaerator; DP – a drain pump; C – a condenser; CP – a condensate pump; HPH-5 – a high-pressure heater; SG – a steam generator; LPH-1, 2, 3, 4 – the low-pressure heaters, FWP – a feed water pump; E – an ejector; EG – an electric generator.

Other options for cycle arrangements are of less interest, since they have much lower efficiency compared to the standard circuit. Figure 5 shows h-s chart of the steam expansion process in a turbine for a standard cycle arrangement and for an arrangement with compression and single-stage fossil-fired superheating

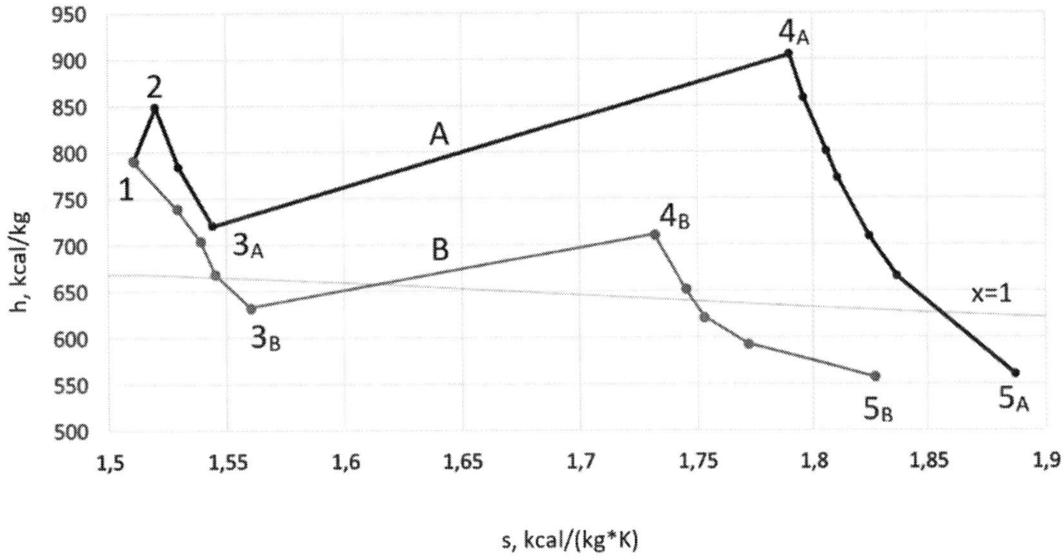

Figure 5. The h-s chart of the steam expansion process in a turbine: A – initial parameters of a cycle arrangement with compression and intermediate fired steam superheating; 2 – initial parameters of the standard BN-1200 cycle arrangement. The numbers in the figure indicate the characteristic points of the expansion process.

The numerical values of the thermodynamic parameters (pressure and temperature) at these characteristic points are given in table 1.

Table 1. Parameters of points on the diagram.

Point number	*P*, kg/cm^2	*T*, °C (*Y*, %)
Cycle arrangement with compression and intermediate fired steam superheating		
1	163	505
2	300	632.53
3_A	41	321.82
4_A	41.2	650.1
5_A	0.031	8.17
Standard BN-1200 cycle arrangement		
1	163	505
3_B	5.93	4.93
4_B	5.81	260.79
5_B	0.052	9.32

It can be seen from the chart that the process of steam expansion in a turbine for a scheme with compression and intermediate fire superheating in the region of much higher enthalpies in comparison with the standard thermal scheme, which makes it possible to significantly increase the thermodynamic efficiency of the cycle.

Figure 5 also shows the following characteristic points of the steam expansion process in the turbine:

- 1 indicates steam parameters at the outlet of the steam generator;
- 2 indicates steam parameters at the compressor outlet (only for cycle arrangement A);
- 3_A and 3_B indicate steam parameters at the end of the expansion process in the high-pressure cylinder;
- 4_A and 4_B indicate steam parameters after superheating;
- 5_A and 5_B indicate steam parameters at the end of the expansion process in the low-pressure cylinder.

5. Conclusion

It can be seen from the results that the application of steam compression (up to 30 MPa) and intermediate fired steam superheating after high-pressure cylinder into the cycle arrangement of the BN-1200 reactor leads to the increasing of the electrical power of the installation by 50%, and the increasing of the net efficiency by 5% in comparison with the standard cycle arrangement of the installation, which leads to a significant increase in the thermodynamic efficiency of the cycle, however, to a lower value compared to the numbers that were achieved when using compression and non-nuclear steam superheating in the VVER-1200 cycle arrangement.

Thus, it can be concluded that the use of compression and non-nuclear steam superheating to obtain ultra-supercritical parameters at a NPP with a BN-1200 reactor is a justified measure, since it leads to a significant increase in the electrical power generated by the NPP, as well as efficiency.

Improvement of the thermodynamic cycle of nuclear power plants with fast reactors requires further studying. The authors are going to consider the effectiveness of the use of compression and fired steam superheating using the example of fast neutron reactors with a lead coolant (BREST-OD-300).

References

[1] Shchepetina T D 2010 O povy`shenii kpd e`nergoblokov s vodo-vodyany`mi reaktorami (VVR) [On Increasing the Efficiency of Power Units with Pressurised Water Reactors (PWR)] *Energiya: Ecomomica, Tehkika, Ekologiya* **12** 21-9 [In Russian]

[2] Beltukov A I, Karpenko A I, Poluyaktov S A, Tashlykov O L, Titov G P, Tuchkov A M and Shheklein S E 2013 *Atomny`e E`lektrostancii s Reaktorami na By`stry`x Nejtronax s Natrievy`m Teplonositelem* [Nuclear power plants with sodium-cooled fast neutron reactors] (Ekaterinburg: Ural Federal University) [In Russian]

[3] Petrov A A, Romenkov A A and Yarmolenko O A 2009 Ispol`zovanie opy`ta beloyarskoj AE`S v oblasti yadernogo peregreva pri sozdanii yaderny`x reaktorov novogo pokoleniya s povy`shenny`mi i sverxkriticheskimi parametrami teplonositelya [Using the Experience of the Beloyarsk NPP in the Field of Nuclear Superheating in the Development of New Generation Nuclear Reactors with Increased and Supercritical Coolant Parameters] *Izvestiya Vuzov. Yadernaya Egengetika* **2** 21-9 [In Russian]

[4] Kasilov V F and Nizovoy A V 2015 Perspektivy` ispol`zovaniya parogazovy`x texnologij v e`nergoblokax atomny`x e`lektrostancij [Prospects for the use of the steam-gas technologies in power units of nuclear power plants] *Nauchnye Issledovaniya: ot Teorii k Praktike* **4** 34-8 [In Russian]

[5] Darwish M A, Al Awadhi F M and Bin Amer A O 2010 Combining the Nuclear Power Plant Steam Cycle with Gas Turbines *Energy* **35** 62-71

[6] Shcheklein S E, Titov G P and Borisova E V 2011 Vliyanie temperaturny`x xarakteristik na glubinu vy`goraniya yadernogo topliva [Influence of the temperature characteristics on the nuclear fuel burn-up] *Vestnik OPU* **2** 104-8 [In Russian]

[7] Shcheklein S E, Tashlykov O L and Dubinin A M 2014 Povy`shenie e`nergoe`ffektivnosti AE`S [Improving energy efficiency of Nuclear Power Plants] *Izvestiya Vuzov. Yadernaya Egengetika* **4** 15-25 [In Russian]
[8] Homenok L A, Kruglikov P A, Smolkin Yu V and Sokolov K V 2012 Opredelenie e`nergeticheskogo e`ffekta modernizacii e`nergooborudovaniya i texnologicheskix sistem AE`S [The determination of the energy effect of modernization of power equipment and technological systems of Nuclear Power Plants] *Teplotehnika* **12** 21-9 [In Russian]
[9] Hrystalyov V A and Hovikova Z Yu 2011 Povy`shenie e`ffektivnosti e`kspluatacii e`nergoblokov ae`s s vve`r-1000 putem modernizacii teplovy`x sxem [Increasing the operational efficiency of npp power units with vver-1000 through modernization of cycle arrangements] *Vestnik SGTU* **3** 166-73 [In Russian]
[10] Kazachkovskii O D 1995 *Reaktory` na by`stry`x nejtronax* [Fast Neutron Reactors] (Obninsk: IATE) [In Russian]
[11] Tashlykov O L, Shcheklein S E, Sesekin A N, Chentsov A, Nosov Y and Smyshlaeva O 2014 Ecological Features of Fast Reactor Nuclear Power Plants (NPPs) at All Stages of Their Life Cycle *WIT Transactions on Ecology and the Environment* **2** 907-18
[12] Filippov G A, Hazarov O I and Belyaev L A 2012 Puti sovershenstvovaniya paroturbinny`x ustanovok AE`S [The ways of inproving of steam turbine installations of Nuclear Power Plants] *Izvestiya TPU* **4** 5-9 [In Russian]
[13] Abdulkadyrov V M, Bogoslovskaya G P and Grabezhnaya V A 2012 Termodinamicheskie cikly` AE`S, rabotayushhix na sverxkriticheskix parametrax vody` [Thermodynamic cycle arrangements of Nuclear Power Plants operating on supercritical water parameters] *Izvestiya Vuzov. Yadernaya Egengetika* **4** 85-91 [In Russian]
[14] Dolgov V V 2002 E`nergobloki na osnove VVE`R s zakriticheskimi parametrami teplonositelya [Power units based on Water-cooled power reactor with supercritical coolant parameters] *Atomnaya Energiya* **92(4)** 277-80 [In Russian]
[15] Baranayev Yu D, Kirillov P L, Poplavskiy V M and Sharapov V N 2004 Yaderny`e reaktory` na vode sverxkriticheskogo davleniya [Nuclear reactors operating on supercritical water coolant] *Atomnaya Energiya* **96(5)** 374-80 [In Russian]
[16] Filippov G A and Yurchevskiy E B 2018 Vozmozhnosti povy`sheniya e`konomichnosti i nadezhnosti parovy`x turbin AE`S [Possibilities of increasing the efficiency and reliability of steam turbines at Nuclear Power Plants] *Teploenergetika* **9** 5-13 [In Russian]

IOP Conf. Series: Materials Science and Engineering 1089 (2021) 012006 doi:10.1088/1757-899X/1089/1/012006

Optimization of refueling times in fast neutron reactors

O L Tashlykov[1], Yu F Dolgii[2,3] and A N Sesekin[2,3]

[1]Department of Nuclear Power Plants and Renewable Energy Source, Ural Federal University, Mira St., 19, Yekaterinburg, 620002, Russia
[2]Department of Applied Mathematics and Mechanics, Ural Federal University, Mira St., 19, Yekaterinburg, 620002, Russia
[3]Institute of Mathematics and Mechanics of the Ural Branch of the Russian Academy of Science, S. Kovalevskaya St., 16, Yekaterinburg, 620108, Russia

E-mail: sesekinl@list.ru

Abstract. The importance of solving issues of increasing energy efficiency and energy saving for the development of the Russian economy was emphasized. The main directions of increasing the energy efficiency of the nuclear power industry (increasing the efficiency of using primary fuel and reducing the shutdown time of the power unit for repair and reloading of nuclear fuel) are highlighted. The importance of route optimization in solving applied problems of nuclear power is updated, in particular, in minimizing the time of nuclear fuel reloading, increasing the utilization factor of the installed capacity of NPP power units. The necessity of periodic reloading of nuclear fuel in reactors has been substantiated. The features of nuclear fuel reloading in thermal and fast neutron reactors are presented. The necessity of using in fast neutron reactors a system for guiding a capture on a fuel assembly using eccentrically located rotary plugs is substantiated. The necessity and importance of optimization of the process of guidance and movement of the overload mechanism in order to reduce the time to stop the power unit is shown. Mathematical models for the mechanisms of reloading fuel assemblies have been built. Time-optimal algorithms for the operation of overload mechanisms with two and three rotating plugs are proposed.

1. Introduction

"Energy efficiency, energy saving, nuclear power" is one of the priority directions of the development of science, technology and technology in the Russian Federation. There are various ways to improve the energy efficiency of nuclear power plants. For NPPs with thermal reactors (VVER, RBMK), the thermal efficiency is approximately 33%, with fast reactors - more than 40%. This means that most of the thermal energy released in the reactor core as a result of the nuclear fission chain reaction is released into the atmosphere. The output of useful energy can be increased, for example, by utilizing low-grade waste heat and nuclear heat supply [1], [2], [3].

Another way is to increase the utilization factor of the installed capacity of the power unit by reducing its downtime for repairs and reloading of nuclear fuel. The duration of scheduled repairs can be reduced by careful planning of work, their optimization, and the introduction of high-performance repair equipment. The critical path determines the duration of the shutdown of the power unit and includes operations for reloading nuclear fuel. Most of these operations are strictly regulated in accordance with nuclear safety rules (for example, the speed of raising and lowering fuel assemblies, moving them in a horizontal plane). But there is a potential in reducing the time of pointing the mechanism to certain coordinates and rearranging the fuel assemblies through route optimization. Work on the use of route optimization as applied to applied problems in nuclear power has been carried out by specialists from the departments of nuclear power, applied mathematics and the IMM UB RAS since the early 2000s. The results of these studies have shown their effectiveness [4], [5], [6], [7].

In VVER reactors, refueling of fuel assemblies is carried out using a refueling machine (RM) under a layer of water in the refueling basin, which performs the functions of radiation protection and removal of residual heat. The RM is a bridge moving over the reactor and the spent fuel pool, along which the cart with the reloading mechanism moves, i.e. guidance of the gripper of the reloading machine is carried out along two mutually perpendicular coordinates.

The features of route optimization in relation to such a scheme for guiding the mechanism of reloading (capturing PM) of VVER-type reactors, which form the basis of nuclear power in Russia and the world, are discussed in detail in the monograph [8]. The problem of reloading nuclear fuel for a VVER-1000 reactor contains a large number of various, poorly formalized restrictions. To solve this problem P.A. Chentsov constructed an efficient greedy algorithm, which was implemented in the form of a standard computer program.

Russia has many years of experience in the construction and operation of sodium-cooled fast reactors (RBR), which represent the innovative development of nuclear power. For 40 and 5 years, the world's most powerful fast-neutron power reactors BN-600 and BN-800 have been reliably and safely operated at the Beloyarskaya NPP. The design of the BN-1200 reactor has been developed, which by its characteristics meets the requirements of the fourth generation of safety [9], [10].

By virtue of the design feature of BN-type reactors, the guidance of the grip of the reloading mechanism on the fuel assembly to be reloaded is carried out by rotating two or three eccentrically located rotary plugs, on the smaller of which the gripper is located.

In the presented work, the ways of solving the optimization problem for the mechanisms of reloading fast neutron reactors with sodium coolant are considered.

2. Features of nuclear fuel reloading

A feature of the operation of a nuclear reactor (the occurrence of a self-sustaining chain reaction of fission) is the presence of a critical mass of nuclear fuel in the core. Since the concentration of fissile isotope nuclei (uranium-235 or plutonium-239) decreases during operation of a nuclear reactor, an excess of fissile material above the critical mass is loaded into the core, which is compensated by special neutron absorbers of the control and protection system (CPS). At present, the loading of fresh nuclear fuel ensures the operation of the NPP power unit with VVER reactors until the next refueling within 12 months, for new power units - 18 months. For the BN-600 reactor, the operating time at rated power between overloads is up to 160 days [11].

In general, the reloading of nuclear fuel is as follows: spent 3-4-5 years fuel assemblies (depending on the duration of the fuel campaign for VVER-type reactors) are unloaded from the central part of the core into the spent fuel pool; partially burnt fuel assemblies move (rearrange) in the core from the periphery to its center; fresh fuel assemblies are loaded (installed) on the periphery of the core. The procedure for rearranging the fuel assemblies is strictly regulated and is carried out taking into account the observance of nuclear safety (preventing the occurrence of an uncontrolled fission chain reaction).

In the core of fast breeder reactors (RBR), the power density is about 500 MW / m3, which is 50 times more than in RBMK reactors and 5 times more than in VVER. Sodium is used to ensure reliable heat dissipation. Being a highly efficient heat carrier, sodium has a number of disadvantages, first of all it is high chemical activity with respect to water and air. The presence of sodium in the reactor vessel requires its reliable isolation from the environment in all operating modes, including refueling. Therefore, in modern fast reactors with a sodium coolant, a system is used to guide the gripper of the refueling mechanism of fuel assemblies to the required coordinates of the core, breeding zone, in-reactor storage or lift slots by rotating two (three for BN-800) eccentrically located rotary plugs (columns), which are at the same time a sealing lid of the reactor and biological shielding (Fig. 1). Rotation tightness is ensured with the help of a water seal filled with a tin-bismuth alloy with a low melting point [11].

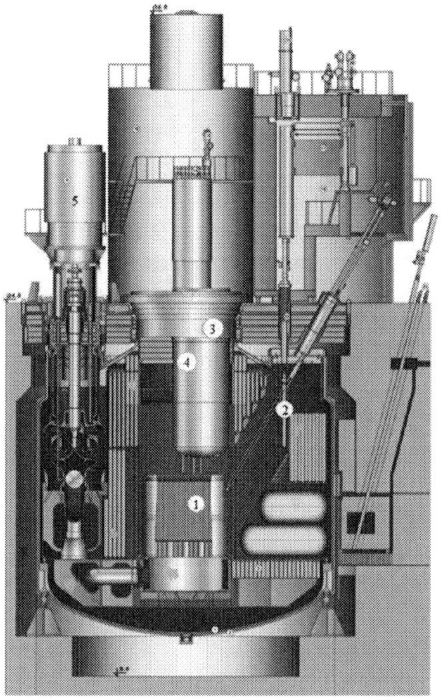

Figure 1. BN-600 reactor: 1 - core; 2 - fuel assembly in the liner of the inclined lift; 3 - large swivel plug; 4 - small swivel plug (central swivel column); 5 - circulation pump.

Some of the first such reactor plants (RU) had a loop layout (for example, BN-350 - [12]), others - integral, when all the equipment of the primary circuit is located in one vessel-vessel of the reactor. Currently, the integral layout is recognized as the most optimal.

In this paper, we consider the problem of the fastest possible guidance of a gripper located on a smaller plug on a given fuel assembly under the assumptions: for the BN-600 reactor, that plugs can rotate simultaneously, and for the BN-800 reactor, that three plugs will rotate sequentially, i.e. only one plug can be turned at a time.

To unload a burned-out fuel assembly from the reactor, the reloading machine must move to its coordinates, lower the gripper, perform coupling, lift the gripper from the fuel assembly, transfer it to the spent fuel pool, lower it, release and raise the gripper. These operations are performed for each unloaded fuel assembly, regardless of the reloading order. In the reverse order, fresh fuel assemblies are loaded. The time spent on the operation of lowering-lifting the gripper of the reloading machine, coupling-uncoupling the gripper from the fuel assembly, can be ignored in the calculation algorithm, since it is a parameter that cannot be optimized in this context. As a result, the problem can be represented as flat, while some of the displacements are not considered [8]. The considered problem of optimizing the sequence of operations for rearranging fuel assemblies consists in minimizing the time for replacing nuclear fuel and, accordingly, reducing the downtime of the NPP power unit.

3. Mathematical models of overload mechanisms and optimal control algorithms

Reloading mechanism for the BN-600 reactor consists from two rotating plugs. The big plug is a disk of radius R_1 with a geometric center fixed during movement. The large plug has an eccentric circular cutout of radius R_2, into which a small plug is placed - a disk of radius R_2, the center of mass of which is at a distance e_2 from its geometric center (Fig. 2). When describing the interaction of plugs, we neglect the friction forces. The movement of the mechanical system is flat. The axial moment of inertia of the large plug is J_1 and of the small plug relative to its center of mass - J_2, the mass of the second plug is m_2. The distance between the geometric centers of the large and small plugs is e_1.

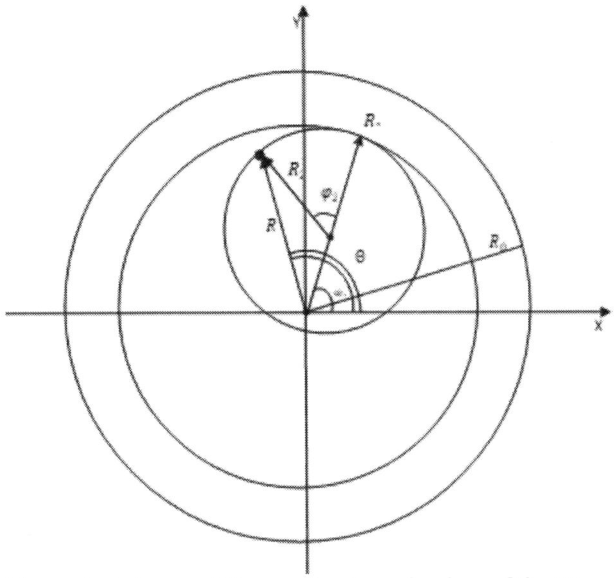

Figure 2. Scheme of the overload mechanism of the BN-600 reactor.

Under the assumption that the geometric center of mass of the small plug coincides with its center of mass (ie, the parameter $e_2 = 0$), according to [13, 14], the mathematical model describing the movement of the overload mechanism will have the form:

$$\ddot{\varphi}_1 = \frac{1}{\Delta} J_2(u_1 - u_2),$$
$$\ddot{\varphi}_2 = \frac{1}{\Delta}(-J_2 u_1 + (J_1 + J_2 + m_2 e_1^2) u_2), \tag{1}$$

where

$$\Delta = J_2(J_1 + m_2 e_1^2).$$

The control moments applied to the large and small plugs have restriction

$$|u_1| \leq \mu_1,$$
$$|u_2| \leq \mu_2. \tag{2}$$

Let

$$a_1 = \frac{1}{J_1 + m_2 e_1^2},$$
$$a_2 = \frac{J_1 + J_2 + m_2 e_1^2}{J_2(J_1 + m_2 e_1^2)}.$$

Then the system of equations (1) takes the form

$$\ddot{\varphi}_1 = a_1(u_1 - u_2),$$
$$\ddot{\varphi}_2 = -a_1 u_1 + a_2 u_2. \tag{3}$$

Let us rewrite system (3) in normal form. As a result, we get

$$\begin{aligned} \dot{\varphi}_1 &= \omega_1 \\ \dot{\omega}_1 &= a_1(u_1 - u_2) \\ \dot{\varphi}_2 &= \omega_2 \\ \dot{\omega}_1 &= -a_1 u_1 + a_2 u_2 \end{aligned} \tag{4}$$

The coordinates of a point in a fixed coordinate system r and θ are related to the variables φ_1 and φ_2 using the formulas

$$R = \sqrt{(R_1 - R_2)^2 + R_2^2 - 2(R_1 - R_2)R_2\cos(\pi - \varphi_2)},$$

$$\theta = \varphi_1 + arccos\frac{R^2 + (R_1 - R_2)^2 - R_2^2}{2(R_1 - R_2)R} sign(\varphi_2).$$

Let the start point have coordinates $x^0 = (\varphi_1^0,\ 0,\ \varphi_2^0,\ 0)^T$,, and the end point have coordinates $x^f = \left(\varphi_1^f,\ 0,\ \varphi_2^f,\ 0\right)^T$. The phase vector $x(t) = \left(\varphi_1(t), \omega_1(t), \varphi_2(t), \omega_2(t)\right)^T$ is described by the system of equations (4). Consider the problem of the fastest movement of the phase point from the initial position to the final one subject to constraints (2). This problem was solved in [13, 14] using the L.S. Pontryagin maximum principle [15]. Let's introduce the functions $\varphi = \varphi_1 + \varphi_2$ and $\chi = \varphi_1 + \frac{\alpha_1}{\alpha_2}\varphi_2$. Let's $\Delta\varphi = \varphi_1^f + \varphi_2^f - \varphi_1^0 - \varphi_2^0$ and $\Delta\chi = \varphi_1^f + \frac{\alpha_1}{\alpha_2}\varphi_2^f - \varphi_1^0 - \frac{\alpha_1}{\alpha_2}\varphi_2^0$. Then, according to [14,15], if the condition

$$a_2|\Delta\chi| \le \frac{\mu_1}{\mu_2} a_1|\Delta\varphi|, \tag{5}$$

then the optimal control solving the time optimal problem will have the form

$$u_1 = \frac{4a_2\left(\Delta\varphi_1 + \frac{\alpha_1}{\alpha_2}\Delta\varphi_2\right)}{a_1(a_2 - a_1)\vartheta_2^2} sign\left(t - \frac{\vartheta}{2}\right),$$

$$u_2(t) = \mu_2 sign\big(\varphi(\vartheta) - \varphi(0)\big) sign\left(\frac{\vartheta}{2} - t\right).$$

if the condition

$$a_2|\Delta\chi| \ge \frac{\mu_1}{\mu_2} a_1|\Delta\varphi|, \tag{6}$$

then

$$u_1(t) = \mu_1 sign\big(\chi(\vartheta_1) - \chi(0)\big) sign\left(\frac{\vartheta}{2} - t\right).$$

$$u_2 = \frac{4\Delta\varphi}{(a_2 - a_1)\vartheta_1^2} sign\left(\frac{\vartheta_1}{2} - t\right).$$

Thus, for all admissible starting and ending points, we have constructed optimal controls. The found optimal controls are relay ones with one switching. The optimal time for case (5) is determined by the formula

$$\vartheta_2 = \sqrt{\frac{|\varphi(\vartheta) - \varphi(0)|}{\mu_2}},$$

and for case (6) is given by the formula

$$\vartheta_1 = \sqrt{\frac{\alpha_2|\chi(\vartheta_1) - \chi(0)|}{\alpha_1\mu_2}}.$$

In [16], a similar problem was considered for the BN-800 reactor. Let us introduce the notation. Let R_1 be the radius of the disk - the big plug, O_1 is the geometric center of the big plug, which remains stationary all the time. The large plug has an eccentric circular cutout of radius R_2 inside which is placed the middle plug - a disk of radius R_2, the geometric center O_2, which during movement remains stationary relative to the large plug. The middle plug has an eccentric circular cut of radius R_3, inside which is placed a small plug - a disk of radius R_3, the geometric center O_3, which during movement remains stationary relative to the middle plug (see Fig. 3).

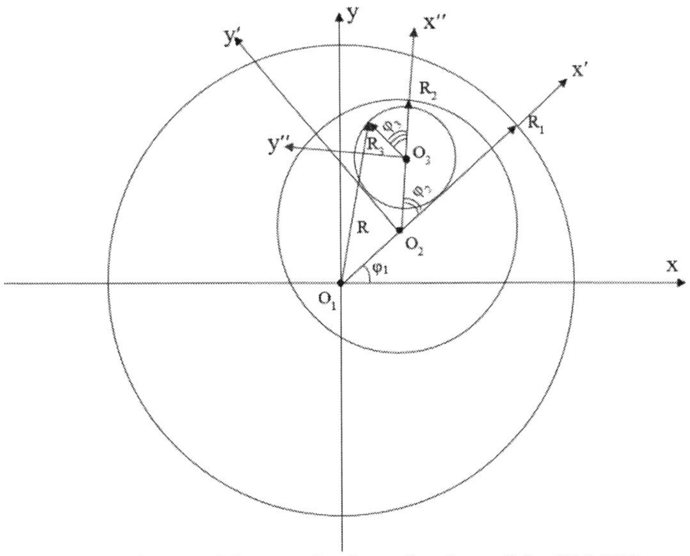

Figure 3. Scheme of the overload mechanism of the BN-800 reactor.

When describing the interaction of plugs, we neglect the friction forces. The movement of the mechanical system is flat. The mathematical model for the transfer mechanism with three plugs is as follows

$$\begin{gathered}
J_1\ddot{\varphi}_1 + J_2(\ddot{\varphi}_1 + \ddot{\varphi}_2) + J_3(\ddot{\varphi}_1 + \ddot{\varphi}_2 + \ddot{\varphi}_3) \\
+m_2(e_2^2\ddot{\varphi}_1 + e_2 a\cos(\varphi_2 + \alpha)(2\ddot{\varphi}_1 + \ddot{\varphi}_2) - e_2 a\sin(\varphi_2 + \alpha)(2\dot{\varphi}_1 + \dot{\varphi}_2)\dot{\varphi}_2 \\
+m_3(e_2^2\ddot{\varphi}_1 + e_3^2(\ddot{\varphi}_1 + \ddot{\varphi}_2) + e_2 e_3\cos(\varphi_2)(2\ddot{\varphi}_1 + \ddot{\varphi}_2) \\
-e_2 e_3\sin(\varphi_2)(2\dot{\varphi}_1 + \dot{\varphi}_2)\dot{\varphi}_2) = u_1, \\
J_2(\ddot{\varphi}_1 + \ddot{\varphi}_2) + J_3(\ddot{\varphi}_1 + \ddot{\varphi}_2 + \ddot{\varphi}_3) + m_2 e_2 a(\cos(\varphi_2 + \alpha)\ddot{\varphi}_1 + \sin(\varphi_2 + \alpha)\dot{\varphi}_1^2) \\
+m_3\left(e_3^2(\ddot{\varphi}_1 + \ddot{\varphi}_2) + e_2 e_3(\cos(\varphi_2)\ddot{\varphi}_1 + \sin(\varphi_2)\dot{\varphi}_1^2)\right) = u_2,
\end{gathered}$$

$$J_3(\ddot{\varphi}_1 + \ddot{\varphi}_2 + \ddot{\varphi}_3) = u_3 \tag{7}$$

As generalized coordinates, we select the angle of rotation of the large plug -φ_1, the angle of rotation of the middle plug - φ_2, with respect to the large plug, and the angle of rotation of the small plug - φ_{13}, with respect to the middle plug. The generalized forces are the control moments u_1, u_2, and u_3, applied to the large, medium and small plugs, respectively.

Note that the mathematical model in the case of three rotating plugs is much more complicated than the model for two rotating plugs. In this regard, it was not possible to obtain an optimal solution under the assumption that all three plugs can rotate simultaneously.

A variant of the problem was considered when the plugs rotate sequentially. According to (7), if only the i-th plug rotates, then the equation of motion will have the form

$$J_i^* \ddot{\varphi}_i = u_i, \text{ i=1,2,3.} \tag{8}$$

here

$$J_1^* = J_1 + J_2 + J_3 + m_2 e_2^2 + 2e_2 a \cos(\varphi_2 + \alpha) + m_3 \left(e_2^2 + e_3^2 + 2e_2 e_3 \cos(\varphi_2)\right),$$
$$J_2^* = J_2 + J_3 + m_3 e_3^2$$
$$J_3^* = J_3.$$

According to [13, 14], the optimal time for the reversal of one i-th plug through the angle $\Delta\varphi_i$ is set by the value

$$\theta_i = 2\sqrt{\frac{J_i}{\mu_i}|\Delta\varphi_i|} \quad i = 1,2,3. \tag{9}$$

Then the total time for the successive turn of three traffic jams is given by the value

$$T = \theta_1 + \theta_2 + \theta_3 = 2\sqrt{\frac{J_1}{\mu_1}|\Delta\varphi_1|} + 2\sqrt{\frac{J_2}{\mu_2}|\Delta\varphi_2|} + 2\sqrt{\frac{J_3}{\mu_3}|\Delta\varphi_3|}. \tag{10}$$

In this case, the values $\Delta\varphi_i$ (i = 1,2,3) are connected by the constraints

$$\begin{aligned} x = R_3 \cos(\varphi_{10} + \varphi_{20} + \varphi_{30} + \Delta\varphi_1 + \Delta\varphi_2 + \Delta\varphi_3) + e_3 \cos(\varphi_{10} + \varphi_{20} + \Delta\varphi_1 + \Delta\varphi_2) + \\ + e_2 \cos(\varphi_{10} + \Delta\varphi_1), \\ y = R_3 \sin(\varphi_{10} + \varphi_{20} + \varphi_{30} + \Delta\varphi_1 + \Delta\varphi_2 + \Delta\varphi_3) + e_3 \sin \varphi_{10} + \varphi_{20} + \Delta\varphi_1 + \Delta\varphi_2 + \\ + e_2 \sin(\varphi_{10} + \Delta\varphi_1). \end{aligned} \tag{11}$$

As a result, we got a nonlinear programming problem: it is required to minimize functional (10) under constraints (11).

The problem of minimizing functional (10) under constraints (11) was solved in [16] numerically. It follows from (11) that in fact $\Delta\varphi_i$ (i = 1,2) depend on $\Delta\varphi_3$. Below are the graphs of T dependence $\Delta\varphi_3$. Software has been developed that allows finding a solution to problem (10) - (11). Below are the graphs of the functions $T = T(\Delta\varphi_3)$.

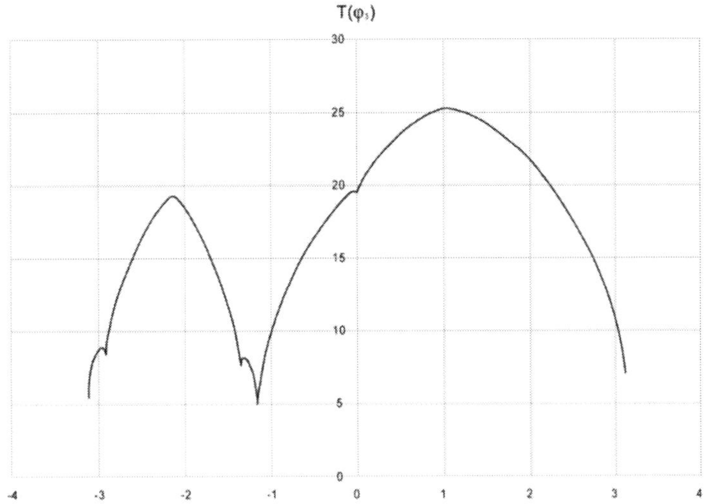

Figure 4. The boundary conditions for the problem, the solution of which is this in Figure, were as follows, $x = 0.4$; $y = 0.6$; $\Delta\varphi_{10} = 1$; $\Delta\varphi_{20} = 1$; $\Delta\varphi_{30} = -1$.

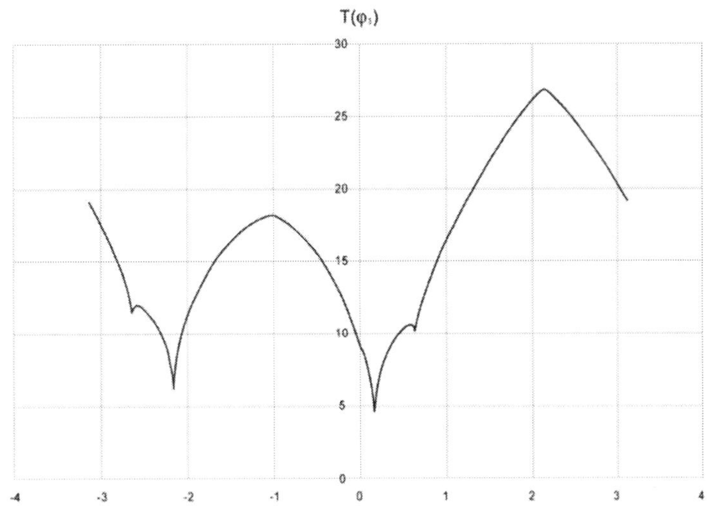

Figure 5. The boundary conditions for the problem, the solution of which is this in Figure, were as follows, $x = 0.5$; $y = 0.5$; $\Delta\varphi_{10} = -1$; $\Delta\varphi_{20} = 1$; $\Delta\varphi_{30} = 1$.

4. Conclusion

An optimal algorithm is proposed for the problem of the fastest guidance of the capture of the fuel refueling mechanism for the BN-600 reactor. For the BN-800 reactor, assuming that the plugs can rotate only sequentially, the time-optimal guidance algorithm is proposed. Application of the proposed algorithms will help to reduce the time of reloading nuclear fuel. And this, in turn, will lead to a reduction in the downtime of a reactor shut down for reloading nuclear fuel.

5. Acknowledgments

The research was supported by Russian Foundation for Basic Research, project no. 20-08-00873.

References

[1] Kostarev V S, Tashlykov O L and Klimova V A 2019 The increasing of the energy efficiency of nuclear power plants with fast neutron reactors by utilizing waste heat using heat pumps. *IOP Conf. Ser.: Mater. Sci. Eng.* **552(1)** 012022

[2] Tashlykov O L, Tolmachev E M, Semenov M Yu and Sapozhnikov B G 2012 Snizhenie teplovy`x nagruzok AE`S na okruzhayushhuyu sredu putem ispol`zovaniya teplovy`x nasosov v sxeme osnovnogo kondensatora paroturbinnoj ustanovki [Reducing heat loads NPP on the environment through the use of heat pumps in the circuit of the main steam turbine condenser] *Al`ternativnaya E`nergetika i E`kologiya* **3** 16-21 [In Russian]

[3] Tashlykov O L, Kovin I V and Kokorin V V 2012 Utilizatsiya nizkopotentsial'noy teploty AES s reaktorom na bystrykh neytronakh s ispol'zovaniyem teplovogo nasosa [Disposal of low-grade heat from the nuclear reactor on fast neutrons with heat pump] *Al`ternativnaya E`nergetika i E`kologiya* **3** 22-5

[4] Sesekin A N, Tashlykov O L, Shcheklein S Y and Chentsov A G 2014 Route optimization in the removal of radiation hazards *WIT Transactions on Ecology and the Environment* **190(2)** 919-26

[5] Petunin A A, Sesekin A N, Tashlykov O L and Chentsov A A 2017 Marshrutnaya optimizaciya na ob``ektax ispol`zovaniya yadernoj e`nergii i v mashinostroenii [Route optimization on the nuclear objects and in mechanical engineering] *Mathematical modeling and information technologies MMIT 2016* (Yekaterinburg: *CEUR Workshop Proceedings*) 69-79 [In Russian]

[6] Chentsov A A, Chentsov A G, Sesekin A N and Tashlykov O L 2019 Application of a generalized bottleneck routing problem to the task of adhering to acceptable doses of radiation during the dismantling of radiation hazardous objects *IFAC-Papers On Line* **52(13)** 2656-61

[7] Grigoryev A M and Tashlykov O L 2019 Rou te optimization during works in non-stationary radiation fields with obstacles *AIP Conf. Proc.* **2174(1)** 020216

[8] Korobkin V V, Sesekin A N, Tashlykov O L and Chentsov A A 2012 *Metody` marshrutizacii i ix prilozheniya v zadachax povy`sheniya e`ffektivnosti i bezopasnosti e`kspluatacii atomny`x stancij* [Routing methods and their applications in improving the safety and efficiency of operation of nuclear power plants] (Moscow: Novy`e texnologii) [In Russian]

[9] Tashlykov O L, Sheklein S E, Sesekin A N, Chentsov A A, Nosov Y V and Smyshlaeva O 2014 Ecological features of fast reactor nuclear power plants (NPPs) at all stages of their life cycle *WIT Transactions on Ecology and the Environment* **190(2)** 907-18

[10] Tashlykov O L, Sheklein S E, Nosov Y V and Smyshlaeva O 2016 Ecological foresight in the nuclear power of XXI century *Int. J. of Energy Production and Management* **1(2)** 133-140

[11] Beltukov A I, Karpenko A I, Poluyaktov S A, Tashlykov O L, Titov G P, Tuchkov A M and Shcheklein S E 2013 *Atomnyye elektrostantsii s reaktorami na bystrykh neytronakh s natriyevym teplonositelem. V 2 chastyakh. Chast' 1* [*Nuclear power plants with fast neutron sodium cooled reactors* In 2 parts, part 1] ed S E Shcheklein and O L Tashlykov (Yekaterinburg: URFU) [In Russian]

[12] Nosov Y V, Rovneiko A V, Tashlykov O L and Shcheklein S E 2019 Decommissioning Features of BN-350, -600 Fast Reactors *Atomic Energy*. 2019 **125(4)** 219-223

[13] Dolgii Y F, Tashlykov O L, Sesekin A N and Zaynullina E Z 2018 Optimal control of the fuel reload mechanism *IFAC-PapersOnLine* **51(32)** 636-41

[14] Dolgii Y F, Petunin A A, Sesekin A N and Tashlykov O L 2018 Optimal control of the system of coupled cylinders *AIP Conf. Proc.* **2048** 020007

[15] Pontryagin L S, Boltyanskii V G, Gamkrelidze R V and Mishechenko E F 1962 *The mathematical theory of optimal processes* (New York: John Wiley & Sons)

[16] Dolgii Yu F, Sesekin A N, Tashlykov O L and Tran K T 2019 Sequential optimal control of the nuclear fuel reload mechanism *AIP Conf. Proc.* **2172** 070015

IOP Conf. Series: Materials Science and Engineering 1089 (2021) 012007 doi:10.1088/1757-899X/1089/1/012007

Organizing the information security in Modbus TCP interfaces for use in the energy complex

F R Ametov, E A Bekirov and M M Asanov

Electrical Energy and Electrical Engineering Department, V.I. Vernadsky Crimean Federal University, Academician Vernadsky Ave., 4, Simferopol, 295007, Russia

E-mail: ametov.ferat@gmail.com

Abstract. A modern automated control system at electric power companies is a complex of many systems, modules and interfaces. One of the most important and, at the same time, vulnerable is the switching module, which is a set of devices and interfaces for transmitting, processing and analyzing data. The Modbus protocol is considered in this work as one of the most common interfaces for control systems at power plants. Its advantages and disadvantages are analyzed. Modern control interfaces are susceptible to cyber attacks, which poses a threat to the entire energy complex. The study of Modbus problems is considered, the main solutions for their elimination are proposed and analyzed in order to increase resistance to cyberthreats. A proprietary solution for modernizing the Modbus protocol was proposed based on the implementation of its own authentication mechanism, as well as data transmission using cryptographically strong encryption algorithms. General functional and structural schemes, as well as schemes that take into account the use of the proposed solution, have been formed. The calculation of the cryptographic strength of the proposed cryptography algorithms is made. Conclusions regarding the potential increase in resilience to cyber threats were formulated based on the data collected during the analysis.

1. Introduction

Automated control systems for the modern energy complex components are an integrated distributed computing system that switches over open protocols, which were developed primarily to provide functional technological advantages and ease of use.

Initially, these systems were meant and built as protected from external influences local subsystems, but their encapsulation and security became ineffective and outdated in the process of communication technologies development.

The control and communication systems architecture is a set of components, systems, subsystems, protocols and algorithms that determines the ways of interconnection between them.

Figure 1 shows a general diagram of a communication system, which consists of a main controller, which is a computational one, a module and a power converter. At the same time, the switching module is used to connect the controller with other modules and controllers [1].

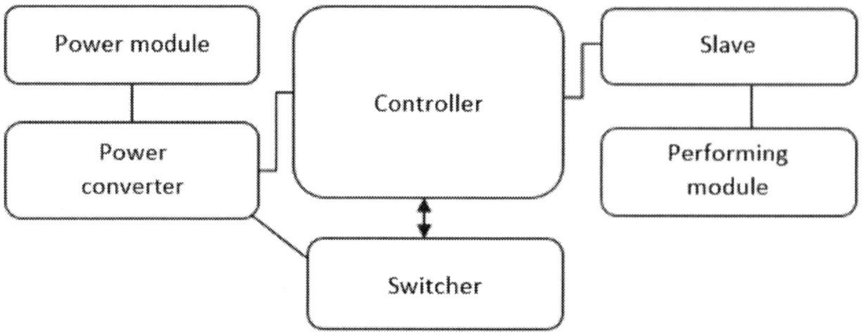

Figure 1. Diagram of the simplest communication system.

1.1. Description of a problem

Most of the components, protocols and interfaces used for organizing control systems were developed and introduced into production a long time ago. Information security of such systems has become outdated over time, and the possibility of modernization is impractical, ineffective and, in some cases, even impossible.

In this regard, an urgent task for electric power information systems is the implementation of system components secure communication protocols, the development and implementation of effective solutions to ensure secure data transmission [2].

The aim of the study is to develop a modernized, safe and efficient interaction and communication process of electric power information systems.

The following tasks have been set to achieve the goal:

1. Analyze the most widely used protocols and interfaces of information system components interaction;
2. Build and review functional and structural interfaces diagrams;
3. Analyze and propose a solution to modernize existing solutions for the components interaction;
4. Implement own solutions to ensure information security of communication system components.

The study will be carried out within the Modbus communication protocol framework. According to statistics, almost all modern ASZU and SCADA systems have Modbus interfaces support, which allows them to be implemented and modernized everywhere with minimal costs.

Modbus is an open communication protocol based on a master-slave architecture. It can be used to transfer data via RS-485, RS-422, RS-232 serial communication lines, and TCP / IP networks.

Let us consider the Modbus TCP protocol, which is widely used to organize the industrial networks control [3].

The protocol data interface has a common approach with other Modbus types, which is the packet transfer of data encoded into a binary format. Figure 2 shows a comparative diagram of the Modbus RTU and Modbus TCP protocols.

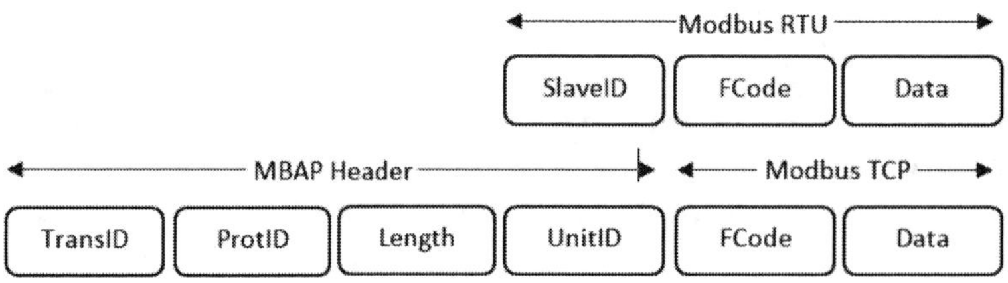

Figure 2. Structural difference between Modbus TCP and Modbus RTU.

The feature of packet transmission in TCP/IP networks allows encoding messages within the network with an additional MBAP Header block containing information to uniquely identify each request [4].

Thus, the signal signature is different:

Modbus RTU: 11 03 006B 0003 7687

Modbus TCP: 0001 0000 0006 11 03 006B 0003

where:

- 11 - the component (slave) unique identifier;
- 03 - the executed function code;
- 006B - the first requested register address;
- 0003 - total number of requested registers;

TCP MBAP Header supplements the request with a few more bits of information:

- 0001 - the request unique identifier;
- 0000 - protocol identifier;
- 0006 - message length.

2. Materials and Methods

2.1. Structural diagram

All devices built on the Modbus interface work at the physical level according to the Master → Slave model. This means that there is a Master control unit and Slave devices controlled by it (Figure 3).

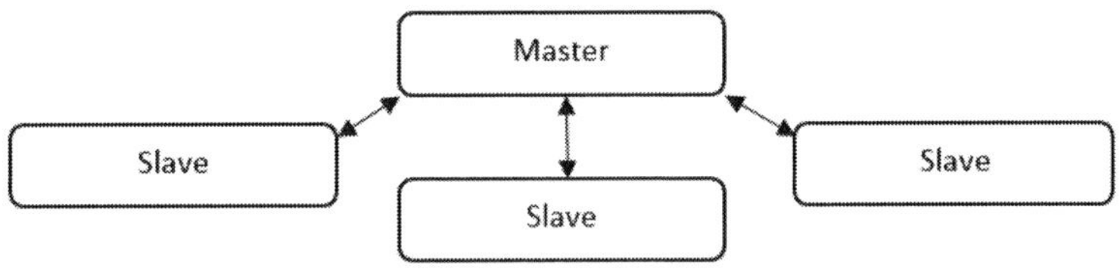

Figure 3. The master → slave model.

The Modbus operation principle is based on the fact that the master device alternately polls the slave devices that respond to it. The latter, in turn, cannot initialize the request themselves [5].

Figure 4 shows a simple functional diagram according to the master-slave principle.

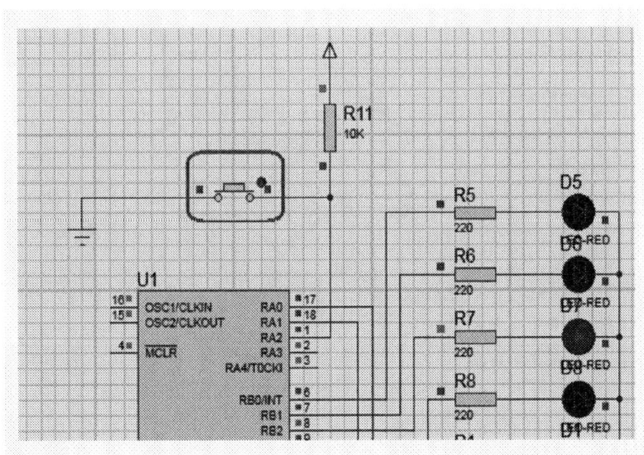

Figure 4. Block diagram according to the master-slave principle.

The diagram shown in Figure 4 contains the main components of the communication interface: the controller (U1) for communication with the final indication devices D1-D20 via R1-R20 resistors, respectively. An input device that generates signals to the microcontroller is separately allocated [6].

An important component of the Modbus TCP protocol is the ability to integrate with external servers via a TCP / IP connection, the structural model of which is shown in Figure 5:

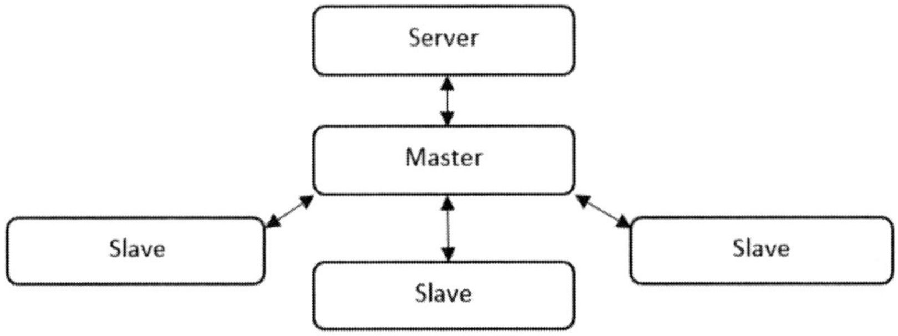

Figure 5. Master → slave model of Modbus TCP protocol.

The master device has the ability to exchange requests with remote servers for various kinds of calculations or checks.

Thus, the master microcontroller generates auxiliary signals to the slave components, measures the analog values of the electrical device required operating parameters and transmits them to the computing servers. In addition, it can receive data from the server and transmit it to the slave output devices [7].

2.2. Functional diagram

The functions and subfunctions graphical systematization of the Modbus interface is shown in the functional diagram in Figure 6:

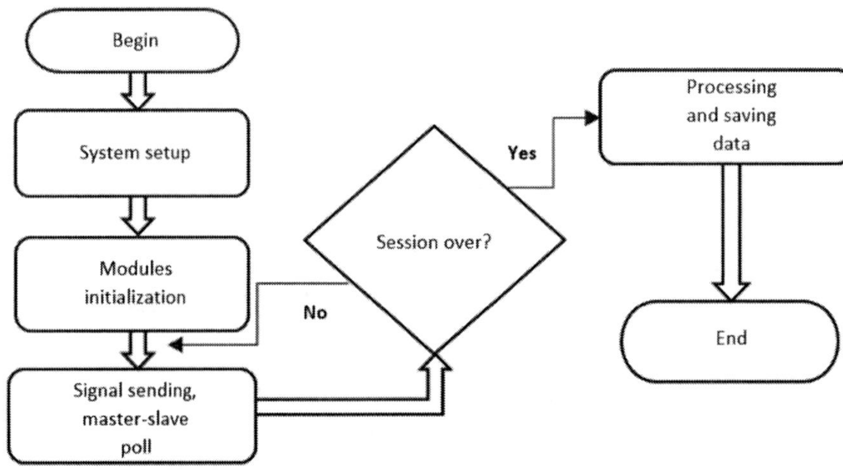

Figure 6. Functional block diagram.

The functional block diagram defines the sequence of command execution, the conditions for data or control flows, and the requirements for operation and performance as required by the baseline.

2.3. *Conclusions on data analysis*
Modbus is a widespread and functional communication interface between information systems components in industrial plants. It has a number of advantages, including: scalability, openness, low demand for resources. However, it has a number of disadvantages related to information security:
- data transmission is made in clear text;
- the ability to authenticate users and devices is not provided.

2.4. *Research problems Analysis*
The Modbus protocol was developed in the 1980s and has gradually been modernized, but safety issues have not been given due attention. Only in 2018, the architects of Schneider Electric USA Inc first proposed the modernized Modbus Secure protocol, which implies the use of Modbus TCP features to implement authentication within the system.

Modbus Secure uses the modern TLS cryptographic protocol for authentication, based on sophisticated encryption algorithms. This technology assumes the use of X.509 trust certificates to identify components among themselves. The protocol features are:
- Each system component is assigned a certificate with a public key
- The server acting as the certificate verification center has a private key to check the validity of the remaining components public keys;
- Checking the trust of certificates is performed before each request to master or from it;
- Data transmission is encrypted according to the algorithms defined on the server for the installed certificate [8].

Modbus Secure solves the main information security problems inherent in Modbus interfaces, and has the ability to be implemented step-by-step without significant changes to the current designed systems.

It is worth highlighting the disadvantages of the protocol:
- Each request to or from master is verified with a trust certificate, which is labor-intensive in terms of encryption / decryption of asymmetric encryption algorithms;
- To identify expired certificates, one must compare their unique identifiers with a special list each time.

3. Results
The analysis shows that the main requirements for a secure interface are the ability to authenticate and encapsulate data. Modbus Secure solves the main problems, but has its own disadvantages in the form of increased labor costs for hiding data.

The current research proposes a solution to improve and modernize algorithms for checking trust and encrypting messages in order to increase the information security efficiency of electric power communication systems

The formulated task of developing a new solution will represent the modernization of Modbus Secure in several aspects.

3.1. *Device authentication within the protocol*
The most time consuming part of authentication is encrypting and decrypting public and private keys using asymmetric encryption algorithms. This procedure occurs every time a signal is sent when using the master device [9].

It is proposed to introduce the term session into the communication process. This protocol is widely used in modern communication infrastructures.

A session has a certain length, during which the exchange of data is considered valid and possible. Each session has its own unique identifier - sessionId [10]. This identifier is stored on the master component, and is transmitted at each connection to the server to check the relevance of the current session.

The parties terminate the connection at the end of the session, and re-check the trust of certificates and register a new session upon reconnection [11].

3.2. Data encryption

Modbus Secure encrypts data with asymmetric encryption algorithms defined on the server. The often used algorithms are RSA, RC4 or AES. These algorithms imply the public keys transfer for decryption from master components to the server [12].

As part of the authentication implementation through the sessionId exchange, it becomes necessary to encrypt this value before transmission. An end-to-end encryption algorithm based on the Diffie-Hellman protocol is proposed to use for this.

3.3. Diffie-Hellman Protocol

This protocol implies a secure exchange of keys, i.e. the key is not transferred from one switching device to another in any form, but is generated on each side separately, by matching some parameters (Figure 7) [13].

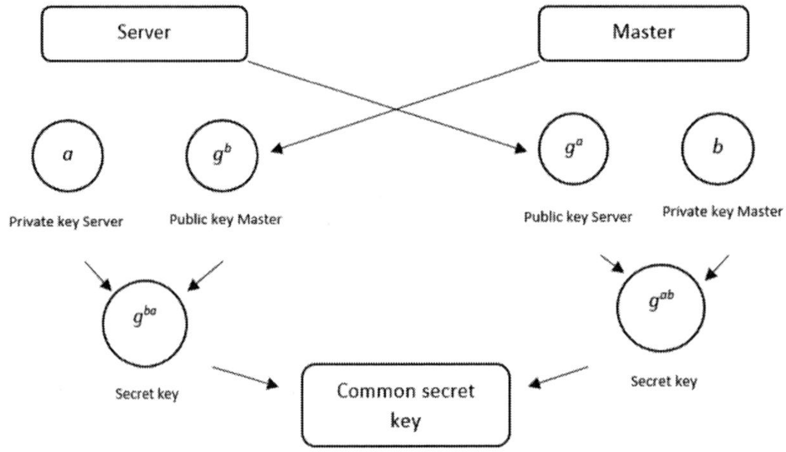

Figure 7. Diffie-Hellman secret key generation algorithm.

Let us suppose that Server (S) and Master (M) want to start exchanging information. They need to implement an algorithm to generate a shared encryption key.

Let p and g are large primes known to both sides. The main primitive of the algorithm is finding the value modulo [14]. General formula is the following:

$$y = g^x \bmod p \tag{1}$$

Function (1) is considered to be unidirectional because the calculation of the y value is simple, and the calculation of the discrete logarithm function (2) for large primes is very difficult.

$$x = \log_g y \bmod p \tag{2}$$

The value a is generated on the S side and passed to M:

$$2 \le a \le p-1$$

$$\text{Server} \to \{S = g^a \bmod p \text{ S}\} \to \text{Master} \tag{3}$$

The value b is generated on the M side and transmitted to S:

$$2 \le b \le p-1$$

$$\text{Master} \to \{M = g^b \bmod p\} \to \text{Server} \tag{4}$$

The following session key is calculated on the Master side:

$$Km = S^b \bmod p \tag{5}$$

The following session key is calculated on the Server side:

$$Ks = M^a \bmod p \tag{6}$$

As a result, we get the same generated keys:

$$Km = Ks \tag{7}$$

4. Discussion

Suppose that the public values are $p = 23$, $g = 5$. The results of calculating the secret key are shown in Table 1.

Table 1. Results of calculating the secret key for each link component

Side	Secret key	Public key	Opponent's public key	Key calculation	Key
Server	6	$S = 5^6 \bmod 23 = 8$	$M = 5^{15} \bmod 23 = 19$	$Ks = 19^6 \bmod 23$	2
Master	15	$M = 5^{15} \bmod 23 = 19$	$S = 5^6 \bmod 23 = 8$	$Km = 8^{15} \bmod 23$	2
Hacker	?	?	?	?	?

Table 1 analysis shows that an attacker can know the common values of p and g, but the secret keys generated by each of the components on their side are not known to him and compromising the secret key becomes a very difficult task, and when using TLS authentication, it is almost impossible.

The functional diagram of the upgraded Modbus protocol is shown in Figure 8.

Figure 8 shows the functional diagram of the switching module, taking into account the upgrades proposed during the study. An authentication process is added before the immediate start of the communication session and after the modules are initialized. This process is modified in Modbus Secure and improved by a different approach to the trust verification process.

In addition, the signal is sent from master to slave and vice versa via the encrypted Diffie-Hellman protocol, due to which the data is transmitted in an encapsulated form. Moreover, "Man-in-the-Middle" attack becomes impossible thanks to the authentication implementation [15].

5. Conclusions

The Modbus interface protocol, which is most in demand in industrial networks, was considered. Its components are determined, based on which functional and structural diagrams are built. The protocol has a number of advantages, among which that it is massive and undemanding to auxiliary systems. However, information security problems make Modbus-based systems susceptible to cyber-attacks.

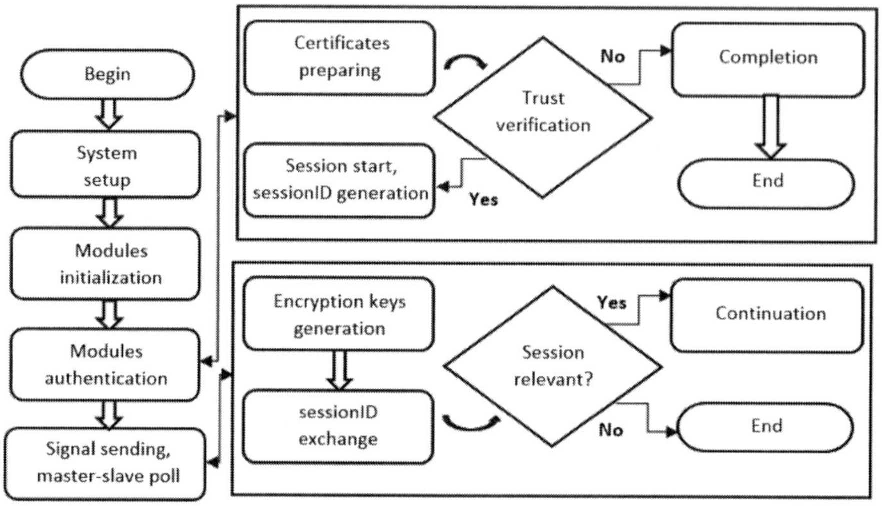

Figure 8. Functional diagram of the system taking into account the proposed modernization.

The goal of the study and the tasks for achieving it were determined as a result of the analysis. The main focus of the research is the protocol modernization in order to increase its safety and stability.

Researches in the field of considered problems are analyzed, innovative proposals and solutions applied in practice are examined.

Based on the considered principles and solutions, our own proposals for modernizing the Modbus protocol in order to increase its stability are identified and suggested, the principles and algorithms proposed for implementation are substantiated.

References

[1] Knapp E D 2014 Industrial Network Security: Securing Critical Infrastructure Networks for Smart Grid, SCADA, and Other Industrial Control Systems (Waltham: Elsevier Science)

[2] Hieb J 2008 Security Hardened Remote Terminal Units for SCADA Networks: PhD thesis (Kentucky: University of Louisville Press)

[3] Goldenberg N and Wool A 2013 Accurate modeling of Modbus/TCP for intrusion detection in SCADA systems Int. J. of Critical Infrastructure Protection **6(2)** 63-75

[4] MODBUS/TCP Security Protocol Specification URL https://modbus.org/docs/MB-TCP-Security-v21_2018-07-24.pdf

[5] Ramirez J P, Garcia E and Alvarez J 2020 Master-slave synchronization via dynamic control Communications in Nonlinear Science and Numerical Simulation **80** 104977

[6] Aslam S, Hannan S, Haider A and Hamza Tariq M 2016 Exploring PIC 24F series Microcontroller using MPLAB and Proteus JCRS **4(2)** 164-76

[7] Zhu Y Fang and Xiong L 2012 Communication platform for energy management system in a master-slave control structure microgrid Harbin: 7th Int. Power Electronics and Motion Control Conf. (Harbin: IEEE) pp 141-5

[8] ICS Security Conference URL https://www.icscybersecurityconference.com

[9] Hayes G and El-Khatib K 2013 Securing modbus transactions using hash-based message authentication codes and stream transmission control protocol Third Int. Conf. on Communications and Information Technology ICCIT (Beirut: IEEE) pp 179-84

[10] Breau J. BelserJ. Doerr R. Rieschick G. 2012 Patent US8213423B1 Unique session ID sharing

[11] Anderson S and Garvin R 1995 Sessioneer: flexible session level authentication with off the shelf servers and clients Computer Networks and ISDN Systems **27(6)** 1047-53

[12] Fovino I N, Carcano A, Masera M and Trombetta A 2009 Design and Implementation of a

Secure Modbus Protocol Int. Conf. on Critical Infrastructure Protection (Arlington: Springer) pp 83-96
[13] Steiner M, Tsudik G and Waidner M 1996 Diffie-Hellman key distribution extended to group communication New Delhi: 3rd ACM Conf. on Computer and Communications Security (New York: ACM) pp 31-7
[14] Kocher P 1996 Timing Attacks on Implementations of Diffie-Hellman, RSA, DSS, and Other Systems Annual Int. Cryptology Conf. (Santa-Barbara: Springer) pp 104-13
[15] Khader A S and Lai D 2015 Preventing man-in-the-middle attack in Diffie-Hellman key exchange protocol 22nd International Conference on Telecommunications ICT (Sydney: IEEE) pp 204-8

Research of operation the parameters in the boiler TGMP-204HL at burning gas

M A Taymarov[1], R V Akhmetova[2], E A Akhmetov[3] and A S Sabirzyanova[4]

[1]Department of Power Engineering, Kazan state power engineering university, Krasnoselskaya Str., 51, Kazan, 420066, Russia
[2]Department of Electric station, Kazan state power engineering university, Krasnoselskaya Str., 51, Kazan, 420066, Russia
[3]Department of Industrial power system and systems of heat supply, Kazan state power engineering university, Krasnoselskaya Str., 51, Kazan, 420066, Russia
[4]Department of Power supply of enterprises and energy resource saving technologies, Kazan state power engineering university, Krasnoselskaya Str., 51, Kazan, 420066, Russia

E-mail: ahmetova_rv@bk.ru

Abstract. The efficiency of fuel combustion in power boilers of thermal power plants largely depends on the layout of the burners in the combustion chamber, on their number and unit capacity. An important condition for the efficiency of the selected combustion mode is the absence of a torch surge from the combustion products onto the screen heating surfaces of the furnace. To prevent the formation of large amounts of harmful emissions of nitrogen oxides into the atmosphere, the necessary requirement for burning fuels is to reduce the local values of the flame temperature in the volume of the furnace. The use of powerful burners in power boilers, with a decrease in their number, leads to an increase in the length of the torch and, with their frontal arrangement, gives a torch thrust onto the rear screen. The article presents the results of an experimental trail of the efficiency of gas burning in the direct-flow supercritical boiler TGMP-204HL of the Surgut GRES-2 PJSC UNIPRO when it is operating on 24 burners at various steam loads. The values of the efficiency of boilers, emissions of nitrogen oxides and the distribution of the radiation intensity of combustion products along the height of the furnace are given. Comparison of the obtained results on efficiency and emissions of nitrogen oxides with the data of the TGMP-204HL boiler using 36 burners is carried out.
Keywords. Boiler, the efficiency factor, natural gas, combustion, burner, intensity, radiation, nitrogen oxide emissions.

1. Introduction

The efficiency of fuel combustion in power boilers of thermal power plants is influenced by many factors, among which the main ones are the design of the burners their location in the furnace, the heat output of the furnace, the temperature and pressure of the feedwater, the method of air heating, the presence of recirculation of flue gases [1-11]. Combustion of fuels in boilers of high power supercritical pressure has its own characteristics as the boilers are once-through. The TGMP-204HL single-shell supercritical boiler selected for experiments with gas combustion is installed at the most powerful in Russia Surgut GRES-2. Boiler TGMP-204HL (station No. 1) is designed to burn fuel using 36 vortex gas-oil burners located in three tiers on the front and rear walls of the furnace. The boiler TGMP-204HL (station No. 1) has been modernized and operates at a load of 800 MW on 24 burners located on the first lower and second tiers of the Surgut GRES-2. The third upper tier of 12 burners was turned off during the experiments. A feature of the boiler operation to reduce nitrogen oxide emissions is the use of a smoke exhauster for recirculation of combustion products directly into the burners. This article presents the results of an experimental research of the operating parameters of the boiler when using such a scheme for burning natural gas.

2. Literature review

In addition to the design of the TGMP-204HL boiler with a counter-wall arrangement of burners the Taganrog Boiler Plant designed and manufactured for the Ryazan GRES a single-body TGMP-204P boiler with a bottom arrangement of gas-oil direct-flow vortex burners of the Ural ORGRES design [12]. The purpose of the development of the TGMP-204P boiler is to reduce the uneven distribution of heat generation along the height of the furnace, which has a TGMP-204HL boiler with a counter wall arrangement of 36 burners [13, 14]. When the bottom is located in the boiler TGMP-204P, the consumption of natural gas through one burner is 16.8×10^3 m^3 / h, and their number is 12 pieces (Figure 1).

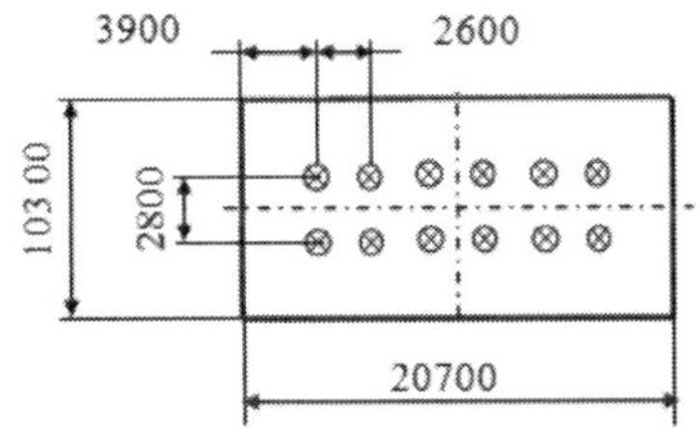

Figure 1. The diagram of the bottom arrangement of burners in the TGMP-204P boiler.

Structurally the bottom burner has a central supply of combustion gas 1 from the collector to the coaxial conical pipe 13 (Figure 2). Typical tangential peripheral vane air swirlers 3.5 with fixed blades have a twist parameter of about unity. To improve mixture formation during fuel oil combustion, an axial air swirler 6 is provided in the bottom burner design.

A characteristic feature of the hearth arrangement is the low radiation values in the lower part of the furnace volume (Figure 3). According to the results of tests of TGMP-204P boilers at the Ryazan GRES, the intensity of flame radiation in the lower part of the furnace decreased by 20 % compared to the TGMP-204HL boiler. As a result, the temperature of the metal of the shield tubes of the lower radiation part decreased by 40 ° C, which made it possible to increase their service life. However, the use of the hearth layout resulted in a decrease in the efficiency of the TGMP-204P boiler, compared to the TGMP-204HL boiler, due to the increased temperature of the exhaust gases [12-13].

To reduce the temperature of the exhaust gases and increase efficiency, the TGMP-204HL boiler (station No. 1) at Surgut GRES-2 was upgraded by increasing the flow capacity of the burners for the burned gas and increasing the air twist. This made it possible to reduce the number of working burners from 36 to 24 and reduce the temperature of the exhaust gases by disabling the third upper tier of 12 burners.

3. Materials and Methods

The single-flow single-body boiler TGMP-204HL (marking according to GOST PP - 2650-25-545 / 542GM) produced by the Taganrog boiler plant is designed to produce supercritical steam with a pressure of 25.5 MPa and a temperature of 545 ° C when burning natural gas and working in a block with a single-shaft condensing turbine LM3 K-800-240-5. The TGMP-204HL boiler with a gas-tight steam capacity of 2650 t / h has a U-shaped layout. The boiler consists of a combustion chamber and a downcomer flue, connected in the upper part by a horizontal rotary flue. The boiler is equipped with two freestanding regenerative rotating air heaters SKHP 32/1850 with a rotor diameter of 14 m.

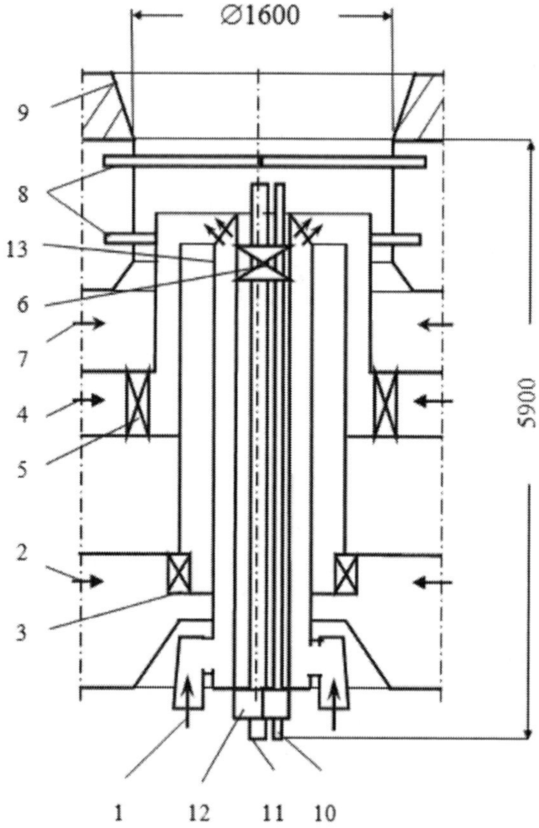

Figure 2. The diagram of the bottom burner of the TGMP-204P boiler at Ryazan GRES with a twist parameter of 1.061: 1 – central natural gas supply; 2 – air supply to the central annular channel, 3 – air-swirling tangential blades of the central register, 4 – air supply to the peripheral annular channel, 5 – air-swirling tangential blades of the peripheral register, 6 – axial air swirler for burning fuel oil; 7 – recirculation flue gas supply, 8 – compensatory lenses, 9 – embrasure, 10 – electric igniter, 11 – oil nozzle, 12 – pneumatic valve, 13 – coaxial conical pipe.

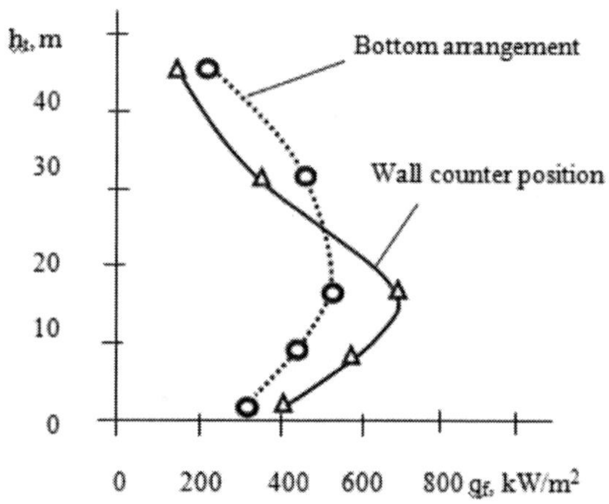

Figure 3. The intensity of the flame radiation q_f along the height h_t of the middle section of the furnace of the TGMP-204P boiler at Ryazan GRES with a bottom arrangement of 12 burners and for the TGMP-204HL boiler of the Surgut GRES-2 with a wall opposite arrangement of 36 burners on three tiers during gas combustion with a recirculation coefficient of the unit equal to 23% and electrical loads 800 MW [12-14].

The air heating temperature is 360 ° C. The design feed water heating temperature is 270 ° C at a pressure of 31 MPa. The boiler combustion chamber is made in the form of a rectangular cross-section with a depth of 10.3 m and a width of 20.7 m, shielded by a bottom screen, lower, middle and upper radiation parts in the form of panels of floating pipes. The boiler does not have its own frame and is suspended from the metal structures of the boiler house. The mark at the top of the boiler is 67.3 m.

The bottom screen is located in the lower part of the firebox and constructively shields a part of the front and rear walls up to the level of 22.8 m. In addition, two middle panels of the first lower radiation part (LRP-1) are located on the side walls of the furnace up to a mark of 22.8 m. The second lower radiation part (LRP-2) is formed from eight extreme panels of the front and rear walls of the lower part of the furnace (along two panels on the side of the side walls) located up to the 22.8 m mark and the four outer panels of the side walls to the 22.8 m mark.

The first medium radiation unit (MRP-1) is a direct continuation of the LRP-2 screens. The second middle radiation part of the MRP -2 screens the central part of the front, rear and side walls of the furnace and is located between the marks of 22.8 m and 35.8 m. The upper radiation part (URP) shields the upper part of the combustion chamber and occupies completely the front, side and rear furnace walls between 35.8 m and 55.3 m marks, and the URP of the rear wall of the furnace screens the aerodynamic protrusion and part of the horizontal flue at the 43.6 m mark. The ceiling of the horizontal flue and convection shaft furnace is screened by a ceiling screen. On the side walls of the reversing chamber between the marks of 43.6 m and 55.3 m, there are screens of the horizontal flue. In the horizontal gas duct in the area of the gas window, a platen superheater (PSH) is installed then the first and second stages of the convective high-pressure superheater CHPSH- (1, 2) and the second stage of the convective low-pressure superheater CLPSH -2 are installed along the gas flow. gases in the convection shaft are installed CLPSH -1 surfaces. The last in the flow of gases are the blocks of the water economizer. The rear screen of the URP forms an aerodynamic protrusion in the upper part of the combustion chamber which protects the screens from direct radiation from the combustion chamber.

The steam-water path of the boiler is done double-flow with autonomous regulation of the medium flow and temperature of the superheated steam. The primary steam temperature is controlled by injections in front of the screens and in the cut of the high-pressure superheater. To regulate the temperature of the secondary steam, two recirculation smoke exhausters GD-26-12-1 are installed which collect the flue gases behind the convection shaft and supply them to the burners. Injections after the first stage of the low pressure superheater are also used. The recirculation gases are injected into the burners and also serves to reduce the rate of high-temperature corrosion of the LRP and suppress nitrogen oxides. In the design, the combustion chamber is equipped with 36 gas-oil burners with a capacity of 60 MW, placed in opposite directions, 18 pieces each on the front and rear walls of the furnace in three at elevations of 12.9 m, 15.9 m and 18.9 m. Air and recirculation gases are supplied to the burners by common ducts. The boiler is designed to operate on a balanced draft and is equipped with two VDN-36 x 29 blower fans and two DOD-43-500 smoke exhausters GM.

In figure 4 shows a diagram of the TGMP-204HL boiler (station No. 1) of the Surgut GRES-2 when using 24 burners during the experiments during gas combustion. The measurement of the light emission from the flame was carried out through 44 hatches, on the walls along the height of the firebox from service areas 3 (Figure 4). In figure 4 shows, as an example, the design arrangement of 36 burners in three tiers (view A). As already noted, the experiments were carried out with 6 burners on the front wall and 6 burners on the back wall turned off on the third tier.

The use of 24 burners for use on three tiers is need, due to the fact that the temperature of the combustion products at the outlet of the furnace when using the third upper tier increases greatly and the heat load on the superheating surfaces increases. This leads to the need to turn on desuperheaters so that the temperature of the superheated steam does not exceed 545 °C. When the burners of the third tier are switched on, the temperature of formation of thermal nitrogen oxides also increases, since the temperature of the combustion products recirculated to the burners from the downcomer flue also increases.

The diagram of the vortex gas-oil burner with a thermal power of 60 MW used on the TGMP-204HL boiler (station No. 1) of the Surgut GRES-2 is shown in figure 5.

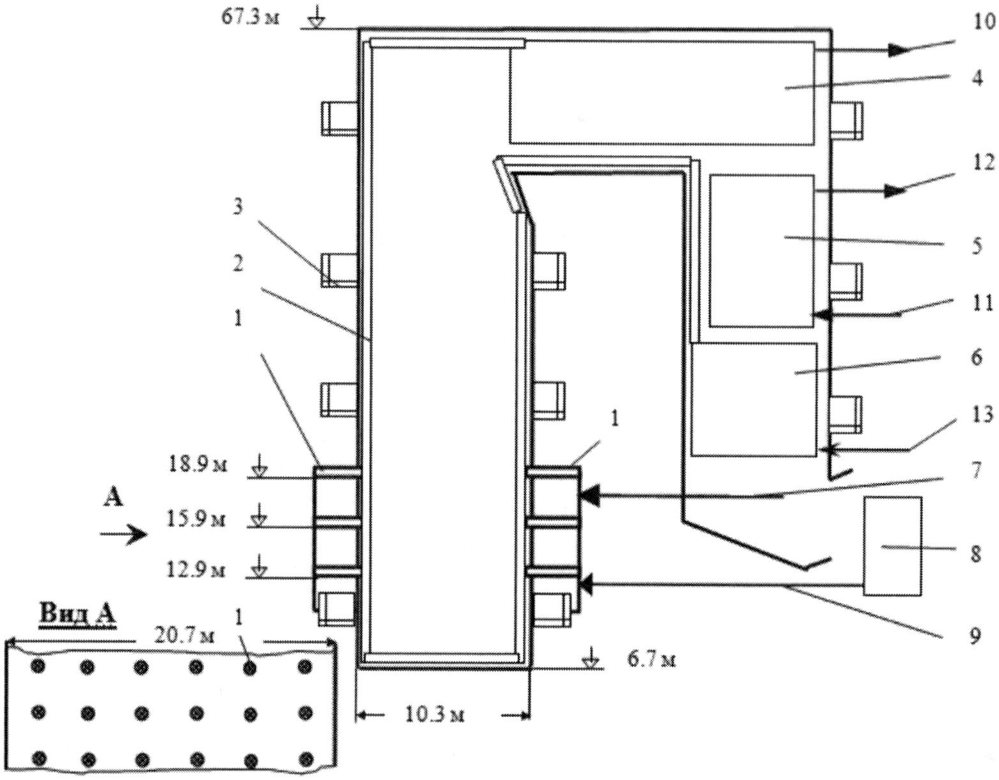

Figure 4. Diagram of the longitudinal section of the TGMP-204HL boiler (station No. 1) of the Surgut GRES-2 using 24 burners at the nominal capacity when burning gas: 1 – burners, 2 – evaporating surfaces, 3 – service platforms, 4 – superheating surfaces, 5 – intermediate superheater, 6 – economizer, 7 – gas recirculation, 8 – regenerative air heater, 9 – heated air, 10 – superheated supercritical steam, 11 – steam for reheat, 12 – steam after reheat, 13 – feed water.

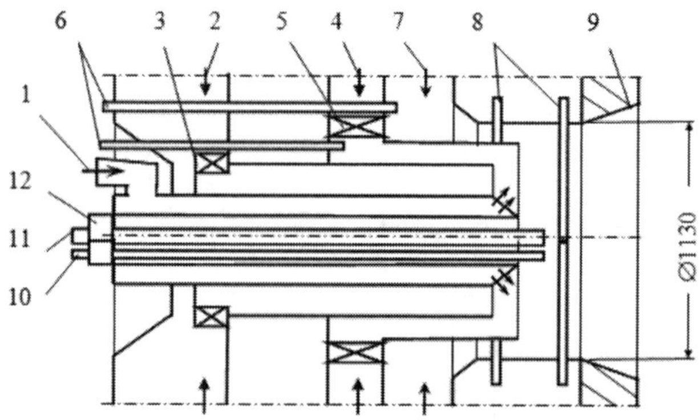

Figure 5. The diagram of the oil-gas burner used on the TGMP-204HL boiler (station No. 1) of Surgut GRES-2 with a capacity of 5.2 t / h for fuel oil or 5.6 x 103 m^3 / h for natural gas with a twist parameter of 1.115: 1 – central supply natural gas; 2 – air supply to the central annular channel, 3 – air-swirling tangential blades of the central register, 4 – air supply to the peripheral annular channel, 5 – air-swirling tangential blades of the peripheral register, 6 – air pressure sensors; 7 – recirculation flue gas supply, 8 – compensatory lenses, 9 – embrasure, 10 – electric igniter, 11 – oil nozzle, 12 – pneumatic valve.

Tangential blades of air swirling registers 3.5 (Figure 5) are non-rotating. Different amount of thermal expansion of the air box with built-in burners and screens of the combustion chamber is eliminated by lens compensators 8. The pneumatic valve 12 prevents the knocking out of combustion products into the boiler room.

The measurement of the radiation intensity of the torch was carried out with a TERA-50 total radiation radiometer with a RK-15 graduation. The temperature of combustion products in the near-wall layer of the furnace was measured using double chromel-alumel thermocouples [2]. The concentration of nitrogen oxides NO_x in the combustion products in the operating section was measured with a DAG-500 gas analyzer. The error in determining the efficiency was ± 0.71%. The operating and balance parameters of the boiler operation were measured using standard instruments with checking their readings by the sensors of the primary calibrated instruments of the set-up and testing group of the Surgut GRES-2. The heat of combustion of the fuel gas was determined in the central laboratory of the Surgut RES-2 by measurements on an ABK-1V electronic bomb calorimeter The Urengoy natural gas with a combustion heat $Q_{н}^{р}$ = 33662 kJ / Nm^3 was burned burners.

4. Results

Table 1 shows the results of a study of the operating modes of the TGMP-204HL boiler (station No. 1) of the Surgut GRES-2 during gas combustion in 24 burners with the third upper tier turned off, on which, according to the design data, 12 more burners should operate. Of these, 6 burners are installed on the front wall and 6 burners - on the rear wall of the furnace.

Table 1. The results of measurements of the operating parameters of the TGMP-204HL boiler (station No. 1) of the Surgut GRES-2 during gas combustion in 24 burners.

Parameter	Electrical load of the unit, MW				
	440	550	650	750	810
Steam capacity, t / h	1282	1587	1908	2250	2465
Superheated steam pressure, MPa	23.5	23.5	23.5	23.5	23.5
Superheated steam temperature, °C	545	545	545	545	545
Intermediate superheat steam temperature, °C	525	527	529	530	531
Feed water temperature, °C	240	250	260	268	272
Combustion gas consumption, thousand nm3/h	127	154	184	214	236
Gas pressure before burners, MPa	0.035	0.048	0.065	0.083	0.093
Exhaust gas temperature, °C	127	130	135	142	145
Heat loss with flue gases,%	5.68	5.81	6.02	6.34	6.38
Recirculation ratio,%	52	45	34	30	1,024
Excess air ratio in the operating section	1.041	1.035	1.029	1.026	287
The content of nitrogen oxides NO_x in combustion products, reduced to α = 1.4, mg / nm^3	80	90	148	196	287
Gross boiler efficiency,%	93.95	93.83	93.75	93.50	93.37

On the TGMP-204HL boiler (station No. 1) of the Surgut GRES-2, using 24 burners instead of 36 burners, the flame radiation intensity q_f increases at altitudes h_t in the range from 10 to 20 m (Figure 6).

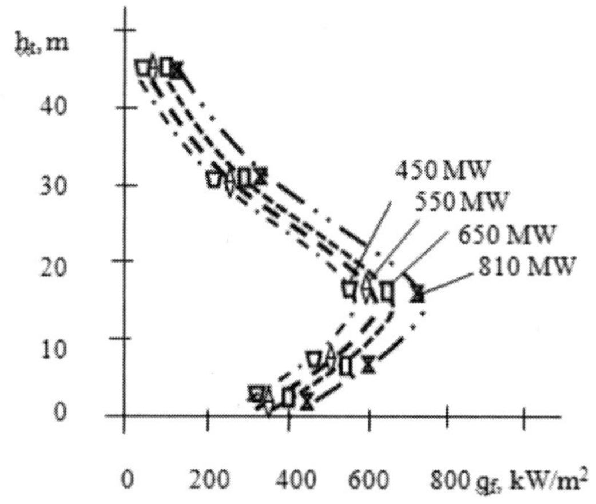

Figure 6. The intensity of the flame radiation q_f along the height h_t of the middle section of the furnace of the TGMP-204HL boiler (station No. 1) of the Surgut GRES-2 with a wall-mounted counter-arrangement of 24 burners on two tiers during gas combustion for electrical loads of the 440, 550, 650 and 810 MW unit.

In comparison with the data in figure 3 for the bottom arrangement of the burners of the TGMP-204P boiler and the wall-mounted opposite arrangement of 36 burners in three tiers on the TGMP-204HL boiler, with an electric load of the 810 MW unit in the TGMP-204HL boiler with 24 burners located in two tiers, the radiation intensity q_f in the altitude area from 10 to 20 m much higher. In this case, the recirculation coefficient was 26 %, which led to a decrease in the flame temperature at the burner level. For the data in figure 3, the recirculation ratio is 23%.

5. Discussions

Combustion of gas in the TGMP-204HL boiler (station No. 1) of the Surgut GRES-2 with 24 burners with an increased twist parameter and a recirculation coefficient made it possible to reduce the concentration of nitrogen oxides in comparison with the TGMP-204HL boiler [14] operating on 36 burners (Figure 7).

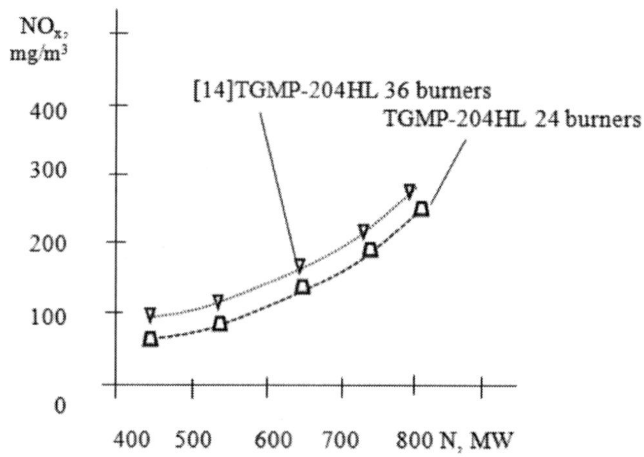

Figure 7. The concentration of nitrogen oxides NO_x in combustion products, reduced to the excess air ratio α= 1.4, when burning gas in the TGMP-204HL boiler (station No. 1) of the Surgut GRES-2 with 24 burners with an increased twist parameter depending on the electrical load N block in comparison with the data [14] for TGMP-204HL with 36 burners located in three tiers.

Combustion of gas in a TGMP-204HL boiler in 24 burners located in two tiers allows obtaining higher efficiency values as compared to combustion in 36 burners located in three tiers [14] (Figure 7).

This is due to the fact that when 36 burners are used and the third tier is used, the temperature of the combustion products at the outlet from the furnace increases and at the same time, heat losses with exhaust gases increase due to an increase in their temperature in the balance section.

To further increase the efficiency of gas combustion in TGMP-204HL boilers with 24 burners located in two tiers it is necessary to replace tangential vane air swirlers with axial peripheral ones, which make it possible to obtain higher values of the swirl parameter [10, 17]. In this case, it is also necessary to replace the central cone-shaped gas supply with a combined one, which includes peripheral gas distribution pipes with rotary nozzles and a coaxial central nozzle, which will improve mixture formation in the flare and reduce the intensity of the flame radiation at the burner level.

6. Conclusions

1.Modernization of the TGMP-204HL boiler for gas combustion in 24 burners located in two tiers, in comparison with combustion in 36 burners located in three tiers with an increase in the air swirl parameter in the burners allows to obtain an increase in efficiency and a decrease in the concentration of nitrogen oxides in the outgoing gases.

2. To further increase the efficiency of gas combustion in the TGMP-204HL boiler in 24 burners located in two tiers it is necessary to replace tangential air swirlers in the burners with axial peripheral ones and use combined gas distribution using peripheral gas supply pipes with nozzles and a central coaxial-conical gas supply which will reduce the local values of the flame radiation intensity at the level of the burners.

References

[1] Taymarov M, LavirkoY and Belyaeva E 2015 Intensivnost` luchistogo teploobmena v topke kotla pri izmenenii parovoj nagruzki [The intensity of radiant heat transfer in the boiler furnace with a change in steam load] *Izvestiya Vy`sshix Uchebny`x Zavedenij. Problemy` E`nergetiki* **7-8** 69-72 [In Russian]

[2] Zhang J Yuan H, Zhao J and Mei N 2017 Viscosity estimation and component identification for an oil-water emulsion with the inversion method *Applied Thermal Engineering* **111** 759-67

[3] Gelderen L, Malmquist L M V and Jomaas G 2017 Vaporization order and burning efficiency of crude oils during in situ burning on water *Fuel* **191** 528-37

[4] Kadota T and Yamasaki H 2002 Recent advances in the combustion of water fuel emulsion *Progress in Energy and Combustion Science* **28(5)** 385-404

[5] Ahrenfeldt J, Thomsen T, Henriksen U and Clausen L 2013 Biomass gasification cogeneration *Applied Thermal Engineering* **50(2)** 1407-17

[6] Sanchez A L and Williams F A 2014 Recent advances in understanding of flammability characteristics of hydrogen *Progress in Energy and Combustion Science* **41** 1-55

[7] Taymarov M A, Akhmetova R V, Akhmetov E A and Sungatullin R G 2020 The efficiency of gas combustion in TGM-84B boilers with a different number of burners *IOP Conf. Ser.: Mater. Sci. Eng.* **791** 012018

[8] Taymarov M Kuvshinov N Akhmetova R Sungatullin R Chiklyaev D 2016 Issledovanie ximicheskix processov obrazovaniya oksidov azota pri szhiganii gaza i mazuta [Research of the chemical processes of the formation of nitrogen oxides during the combustion of gas and fuel oil] *Vestnik KNITU* **19(20)** 80-3 [In Russian]

[9] Taymarov M A, Akhmetova R V, Chiklyaev Ye G, Lavirko Y V, Akhmetov E A and Garifullina A O 2019 Study of the speed of flame distribution in the combustion of methane-hydrogen fractions *E3S Web Conf.* **124** 05065

[10] Taymarov M A, Akhmetova R V and Akhmetov E A 2019 Efficiency of application of various layout arrangements of oil-gas burners in thermal power plant boilers *IOP Conf. Ser.: Mat. Sci. Eng.* **552** 012008

[11] Krishna N A and Mishra S C 2006 Discrete transfer method applied to radiative transfer in a variable refractive Index semi-transparent medium *J. of Quantitative Spectroscopy and Radiative Transfer* **102(3)** 432-40
[12] Taymarov M A, Akhmetova R V, Margulis S M and Kasimova L I 2018 Vliyanie kavitacionnoj obrabotki na vy`goranie chasticz mazuta v topkax kotlov [The effect of cavitation treatment on the extinguishing of particles of oils in fuel boilers] *Izvestiya Vy`sshix Uchebny`x Zavedenij. Problemy` E`nergetiki* **20(9-10)** 52-9 [In Russian]
[13] Taymarov M A, Akhmetova R V, Sungatullin R G, Lavirko Y V and Zheltuhina E S 2017 Snizhenie vredny`x vy`brosov v atmosferu oksidov azota kotlami TE`S [Decrease in harmful emissions in the atmosphere of nitrogen oxides by boilers of thermal power plant] *Izvestiya KGACU newsvol* **11(9)** 180-8 [In Russian]
[14] Roslyakov P V and Yegorova L E 1996 Vliyanie osnovny`x xarakteristik zony` aktivnogo goreniya na vy`xod oksidov azota [Influence of the main characteristics of a zone of active burning on nitrogen oxides output] *Teploe`nergetika* **9** 22-6 [In Russian]
[15] Burke M P, Chaos M, Ju Y, Dryer F L and Klippenstein S J 2012 Comprehensive H2/O2 kinetic model for high-pressure combustion *Int. J. of Chemical Kineticsvol* **44(7)** 444-74
[16] Shimizu K, Hibi A, Koshi M, Morii Y and Tsuboi N 2011 Updated Kinetic Mechanism for High-Pressure Hydrogen Combustion *J. of Propulsion and Powervol* **27(2)** 383-95
[17] Lu T and Jin J-M 2017 Electrical-thermal co-simulation for analysis of high-power rf/microwave components *IEEE Transactions on Electromagnetic Compatibility* **59(1)** 93-102

IOP Conf. Series: Materials Science and Engineering 1089 (2021) 012009 doi:10.1088/1757-899X/1089/1/012009

Transfer of the steam turbine PT-60-130/13 to the adventure T-27/33-1,28 with the replacement of electric motors of feeding electric pumps with a steam turbine drive

D S Balzamov[1], E Yu Balzamova[2], V V Bronskaya[3] and O S Kharitonova[4]

[1]Department of Power Supply of Enterprises and Energy Resource Saving Technologies, Kazan State Power Engineering University, Krasnoselskaya St., 51, Kazan 420066, Russia
[2]Department of Economics and Organisation Production, Kazan State Power Engineering University, Krasnoselskaya St., 51, Kazan 420066, Russia
[3]Department of Chemical Process Engineering, Kazan National Research Technological University, Karl Marx St., 68, Kazan, 420015, Russia
[4]Department of Chemical Technology of Petroleum and Gas Processing, Kazan National Research Technological University, Karl Marx St., 68, Kazan, 420015, Russia

E-mail: olga.220499@mail.ru

Abstract. For the production of thermal and electric energy, generating enterprises consume fuel, electricity, water for various purposes, thermal energy, etc. The consumption of electricity which is used mainly on the motor of pumping equipment, in the cost of the released energy resource is a significant share. Nowadays, reducing the consumption of energy for the auxiliary needs of generating enterprises is a crucial task to ensure competitiveness according to specific indicators on the wholesale electric power market. This article provides the analysis and justification of the technical decision to replace the electric motor of pumps with a steam turbine drive. A scheme for connecting a steam turbine drive to the feed pumps PE-500-180 (№1) and PE-500-180 (№2) is proposed, taking into account the characteristics of the thermal diagram of the boiler and turbine section. The work schedules of feed pumps have been analyzed, the required steam consumption has been determined, taking into account the loading of the production steam extraction of operating turbines for 200 t/h. According to the results of the analysis of the performance indicators of the station, it was determined that in the condensation modes during the period with a low supply of the thermal energy, the thermal economy of the station decreases, which leads to an increase in the consumption of fuel equivalent in the amount of 1660806 rubles. At the same time, the increase in the volume of sales of electric energy and capacity on the open electric capacity market after the reconstruction will amount to 75493070,81 rubles. Thus, the proposed technical solution for the reconstruction of the feed pump drive is a rather promising direction with the correct construction of the scheme and the presence of consumers of the heat energy.

1. Introduction

Today, it is considered that the power consumption for auxiliary needs of the combined heat and power plant (CHPP) is 5-15% of the installed capacity of the CHPP. The auxiliary needs of station are a set of auxiliary equipment that ensures the smooth operation of its main elements. These elements include lighting systems, emergency power supplies, electric motors of all mechanisms, battery installations, etc. Such factors like fuel, location and mode of operation, equipment condition influenced on the smooth operation, but analysis of research shows that actual consumption for auxiliary needs of CHP plants can in some cases reach 35% of their production [1].

Figure 1 presents the structure of electricity losses at CHP plants.

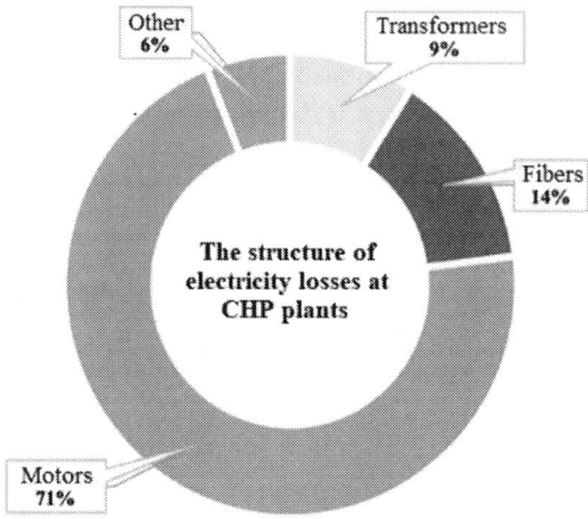

Figure 1. The structure of electricity losses at CHP plants

According to the estimates of the studies carried out over the past 10 years, the average efficiency coefficient of Russian CHPPs is 35-37%. It should be noted that the specific fuel consumption for the generation of one kWh of electricity in recent years has decreased by only 1,5% [2].

The main problems of the modern energy complex of Russian CHPPs include:

- Decrease of the performance of old heat and power equipment, which leads to frequent shutdowns and downtime;
- Decrease in the electricity generation and increase in the specific fuel consumption;
- Low rates of commissioning new capacities using modern technologies;
- Decrease in economy of CHPP operation due to increased energy consumption for auxiliary needs.

As it follows from figure , the main consumers of electricity for the auxiliary needs of CHP plant are engines, mainly for driving injection equipment.

This is especially critical for heating CHP plants operating according to the thermal load schedule during the non-heating period of the year. A characteristic feature of the operation of such CHP plants during the non-heating period is that their electrical load is more dependent on the heat load of the plant. This is due to the fact that such combined heat and power plants use cogeneration turbine. The operation of the main equipment of the CHPP at partial loads leads to the need to regulate the performance of the mechanisms of the auxiliary needs of the plant, as a result of which they operate in the zone of nonoptimal value of the efficiency coefficient. Bypassing and throttling are the main ways of regulating the performance of auxiliary mechanisms, which were laid down in those years when no one thought about the energy efficiency. But, today, these methods are not effective, since the regulation of productivity is achieved by increasing the resistance of the network, and this leads to additional losses on the gate or valve of the mechanism. The most effective way to control the performance of centrifugal mechanisms is to change the rate of rotation of the impeller, which can be achieved by installing variable speed drives, as well as replacing the electric motors by a steam turbine drive, the performance of which can be regulated by changing the supply of steam. [1].

IOP Conf. Series: Materials Science and Engineering 1089 (2021) 012009 doi:10.1088/1757-899X/1089/1/012009

The analysis of the condition of the equipment of the research object (thermal power station (TPS)), its thermal diagrams and heat consumption in the steam and hot water showed that to improve the efficiency of TPS, one of the most effective actions are [2-3]:

- reconstruction of turbine PT-60-130/13;
- installation of the steam turbine motor on feeding pumps.

It should be noted that the TPS was operated by the turbine PT-60-130/13, which has passed 3 life extensions (at the time of the survey the lifetime was 259730 hours), which has a number of observations requiring significant capital investment. The internal relative efficiency factor of the high-pressure cylinder of the turbine PT-60-130/13 was lower than the normative values due to significant wear and tear of the turbine blading, especially the operating blades of the control stage, in the result of cracking the bottom of the stator of the high-pressure cylinder, as well as insufficient rotor geometry and end seal gaps. Considering this condition of the high-pressure cylinder, it is evident that their repair is inadvisable.

For the motor of feeding pumps, the electric motors of the brand AC-4000/6000 with a power of 3150 kW were used. However, losses in turbine generators, losses in transformers of auxiliary needs, losses in transmission of electric energy to the motor and losses in the electric motor are obvious. There is also an overconsumption of electricity on the motor of feeding pumps when operating on partial loads [5-7].

Thus, the following objectives were pursued when implementing a technical solution:

- increasing the efficiency of the production process;
- improving the safety of the production process;
- Significant reduction of electricity losses for auxiliary needs;
- Additional production of electrical energy;
- Optimization of loading of turbines type P;
- Stable maintenance and stepless control of feedwater pressure;
- Increase the operating life, decrease in the number of repairs of shut-off and control valves.

2. Materials and Methods

For the reconstruction of the turbine PT-60-130/13 due to the unsatisfactory condition of passage, it was proposed to dismantle the high-pressure cylinder TG-1 with the replacement of the low cylinder TG-1 pressure on the operation of the stream of 13 ata (the turbine was included). Reconstruction involved dismantling of the rotor stages of the high-pressure cylinder, reconstruction of the steam distribution of the low-pressure cylinder with the organization of new steam starting and installation of new valves [8, 9, 11].

The passage of the low-pressure cylinder in part of the mean pressure is also undergoing reconstruction, namely:

- reconstruction of the steam distribution of the low-pressure cylinder with the organization of new steam starting and installation of new valves;
- dismantling pressure stages;
- installation of mounted discs with new blading to ensure the required throughput capacity
- installation of new vane carriers;
- Low-pressure compartment (pressure stages behind the turning carrier) and condenser are beyond reconstruction.

Steam in the amount of 225 t/h to the reconstructed turbine is supplied from the collector of 13 ata. The installed electrical capacity of the turbine is about 35 MW. After reconstruction, the turbine PT-60-130/13 should be labeled as T-35-13. The additional steam flow rate of 13 ata to the turbine T-35-13 can come from the turbines P-50/60-130/13, PT-135-130 and from the motor turbines of feeding pumps. This solution will increase the loading of turbines type P, increase the thermal energy supply from the rebadged turbine type T-35-13, as well as increase its internal relative efficiency factor.

To avoid losses related to energy losses in the generator windings, electric motor and transformer, it is proposed to use steam turbines for driving feeding pumps of type PE-500-180 (№1) and PE-500-180 (№2).

Turbines to drive feeding pumps compared to traditional axial turbines have a more economical (2%) passage, fewer stages (3 times), fewer nozzle and rotating blades. These benefits reduce the cost of repair works. Moreover, these turbines are cheaper than axial. The use of a turbine to drive feeding pump instead of an electric motor is also advisable, based on the possibility of regulating the flow of feedwater under partial modes (this is not provided by the electric motor).

The turbo drive construction is a single-cylinder turbine installed on the existing foundation frame of the pump unit within the electric motor layout. The weight load from the turbo drive does not exceed the weight load of the replaced electric motor, which allows to install the turbo drive to the place of the dismounted electric motor with the current position of the feeding pump and without the reconstruction of the foundation [4].

Steam distribution of the throttle turbine. Block and control valves are installed in front of the turbine consequentially. The steam supply from the control valve to the steam inlet zone of the front of the cylinder is made by two tubes. Steam exhaust from the back of the cylinder is carried out up by two sides symmetrically [9].

The main characteristics of the passage are presented in table 1.

Table 1. The characteristics of the passage of the turbo motor.

Parameter	Value
Minor diameter	800 mm
Rotational speed	2985 rpm
Steam flow rate	100 t/h
Internal efficiency factor	0,73
Height of nozzle blades	20/24/28 mm
Number of nozzle blades	70/70/70
Height of rotating blades	20/24/28 mm
Number of rotating blades	160/160/160

The motor turbine is activated between the turbine exhaust pipeline P-20/40-130/30 and the collector of 13 ata.

Installation of turbines as a feeding pump motor will significantly reduce the electricity consumption for auxiliary needs and ensure the optimum loading of TG-2,5. Also, the regulation of the feedwater rate due to the change in the number of rotators of the motor turbine will reduce the specific consumption of electricity for the pumping of the medium [5,10,12].

Turbo motors of the feeding pumps are proposed to be installed in the feedwater diagram of the I stage. At the same time, the steam consumption per feeding pump is 100 t/h, which leads to the loading of the production selection TG-5 at 200 t/h and, accordingly, to the increase of the electrical load on it by 18 MW.

To ensure the required steam temperature released to the consumer, it is important to provide a steam cooler on the exhaust of the turbo motor [13,14].

Figure 2 shows the diagram of connecting a steam turbine to the steam pipelines of a combined heat and power plant.

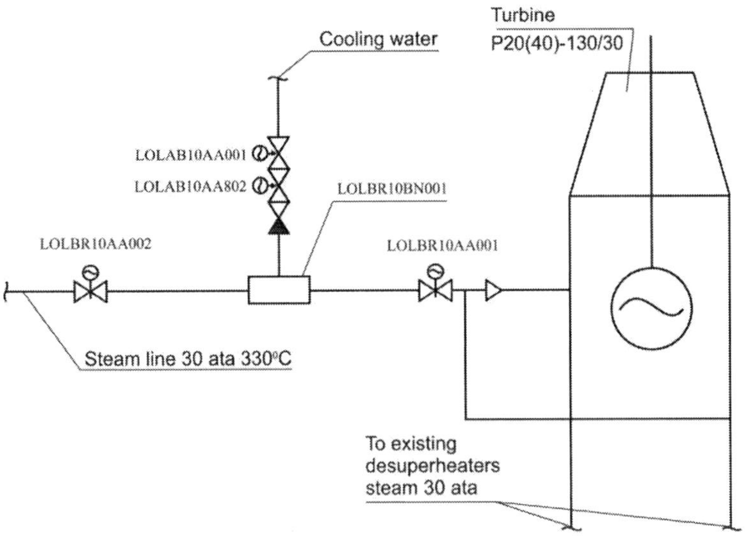

Figure 2. The diagram of steam lines 30 ata №5 and 6 P=28.5±1,5 kgs/cm^2, T=430°C.

Figure 3 shows the operation schedule of the feed pump in the power unit of a combined heat and power plant.

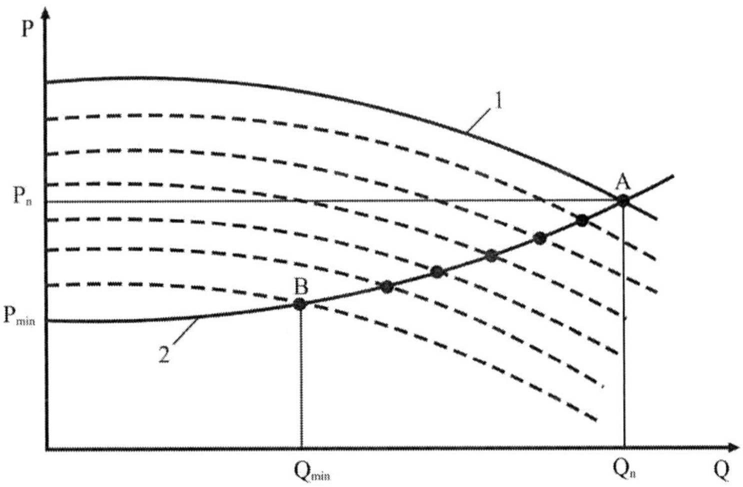

Figure 3. Schedule of the feed pump operation.

It can be seen on figure 3 that the operating point A corresponds to the nominal flow rate Qn and the pressure Pn. With a decrease in pump speed, the operating point shifts along the curve of the network characteristic to the position of point B, at which the flow becomes the minimum permissible and the recirculation line is turned on. The unloading of the power unit is mainly limited by the reliability of the boiler operation and is 50% of the nominal power. Less load is observed only when it starts and stops. As a result, at the CHP plant, the smooth regulation of revolutions and change in the performance of the main feed pump can be carried out in the entire operating range of loads with the complete exclusion of throttling [17, 18].

3. Results

Analysis of the technical solution of the modernization of the turbine PT-60-130 by replacing the high-pressure cylinder and the passage of the low-pressure cylinder showed that the nominal capacity of the upgraded turbine PT-70-130/13 will be 67.2 MW, which is higher than the current nominal capacity of the turbine PT-60-130 by 7,2 MW. The maximum capacity of the upgraded turbine PT-70-130/13 will be 86,4 MW, which is higher than the current maximum capacity of the turbine PT-60-130 by 11,4 MW. Thus, this action does not allow to increase the actual value of the electrical power of the plant by 20 MW, as specified in the technical solution.

It should also be noted that a year before the energy survey the total overhaul of the TG was carried out after which the indicators of the economical operation significantly improved and at the moment have slightly conceded indicators of the economical operation of the upgraded turbine. In connection with it, the modernization of the turbine PT-60-130 with the replacement of the high-pressure cylinder and the passage of the low-pressure cylinder seems impractical.

As a result of the analysis of the technical solution, the turbine of the type T-45-15 was selected as included, however, according to the conditions of loading the heat selection and the permissible steam load of the condenser, the turbine will not be able to be loaded over 25 MW. The installed electrical capacity of the turbine is approximately 25 MW.

Steam in the amount of 236 t/h is supplied from the collector of 13 ata to the reconstructed turbine. The additional steam flow rate of 13 ata to the T-45-13 can come from the turbines P-50/60-130/13 and PT-135-130. This solution will increase the loading of turbines type P, increase the thermal energy supply from the relabeled turbine type T-35-13, as well as increase its internal relative efficiency factor.

After reconstruction the thermal economy of the station will decrease due to low economical operation of the reconstructed turbine T-25-13 on condensation modes during the period with low release of the thermal energy. The increase in the consumption of oil equivalent for the year will be 501,86 TW that is equivalent to a cash amount of 1,660,806 RUB.

The increase in the volume of electricity sales in connection with the reconstruction will be 193231,47 kW·h that is equivalent to 23241098,52 rubles.

The increase in the volume of the power sales due to the reduction of the installed capacity restrictions after reconstruction will be 486,03 MW which is equivalent to a cash amount of 52251972,29 rubles.

As a result, the total cost estimate of the economic efficiency of the solution will be 73832264,83 rubles and the payback period will be 9,06 years.

4. Conclusion

Implementation of the technical solution allowed to reduce electricity consumption for auxiliary needs, optimize the loading of the turbine P-20/40-130/31, maintain stable and stepless control pressure of feedwater, improve the safety of the production process and reliability of the provision of thermal energy in the hot water.

Thus, the implementation of technical solutions for the introduction of steam turbine drives at CHP plants, taking into account the operating conditions of the thermal scheme, is a rather promising direction along with the introduction of variable speed drives. Thus, according to the ARCAdvisory Group (USA), the global market for variable speed drives is estimated at $ 2.5 billion and will only grow in the coming years. In Russia, variable speed drives make up no more than 2.5% of the entire drive market, which opens up the great prospects for the introduction of both variable frequency drives and drives based on the steam and gas turbines.

5. Acknowledgments

The study was carried out within the framework of a scientific project of the Russian Science Foundation (RSF) № 18-79-10136.

References

[1] Lazurenko A P, Cherkashina G I and Krugol N M 2017 Povy`shenie e`nergoe`ffektivnosti raboty` te`cz za schet snizheniya potrebleniya e`lektroe`nergii na sobstvenny`e nuzhdy` [Increasing the energy efficiency of the CHPP operation by reducing the consumption of electricity for its own needs] *E`nergetika: E`konomika, Texnologii, E`kologiya* **2** 82-90 [in Russian]

[2] Lazarev G B 2012 Upravlenie e`ffektivnost'yu mexanizmov sobstvenny`x nuzhd TE`S [Efficiency management of own needs mechanisms of TPP] *E`nergiya Edinoj Seti* **5(5)** 58-67 [in Russian]

[3] Balzamov D S, Sabitov L S, Timershin B F and Balzamova E Yu 2018 Increase of efficiency of heat sources work due to application of condensation economizers on an example of a boiler PTVM-180 *IOP Conf. Series.: Mater. Sci. Eng.* **412** 012007

[4] Balzamov D S, Balzamova E Yu, Ibatullin I S and Sabitov L S 2019 Efficiency increase of gas turbine work in the summer period *IOP Conf. Series.: Mater. Sci. Eng.* **570** 012008

[5] Balzamov D S, Akhmetova I G, Balzamova E Y, Oykina G I and Coman G 2019 An analysis of the viability of implementing steam screen machines at the facilities generation enterprises to reduce the energy costs for their own need *E3S Web of Conf.* **124** 01016

[6] Remizov D V, Egorov V A, Greenman M I, Egorov V A, Ushakov S L and Plakhin A A 2009 Puti uvelicheniya teplovoj i e`lektricheskoj moshhnosti TE`Cz [Ways to increase the thermal and electrical power of CHP] *Vserossijskaya Konferenciya Rekonstrukciya E`nergetiki-2009* (Moscow: INTEXE`KO) pp 7-11 [in Russian]

[7] Balzamov D S, Balzamova E Yu, Bronskaya V V, Ignashina T V and Kharitonova O S 2020 Analysis of the possibility of modernization of the state district power station by building the combined cycle plant *J. Phys.: Conf. Ser.* **1515** 042100

[8] Solovev S A, Soloveva O V, Gilmurahmanov B Sh and Lamberov A 2020 Numeral simulation of a flow mixer for a radial-type chemical reactor *IOP Conf. Ser.: Earth Environ. Sci.* **421** 072017

[9] Balzamova E Yu, Balzamov D S, Bronskaya V V and Kharitonova O S 2020 Modernization of the scheme for the delivery of thermal power from the state district power station *IOP Conf. Ser.: Mater. Sci. Eng.* **734** 012194

[10] Zatsarinnaya J N, Logacheva A G and Solovyova A A 2019 Analysis of thermodynamic efficiency of the fuel preparation systems with an intermediate hopper at thermal power plants *IOP Conf. Ser.: Earth Environ. Sci.* **288** 012130

[11] Remizov D V and Fomin V A 2009 E`nergosberegayushhie rekonstrukcii uzlov parovy`x turbin. Novy`e remontny`e texnologii i ix e`ffektivnost` [Energy-saving reconstruction of steam turbines. New repair technologies and their efficiency] *Vserossijskaya Konferenciya Rekonstrukciya E`nergetiki-2009* (Moscow: INTEXE`KO) pp 17-21 [in Russian]

[12] Zinnatullina G N, Davydov A V, Bronskaya V V and Zinnatullina N H 2014 Raschet osnovny`x texnologicheskix parametrov centrobezhny`x plenochny`x teplomassoobmenny`x apparatov [Calculation of main technological parameters of centrifugal film heat and mass] *Vestnik KTU* **17(3)** 89-92 [in Russian]

[13] Balzamov D S, Akhmetova I G, Balzamova E Yu, Oykina G I and Bronskaya V V 2019 Options for organising own sources of energy supply at the facilities of generating companies based on steam screen machines 2019 *J. Phys.: Conf. Ser.* **1399** 055018

[14] Kharitonova O S, Bronskaya V V, Ignashina T V, Al-Muntaser A and Khairullina L E 2019 Modeling of absorption process using neural networks *IOP Conf. Ser.: Earth and Environ. Sci.* **315** 032025

[15] Konakhina I A, Konakhin A M and Shinkevich O P 2011 Termodinamicheskij analiz sistemy` sbora i povtornogo ispol`zovaniya kondensata [Thermodynamic analysis of the system for collection and re-use of condensate] *Vestnik IGEU* **2** 11-4 [in Russian]

[16] Balzamov D S and Konakhina I A 2010 Sistema e`nergotexnologicheskogo kombinirovaniya

vy`sokotemperaturnogo uchastka degidrirovaniya izoamilenov v izopren [System of energy technical combinations of a high temperature section of isoamilenes in izopren dehydration] *Vestnik KGEU* **1** 16-25 [in Russian]

[17] Mamontov N I, Kobtsev O M and Panov V V 2012 E`ffektivnost` i celesoobraznost` primeneniya turooprivodov pitatel`ny`x nasosov na TE`Cz i e`nergoblokax 150 - 200 MVt TE`S [Efficiency and feasibility of using turbo drives for feed pumps at CHPPs and power units of 150-200 MW TPPs] *Kompressornoe i E`nergeticheskoe Mashinostroenie* **1** 9-14 [in Russian]

[18] Kobtsev O M, Panov V V and Pugacheva T N 2012 primenenie turboprivoda pitatel`nogo nasosa na e`nergoblokax 150–200 MVt TE`S [Application of a turbo drive of a feed pump at power units of 150–200 MW *TPP] Vestnik NPI* **7** 117-123 [in Russian]

Algorithm of the automatic synchronizer operation

B O Grigorkin and G S Safronov

Department of automated electric power systems, Novosibirsk state technical University, Karl Marx Ave., 20, Novosibirsk, 630073, Russia

E-mail: gbo@ngs.ru

Abstract. When connecting powerful synchronous electric machines (SM) to the power system uses automatic synchronizers with constant lead time. It is necessary to improve their accuracy and performance for the success of synchronization on the technological requirements of production. METHODS. For SM change the angle of the rotor is determined in the three phases of the synchronized voltages. This allows to increase the number of measurements per unit of time. The least-squares method is used for extrapolating the function values at the interval forecast. REZULTS. The proposed algorithm for automatic synchronizer uses to predict the set of values of the functions change the angle of the rotor. This distinguishes it from the known algorithms, where the forecast is based on the three rotor angle values determined over the same time interval. This property improves the accuracy, reduces the time interval determine the forecast considers possible change of the acceleration slip angle of the rotor. Using the least-squares method allows to abandon the traditional derivative calculation of the angle of the rotor, which is associated with increased interference. In addition, this method characteristic filter properties. CONCLUSIONS. The developed algorithm allows to quickly define the time of synchronization and improves the conditions of the SM by keeping dynamic stability.
Keywords: automatic synchronizer, exact synchronization method, least squares method, synchronous motor.

1. Introduction

When powerful SM and power systems are switched on in parallel, it is often critically important to increase the speed and accuracy of the automatic synchronizers (SA) to determine the switching angle [1]. This is especially true when the backup power supply of SM is switched on by the method of precise synchronization in industries with a continuous technological process [2]. Here, an interruption in the work of the SM leads to large economic and environmental losses (metallurgy, petrochemistry, transit of hydrocarbons, treatment facilities, etc.). The current-free pause, admissible under the condition of maintaining the dynamic stability, namely, the rotation of the SM rotor at an angle of 360 degrees, can be 450 - 550 ms and depends on the operating parameters and characteristics of the motors. [3, 4]. Considering the proper switching time of the operating and backup power switches, as well as the time spent on detecting an accident by relay protection, the interval for determining the moment of switching on using the SA can be reduced to three periods of industrial frequency. This raises the requirements for the performance of the extrapolation function of the rotor angle motion. Also, for the success of synchronization and preservation of the dynamic stability of the SM, it is necessary to predict the moment of switching on for the time of operation of the backup power switch. This is explained by the large value of the angular slip frequency during deceleration of the SM rotor and, consequently, by the significant phase difference during synchronization due to an error during switching on.

2. Materials and Methods

2.1. Automatic synchronizers

At present, the production uses SA with a constant lead time: AST-4, SA-1, AS-M1, AS-M2, SPRINT-M [5] and a constant lead angle: UBAS [6]. The latter devices have characteristics that are unacceptable for SM synchronization:

- low accuracy due to the dependence of the lead time value (forecast of the synchronization moment) on the value of the angular slip frequency;
- limited in magnitude lead angle, no more than 180 degrees, and therefore will not be considered further. To predict the movement of the rotor angle in synchronizers, the Fourier series expansion is used [7]:

$$\delta_{ON} = \delta_{AD} + \omega_S \Delta t_{ON} + \frac{a_S \Delta t_{ON}^2}{2}, \quad (1)$$

where:
δ_{ON} - is the angle of inclusion 360 degrees;
δ_{AD} - lead angle;
ω_S - angular slip frequency;
Δt_{ON} - lead time; a_S - sliding acceleration.

The determination of the values of ω_s and as is carried out by calculating the first and second derivatives of the angle of motion of the rotor δ (t), respectively. The limitation by three terms of the series in expression (1) is explained by the assumption of uniformly slowed rotation of the rotor of a SM in the process of running. The mechanical moment on the SM shaft can have a fan characteristic, as a result of which the magnitude of the sliding acceleration changes over time, and in equation (1) it is necessary to consider the third derivative, and sometimes the subsequent ones.

However, many synchronizers are limited to the definition of the first derivative and cannot be used for the problem under consideration [8-16].

In analog synchronizers, for example, SA-1, when measuring the value of δ (t) with a phase meter, the time interval proportional to the phase difference is converted into a voltage amplitude using a second-order Butterworth low-pass filter (LPF).

At the moment of turning off the accident by the switch of the working power, a sharp change in the phases of the measured voltages may occur. A transient oscillatory process will begin in the analog low-pass filter. It will increase in series connected aperiodic links of the differentiating circuits used to determine the values of ω_s and a_s. As a result, the solution of equation (1) will be impossible, and the synchronizer is unusable for a time comparable to the interval of the synchronization process.

In digital synchronizers, for example, SPRINT-M, the time interval proportional to the phase difference is measured twice per period. The values of ω_s and as are calculated by the increments of the current values δ (t), which is theoretically equivalent to the extrapolation of the function of changing the rotor angle by three values and occupies a fixed interval equal to at least two periods of the power frequency.

However, the increments of the measured values over such a short time interval are small. The implementation of the operation of double differentiation δ (t) leads to a sharp deterioration in the signal-to-noise ratio due to the presence of interference and harmonic components of the measured signal.

At the same time, it is not possible to eliminate this disadvantage of the SA, as it is done in analog versions with the help of filtering, in a digital device based on one value of each measured value. This explains the inaccuracy of predicting the time of synchronization of the SA.

To increase the speed and, above all, the reliability of extrapolation, it is proposed to measure the phase difference of synchronized voltages in three phases, which will increase the amount of information received per unit of time three times.

In addition, since the run-out of the SM is monotonically slow and the switching to the backup power is limited to a short time interval, then all the measured values of δ (t) recorded in the memory of the CA are used for extrapolation. Thus, the number of these values can be large, which makes it possible to consider the nature of the load on the SM shaft, other features of the run-out, namely the exchange of electrical power in the run-out process and build an accurate forecast for the lead time.

The extrapolation process itself is carried out using the least squares method, which not only does not increase the amount of noise in the measured signals, as in the known SA, but, on the contrary, reduces their influence on the synchronization accuracy.

2.2. *The algorithm of the automatic synchronizer*

The operation of the proposed device is based on the representation of the functional dependence δ (t) in the form of a regression-type model with respect to the basis vector polynomial in the time factor:

$$\delta(t) \equiv \delta(b,t) = b^T \cdot \phi^T(t) = b_0 + b_1 \cdot t + b_2 \cdot t^2 + \ldots + b_n \cdot t^n, \quad (2)$$

where:
$b^T = (b_0, b_1, \ldots + b_n)$ – vector of independent coefficients of the model;
$\phi^T(t) = (1, t, \ldots, t^n)$ – the basis vector of the model;
T – the transposition symbol.

Finding the numerical values (or estimates) of the coefficients of the vector b makes it possible to predict the function δ (t) for a certain time interval due to the intrinsic time of switching on the switching equipment.

To find estimates of the parameters b in model (2), the method of least squares (MLS) was used [17]. Formally, the MLS estimates $\tilde{b}$ of the parameters b of the model δ (b, t) are in accordance with the dependence:

$$\hat{b} = (\Phi^T \cdot \Phi)^{-1} \cdot \Phi^T \cdot \tilde{\delta}. \quad (3)$$

The matrix Φ can be represented as:

$$\Phi = \begin{pmatrix} \varphi^T(t_1) \\ \varphi^T(t_2) \\ \ldots \\ \varphi^T(t_N) \end{pmatrix},$$

where $t_1, t_2, \ldots, t_N$ – moments of measurements of the angle δ (t).

The vector of angle measurements δ (t) at times

$$t_1, t_2, \ldots, t_N, N \geq (n+1),$$

$$\tilde{\delta}^T = (\delta(t_1), \delta(t_2), \ldots, \delta(t_N)).$$

The filtering properties of estimates of the form (3) in the model are known: $\delta(t) \equiv \delta(\hat{b}, t)$.

In the device under consideration, for N> (n + 1), they have a positive effect on increasing the accuracy and reliability of the choice of the synchronization time point, which is found from the solution of the equation with respect to t_{ON}:

$$\tilde{\delta}(\hat{b}, t_{ON}) = \delta(t_{ON}).$$

Matrix $(\Phi^T \Phi)^{-1} \Phi^T$ is calculated once in advance before the start of operation of the device with a predetermined accuracy. The values of its elements are stored in the ROM of the computing device that implements the operation of this algorithm.

In order to improve the accuracy of estimates $\tilde{b}$ of model (2) and reduce the total time for obtaining the values δ (t_i) in the proposed device, joint measurements of the angles δ (t) are made from three phases of voltages of the synchronized running out SMs and the backup power source. In this case, it is assumed that models of the form (2) for each (i-th) phase have the form:

$$\delta_{(i)}(t) \equiv \delta_{(i)}(b_{(i)}, t) = b_{(i)}^T \phi(t) + c_i = (b_0 + c_i) + b_1 t + \ldots + b_n t^n = \tilde{b}_{i0} + b_1 t + \ldots + b_n t^n,$$

$$i = 1,2,\ldots,3;\ c_i = 0;\ \tilde{b}_{i0} = b_0.$$

By combining the results of measurements of the angle δ (t) at times $t_1, t_2, \ldots, t_N$, (N is a multiple of 3) from three phases, we find the estimates of the coefficients:

$$\tilde{b}_{(3)}^T = (\tilde{b}_{10}, \tilde{b}_{20}, \tilde{b}_{30}, b_1, \ldots, b_n).$$

These estimates:

$$\hat{\tilde{b}}_{(3)} = (\Phi_{(3)}^T \cdot \Phi_{(3)})^{-1} \cdot \Phi_{(3)}^T \cdot \tilde{b}_{(3)}, \quad (4)$$

where, in contrast to formula (3), the matrix $\Phi_{(3)}$ has the form:

$$\Phi_{(3)} = \begin{pmatrix} 100\ \tilde{\varphi}^T(t_1) \\ 010\ \tilde{\varphi}^T(t_2) \\ 001\ \tilde{\varphi}^T(t_3) \\ \ldots \ldots \ldots \\ 100\ \tilde{\varphi}^T(t_{N-2}) \\ 010\ \tilde{\varphi}^T(t_{N-1}) \\ 001\ \tilde{\varphi}^T(t_N) \end{pmatrix} \quad (5)$$

- 1st block of matrix $\Phi_{(3)}$

- the l-th block of the matrix $\Phi_{(3)}$,

$$\tilde{\phi}^T(t) = (t, t^2, \ldots, t^n).$$

Then the vector of angle measurements δ (t) in three voltage phases:

$$\tilde{\delta}_{(3)}^T = (\delta_{11}, \delta_{21}, \delta_{31}, \delta_{12}, \delta_{22}, \delta_{32}, \ldots, \delta_{1l}, \delta_{2l}, \delta_{3l}),$$

$$N = 3l,$$

where δ (t_i, j) – the value of the angle in the i-th phase in the j-th block of measurements, at time t_p, p = 3 (j -1) + i.

The algorithm implemented in this device allows, at a fixed time interval allotted to the computing device for finding estimates $\tilde{b}$ (3) by formula (5), to increase the number of measurements N. For this, it is necessary that the interval between measurements:

$$\Delta t = t_i - t_{i-1} \geq \Delta t_{min},$$

where Δt_{min} – the time interval characterizing the speed of the measurement and computational procedures of the device.

This increases the accuracy of finding estimates $\tilde{b}(3)$, since the variance of the estimate for model (4) for i = 1 is

$$D(\delta_1(\hat{b}_{(1)}, t)) = \phi^T(t)(\Phi_{(3)}^T \cdot \Phi_{(3)})^{-1} \cdot \phi(t) \frac{\hat{\sigma}^2}{N-1},$$

$$\phi^T(t) = (1, t, \ldots, t^n),$$

where:
$\Phi_{(3)}$ – the matrix, which has the form (5);
$\tilde{\sigma}^2$ is an estimate of the variance of the noise component of measurements

$$\tilde{\delta}\, ij, i = 1,2,3;\ j = 1,2,\ldots,l.$$

IOP Conf. Series: Materials Science and Engineering 1089 (2021) 012010 doi:10.1088/1757-899X/1089/1/012010

However, in this case, the material costs for the construction of the measuring system and the amount of memory of the computing device allocated for the storage of the matrix $(F^T{}_{(3)}\,\Phi_{(3)})^{-1}\,FT_{(3)}$ increase. Matrix elements are calculated in advance with a predetermined accuracy.

The moment when the signal is applied for synchronization is determined by solving the equation:

$$\delta(t)(\hat{b}_{(1)}, t) = \delta(t_{ON}).$$

2.3. Full-scale tests of the synchronizer.

The purpose of the tests was to determine the possibility of switching on the backup power supply of SM by the method of precise synchronization. And also, a comparison of several synchronizer algorithms in terms of speed and accuracy. The experiments were carried out at the «Bashneft-UNPZ» plant, Ufa.

In figure 1 shows a simplified diagram of the experiment. It contains a 10 kV network, operating power switches Q1, Q2 (disconnection time 100 ms), backup power switch QB (turn-on time 220 ms), SM-M1, M2. The tests involved STD-6300 engines with a power of 6.3 MW; LF = 0.95; PF = 1.

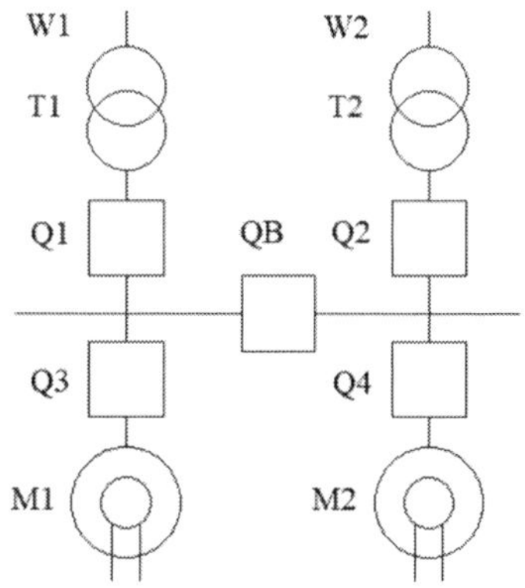

Figure 1. Simplified diagram of the STD-6300 motor synchronization.

Symbols in Figure 1: W1, W2- supply-lines voltage; Q1, Q2, Q3, Q4, QB - switches; M1, M2 - STD-6300 engines.

The test results are shown in Figures 2 – 6. To simulate the emergency mode, the power line W1 of the working power source of the motor M1 was turned off (time t_0). The relay protection detected the loss of power, opened the Q1 switch, made it possible for the synchronizer to work and switch the QB switch. The synchronizer predicted the time t_1 and generated the switch control signal QB with a lead time of 220 ms. This switch was turned on when the phases of the synchronized voltages were equal (Figure 2). Synchronization was successful.

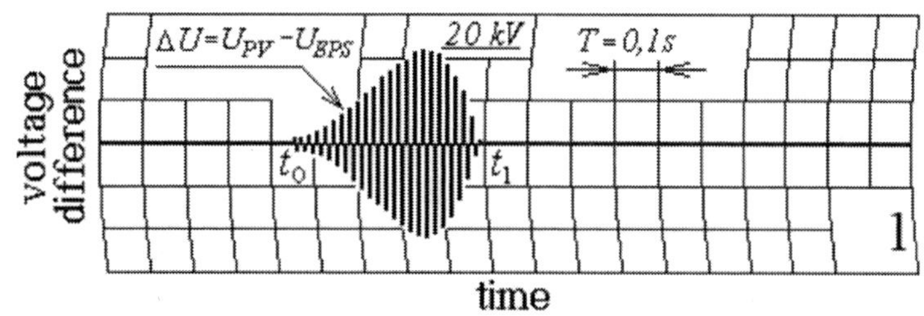

Figure 2. Registration of voltage between the stator of the motor STD-6300 and the backup power supply: T- time interval, s; ΔU - voltage difference, kV; U_{PV} - primary voltage M1, kV; U_{BPS} - voltage power backup, kV.

The synchronizer recorded the nonlinear nature of the change in the rotor angle of the motor M1 (Figure 3). The no current pause was about 450 ms. Considering the time interval that turns on the action of the relay protection and switches Q1, QB, 60 ms is left for the synchronizer to work. Therefore, the rotor angle values were measured between the three phases of the stator voltage of the motor and the backup power supply.

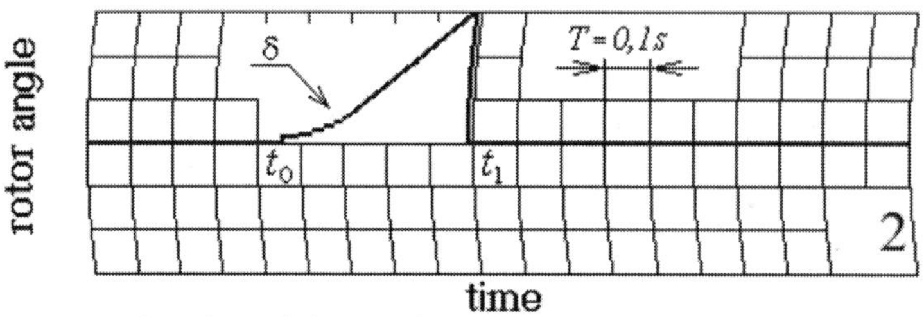

Figure 3. Registration of changes in the rotor angle of the STD-6300 engine: T– time interval, s; δ – the STD-6300 rotor angle.

When the backup power was turned on, the STD-6300 rotor rocked without turning it (Figure 4). The engine retained its dynamic stability. The transient process lasted approximately 820 ms.

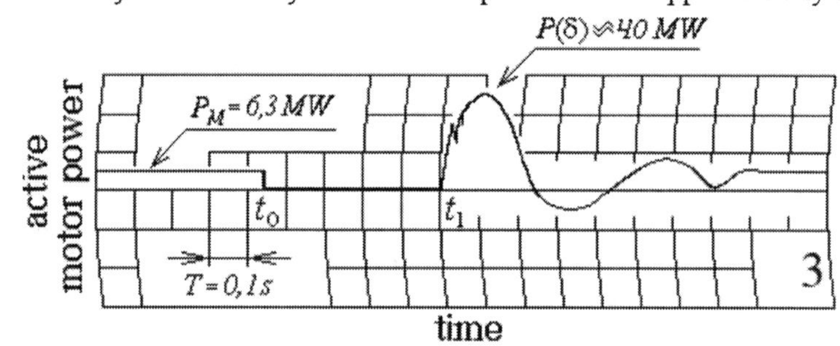

Figure 4. Registration of the active power of the STD-6300 engine: T– time interval, s; P_M – active power consumed by STD-6300 motor in normal mode, MW; P(δ) is the active power consumed by the STD-6300 motor in the transient process, MW.

Since the synchronization error in the switching angle is insignificant, there was no equalizing current (Figure 5). The increase in current during the rotor swing of the motor did not lead to a false operation of the relay protection.

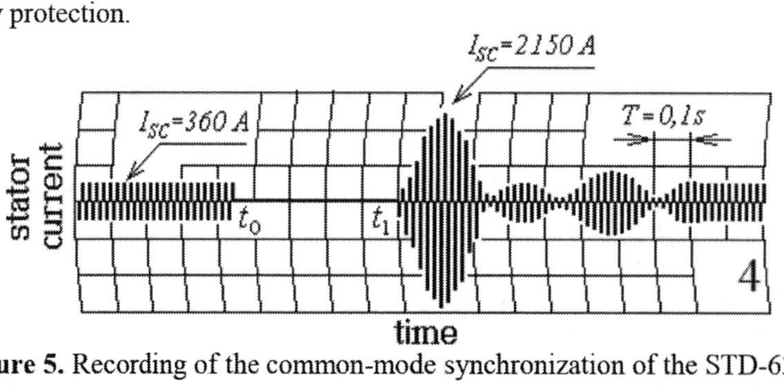

Figure 5. Recording of the common-mode synchronization of the STD-6300 synchronous motor: T– time interval, s; I_{SC} – stator current of STD-6300 motor, A.

When the STD-6300 runs out, there was a residual voltage on its stator (Figure 6). This ensured the possibility of operation of consumers whose power source was the power line W1 (Figure 1) and, therefore, the continuity of the technological process. The voltage drop during synchronization was negligible. Therefore, the M1 and M2 engines remained stable.

The synchronizer has also been tested for STD-12500, STD-3150, STD-1600, STM-1500, DSP-116 motors.

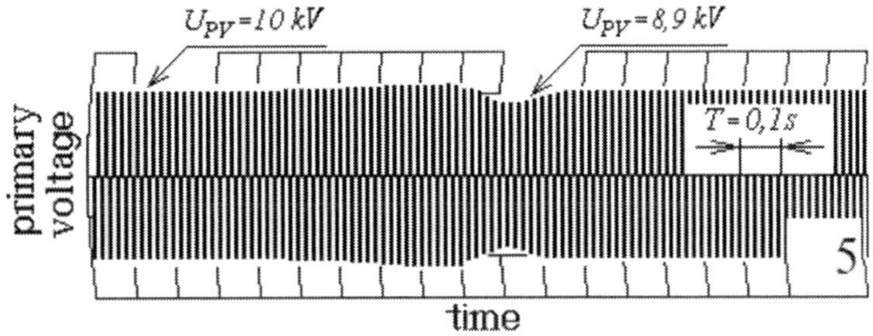

Figure 6. Recording the voltage on the stator of the STD-6300 motor: T– time interval, s; U_{PV} – primary voltage, kV.

3. Results

An algorithm for the operation of an automatic synchronizer to turn on the backup power supply of the SM has been developed. He has the necessary characteristics to solve this problem. Namely: high speed due to the lack of time for making a decision and high accuracy, since synchronization takes place at a large slip value.

Such characteristics are obtained by increasing the number of measurements of the value of the rotor angle of the SM in a unit of time. This reduces the time interval for extrapolation and allows one to take into account the nonlinearity of changes in the mechanical moments on the motor shaft. In addition, the used mathematical apparatus reduces the influence of measurement errors.

4. Conclusion

The accuracy of the trajectory extrapolation δ (t) mainly depends on the type of this function and the characteristics of the measuring means of the device under consideration. To predict the values of δ (t), the angles between the vectors of the same phases of synchronized voltages are measured. This makes it possible to increase the number of measurements per unit of time and ensure high performance of the proposed automatic synchronizer.

In addition to the field tests presented in the article, experiments were also carried out for various sets of loads. They showed that a group of SMs of different power during synchronization can be considered as an equivalent SM. For this mode, the proposed synchronizer can also be used.

In addition, due to the high voltage on the SM stator (Figure 6), the asynchronous load is switched to a backup source, receiving power from a SM operating in the generation mode. This load includes asynchronous motors, 0.4 kV busbars (own needs), etc. This allows you to maintain the SM in operation, as well as ensure the continuity of the technological process.

References

[1] Colak I and Yılmaz E N 1999 Simulation of the parallel connection of electrical power stations *Int. J. of Electrical Eng. & Education* **36(4)** 332-41

[2] Prokopchik V V and Shirokov O G 1997 Problemy elektrosnabzheniya predpriyatij s nepreryvnym tekhnologicheskim processom [Problems of ensuring a continuous technological process] *Elektrofikaciya metallurgicheskih predpriyatij Sibiri* [electrophication of metallurgical enterprises in Siberia] (Tomsk: TGU) pp 56-71[In Russian]

[3] Grigorkin B O 2017 *Issledovanie vliyaniya harakteristik mekhanizmov na vybeg sinhronnogo dvigatelya* [Study of the effect of mechanisms on the run-out of a synchronous motor] (St. Petersburg: KultInformPress) [In Russian]
[4] Fitzgerald A E, Kingsley C Jr and Umans S D 2007 *Electric machinery* (New York: McGraw-Hill)
[5] Dyakov A F and Ovcharenko N I 2000 *Microprocessornaya avtomatika i releinaya zashchita elektricheskish system* [Microprocessor-based automation and relay protection of power systems] (Moscow: MEI) [In Russian]
[6] Kireeva E A and Cyruk S A 2010 *Relejnaya zashchita i avtomatika elektroenergeticheskih sistem* [Relay protection and automation of electric power systems] (Moscow: Akademiya) [In Russian]
[7] Shewhart W A and Wilks S S 2019 *Time Series Analysis with Long Memory in View* (USA: Wiley)
[8] Thompson M J 2012 Fundamentals and advancements in generator synchronizing systems *65th Annual Conf. for Protective Relay Engineers* (College Station: IEEE) pp 203-14
[9] Seeley N C, Craig C and Rainey T 2012 Advances in generator control and automatic synchronization - Eliminating the need for standalone synchronization systems *Petroleum and Chemical Ind. Conf.* (Chicago: IEEE) 6549648
[10] Ingle R and Halmare A 2014 Automatic Synchronisation of Alternator for Small Power Plant *Int. Conf. on Industrial Automation and Computing* (Lancaster: IJERA) pp 44-6
[11] Emam S E A 2004 Automaic digital synchronization *Int. Conf. on Electrical, Electronic and Computer Engineering* (Cairo: IEEE) pp 778-84
[12] Bayındır C R, Demirbau S and Bal G 2006 *3rd Int. Conf. on Technical and Physical Problems in Power Eng.* (Turkey: Istanbul) pp 134-7
[13] Khrushchev Yu V, Khrushchev I Yu, Prokhorov A V, Belyaev N A and Vasiliev A S 2020 Programmed Trajectory motion Control for Synchronous Generators *Int. J. of Electrical Power & Energy Systems* **119** 105884
[14] Bekiroglu E and Bayrak A 2009 Automatic synchronization unit for the parallel operation of synchronous generators *IEEE EUROCON* (St.-Petersburg: IEEE) pp 766-71
[15] Kundur P 1993 *Power System Stability and Control* (New York: McGraw-Hill) pp 169-98
[16] Anderson P M and Fouad A A 2003 *Power System Control and Stability* (New York: Wiley) pp 82-149
[17] Lawson C L and Hanson R J 1974 *Solving Least Squares Problems* (New Jersey:Prentice-Hall)

Complex power control system for the installation of transverse capacitive compensation in the system with distributed load

L A German[1], A S Serebryakov[2], V L Osokin[2], and D E Dulepov [2]

[1]Department of Engineering and Technology of Railway Transport, Samara State University of Railways, Branch in Nizhny Novgorod, Komsomolskaya Sq., 3, Nizhny Novgorod, Nizhny Novgorod region, 603000, Russia
[2]Department of Electrification and Automation, Nizhny Novgorod State Engineering and Economic University, Oktyabr'skaya St., 22A, Knyaginino, Nizhny Novgorod region, 606340, Russia

E-mail: lagerman@mail.ru

Abstract. The operation of transverse capacitive compensation installations (CI) when they are switched on away from the power source (for example, at partitioning posts in the traction network or in general-purpose networks, for example, at branch points from main networks) under distributed load has a number of features. These features are considered on the example of an electrified railway section. In particular, the problem of controlling the power of the CI at medium and maximum loads is analyzed. Power control schemes of the CI have been known for a long time, usually regulation is performed to minimize power losses in the supply network, to reduce the maximum reactive power of the consumer, to stabilize the voltage mode in the power system node, etc. For electrified sections, regulation is typical to reduce power losses in the traction network or to stabilize the voltage at the AC traction network partitioning post. However, with the growth of heavy traffic on the railways, the requirements for increasing the capacity of railway sections while reducing electricity losses in the traction network have increased, that is, there are requirements for an integrated control system for CI. The complexity of solving this problem is determined by the fact that the CI is located at the partitioning post. The article offers an algorithm for solving this problem using the harmonics of the electric rolling stock current.

1. Introduction

Operation of transverse capacitance compensation installations (CI) when they are switched on away from the power source (for example, at partitioning posts in the traction network or in General-purpose networks, for example, at points of branch from the main networks) it has a number of features. Let's look at these features on the example of an electrified railway section. In particular, we will analyze the well-known problem of controlling the power of the CI.

Until recently, specialists were guided by well-known regulatory documents when calculating the capacity of transverse capacitive compensation (CI) installations and developing principles for its regulation. However, it is known [1-6] that CI is a powerful technical means of increasing the capacity of railway sections. Therefore, it is no coincidence that a number of measures have recently been implemented for the CI to increase throughput.

Research goal is to develop methods and technical tools for a comprehensive power control system for installing transverse capacitive compensation in distributed load systems, which provides energy savings for small and medium loads and a set voltage value for large loads.

2. Materials and Methods

The control and automation of existing installations of transverse capacitive compensation (CI) are considered in [7-15]. And in [1] it is specified it is indicated that the third harmonic of the traction current is used to control the power of the CI of the partitioning post (PP). This is because the traction load measurements at the partitioning post do not reflect the actual load related to the PP. Indeed, if the same currents are located on both sides at the same distance from the PP, then the traction current attributed to (accounted for by) to the PP, is zero (no current flows through the PP), and the measured current of the PP supply line is zero. The power control CI occurs only for reducing power losses in the traction network, and does not take into account the main indicator of the mode - carrying capacity of the site.

To increase the capacity of the railway section, the voltage level in the traction network is increased, and therefore, the power of CI installations is currently being increased for this purpose. The most common CI installations on domestic railways are condenser installations that are switched on by a third - harmonic filter with sequential switching on of the main capacitor Bank and reactor to prevent resonant phenomena. An increase in the power of the CI is accompanied by an additional installation of the capacitor section, which allows to regulate the power of the CI, while the additional section can be switched on in parallel or in series with the main capacitor bank. Next, we will consider the inclusion of an additional section of the CI_{add} in parallel with the main capacitor bank.

Thus, the main goal of the research carried out by the authors is to regulate the power of the CU in such a way as to increase the capacity of the electrified section of the railway.

We propose the following scheme for regulating the power of the CI (Fig.1), where the following designations are made:

1. Partitioning post (PP) with supply lines of the contact network on the switches;
2. PP tire;
3. Main switch of CI;
4. CI current transformer;
5. The reactor CI;
6. The main capacitor bank CI;
7. Additional capacitor bank CI_{add};
8. A switching device of the additional capacitor Bank CI_{add};
9. Unit additional capacitor Bank CI_{add} with the switching apparatus;
10. The first resonant filter for the third harmonic in the secondary winding of the current transformer CI;
11. Relay for monitoring the third harmonic of the CU current;
12. Switch of the supply line PP;
13. The current transformer of the supply line PP;
14. The second resonant filter of the third harmonic in the secondary winding of the current transformer of the supply line;
15. The second band stop filter of the third harmonic in the secondary winding of the current transformer of the supply line;
16. Ammeter $A_{(3)}$ of the third harmonic of the supply line current;
17. Ammeter $A_{(1)}$ of the first harmonic of the supply line current;
18. The estimated unit;
19. Comparison unit $I^*(1) > I$ installation;
20. Block for enabling an additional section CI_{add};
21. Block for disabling an additional section CI_{add};
22. The first barrier filter for the third harmonic in the secondary winding of the current transformer CI.

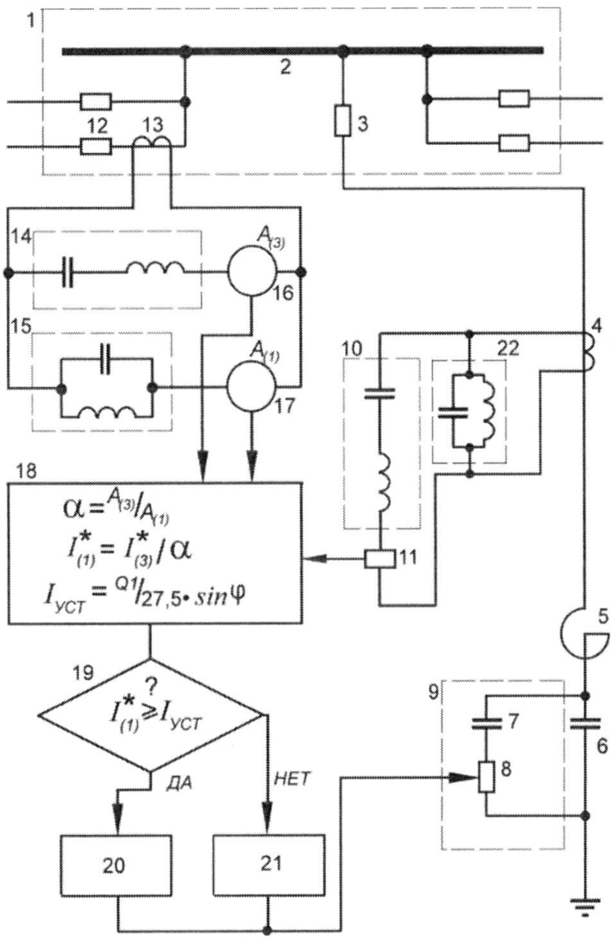

Figure1. CI power control scheme.

3. Results

Description of the operation and characteristics of blocks in the diagram.

Power of additional section CI, 9. Based on [1] power CI (Q1) of sectioning by the actual loads is usually equal to 4...5 MVAr and is determined by the need to reduce reactive power inter-station zone area to normalized values according to the formula "difference of tangents". In addition, [1] presents regulatory documents that require an increase in the power of the CI (up to Q2) to meet the requirement to ensure the voltage on the current collector of an electric rolling stock (ERS) of at least 21(24) kV. According to design experience, usually the implementation of [1] requires an increase in the power of the CI compared to the requirement ($Q1$) by 1.5 times. If the requirement to increase the throughput is met, it will be necessary to further increase the capacity of the CI ($Q3$). According to the design experience, the implementation of requires an increase in the power of the CI ($Q3$) in comparison with the variant ($Q2$) by another 1.5-2 times. Currently, the AC railway network has more than 200 such CIs, almost all of them require major repairs and corresponding modernization in terms of their terms and physical condition. Therefore, it is logical to pay attention at the present time to the modernization of existing CI taking into account the increase in the capacity of sections, which is considered in this article.

The total power of the CI together with the additional section of the CI_{add} is determined by

$$Q3 = (U_d - U_{min})\, U_{sgdm}/X_{in}, \tag{1}$$

and then the power of the additional section of the CI_{add}, 9, is equal to

$$Q\, CI_{add} = Q3 - Q1, \tag{2}$$

where: U_d – the desired voltage on the current collector, to increase the capacity of the section, depending on the nature of the load U_d = 25-28 kV;
U_{min} – minimum voltage on the heavy-duty current collector according to the calculation;
U_{sgdm} – rated voltage of a static generator of reactive power of a static gas distribution mechanism (SGDM);
X_{in} – input inductive resistance of the sectioning post;
$Q1$ and $Q3$ of the CU capacity, respectively, according to the standard and the condition for increasing the capacity of the section [1].

On the necessity of forming filters of harmonic components of the traction current, 10,14,15. To get the current assigned to the PP, unfortunately, it is impossible to determine it from measurements. Indeed, at the same ERC currents on both sides and at the same distances from the PP, the PP current is zero. Under the current referred to the PP, we understand the current of the PP in this instantaneous scheme, obtained as a result of decomposition of the instantaneous currents of the inter-station zone network between traction substations and the PP. Therefore, to calculate the PP current, the third harmonic current $I_{(3)}$ is used, which is characteristic of the rectifier electric rolling stock (ERS). No matter how placed ERS and as they were not all currents third harmonic of ERS are snapped via CI PP because the resistance of CI for the third harmonics of all the locomotives under review inter-station zone will be practically equal to zero. That is why a resonant filter for the third harmonic, 10, is connected to the current transformer of the CI PP, 4, with the third harmonic current monitoring relay, 11, in series in the CI PP. Note that the need to incorporate a band stop filter of the third harmonic is dictated by the observance of the mode of operation of the current transformer: as it always flows a current of the first harmonic, defined by the voltage on the PP tires, which excludes the no current in the secondary winding of the current transformer installation transverse capacitive compensation.

In CI PP will be to focus the whole current of the third harmonic of the traction network related to PP, and its maximum value will then define the current first harmonics of traction current, which should include an additional section CI.

However, to control the actual current first and third harmonics of the traction current should perform their measurements on a power line contact network (on any supply, such as switch 12 (Fig.1).

The need to obtain currents of the first and third harmonics is dictated by determining the ratio of these currents in the traction network during normal operation of the inter-substation zone. The highest values of the first and third harmonic components prevail in the traction current. The current sensors of the third and first harmonics of current CI are formed on the secondary side of the current transformer 13 of the supply line for the circuit of the current transformer closes on resonant third harmonic filter 14 and band-stop filter of the third harmonic (to obtain the first harmonic current in the circuit of the filter 15), in a circuit which included the ammeter $A_{(1)}$,17, and $A_{(3)}$,16. For experimental determination of the coefficient α ($\alpha = A_{(3)}/A_{(1)}$) a one-way power supply scheme should be performed at the time of measurement with the neighboring substation switched off (for increased values of traction currents and $I^*_{(3)}$ and excluding equalizing currents).

As a result of measurements, we get

$$\alpha = A_{(3)}/A_{(1)} \tag{3}$$

For calculations, we take the average value of α from the measurements made. Using the coefficient α (3), the current of the first harmonic is determined by the formula (4) $I^*_{(1)}$

$$I^*_{(1)} = I^*_{(3)}/\alpha\,, \tag{4}$$

where $I^*_{(3)}$ maximum value of the measured third harmonic of the traction network current when the additional section is disabled CI_{add}, and $I^*_{(1)}$ – the calculated value of the first harmonic of the traction current, at which the additional power of the CI_{add} is turned on, provided (5),

$$I^*_{(1)} \geq I \text{ installation}, \tag{5}$$

and when $I^*_{(1)} \geq$ I installation additional section CI_{add} is switched on, and when $I^*_{(1)} < I$ installation CI_{add} is switched off. If it is necessary to control the capacitive component of the traction load, instead of $I_{(1)}$ and I installation, $I_{(1)}\sin\varphi$ and I installation $\sin\varphi$ are determined, where $\sin\varphi$ is determined by measurements in the section under consideration.

The estimated unit 18 determines:

- the third harmonic coefficient $\alpha = A_{(3)} / A_{(1)}$ with the maximum value of $A_{(3)}$ and the disabled additional section of the CI using 16 and 17;
- the current of the first harmonic of the traction current, at which an additional section of the CI_{add} (4) and (5) is connected);
- set point current I installation for switching to the capacity increase mode (see below (6).

Thus, there are two modes of operation of the CI:

1) mode for reducing power losses in the traction network using the power control algorithm from [2].
2) with an increase in the current of the traction network assigned to the PP (5), you should switch to the second mode – the mode of increasing the capacity of the section.

We propose to pre-set the set point current to enable an additional section for the power of the $Q1$ CI:

$$I \text{ installation} = Q1/(27.5 \sin\varphi) \tag{6}$$

Here, the introduction of $\sin\varphi$ in the denominator determines the setting for the full value of the traction current. For each real railway section, the preliminary I installation value is specified based on the actual load of the section, the nature of train traffic, and the parameters of the power supply system. Further, according to the maximum current of the third harmonic in CI $I^*_{(3)}$ measured with disconnected additional section of the CI_{add} and the average value of the coefficient α determines the current of the first harmonic of traction load $I^*_{(1)}$ and the execution (5) inclusion of additional sections CI_{add}.

So, the novelty of the proposed technical solution is as follows:

1. The power control of the CI occurs not only to reduce power losses in the traction, but also to increase the capacity of the railway section, that is, a comprehensive power control is introduced.
2. The need for the introduction of power control to increase throughput is determined by the increased value of traction current, are related to PP, and when the traction current of the first harmonic more I include an additional section CI_{add}.
3. The traction current of the first harmonic assigned to the PP is determined by measuring the third harmonic of the current in the CI and then calculated using the formula (4) with the measured harmonic coefficient α.

The third harmonic coefficient α for calculating the first harmonic of the traction current is determined by forming the resonant and blocking filters of the third harmonic on the secondary winding of the current transformer of the supply line of the contact network of the PP.

4. Installation work.

When increasing the traction current assigned to the PP before setting I installation and above, a command is given to turn on the additional section of the cud. As a result, the total power of the CI increases, therefore, the voltage in the traction network increases and the conditions for passing heavy trains improve. As soon as the traction current decreases (*I*) $I^*_{(1)}$, which indicates a decrease in the traction load, the additional section of the CI_{add} is switched off. The installation current for switching on the additional power of the CI_{add} is determined by the expression (6).

Thus, a comprehensive system for regulating the power of the CI is implemented: before the traction current $I^*(1)$, the power of the CI is regulated to reduce power losses in the traction network, and during periods when the traction loads $I^*_{(1)}$ exceeds I installation (that is, $I^*_{(1)} \geq I$ installation), the power is regulated according to the law of increasing the capacity, that is, the AC is connected.

5. Conclusion

The idea of a comprehensive power control system for the CI should be implemented on all existing CI (first of all, on the PP), where the capacitor bank with the reactor is connected by a third-harmonic filter. However, the integrated power control system can also be used in other regulated CI, for example, in SGDM.

So, a comprehensive system of power control of the CU allows you to increase the capacity of the railway and reduce power losses.

References

[1] German L, Serebryakov A and Osokin V 2019 Three-stage Filter Compensating Installation of Traction Network of Alternating Current *Int. Ural Conf. on Electrical Power Eng. UralCon* (Chelyabinsk: IEEE) pp 229-33

[2] Menter F E and Kaltenbach M 1999 Considerations for AC railway traction power equipment-cost or performance-contradiction or challenge? *IEEE Power Eng. Society Summer Meeting. Conf. Proc.* (Edmonton: IEEE) pp 463-5

[3] Plakhova M, Mohamed B and Arboleya P 2015 Graph theory approach for a 2×25kV AC bivoltage traction system *6th Int. Conf. on Power Electronics Systems and Applications PESA* (Hong Kong: IEEE) pp 1-7

[4] Szeląg A and Patoka M 2014 Some aspects of impact analysis of a planned new 25kV AC railway lines system on the existing 3 kV DC railway system in a traction supply transition zone *Int. Symp. on Power Electronics, Electrical Drives, Automation and Motion* (Ischia: IEEE) pp 796-800

[5] Kabalci E 2017 Reactive Power Compensation in AC Power Systems *Reactive Power Control in AC Power Systems* ed N M Tabatabaei, A J Aghbolaghi, N Bizon, F Blaabjerg (Cham: Springer) pp 275-315

[6] Steimel A 2008 *Electric Traction-Motive Power and Energy Supply: Basics and Practical Experience* (Munich: Oldenbourg Industrieverlag)

[7] Faraji D, Rabiei A, Mohammadi B and Hoseynpoor M 2011 Reactive power generation management to improve voltage stability margin of power systems Australian *J. of Basic and Applied Sciences* **5(6)** 957-63

[8] Benidris M, Sulaeman S, Tian Y and Mitra J 2016 Reactive power compensation for reliability improvement of power systems *IEEE/PES Transmission and Dist. Conf. and Exposition (T&D)* (Dallas: IEEE) 7519910

[9] Patel V and Patel R 2019 Reactive Power Management through Dispatch Analysis using Evolutionary Algorithms *Int. J. of Innovative Technology and Exp. Eng.* **8** 759-65

[10] Dixon J, Moran L, Rodríguez J and Domke R 2005 Reactive power compensation technologies, state-of-the-art review *Proceedings of the IEEE* **93(12)** 2144-64

[11] Mujawar I, Patil D and Mujawar I 2013 Reactive Power Compensation and Harmonic Mitigation of Distribution System using SAPF Compensator *Int. J. of Eng. Science and Innovative Technology* **2** 494-02

[12] Lam C, Choi W, Wong M and Han Y 2012 Adaptive DC-Link Voltage-Controlled Hybrid Active Power Filters for Reactive Power Compensation *IEEE Transactions on Power Electronics* **4(27)** 1758-72

[13] Luo A, Shuai Z, Zhu W and Shen Z J 2009 Combined System for Harmonic Suppression and Reactive Power Compensation *IEEE Transactions on Industrial Electronics* **2(56)** 418-28
[14] Rahmani S, Hamadi A, Al-Haddad K and Dessaint L A 2014 A Combination of Shunt Hybrid Power Filter and Thyristor-Controlled Reactor for Power Quality *IEEE Transactions on Industrial Electronics* **5(61)** 2152-64
[15] Luo A, Peng S, Wu C, Wu J and Shuai Z 2012 Power Electronic Hybrid System for Load Balancing Compensation and Frequency-Selective Harmonic Suppression *IEEE Transactions on Industrial Electronics* **2(59)** 723-32

Decision-making algorithm for the development of the district of electrical grids

N G Semenova[1], A D Chernova[2] and L A Vlatskaya[2]

[1]Department of Automatic Electric Drive, Electromechanics and Electrotechniques, Orenburg State University, Prospect Pobedy, 13, Orenburg, 460018, Russia
[2]Department of Heat Technology and Power Engineering, Orenburg State University, Prospect Pobedy, 13, Orenburg, 460018, Russia

E-mail: ng_sem@mail.ru

Abstract. The article presents the stages and operational actions of the formation of a priority alternative for the development of the district of electrical grids. It is proposed to make a priority alternative from the preferred variants for the development of electrical grids objects. The selection and ranking of variants is influenced by the financial constraint, the technical condition of the equipment, the consequences of failure. The article presents two-level block diagram of the decision-making process at the local and global levels. It is proposed at the local level to select the variant for the development of an object of electrical grids by multi-criteria assessment and using neural network technology. At the global level, it is proposed to use the integration of an algorithm for constructing a decision tree and an algorithm for solving an optimization problem.

1. Introduction

The technical condition of electrical grids is one of the most important indicators of the country's power industry. Assessment of alternatives for the development of Russian electrical grids should take into account the following features of the technical condition of equipment:

1) equipment deterioration degree (the worn-out equipment park in some district is more than 50 %);

2) a large amount of equipment with different technical and economic characteristics;

3) growth of socio-economic requirements and financial constraints;

4) many possible alternatives for the development of electrical grids.

Therefore, the development of automated systems for the selection of a priority alternative for the development of power grids is becoming significant [1]. This allows to process large amounts of information; rank alternatives; reduce the time for making a decision; improve the validity of the decision, analyze and evaluate every possible alternative for the development of power grids.

Analysis of scientific in the development of automated systems in the energy sector has identified the following areas:

- optimization of the power supply system under uncertainty [2] using fuzzy logic techniques for the determination of development strategies, multi-criteria optimization and expertise to select the best variant;
- selection of points for placement of deep input substations under load uncertainty [3] by multi-criteria optimization and reducing several criteria to one;

- selection of the electrical grid configuration by discrete optimization methods according to the criterion of minimum losses with the possibility of using it in multi-agent systems [4,5];
- calculation, analysis and optimization of electricity losses in the power supply networks of an industrial enterprise, taking into account multi-criteria assessments of measures [6];
- optimization of the functioning and development of city power supply systems [7,8,9], based on the method of weighted average normalized assessment;
- assessment of the state of equipment of power supply systems [10,11] using an artificial neural network.

Abroad, DSSs are being developed in hydropower [12], Smart Grid [13], projects for the repair of electrical networks [14], reconfiguration of the energy distribution system [5], and energy management.

The district of electrical grids have a hierarchical structure, multi-connectivity and nesting of objects. As noted in [15], the hierarchical structure in any complex system allows to divide the process of selecting the best alternative into such a number of levels that "the solution of the optimization problem at each of them was not difficult".

In the paper, the district of electrical grids (DEG) is a set of technologically interconnected objects located on the same territory and served by one structural unit. An electrical grid object (EGO) is a collection of a substation and power supply lines.

In connection with the above, the study proposes a two- level decision-making structure. The preferred variant for the development of the object is selected at the first (local) level. The priority alternative for the development of the DEG is formed from the preferred variants for the development of object, taking into account the financial constraints of the investment program. This is the second (global) level.

Analysis of the literature in the field of decision-making theory revealed the following main stages of decision-making [15]: formation of alternatives; assessment of alternatives; selection of a priority alternative. The authors proposed to adapt the classical decision-making process in accordance with the purpose of the study. Figure 1 depicts the operational steps at each stage of decision making.

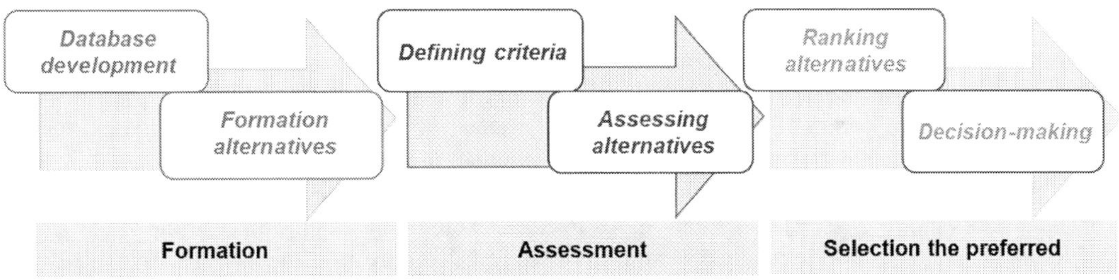

Figure 1. Stages and operational actions of the decision-making process for the development of electrical grids.

Below is the proposed structure of the decision-making process for each highlighted level.

2. Structure of the decision-making process

2.1. Selection the preferred variant at the local level

1. Formation set of variants.

1.1 Database development.

A relational database has been developed for structuring the nomenclature of equipment by their parameters. As a result, semantic links were established between such basic entities as: «Number of lines» $L\{l_1,l_2,\ldots l_m\}$; «Distance to power source» $D\{d_1,d_2,\ldots d_z\}$; «Wire type» $W\{w_1,w_2,\ldots w_n\}$; «Number of chains» $C\{c_1, c_2,\dots c_k\}$; «Pillow» $P\{p_1,p_2,\dots p_j\}$; «Wire section» $F\{f_1,f_2,\dots f_g\}$; «Substation scheme number» $S\{s_1,s_2,\dots s_i\}$; «Breaker type» for the high and low voltage sides of the substation $BHV\{b_{HV1},b_{HV2},\dots b_{HVh}\}$ и $BLV\{b_{LV1},b_{LV2},\dots b_{LVh}\}$; « Transformer power» $T\{t_1,t_2,\dots t_v\}$ etc. [16].

1.2 Formation of variants consists in generating tuples from variable sets of parameters of electrical grids equipment.

It is proposed to use a cyclic search with matching equipment parameters and checking the level of prospective load. The set of variants is formed by the Cartesian product of the nomenclature data:

$$A\{A1, A2,\dots Aq\} = L\times D\times W\times C\times P\times F\times S\times B_{HV}\times T\times B_{LV}\,A \tag{1}$$

Each of the variants is a tuple of length 10:

$$\begin{aligned} &Aq\{d_z, w_n, l_m, c_k, f_g\, p_j, s_i, b_{HVh}, t_v, b_{LVh}\} \,|\, l_m \in L \wedge d_z \in D \wedge w_n \in W \wedge c_k \in C \wedge \\ &\wedge p_j \in P \wedge f_g \in F \wedge s_i \in S \wedge b_{HVh} \in B_{HV} \wedge t_v \in T \wedge b_{LVh} \in B_{LV} \end{aligned} \tag{2}$$

Formation of variants takes into account the technical condition index of each piece of equipment [17] with the subsequent determination of the development scenario: technical re-equipment or reconstruction.

2. Assessment variants.

The second stage of the local level consists of the following operational steps:

2.1 Selection and substantiation of decision-making criteria and development of their mathematical models.

Electrical grids are characterized by a continuous increase in loads due to the appearance of new consumers. Therefore, they are dynamically developing systems. In addition, it is necessary to take into account economic, technical, social, environmental and other requirements.

In this regard, it is proposed to assess the development of an object of electrical grids by multi-criteria assessment model.

As a system of particular criteria for evaluating variants for the development of objects, it is proposed to use the following criteria: economic (total discounted costs); technical (annual loss of electricity); technical and economic (damage from power failure); socio-ecological (area of land withdrawn); technical and operational (unification of equipment, cenology) [2,16].

The development of mathematical models of particular criteria for evaluating variants is proposed to be carried out on the basis of regulatory documents in the field of electric power, as well as taking into account the typification of the nomenclature parameters of equipment. The development of mathematical models of particular criteria is detailed in the works of the authors [2,16].

2.2 Multi-criteria assessment of each variant.

The authors propose to evaluate alternatives for the development of an object of electrical grids by the theory of fuzzy sets or an artificial neural network (ANN). Application of the theory of fuzzy sets is described in detail in the works [2,3]. In this study, we will consider the use of ANN.

To train the ANN, a set of vectors of input and output values was used. The input vector contains the criteria for evaluating the alternative X. The value of the output vector corresponds to them. It contains the preference factor T:

$$(X^i, T^i), \tag{3}$$

where:
i is the number of the training alternative;
X is input vector containing the criteria reduced to unit power. These criteria assess the effectiveness of decision making;
T is a unit output vector that is the preference factor for the alternative. It characterizes the belonging of alternatives to the class of preferable or inexpedient solutions, varies from 0 to 1.

$$X^i = \left\{ \frac{C^i}{P}, \frac{\Delta W^i}{P}, \frac{D^i}{P}, \frac{S^i}{P}, N^i \right\}, \quad (4)$$

where:
C is the total discounted costs for the i-th alternative of object;
ΔW is annual loss of electricity;
D - damage from power failure;
S is area of land withdrawn;
N is coefficient unification of equipment;
P is level of prospective load.

In the work for training the ANN, the total number of examples was 2180. The number of actually existing substations in the Orenburg region, referred to the set of preferred solutions, was 250. The number of examples from the set of inappropriate solutions was 1930.

3 Selecting the preferred variant

3.1 Ranking variants.

The variants for electrical grids objects are ranked by the value of the preference coefficient, which is determined by the ANN.

3.2 Decision-making.

The preferred variant for the development of an object has the highest preference coefficient.

2.2. Selection a priority alternative at the global level

1. Formation set of alternatives.

1.1 Database development.

The proposed operational action limits the number of variants for the development of the objects included in the district of electrical grids. The decision-maker fixes the threshold value of the coefficient of preference T_{cr} for alternatives and the index of the technical condition TCI_{cr}:

$$\begin{cases} T_i \geq T_{cr} \\ TCI_{EGOi} \leq TCI_{cr} \end{cases} \quad (5)$$

where:
TCI_{EGOi} is the index of the technical condition of the i-th EGO;
T_i is the value of the preference coefficient of the i-th variants, the calculation of this coefficient is performed by the ANN in operational action 2.2.

1.2 Formation alternatives of the district of electrical grids

1.2.1 Formation of preliminary alternatives for the district of electrical grids was implemented by the algorithm for constructing a decision tree (figure 2) and by traversing it in a forward order with a depth-first search [18]. The nodes of the tree are variants of electrical grid objects (the subscript indicates the serial number of the variant, and the upper one - the object number).

The formation of a branch is completed when the amount of costs for the implementation of development measures at the district objects $\sum C^{EGOi+1}$ exceeds the volume of the investment program $C_{\max}$:

$$\sum_{i=0}^{n} C^{EGO_{i+1}} > C_{\max} \quad (6)$$

All variants for EGO at the nodes of the branch were added to the preliminary development alternatives for the district of electrical grids.

1.2.2 The final formation of alternatives for the district of electrical grids.

If the decision branch included all EGO, then this is a finally formed alternative for the development of the district of electrical grids A^{DEGi}.

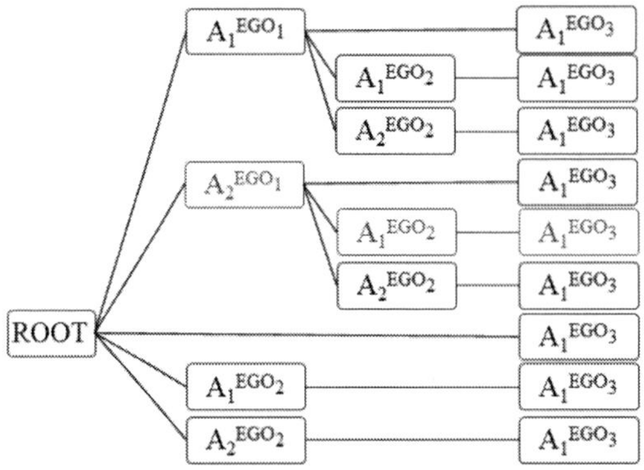

Figure 2. In this case simply justify the caption so that it is as the same width as the graphic.

If the decision branch does not include all EGO, then the formation of final alternatives depends on the amount of costs $\sum C^{EGOi+1}$ for the implementation of development measures at the district objects:

- if the unspent funds of the investment program C_{res} are sufficient for the spot replacement of emergency equipment at the object that were not included in the preliminary alternative:

$$C_{\max} - \sum_{i=0}^{n} C^{EGO_i} - C_{res} \geq 0\,, \tag{7}$$

then the preliminary alternatives were complemented by equipment replacement options;
- if the unspent funds of the investment program are not sufficient for the replacement any equipment at any object, then the preliminary alternatives were considered finally formed.

2. Assessment of alternatives.

2.1 Defining criteria.

It was proposed to use the value of repair cost C_R and damage from equipment accidents C_D at EGO that were not included in the alternative development of the DEG as an assessment criterion [19].

Making a decision on the selection of a priority alternative for the development of the DEG is an optimization task. Therefore, its mathematical model is proposed on the basis of regulations of electrical grid enterprises of the Orenburg region. It consists:

- Objective function – minimization of repair costs and damage from equipment accidents (formula 8);
- Constraints – Financial volume of the investment program $C_{\max}$ (inequality 9, where C^{EGOi} is the cost of implementing development measures at the i-th EGO, $C^{EGi'j}$ is the cost of spot replacement of equipment at the j-th EGO) ;
- Boundary conditions: non-negative values of the determined quantities; belonging to natural numbers of the object serial number (formulas 10-11):

$$\begin{cases} \sum\limits_{j=1}^{n} (C_{R_j} + C_{D_j}) \to \min & (8) \\ \sum\limits_{i=1}^{n} C^{EGO_i} + \sum\limits_{m=1}^{l} C^{EGO'_j} \leq C_{\max} & (9) \\ C_{R_j}, C_{D_j}, C^{EGO_i}, C^{EGO'_j}, C_{\max} \geq 0, & (10) \\ i, j = 1,2..n & (11) \end{cases}$$

3. Selection the best alternative

3.1 Ranking of alternatives for the development of the district of electrical grids.

3.1.1 The alternatives for the development of the DEG were ranked according to the value of the objective function of the optimization task if they did not contain all objects. When the values of the objective function for several alternatives were equal, they were sorted according to the increase in capital costs.

3.1.2 If all objects were included in the investment program, then the alternatives for the development of the DEG were ranked according to the value of capital costs, because the objective function took a value equal to 0.

3.2 Decision-making.

Automated decision support systems are used to generate a variety of alternatives, to evaluate and rank them. The main purpose of such programs is to provide reasoned information to the head of the department for technical re-equipment and reconstruction of electrical grids to select an adequate and effective option for the development and functioning of electrical grids. Automated decision support systems do not replace the manager, they only provide technical support. The final decision is made by the manager.

3. Results of the software implementation of the algorithm for selection the priority alternative for the development of the district of electrical grids

In the paper, the developed algorithm is implemented in the C# [20]. Visual Studio was selected as the development environment. The program has been tested and validated on the example of power lines and substations in the Central District of Electrical Grids of the Orenburg Region.

The results of the selection of the preferred variant for the development of the EGO at the local level are shown in table 1.

Table 1. Results of decision-making on objects of the Central District.

Object	Measure	Composition of alternatives and their number	
Substation 35/10 kV "Stepanovskaya"	Technical re-equipment due to increased load	Replacement of 2 transformers 6.3 MVA with 10 MVA; replacement of circuit breakers on the 35 and 10 kV side; change of the typical scheme of the substation.	18 pcs
Substation 110/35/10 kV "Selskaya"	– " –	Replacement of 16 and 10 MVA transformers with 25 MVA transformers; installation of circuit breakers on the 110 kV side, replacement of circuit breakers on the 35 and 10 kV side; change of the typical scheme of the substation.	18 pcs
Substation 110/10 kV "Stepnaya"	– " –	Replacement of the 25 MVA power transformer with 40 MVA.	1 pcs
Substation 110/10/6 kV "Yugo-Vostochnaya"	Reconstruction due to equipment wear	Replacement of transformers and circuit breakers	18 pcs

The results of the selection of the priority alternative for the development of the DEG at the global level are shown in table 2. The investment program was limited to 700 million rubles. The priority alternative includes all considered objects in the Central District of Electrical Grids the Orenburg Region.

The results of the decision support system are consistent with the proposals for substations of the Central District specified in the "Comprehensive program for the development of electrical grids with a voltage of 35 kV and above in the Orenburg region for the period 2018-2022", as well as in the "Scheme and program for the prospective development of the Orenburg region for 2019-2023".

Table 2. Priority alternative for the development of the Central District.

Object	Equipment replacement results					
	HV circuit breakers		MV and / or LV circuit breakers		Transformers	
Substation "Yugo-Vostochnaya"	SF6	3 pcs	vacuum	6 pcs	40 MVA	2 pcs
Substation "Stepanovskaya"	vacuum	3 pcs	vacuum	3 pcs	10 MVA	2 pcs
Substation "Selskaya"	SF6	3 pcs	vacuum	6 pcs	25 MVA	2 pcs
Substation "Stepnaya"	–	–	–	–	40 MVA	1 PC

4. Conclusions

1. A two-level block diagram of decision-making on the development of an electric grids region is proposed. It contains the local level of decision-making on the selection of preferred variants for the development of electrical grid object; and the global level of decision-making on the selection of the priority alternative for the development of the district of electrical grids, it is formed from the preferred variants for the development of objects, taking into account the financial constraints. The following stages are highlighted at each level: formation of an alternative; assessment of alternatives; selection the best alternative.

2. It is proposed to use particular criteria to assess variants for the development of objects: economic (total discounted costs); technical (annual loss of electricity); technical and economic (economic losses from power failure); socio-ecological (area of land withdrawn); technical and operational (unification of equipment, cenology).

3. It is proposed to carry out multicriteria estimation and ranking of variants of electrical grids objects by a neural network trained by the error backpropagation algorithm.

4. A mathematical model has been developed for selection a priority alternative for the development of the district of electrical grid. It is represented by a system of equations for conditional optimization, including: objective function - minimization of repair costs and damage in the event of equipment failure that was not included in the investment program; limitation - the financial volume of the investment program.

5. The selection of the priority alternative for the development of the district of electrical grid at the global level is proposed to be implemented by integrating the algorithm for constructing a decision tree and an algorithm for solving the optimization problem. Preliminary and final alternatives for the development of the district of electrical grid are being formed in the context of the limited volume of the investment program. Alternatives are evaluated according to the criterion of repair costs and damage in case of equipment failure. The alternatives of the district of electrical grid are ranked in ascending order of the values of the objective function.

6. The developed automated decision support system was tested and approved. The results obtained are consistent with the schemes and programs for the development of energy for the electric grids of the Orenburg region.

References

[1] Ada S and Ghaffarzadeh M 2015 Decision making based on management information system and decision support system *European Researcher* **93(4)** 260-9

[2] Berdin A S and Semenova L A 2010 Integraciya tekhnocenologicheskogo podhoda i teorii nechetkih mnozhestv v zadachah optimizacii sistem elektrosnabzheniya [Integration of the technocenological approach and the theory of fuzzy sets in the problems of optimization of power supply systems] *Izvestiya Vuzov. Problemy energetiki* **3-4** 151-6 [In Russian]
[3] Leshchinskaya T B 2004 Algoritm resheniya mnogokriterial'nyh zadach optimizacii s ne-opredelennoj informaciej na primere vybora optimal'noj moshchnosti glubokogo vvoda vy-sokogo napryazheniya [Algorithm for solving multicriteria optimization problems with uncertain information on the example of choosing the optimal power of high-voltage deep input] *Elektrichestvo* **10** 8-14 [In Russian]
[4] Liao J, He L, Yuan X and Li H 2008 Application of decision support system to three gorges cascade hydropower station *Int. Conf. on Genetic and Evolutionary Computing* (Hubei: IEEE) pp 500-3
[5] Andervazh M-R, Javadi S and Aliabadi M H 2018 Decision support system for multicriteria reconfiguration of power distribution systems using CSO and efficient graph traversal and repository management techniques *Int. Trans. Electr. Energy Systems* **28(8)** e2579
[6] Flores W C, Mombell E E, Jardini J A, Ratta G and Corvo A M 2011 A Expert system for the assessment of power transformer insulation condition based on type 2 fuzzy logic systems *Expert Systems with Applications* **38(7)** 8119-27
[7] Khalyasmaa A I, Dmitriev S A, Kokin S E and Eroshenko S A 2014 Fuzzy neural networks' application for substation integral state assessment *WIT Transactions on Ecology and the Environment* **190** 599-605
[8] Ram S, Chandel A, Singh G and Mondal M 2012 UV Spectrophotometer Based AI Techniques for Remnant Life Estimation of Power Transformers *IOSR J. of Electrical and Electronics Eng.* **3** 46-55
[9] Khal'yasmaa A I, Dmitriev S A and Kokin S E 2016 An automated system for taking decisions to assess the actual state of electrical equipment *Power Technology and Eng.* **49(5)** 389-92
[10] Khalyasmaa A I and Dmitriev S A 2014 Power equipment technical state assessment principles *Applied Mechanics and Materials* **492** 531-5
[11] Khalyasmaa A I, Dmitriev S A, Kokin S E, Giushkov D A and Kuzin P 2014 Development of 110-220 kV power transformer model for equipment functional state assessment system *Advanced Materials Research* **950-951** 653-8
[12] Barriquello C H, Garcia V J, Schmitz M, Bernardon D and Fonini J S 2017 A decision support system for planning and operation of maintenance and customer services in electric power distribution systems *System Reliability* (London: IntechOpen) pp 355-70
[13] Jones L E 2012 *Strategies and decision support systems for integrative variable energy resources in control centers for reliable grid operations* (Charlotte: Areva Federal Services LLC)
[14] Oatley C J, Ramsay B, McPherson A, Eastwood R and Ozveren C S 1997 A decision support system for electricity distribution network refurbishment projects *Electric Power Systems Research* **40(1)** 27-35
[15] Mazurek M 2013 Concepts of distributed database for decision support in smart grid *Biuletyn Wojskowej Akademii Technicznej* **62(1)** 115-28
[16] Semenova N G and Chernova A D 2018 Vy`bor predpochtitel`nogo resheniya po razvitiyu e`lektricheskix setej na osnove nejrosetevy`x texnologij [Selecting the Preferred Solution on Development of Electrical Supply System Based on Technology of Neural Networks] *Vestnik Y-UGU Seriya: E`nergetika* **18(3)** 38-43 [In Russian]
[17] Yang C, Su Z, Wang X, Liu Y and Qi Y 2017 Design of investment management optimization system for power grid companies under new electricity reform *AIP Conf. Proc.* **1820(1)** 784-91
[18] Levitin A 2012 Introduction to the Design and Analysis of Algorithms (New Jersey: Pearson Education)

[19] Semenova N G and Chernova A D 2020 The selection of a priority package of measures for technical re-equipment and reconstruction of the district of electrical grids *IOP Conf. Ser.: Mater. Sci. Eng.* **791** 012035
[20] Chernova A D 2019 *Sistema podderzhki priyatiya resheniya po razvitiyu rajona elektricheskih setej Patent RU 2019660150* [Decision-making support system for the development of the electric grid region Patent RU 2019660150] (Moscow: Rospatent) [In Russian]

Development and modeling of LED indicator of high voltage insulator status

R N Balobanov[1], D I Amirov[1] and A G Razina[2]

[1]Department of Electric Stations named after V. K. Shibanov, Kazan State Power Engineering University, Krasnoselskaya Str., 51, Kazan, Republic of Tatarstan, 420066, Russia
[2]Department of Applied Physics and Nanoelectronics, Chuvash State University named after I.N. Ulyanov, Moskovsky prospect 45, Cheboksary, Chuvash Republic, 428015, Russia

E-mail: rassel_ipek@mail.ru

Abstract. The article discusses a non-contact method for monitoring the state of high-voltage insulation. The principle of operation of circuits of devices built using various electronic components is described. The results of modeling the operation of the device in the software complex of finite element modeling COMSOL Multiphysics are presented. The modeling was carried out by constructing and analyzing three-dimensional models of insulators of the LK70/35 and LK 70/110 brands with indicators installed on them. The presented method of insulation control will allow, during visual inspection of overhead lines, to additionally receive information about the change in the electric field and the appearance of partial discharges in the insulation. The main results of this article are the determination of the parameters and requirements for the electronic components of the considered circuits of indicators of the state of high-voltage insulators.
Keywords: polymer insulator, indicator, defects, LEDs, power line, insulation condition, assessment, outdoor switchgear, contactless method, dynistor.

1. Introduction
Insulators are one of the most important structural elements of power lines and open switchgears. Improvement and modernization of insulator designs occurs throughout the history of the development of the electric power industry. Insulators are electrical insulating structures designed for fastening and insulating live parts of installations under potentials. Insulators are classified according to voltage, type of installation, purpose and design. The disadvantage of all designs of insulators is the lack of available methods of operational monitoring of the state of internal insulation. An exception is tempered glass suspension insulators. The internal breakdown of such insulators can be easily identified by the breakdown of the glass insulating piece. Due to this, these insulators are widely used in the power industry. However, they have many disadvantages: weak hydrophobicity and tracking resistance in comparison with silicone rubber of polymer insulators, low moisture-discharge characteristics in comparison with polymer insulation, the impossibility of manufacturing large-size support insulation for substations, etc.

Recently, active developments have been carried out in the direction of creating systems for monitoring the state of high-voltage insulating structures. One of the promising directions is the creation of defect indicators installed on an insulating structure. Various options for insulating structures have been developed, consisting of solid or composite insulators, with additional or built-in elements for assessing their condition [1 - 18]. These elements are called defect indicators.

2. Materials and Methods

During the design, two electrical circuits of the indicator were developed. The first circuit is based on the MCP 65R41 comparator (Fig.1).

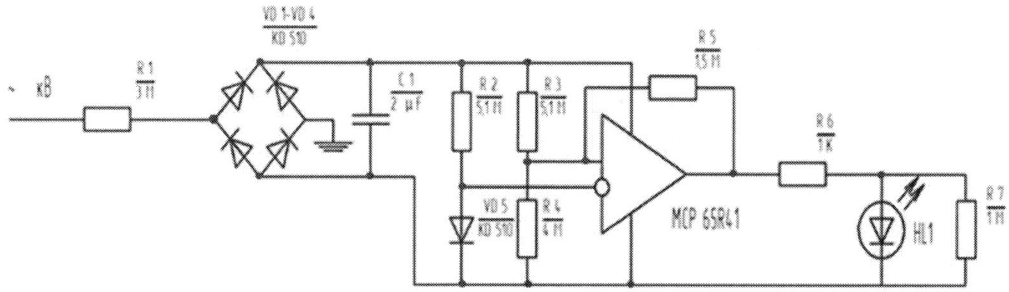

Figure1. Comparator indicator circuit.

In this circuit, the current induced on the electrode, which serves as a capacitor plate, is rectified by the diode bridge and, flowing through the capacitor C1, charges it. The circuit, consisting of R2 and VD5, creates a reference voltage that is applied to the inverse input of the comparator and thus sets the threshold. The voltage generated by the divider built across R3 and R4 is applied to the direct input of the comparator, which is proportional to the voltage across C1. In the process of operation, the voltage on C1 gradually increases to a certain value. At this moment, the voltage at the direct input, proportional to the voltage on C1, reaches the operating threshold and the comparator switches, opening its output transistor and thereby supplying voltage to the LED HL1. The LED lights up. The current flowing through the LED discharges C1 and the voltage across the capacitor begins to drop rapidly, since the capacitor charge current from the electrode is many times less than the discharge current. The voltage at the direct input of the comparator also drops. As soon as it falls below the trip threshold, the comparator switches, interrupting the current through the LED. The LED goes out. Capacitor C1 starts recharging. Thus, the cycle of charging and discharging (turning the LED off and on) is continuously repeated. Resistor R5 in the circuit serves to set the required hysteresis value at the comparator input and determines the time of the LED on. Resistor R6 limits the current through the LED and also determines the on time of the LED. Resistor R7 shunts the LED, allowing C1 to discharge even if the voltage across the LED is below its turn-on threshold.

The second scheme is based on dinistors (Fig. 2). In Figure 2, the dotted lines indicate the connection of the indicator with the high-voltage and grounded part of the structure. A feature of this scheme is the ability to display discharges that occur in the insulation during damage and / or contamination. The circuit consists of two loops. The first one includes the following elements: capacitance C1, resistances R1 and R2, dinistor VD6 and green LED HL1. The second one consists of the capacitance C2, the resistance R3, the dynistor VD7 and the red LED HL2.

Let's consider the principle of the indicator. The distribution of potentials along an insulating structure changes in violation of the integrity of its individual parts. The potential difference in the damaged area decreases, which causes an increase in voltage on the intact part of the structure. The defective condition of the insulator can be detected, for example, by increasing the intensity or frequency of the pulses of the light emitted by the indicator mounted on the portion of the insulating part of the structure or support, as this increases the voltage on it. One of the electrodes of the radiator can be earthed through the grounded part of the insulator or support, the second electrode can be positioned freely.

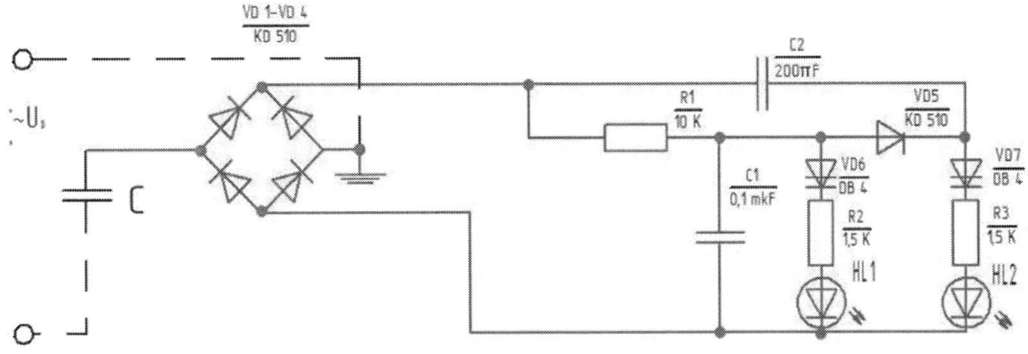

Figure2. Diagram based on dynistors.

To ensure reliable operation of the radiator with a constant installation, it is necessary to connect a discharger in parallel to it. In the event of a surge overvoltage, the discharger shunts the radiator, protecting it from failure.

The scheme works as follows. The current from the diode bridge VD1-VD4 charges the capacitance C1. As charging C1 voltage on it increases. The dynistor VD6 does not pass the current until the voltage on its terminals (on C1) reaches 40 V.As soon as the voltage reaches the desired value, the dynistor opens and the charge from the capacitor C1 is discharged to the LED HL1, as a result of which the LED emits a flash of green light. Simultaneously with the flash, the voltage on the dynistor VD6 drops, and it closes. Next, the capacitance C1 is again charged until the opening voltage of the dynistor and the cycle repeats. Thus, the green LED is pulsed.

If there are no discharges on the insulation, only the green HL1 LED will always be triggered, because due to the voltage drop across the VD5 diode, the dynistor VD7 will always be closed.

At a certain intensity of the discharges on the insulator, the pulses, passing through the second (high-frequency) circuit with a capacitance C2, will increase the voltage on the VD7 dynistor to a level higher than that on the VD6 dynistor. Then VD7 will open first and the red LED HL2 will light up. The value of capacitance C2 is selected in such a way as to react to the appearance of short pulses (≤ 1 μs) caused by discharges on the insulator. More details ot the work of the LED indicator are considered in the thesis.

3. Results

The indicator works without using a separate power source on the principle of accumulating electric field energy. For the efficiency of such a device, it is necessary that during the accumulation of energy, as little of it as possible is lost. Therefore, the development of criteria for the selection of elements of the indicator circuit with minimum leakage currents is fundamental. This was the first challenge in the simulation.

The second task in computer calculations was to assess the possibility of visually detecting the appearance of an insulation defect on various types of insulators by changing the voltage on the indicator, and hence changing the LED blinking frequency.

The calculations were carried out in the COMSOL Multiphysics program [23-24] by constructing and analyzing three-dimensional models of insulators LK70 / 35 and LK70 / 110 with the same indicator installed on them.

At the beginning of the modeling of the LK70 / 35 insulator, a three-dimensional model of the insulator was drawn in the COMSOL program.

Figure 3 shows a separate section of the insulator with an indicator installed on the end piece.

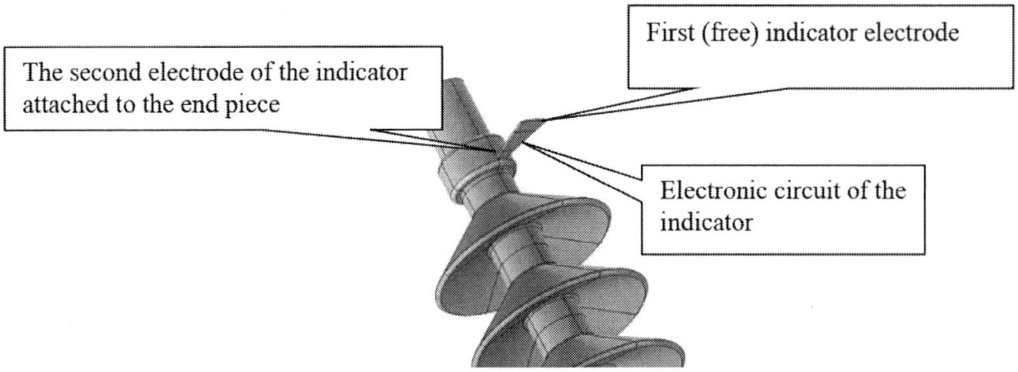

Figure3. Indicator attached to the end cap.

The situation was simulated when the insulator is in air, a phase voltage of 20 kV (35 kV / 1.7) is applied to the ends of the insulator, the amplitude value of which is U0 = 28 kV. A parametric calculation was carried out, where the resistance value of the indicator circuit Ri was used as a parameter. The electronic circuit of the indicator itself was modeled by a strip of semiconducting material, the ends abutting against two electrodes, one of which is fixed to a grounded end cap. The variable parameter Ri was converted into the value of the specific conductivity of the material of the semiconducting strip G from the well-known formula:

$$Ri = \frac{1}{G}\frac{l}{S}, \tag{1}$$

where l is the length of the circuit, m;
S — cross-sectional area of the strip in the middle part, m2.

The equation used by COMSOL Multiphysics to calculate is derived from Ohm's Law, Continuity Equation, and Gauss's Theorem. For a sinusoidal current with an angular frequency ω = 2πf and an isotropic medium, these expressions can be written as:

$$\begin{gathered} J = \sigma \cdot E; \\ \nabla \cdot J = \nabla \cdot (\sigma \cdot E) = -j \cdot \omega \cdot \rho; \\ \nabla \cdot D = \rho; \\ E = -\nabla \cdot V; D = \varepsilon \cdot \varepsilon_0 \cdot E, \end{gathered} \tag{2}$$

where:
J, E, D are vectors of current density (A / m2), electric field strength (V / m) and electric displacement (C / m2), respectively;
σ - specific conductivity (S / m);
ρ is the volumetric charge density (C / m3);
ε is the dielectric constant of the material;
$\varepsilon_0 = 8{,}85*10^{-12}$ F/m – electrical constant; j is the imaginary unit.

After substituting the expressions for E and ρ into the continuity equation, a Poisson equation of the form $\nabla \cdot (-c \cdot \nabla \cdot V) = f$ is obtained from a set of mathematical equations solved by the COMSOL Multiphysics software.

$$\nabla \cdot (-(\sigma + j \cdot \omega \cdot \varepsilon \cdot \varepsilon_0) \cdot \nabla \cdot V) = 0 \tag{3}$$

Table1. Initial data for calculation in COMSOL Multiphysics software.

Name	Value	Size	Description
l	14e-3[m]	0.014 m	Circuit length
S	1e-5[m^2]	1E−5 m²	Cross-sectional area of the strip (diagram) in the middle
R	1e6[1/S]	1E6 Ω	Leakage resistance of indicator circuit
r	R*S/l	714.29 Ω·m	Resistivity of the indicator circuit
G	1/r	0.0014 S/m	Specific conductivity of the indicator circuit
sigma_air	0[S/m]	0 S/m	Specific air conductivity
epsilon_air	1	1	Dielectric constant of air
sigma_copp	5.99e7[S/m]	5.99E7 S/m	Specific conductivity of copper
epsilon_copp	1	1	Dielectric constant of copper
sigma_al	3.77e7[S/m]	3.77E7 S/m	Specific conductivity of aluminum
epsilon_al	1	1	Dielectric constant of aluminum
sigma_sig	1e-14[S/m]	1E−14 S/m	Specific conductivity of rubber
epsilon_sig	2.09	2.09	Dielectric constant of rubber
sigma_gla	G	0.0014 S/m	Specific conductivity of the circuit
epsilon_gla	1	1	Dielectric constant of the circuit
U_0	28[kV]	28000 V	Insulator voltage amplitude
f	50[Hz]	50 Hz	Frequency

Further, the electrical characteristics of materials for each of the areas were established using the example of the isolation area. Also, boundary conditions were set for the potential, "land" and the boundaries of the computational domain.

The parametric calculation was performed for the values of the leakage resistance R of the indicator electrical circuit: 1 MΩ, 10 MΩ, 100 MΩ and 500 MΩ. The most important result is shown in Figure 4.

Shown here is the potential difference between the electrodes of the circuit as the circuit leakage impedance changes from 1 MΩ to 500 MΩ. It can be seen from the graph that the developed circuit based on dinistors (figure 2) will not work until the potential difference does not exceed the response voltage of the dinistors (~ 40 V), which can only happen if the total resistance of the indicator exceeds 20 megohms. This important conclusion formed the basis for the choice of components and the design of the indicator.

The constructed model served as the basis for calculating the operation of an insulator for 110 kV with an indicator, where the capabilities of the indicator were modeled and evaluated in terms of functional purpose.

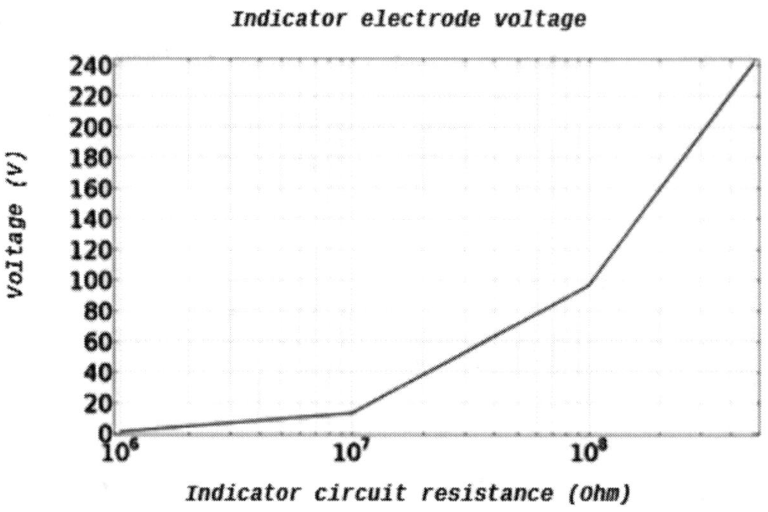

Figure4. Dependence of the amplitude value of the circuit voltage on the resistance of the indicator circuit.

Modeling a 110 kV suspended polymer insulator with a defect. In the case of a 110 kV suspended polymer insulator with a defect, it is necessary to assess the change in the repetition rate of the indicator light pulses in the presence of a defect in the insulating structure shunting part of its length.

For this, a model of a 110 kV insulator of the LK70 / 110 brand was created. A feature of this insulator, in contrast to LK70 / 35, is the presence of screens at the ends, which serve to equalize the electric field. The presence of screens determines the requirements for optimal fixing of the indicator on the insulator. We considered two possible options for attaching the indicator: on the terminal behind the screen and on the screen itself.

The insulation defect was modeled as a conductive channel 1 mm in diameter located inside the insulator between the rubber sheath and the fiberglass rod, as shown in Figure 5. The defect shunts 30% of the insulating structure.

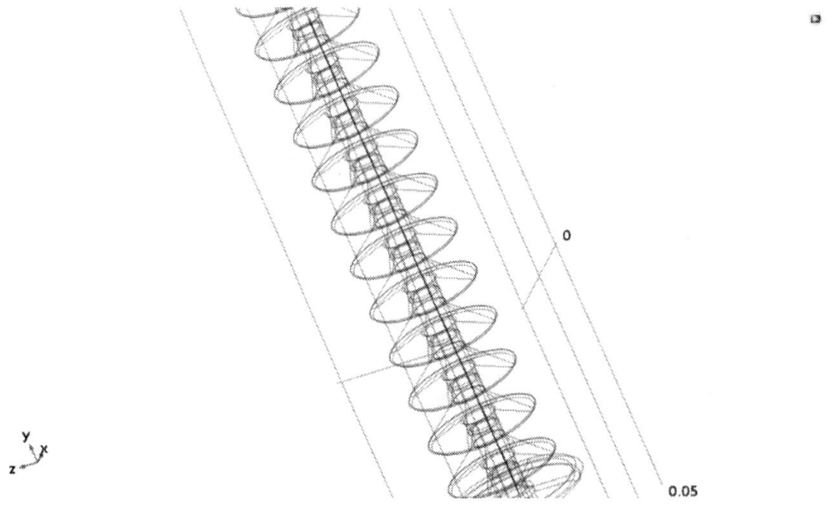

Figure5. Section of the insulator with a highlighted defect.

Calculations in the COMSOL Multiphysics program were carried out according to the initial data (table 1) with the changes and additions reflected in table 2 according to the formula (3).

Table 2 Initial data for calculation with changes and additions

Name	Value	Size	Description
R	40e7[1/S]	4E7 Ω	Indicator circuit resistance
sigma_D	1e-12[S/m]	1E−12 S/m	Defect specific conductivity
epsilon_D	1	1	Defect dielectric constant
U_0	90[kV]	90000 V	Insulator voltage amplitude

Figure 6 shows the results of a parametric calculation in the form of graphs of the dependence of the amplitude of the potential difference between the indicator electrodes on the conductivity of the defect for cases where the indicator is installed behind the screen and on it, respectively. As expected, in absolute terms, the voltage is higher in the second case. However, the change in voltage as the defect develops is negligible. In the first case (figure 6, 1) the voltage increases from about 64 V to 83 V and the ratio is 1.3, and in the second case (figure 6, 2) from 112 V to 145 V and the ratio is also 1.3.

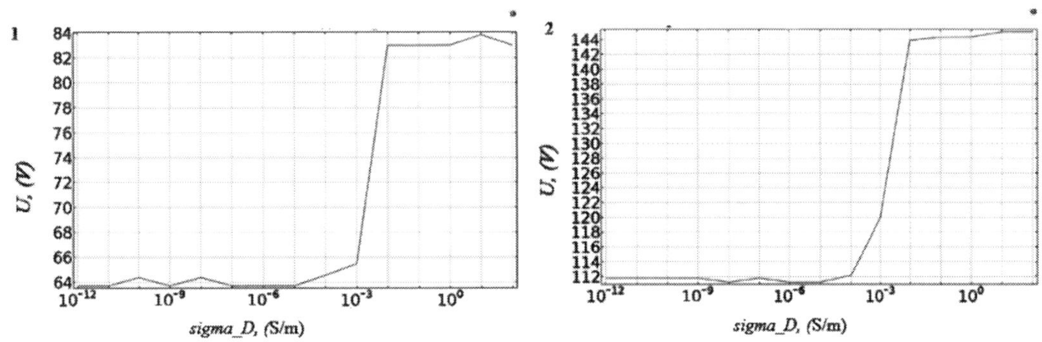

Figure6. 1 - Change in the potential difference from the conductivity of the defect when the indicator is located behind the screen; 2 - Change in the potential difference from the conductivity of the defect when the indicator is located on the screen.

Figure 7 shows the voltage distribution along the indicator from the free electrode (antenna) to the attachment point for cases without a defect (figure 7, a) and with a defect (figure 7, b) with high conductivity, respectively.

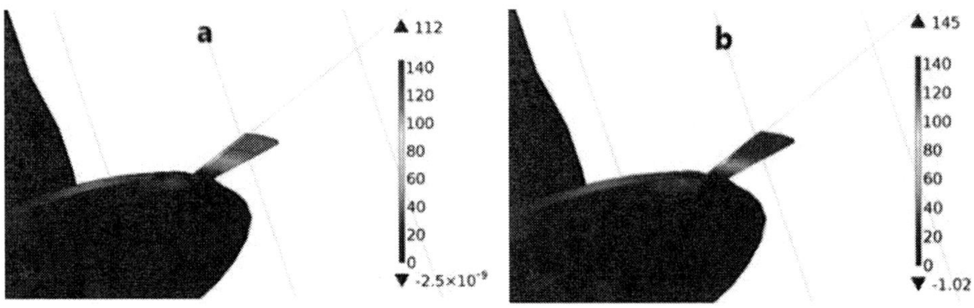

Figure7. Change of potential on the indicator from antenna to ground: a - no defect in insulation (sigma_D=1e-12 S/m) b - defect with high conductivity (sigma_D=100 S/m).

Further, it is important to find out whether the increase in the voltage on the indicator in the event of a defect will be sufficient to change the repetition rate of light pulses of the indicator LEDs when visually distinguishing between the defective and normal states of the insulator.

The principle of operation of the indicator according to the diagram in Figure 2 is based on the accumulation of energy on a capacitor through an integrating circuit.

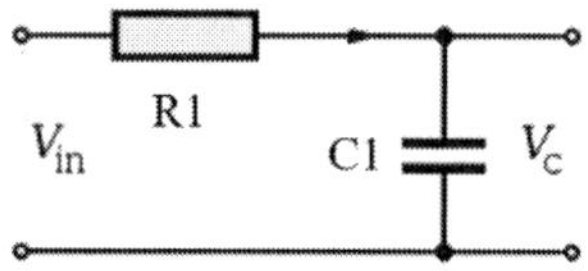

Figure8. Equivalent circuit of the indicator input circuit.

The voltage on C1 will increase over time according to the law:

$$U_c(t) = U_{in} \cdot (1 - e^{-\frac{t}{\tau}}) \tag{4}$$

In the circuit (figure 2), the voltage Uc (t) can only increase up to the switching voltage of the dynistor. The time to reach this value is the shorter, the higher the input voltage Uin. As shown by modeling, with a defect in the polymer insulator LK70 / 100, covering 30% of the insulation, the voltage on the indicator increases 1.3 times.

Let t1 and t2 be the time for the voltage on C1 to rise to the turn-on value of the dinistor $U_c(t_1)$=$U_c(t_2)$ for the case without a defect and with a given defect, respectively. Then, setting $U_c(t_1)$=$U_c(t_2)$=36 V, and U_{in1}=112 V and U_{in2}=145 V from formula (4) we get:

$$\frac{t_2}{t_1} = \frac{\ln(1 - \frac{U_C}{U_{in1}})}{\ln(1 - \frac{U_C}{U_{in2}})} = 0.736 \tag{5}$$

Thus, the presence of a defect will cause an increase in the repetition rate of light pulses by 1 / 0.736 = 1.36 times. As laboratory and field experiments have shown, this increase in the pulse repetition rate on one indicator is difficult to assess. However, when observing simultaneously two or more indicators located on adjacent phases of the same support, the identification of a defective insulator is simplified.

4. Conclusion

Modeling in the COMSOL Multiphysics program formulated the requirements for the electronic components of the developed circuit: the choice of elements should be such that the total resistance of the indicator circuit was at least 20 megohms.

Modeling the operation of the indicator on a polymer insulator in the presence of a defect in the form of a conductive channel showed that an increase in the voltage on the indicator caused by damage to the insulation and, accordingly, an increase in the frequency of light pulses of the indicator is visually detected by the operator when simultaneously observing two or more indicators located on adjacent phases of one and the same support.

References

[1] Aouabed F, Bayadi A, Rahmani A E and Boudissa R 2018 Finite element modelling of electric field and voltage distribution on a silicone insulating surface covered with water droplets *IEEE Transactions on Dielectrics and Electrical Insulation* **25(2)** 413-20

[2] Banik A, Mukherjee A, and Dalai S 2018 Development of a pollution flashover model for 11 kV porcelain and silicon rubber insulator by using COMSOL multiphysics *Electrical Engineering* **100** 533-41

[3] Kuznetsov A A, Kuzmenko A Yu, Rodchenko A D 2014 *Patent RU 148957 Portativnyy pribor dlya udalennoy diagnostiki izolyatorov kontaktnoy seti* [Patent RU 148957 Portable device for remote diagnostics of catenary insulators] (Moscow: Rospatent) [In Russian]

[4] Majzoobi A, Joneidi I A, Mohajer S, Mohseni H and Shayegani AA 2011 3D modeling of electrical field and electrical potential in different contamination condition in Polymeric Insulator *XVII Int. Symp. on High Voltage Eng.* (Hannover) pp 122-6

[5] Bykadorov V F, Pirozhnik A A and Sklyarov P A 2009 *Patent RU 89245 Perenosnoye ustroystvo dlya udalennogo obnaruzheniya povrezhdeniy izolyatorov vozdushnykh liniy* (varianty) [Patent RU 89245 Portable device for remote location of damage to overhead line insulators (options)] (Moscow: Rospatent) [In Russian]

[6] Borghei M, Ghassemi M, Rodriguez-Serna J M and Albarracin R 2020 A Finite-Element-Analysis and an Improved Induced Charge Concept for Partial Discharge Modeling *IEEE Transactions on Power Delivery*

[7] Shumilov Yu N 2011 Indikator proboya vysokovol'tnykh polimernykh izolyatorov [Breakdown indicator for high voltage polymer insulators] *Enerhetyka Nadiynist' ta Enerhoefektyvnist'* **3** 164-71 [In Ukraine]

[8] Benguesmia H, Bakri B, Khadar S, Hamrit F and M'ziou N 2019 Experimental study of pollution and simulation on insulators using COMSOL® under AC voltage *Diagnostyka* **20(3)** 21-9

[9] Lyubimov V A and etc 2010 *Patent RU 2392678 Polimernyy izolyator s kontrolem sostoyaniya izolyatsii* [Patent RU 2392678 Polymer insulator with insulation condition monitoring] (Moscow: Rospatent) [In Russian]

[10] Dar K Y, Kalmykov V L, Rozov V A and Demidov O A 2009 Steklyannyy izolyator kak indikator vnutrennego elektricheskogo sostoyaniya polimernogo izolyatora [Glass insulator as an indicator of the internal electrical state of a polymer insulator] *Energetika I Electroficatsiya* **4** 29-33 [In Russian]

[11] Gallo-Villanueva R C, Perez-Gonzalez V H, Cardenas-Benitez B, Jind B, Martinez-Chapa S O and Lapizco-Encinas B H 2019 Joule heating effects in optimized insulator-based dielectrophoretic devices: An interplay between post geometry and temperature rise *Electrophoresis* **10** 1408-16

[12] Florkowski M, Florkowska B and Zydron P 2019 Partial discharges in insulating systems of low voltage electric motors fed by power electronics - Twisted-pair samples evaluation *Energies* **12** 768

[13] Hussain M M, Chaudhary M A and Razaq A 2019 Mechanism of saline deposition and surface flashover on high-voltage insulators near shoreline: Mathematical models and experimental validations *Energies* **12** 368

[14] Fofana I, N'cho J S, Betie A, Hounton E, Meghnefi F and Yapi K M L 2020 Lessons to Learn from Post-Installation Pollution Levels Assessment of Some Distribution Insulators *Energies* **16** 4064

[15] Ouchen L, Bayadi A and Boudissa R 2019 Dynamic model to predict the characteristics of the electric arc around a polluted insulator *IET Science, Measurement & Technology* **1** 83-90

[16] Zhao S, Zhang C, Zhang Y and Wang S 2019 Influence of Partial Arc on Electric Field Distribution of Insulator Strings for Electrified Railway Catenary *Energies* **17** 3295

[17] Zhao S, Zhang C, Zhang Y and Zhang D 2020 Collision Mechanism and Deposition Characteristics of Particles on the Catenary Insulator Surface *Coatings* **7** 697

[18] Kuznetsov V A and Trubaev P A 2018 Resources and Problems of the Mathematical Simulating Thermo-Technological Processes *J. Phys.: Conf. Ser* **1066** 012024

IOP Conf. Series: Materials Science and Engineering 1089 (2021) 012014 doi:10.1088/1757-899X/1089/1/012014

Electromechanical speed control system for steam turbine generator

S V Stelmashchuk and E D Sherov

Departament of Electric Drive and Automation of Industrial Plants, Komsomolsk-on-Amur State Technical University, 27, Lenin St., Komsomolsk-on-Amur, 681013, Russia

E-mail: rukdpsv@rambler.ru

Abstract. The paper considers the application of a turbine speed electromechanical control system. The range of questions includes a description of the principle of operation and mathematical description of the turbine unit with electromechanical control system, as well as the close motoring for electromechanical control system. The mathematical description is presented in the form of transfer functions based on the linearization of the motion equation of turbine in the relative deviations of the parameters of the turbine unit and the electric drive of the electromechanical control system. The linearized description will allow us to apply classical methods of synthesis of controllers of an electromechanical control system. The use of an electric drive instead of a hydraulic actuator, which are used for centrifugal and hydroelectric controllers, will lead to an improvement in the quality of regulation in terms of precision and speed. This will increase the reliability of synchronization of generators. This will allow you to adjust the rotation speeds of all generators in a coordinated manner, rather than individually.

1. Introduction

During rotation, the generator rotor is acted by external disturbances that affect the rotor speed. This leads to a change in the rotor speed of the generator. Consequently, this leads to desynchronization of parallel operating generators. To maintain the generator rotor speed, it is necessary to adjust the rotor speed.

The existing systems for automatic control of the rotational speed of turbine generators use indirect centrifugal controllers. The centrifugal regulator pendulum parameters are set to the fixed value of the turbine generator speed corresponding to the standard value of 50 Hz.

External disturbances cause difference between rotation angles of generator rotors. Then, for synchronized operation, it becomes necessary to change the rotational speed of the generator rotor within the permissible variations of ± 0.4 Hz. This is enough to synchronize the generators for 20 seconds. However, centrifugal controllers do not have the ability to change the value of the rotation speed demand. Their parameters are calculated for only one fixed speed value.

Electrohydraulic controllers are also used, which have the ability to change the speed demand. In electrohydraulic and centrifugal controllers, the actuator is a hydraulic drive. It is characterized by the limitation of hydraulic systems: low accuracy when processing control signals and high labor intensity in service [1].

The most advanced is the use of electromechanical controllers [2]. They are based on the use of a controlled linear electric drive. The use of an electric drive provides high reliability and accuracy of processing control signals and low labor costs in service.

2. Mathematical description of a steam turbine

Assume that a steam turbine is used without reheating the steam. Figure 1 shows a technique for electromechanical control of the position of the steam turbine valve stem. The executive device in the electromechanical control system is a linear actuator A, which moves the rod by the value y. The actuator rod moves the valve rod V by a value of x through a crank-rocker mechanism, which regulates the steam supply to the turbine T. Thus, the rotor speed ω is regulated, and, therefore, the power P_t of the turbine. The turbine rotates a synchronous generator G, which generates electricity with a power of P_e. The rotor speed is measured by a sensor S and compared with the demand signal ω^*. As a result of comparison, a loop error signal occurs at the input of the speed controller C. The control signal for the actuator occurs at the output of the controller ω_o.

Depending on the type of electric motor, the control signal can be any controlled parameter that determines the idle speed value (for a DC motor) or the synchronous speed value (for an AC motor). For example, for a DC motor, this may be the armature voltage, and for an asynchronous motor, the stator voltage frequency.

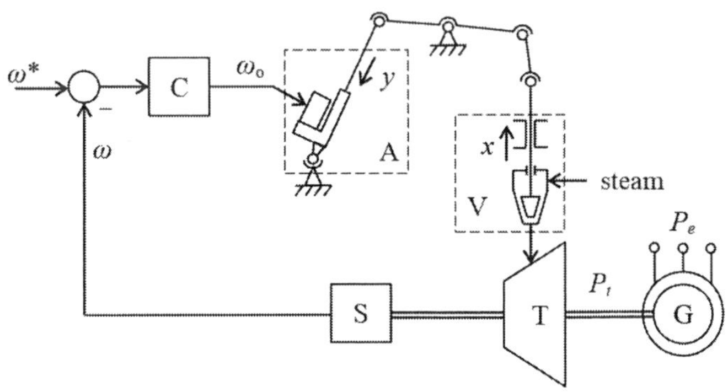

Figure 1. Steam turbine generator with electromechanical speed control.

Differential equation of turbine shaft rotation is used for mathematical simulation of steam turbine operation. The equation has the form of a dynamic balance equation [3]:

$$J\frac{d\omega}{dt} = M_t(x,\omega) - M_g(P_e,\omega),$$

where:

J – total moment of inertia of rotating masses of steam turbine unit (turbine and generator);
M_t – turbine torque;
M_g – load torque on the turbine shaft;
P_e – electrical load power.

The value of the turbine torque is determined by the position coordinate x of the valve and the rotation speed ω of the unit shaft. The load torque on the turbine shaft depends on the value of the electric load P_e and the speed of rotation ω of the unit.

The dependence of torques on the turbine shaft on the valve position, shaft speed and electrical load are nonlinear. Therefore, we linearize nonlinear dependencies. Then the dynamic balance equation for small deviations from the base values corresponding to the steady state (x_0, ω_0, P_0) is presented in the following form:

$$J\frac{d(\omega-\omega_0)}{dt}=\left(\frac{\partial M_t}{\partial x}\right)_0(x-x_0)-\left(\frac{\partial M_g}{\partial P}\right)_0(P_e-P_0)-\left[\left(\frac{\partial M_g}{\partial \omega}\right)_0-\left(\frac{\partial M_t}{\partial \omega}\right)_0\right](\omega-\omega_0).$$

Let's introduce the relative deviations of variables. In this case, base-line values of variables are equal to rated values of the valve position of the turbine x_r, rotation speed of the turbine shaft ω_r, electrical load power P_r and torque on the turbine shaft M_r. Let's introduce the following symbols for the equation variables:

$\delta\omega=\frac{\omega-\omega_0}{\omega_r}$ – is the relative deviations of rotation speed of the turbine shaft;

$\delta P_e=\frac{P_e-P_0}{P_r}$ – is the relative deviations of electrical load power;

$\delta x=\frac{x-x_0}{x_r}$ – is the relative deviations of valve position of the turbine.

As a result, the dynamic balance equation in relative deviations is presented in the following form:

$$J\omega_r\frac{d\delta\omega}{dt}=\left(\frac{\partial M_t}{\partial x}\right)_0 x_r\delta x-\left(\frac{\partial M_g}{\partial P}\right)_0 P_r\delta P_e-\left[\left(\frac{\partial M_g}{\partial \omega}\right)_0-\left(\frac{\partial M_t}{\partial \omega}\right)_0\right]\omega_r\delta\omega.$$

Divide the left and right sides of the equation by the rated torque M_r. The result is:

$$\frac{\omega_r}{M_r}\frac{d\delta\omega}{dt}=\left(\frac{\partial M_t}{\partial x}\right)_0\frac{x_r}{M_r}\delta x-\left(\frac{\partial M_g}{\partial P_e}\right)_0\frac{P_r}{M_r}\delta P_e-\left[\left(\frac{\partial M_g}{\partial \omega}\right)_0-\left(\frac{\partial M_t}{\partial \omega}\right)_0\right]\frac{\omega_r}{M_r}\delta\omega. \quad (1)$$

Let's introduce the following symbols for the equation parameters:

$J^*=J\frac{\omega_r}{M_r}$ – is the relative value of inertia moment;

$k_x=\left(\frac{\partial M_t}{\partial x}\right)_0\frac{x_r}{M_r}$ – is the gain of the torque to the deviation of the valve position;

$\Delta\beta^*=\left[\left(\frac{\partial M_g}{\partial \omega}\right)_0-\left(\frac{\partial M_t}{\partial \omega}\right)_0\right]\frac{\omega_r}{M_r}$ – is the coefficient of inherent stability;

$k_P=\left(\frac{\partial M_g}{\partial P_e}\right)_0\frac{P_r}{M_r}$ – is the gain of the load torque to the deviation of the electrical load power.

Thus, equation (1) in relative deviations will be the expression

$$J^*\frac{d\delta\omega}{dt}+\Delta\beta^*\delta\omega=k_x\delta x-k_p\delta P_e.$$

Here you can see that the controlled part of the turbine torque is the relative deviation of the torque:

$$\delta M_{tx}=k_x\delta x.$$

The torque M_{tx} is that part of the turbine torque generated by the hot steam pressure that is supplied to the rotor blades of the turbine cylinder. The steam supply to the cylinder, therefore the torque on the turbine shaft, is controlled by the relative deviation of the turbine valve position δx.

The load torque on the turbine shaft is a part of the electromagnetic torque of the generator, which is determined by the electrical load

$$\delta M_g=k_p\delta P_e.$$

Equate the base values of the linearized equation to the rated values, i.e. $x_0=x_r$, $\omega_0=\omega_r$, $P_0=P_r$, $M_0=M_r$. Then there is an expression [3]:

$$k_P=\frac{P_r}{M_r}\left(\frac{\partial M_g}{\partial P_e}\right)\Bigg|_{\substack{M_{g0}=M_r\\P_0=P_r}}=1.$$

Then the load torque in relative deviations is equal to the electric load power $\delta M_g = \delta P_e$.

When using relative deviations, approximate substitution of torques δM with corresponding powers δP in relative values is allowed [4]. Therefore, it can be assumed that the relative deviations of the electric power δP_e and the load torque δM_g on the turbine shaft caused by the electric load are equal.

Then the dynamic balance equation (1) in relative deviations is presented in the following form:

$$J^* \frac{d\delta\omega}{dt} + \Delta\beta^* \delta\omega = k_x \delta x - \delta P_e. \quad (2)$$

Let's apply the Laplace transform and present equation (2) in operator form:

$$(T_s p + 1)\delta\omega(p) = K_s(k_x \delta x(p) - \delta P_e(p)),$$

where $T_s = \frac{J^*}{\Delta\beta^*}$ – is the time constant of unit; $K_s = \frac{1}{\Delta\beta^*}$ – is the gain of unit.

Figure 2 shows a block diagram of the steam turbine unit.

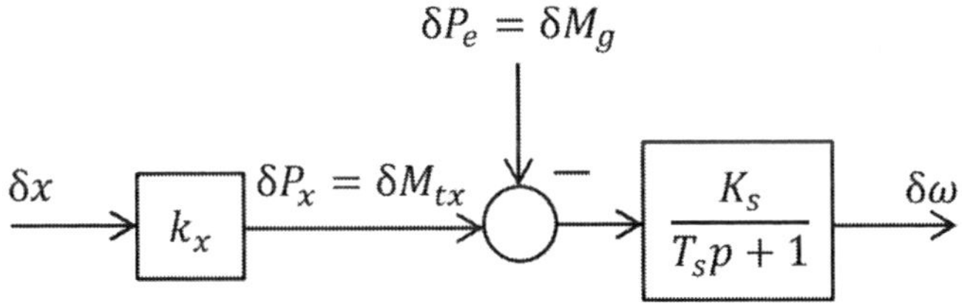

Figure 2. Block diagram of the steam turbine unit.

3. Mathematical description of a electromechanical control

The executive device of the electromechanical control system is a linear electric drive (linear actuator). The linear actuator consists of a DC or AC motor. The electric motor is connected with screw or ball gear through cylindrical, bevel or worm gear. The screw or ball drive nut is connected to the retractable rod. As a result, the rotational motion of the motor is converted to the translational motion of the actuator rod [5].

To date, non-controlled linear actuators are used, which are a ready-made position control system of the actuator rod [6]. These actuators provide movement of the rod with a large force of push and pull of the rod. These forces are sufficient to move the valve of a powerful steam turbine. The linear actuator in the uncontrolled version corresponds to an on-off control system. This makes it difficult to design the controller. It is necessary to take into account the self-oscillating mode of the control system.

A controlled linear actuator must be used. This provides high-quality positioning of the turbine valve with high response time. Stepless control is provided by voltage converters (for DC drives) or frequency converters (for AC drives). In both cases, the controlled parameter is idle speed (for DC drives) or synchronous speed (for AC drives). The parameter designation is ω_0. Figure 3 shows a block diagram of a controlled linear actuator in deviations. The electromagnetic inertia of the drive and the inertia of the converter are ignored because they are small in compare with the inertia of the steam turbine unit.

The electric part of the actuator drive is presented as a proportional block with a gain equal to the stiffness coefficient β_{dr} of the speed-torque characteristic. This link is applicable when the electromagnetic time constant is ignored. Otherwise, the electrical part of the actuator drive must be presented as a first-order aperiodic block with an electromagnetic time constant. It would increase the order of the electromechanical control system. Ignoring the electromagnetic time constant makes it possible to simplify the research of the electromechanical control system of turbine speed and not to take into account the small components of transients.

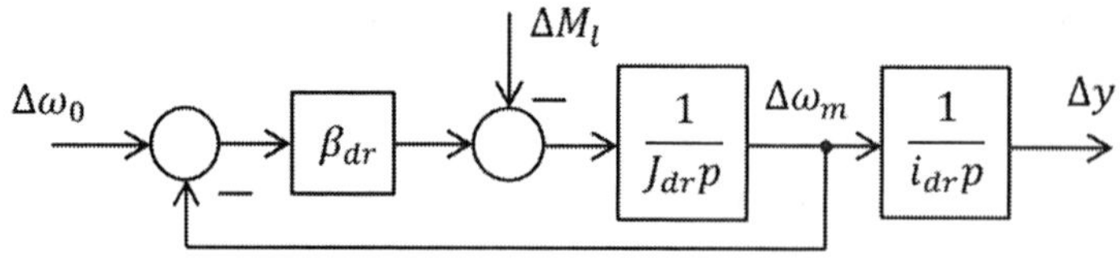

Figure 3. Block diagram of a controlled linear actuator.

The mechanical part of the actuator drive is presented as an integrator with a coefficient equal to the inertia J_{dr} of the actuator drive. It consists of the inertia of the motor, gears, and retractable rod. The mechanical and electrical parts of the drive are covered by the internal feedback loop of the motor. The loop output is the deviation $\Delta\omega_m$ of the motor shaft rotation speed. The control input of the loop is the deviation of the drive control signal $\Delta\omega_0$ (synchronous or idle speed). The disturbance input is the deviation of the load torque ΔM_l on the drive shaft that is produced by the steam pressure force on the turbine valve.

An integrator with a coefficient i_{dr} shows the dependence of the deviation Δy of the rod movement on the deviation $\Delta\omega_m$ of the motor shaft rotation speed. The integrator coefficient is equal to the total gear ratio i_{dr} of the actuator.

According to the block diagram, we will compose a system of differential equations for the actuator operation:

$$J_{dr}\frac{d\Delta\omega_m}{dt} = \beta_{dr}(\Delta\omega_0 - \Delta\omega_m) - \Delta M_l; \qquad i_{dr}\frac{d\Delta y}{dt} = \Delta\omega_m.$$

We present the deviations of variable of differential equations relative to the basic rated values at the inputs (rated motor torque M_{m_r} and rated control ω_{0_r}) and output (position of the rod y_r) of the block diagram of the actuator drive.

$$J_{dr}^*\frac{d\delta\omega_m}{dt} = \beta_{dr}^*(\delta\omega_0 - \delta\omega_m) - \delta M_l; \qquad i_{dr}^*\frac{d\delta y}{dt} = \delta\omega_m,$$

where:

$\delta\omega_{dr} = \frac{\Delta\omega_m}{\omega_{0_r}} = \frac{\omega_m - \omega_{m0}}{\omega_{0_r}}$ – is the relative deviations of rotation speed of the actuator drive shaft;

$\delta\omega_0 = \frac{\Delta\omega_0}{\omega_{0_r}} = \frac{\omega_0 - \omega_{m0}}{\omega_{0_r}}$ – is the relative deviation of the actuator drive control signal;

$\delta M_l = \frac{\Delta M_l}{M_{m_r}} = \frac{M_l - M_{m0}}{M_{m_r}}$ – is the relative deviation of the load torque on the motor shaft of the actuator;

$\delta y = \frac{\Delta y}{y_r} = \frac{y - y_0}{y_r}$ – is the relative deviation of the actuator rod position;

$J_{dr}^* = J_{dr}\frac{\omega_{0_r}}{M_{m_r}}$ – is the relative inertia of the actuator drive;

$\beta_{dr}^* = \beta_{dr}\frac{\omega_{0_r}}{M_{m_r}}$ – is the relative stiffness coefficient of the speed-torque characteristic of the actuator drive;

$i_{dr}^* = i_{dr}\frac{y_r}{\omega_{0_r}}$ – is the relative value of the total gear ratio of the actuator drive;

y_r – is the actuator stem position corresponding to the rated value of the turbine valve position x_r;
ω_{m_0}, M_{m_0}, y_0 – is the steady state of motor speed, motor torque, and rod position, relative to which the mathematical description of the actuator drive is linearized.

The rated control signal ω_{0_r} does not correspond to the rated motor mode, but to the rated load of the power grid. The control signal ω_0 must vary both above and below the nominal value ω_{0_r}.

The actuator drive operates in position mode. Therefore, the following values must be taken as steady state values:

$$\omega_{m_0} = 0,\ M_{m_0} = M_{m_r},\ y_0 = y_r.$$

In practice, an approximate equality between the relative position of the actuator stem and the relative position of the turbine valve is allowed, i.e. $\delta y = \delta x$. Then the block diagram of a controlled linear actuator in relative deviations is shown in figure 4.

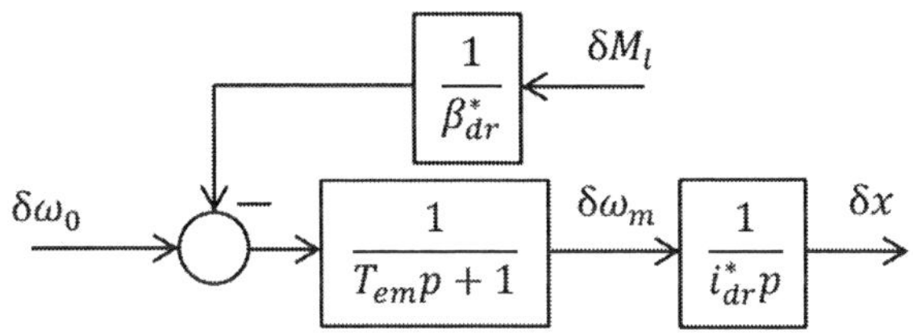

Figure 4. Block diagram of a controlled linear actuator in relative deviations.

Here T_{em} is an electromechanical time constant:

$$T_{em} = \frac{J^*_{dr}}{\beta^*_{dr}} = \frac{J_{dr}}{\beta_{dr}}.$$

The electromechanical time constant of the actuator drive T_{em} is determined depending on the type of drive used, DC or AC, as well as its control system [7, 8].

4. Mathematical description of a steam turbine with electromechanical control

The turbine torque M_{tx} affects actuator drive shaft. It is the load torque on the actuator motor shaft. Then, taking into account the gear ratio of the actuator drive, we can assume that

$$M_l = \frac{M_{tx}}{i_{dr}},$$

or in relative values

$$\delta M_l = \frac{\delta M_{tx}}{i^*_{dr}}.$$

Figure 5 shows a block diagram of a steam turbine unit with electromechanical speed control. It takes into account the specific effect of steam pressure on the turbine valve on the operation of the actuator.

The transfer function of the actuator drive after structural transformations is:

$$W_{dr}(p) = \frac{K_{dr}}{T_{dr}T_{em}p^2 + T_{dr}p + 1},$$

where K_{dr} – is the actuator drive gain

$$K_{dr} = \beta^*_{dr} i^*_{dr} = \beta_{dr} i_{dr} \frac{x_r}{M_{m_r}};$$

T_{dr} – is the mechanical time constant of actuator drive

$$T_{dr} = \frac{\beta^*_{dr} {i^*_{dr}}^2}{k_x} = \frac{\beta_{dr} {i_{dr}}^2}{\left(\frac{\partial M_t}{\partial x}\right)_r} \cdot \frac{x_r}{\omega_{0_r}}.$$

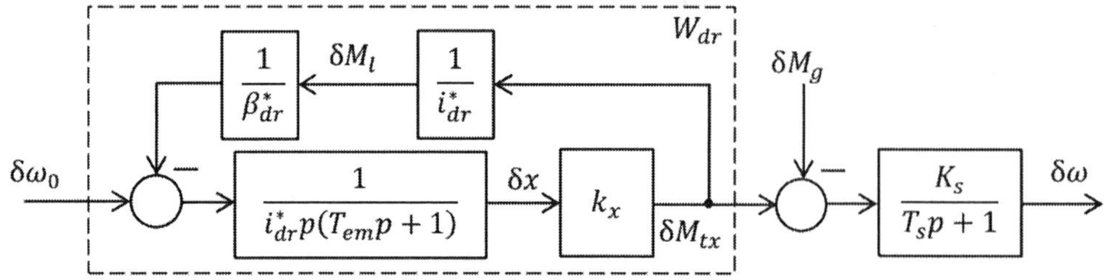

Figure 5. Block diagram of a steam turbine unit with electromechanical speed control.

5. Speed-torque characteristic of actuator drive

The actuator control system determines the relative deviation of the drive control signal. For a DC electric drive, the control action can be the value of the motor armature voltage. For a frequency-controlled AC electric drive, the control action is the frequency of the stator voltage. Perspective electric drives based on synchronous machines can also be used, where the control signal can be the magnitude of the stator voltage. Either way, the control signal must be an electric drive parameter that determines the idle speed (for a DC drive) or synchronous speed (for an AC drive).

The value of the drive control signal is determined such that the created motor torque is balanced by the load torque at zero speed of the positioning drive of the actuator (figure 6).

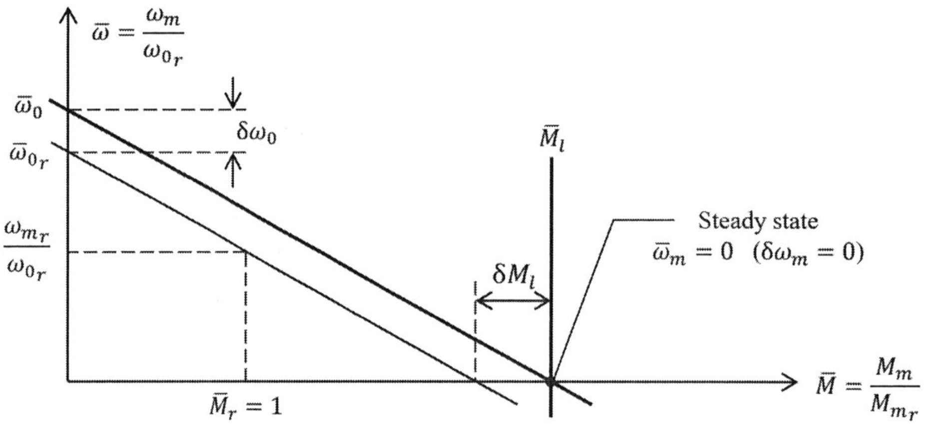

Figure 6. Speed-torque characteristic of the actuator drive.

If the steam turbine unit is operating in rated mode, then the electrical load is rated ($\delta M_g = 0$) and the actuator drive load torque is also equal to the rated value ($\delta M_{tx} = 0$ and $\delta M_l = 0$). To provide the rated mode, the drive control signal is equal to the rated value $\omega_0 = \omega_{0_r}$, i.e. $\delta\omega_0 = 0$.

With increased consumption of the power grid, the electric load increases ($\delta M_g > 0$), which leads to a decrease in the rotation speed of the turbine rotor ($\delta\omega = 0$). To do this, you need to increase the control signal ($\delta\omega_0 > 0$), which moves the actuator rod and, therefore, the turbine valve rod. As a result, the steam pressure in the turbine increases. This increases the load torque on the actuator drive shaft ($\delta M_{tx} > 0$ and $\delta M_l > 0$). When the $\delta M_{tx} = \delta M_g$ position of the valve stem is fixed ($\delta x = const$) and the motor speed ($\omega_m = 0$).

6. Actuator motor power

To move the turbine valve, it is necessary to overpower the force of live steam F_s acting on the valve surface area S_v.

The F_s force depends on the pressure of the live steam p_s that enters the turbine chamber (figure 7):

$$F_s = p_s S_v.$$

Actuator motor power is determined based on the required maximum speed of valve movement V_v:

$$P_m = k\frac{F_s V_v}{\eta_{dr}} = k\frac{p_s S_v V_v}{\eta_{dr}},$$

where:
η_{dr} – is the efficiency of the actuator drive including actuator gears efficiency and the efficiency of the rocker mechanism to transfer motion from the actuator rod to the turbine valve rod;
$k = 1{,}1 \div 1{,}3$ – is the coefficient that takes into account friction torque in drive gears and dynamic components in transient processes.

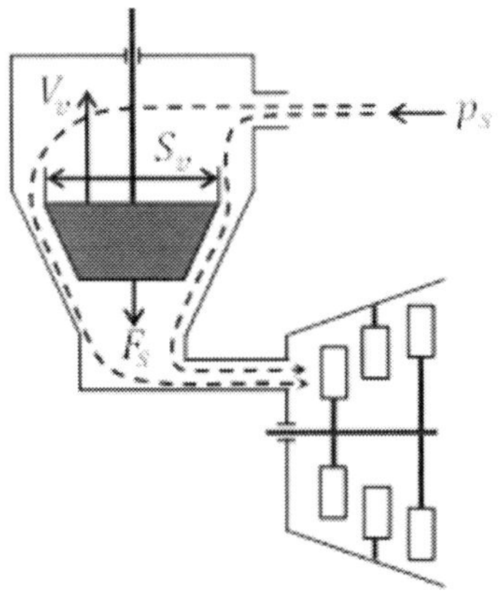

Figure 7. Parameters for determining the power of the actuator motor.

Analysis on calculations of required drive power for turbines with capacity from 250 to 1000 MW [9, 10, 11] showed that live steam pressure reaches no more than 25 MPa. This means that the actuator drive power required is in the range of 15-20 kW. This is technically feasible using modern automation tools. This allows to achieve high speed and position of valve with high quality of control [12].

7. Conclusion

The key point of electromechanical control system is the replacement of the hydraulic servomotor used in the centrifugal and electrohydraulic controllers with an electric drive. The use of a hydraulic servomotor is accompanied by problems associated with the presence of Coulomb friction of the slide valve and positive overlap. Coulomb friction of slide valve leads to start-stop control [1]. The problem of positive overlaps leads to large self-oscillations [13]. The automatic electric drive of the linear actuator does not have these deficiencies. The use of a controlled linear actuator will increase the reliability and quality of the system for automatic control of the turbine generator speed.

The model in figure 5 is a control object in the steam turbine rotor speed control system. Design of the controller C (see figure 1) is performed by approach of control systems. Many publications are devoted to this issue [14, 15].

References

[1] Kirilov I I 1988 *Avtomaticheskoe regulirovanie parovykh turbin i gazoturbinnykh ustanovok* [Automatic control of steam turbines and gas turbine plants] (Leningrad: Mashinostroenie) [In Russian]

[2] Sadovoy A V and Krupnik A A 2011 *Elektromehanicheskaya sistema regulirovaniya skorosti parovoy turbiny* [Electromechanical system of steam turbine speed regulation]

Elektrotehnicheskie i kompyuternye sistemy **79** 198-9 [In Russian]
[3] Meleshkin G A and Merkuriev G V 2006 *Ustoychivost` energosistem* [Stability of power systems] (St. Petersburg: NOU "Tsentr podgotovki kadrov energetiki") [In Russian]
[4] Weinstein R A, Shestakova V V and Katz I M 2013 *Avtomaticheskoe upravlenie elektroenergeticheskimi sistemami v normal`nykh i avariynykh rezhimakh* [Automatic control of electric power systems in normal and emergency operation modes] (Tomsk: Tomsk Polytechnic University) [In Russian]
[5] Aktuatory. Lineynye aktuatory [Actuator. Linear actuators] URL https://lmotion.ru/actuatorlin.shtml [In Russian]
[6] Lineynye aktuatory – osnovnye polozheniya [Linear actuators - basics] URL http://www.linearmotion.ru/article-basics.html [In Russian]
[7] Anuchin A S 2015 *Sistemy upravleniya elektroprivodov* [Electric drive control systems] (Moscow: MPEI Publishing House) [In Russian]
[8] Terekhov V M and Osipov O I *Sistemy upravleniya elektroprivodov* [Electric drive control systems] (Moscow: Academy) [In Russian]
[9] Dobroumov L A and Telyakova T V 1995 *Paroturbinnye ustanovki dlya TES i AES* [Steam turbine plants for thermal power plants and nuclear power plants] (Moscow: Energoatomizdat) [In Russian]
[10] Valamin A E, Kultyshev A Yu, Shibaev T L, GolDberg A A, Bilan V N, Paneque Aguilera H C, Sakhnin Y A, Shekhter M V, Stepanov M Y and Polyaeva E N 2013 Steam turbines produced by the Ural turbine works for combined-cycle plants *J. Thermal Engineering* **60** 533-40
[11] Motorin A V, Raspopov I V and Fursov I D 2004 *Parovye turbiny* [Steam turbines] Vol 2 (Barnaul: AltGTU Publishing) [In Russian]
[12] Aksenov M I, Grekhov V P, Zaritskii M N, Kuprikov A V and Nitievskaia A I 2001 *Avtomatizirovannyy elektroprivod promyshlennykh ustanovok* [Automated electric drive of industrial plants] ed G B Onishchenko (Moscow: RASKHN) [In Russian]
[13] Fragin M S and Yegorshin V P 2017 *Issledovanie vliyaniya real`nykh perekrytiy zolotnikov servomotorov na kolebaniya v sistemakh regulirovaniya parovykh turbin* [Research of influence of real overlaps of spool servomotors on fluctuations in control systems of steam turbines] *J. Nauchno-tekhnicheskie vedomosti SPbGPU* **23** 48-60 [In Russian]
[14] Utkin A V 2018 Synthesis of a control system for a steam turbine *J. Automation and remote control* **79** 2186-202
[15] Kim K K, Ivanov S N and Gorbunov A V 2020 Synthesis of the control device of the electromechanical drive of the main valve *Conf. International conference on industrial engineering, applications and manufacturing, ICIEAM 2020* (Sochi: IEEE) 9112086

IOP Conf. Series: Materials Science and Engineering 1089 (2021) 012015 doi:10.1088/1757-899X/1089/1/012015

New device for the quality control of the high-voltage electrical insulator

A S Serebryakov, V L Osokin, D A Semyonov and D E Dulepov

Department of Electrification and Automation, Nizhny Novgorod State Engineering and Economic University, Oktyabr'skaya St., 22A, Knyaginino, Nizhny Novgorod region, 606340, Russia

E-mail: a.sereb@mail.ru

Abstract. During the operation of high-voltage electrical equipment, its insulation ages, its properties deteriorate, and the electrical strength decreases. To avoid sudden breakdowns of insulation and maintain the necessary level of reliability of high-voltage electrical equipment, the state of its insulation must be periodically monitored. The device described in the article is intended for preventive testing and diagnostics of heterogeneous high-voltage insulation of electric machines, power oil transformers and cables with paper-oil insulation. In particular, the integral assessment of high-voltage insulation aging, its humidification and remaining operational life can be based on the use of the absorption phenomenon (the accumulation of internal absorbed charge). Self-discharge and return voltage are measured to assess the insulation condition. A block diagram of the device for monitoring the electrical insulator quality is given as well as the formulae for determining the spent and remaining operational life of the tested insulation. The proposed device will increase the operational reliability of high-voltage electrical equipment.

1. Introduction

Electrical equipment and its insulation should be reliable for accident free work. During the exploitation process electric insulator ages, its properties deteriorate, its electric strength lowers [1-3]. To avoid sudden breakdowns of insulation and maintain the necessary reliability level of electrical equipment, the insulation state should be periodically monitored [4-6].

The issue of monitoring the state of insulator is particularly acute due to the fast growing of the share of electrical equipment that has worked its standard term. In this case, it is necessary to determine the remaining insulator resource for reasonable extending of the electrical equipment operational life [7, 8]. As the practice shows, the remaining life of electrical equipment can and should be determined not by the time worked, but by the actual technical state of its insulator based on the measurements. One can see how to do this below.

Objective – development of new technical equipment to objectively control the state of heterogeneous high-voltage insulator and determining its remaining operational life using the absorption phenomenon.

2. Materials and Methods

The developed device described in the article is intended for preventive testing and diagnostics of high-voltage insulation of electric machines, power oil transformers and cables with paper-oil insulation. In particular, it may be used for integral assessment of high-voltage insulation aging, its humidification and remaining operational life based on measurement of self-discharge and return voltage.

One can objectively assess the quality of insulation, the level of its aging and remaining operational life by measuring parameters caused by internal absorbed charge within heterogeneous insulator, which is insulation of high voltage electrical machines, power oil transformers and cables with paper-oil insulation [9, 10].

To create effective non-destructive methods for testing the state of the main insulation of high-voltage power electrical equipment, absorption phenomenon is used (accumulation of internal absorbed charge). The state of insulation and the degree of its aging can be assessed according to leakage current [10, 11], absorption current as well as the absorption coefficient which is the ratio which is defined as the ratio of the one-minute value of the insulator resistance to its fifteen second value [9, 12].

The absorption coefficient (or polarization index in the United States) provides an objective assessment of the insulation state since it takes into account the absorption charge in the insulation system. However, monitoring the absorption charge according to the absorption current is inconvenient because the absorption current is small and industrial interference greatly distorts it. Therefore, it is more convenient to use other methods for detecting the absorption phenomenon. For example, in practice, it is possible to use the method of measuring the self-discharge and return voltage [9, 12].

The self-discharge voltage is measured at the test object after it is charged from a constant voltage source and its disconnection from this source. In this case, the electrical capacitance of the insulator is discharged over time to its internal resistance. Figure 1 shows the dependence of the self-discharge voltage U_S on time for a 1000 kV/A distribution transformer with different exploitation terms. Power supply voltage is 2500 V. Each curve of the graph represents the sum of damped exponentials with the same signs and different time constants. It is shown in [9, 12] that the most informative is the self-discharge voltage measured at the 15th second after the start of the measurement. It is defined as U_{S15} and used as an assessment of aging and the state of isolation. The dependence of U_{S15} on the operating time changes almost linearly, decreasing by an average of 30 V per year. By the end of the operational life, the speed of reduction of the self-discharge voltage lowers slightly. The dependence of the self-discharge voltage U_{S15} measured at the 15th second of the operating time of the transformer τ can be expressed by the equation of a straight line: $U_{S15} = 1000\text{-}30\tau$. Here, τ is the operating time, years.

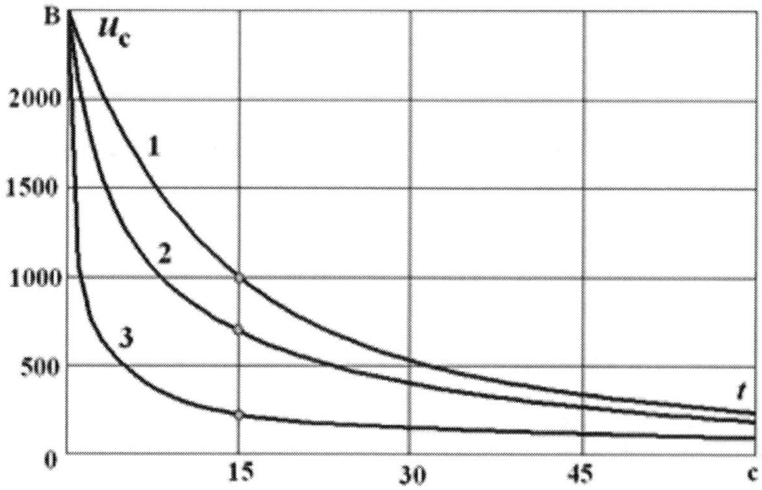

Figure 1. Time dependence of self-discharge voltage for distribution transformers with different operational life: 1 – new transformer at installation; 2 – after 10 years of operation; 3-after 28 years of operation.

Spent time resource is

$$\tau = \frac{1000 - U_{S15}}{30}.$$

Remaining resource is

$$\tau_0 = \tau_{stand} = \frac{1000}{30} - \frac{1000 - U_{S15}}{30} = \frac{U_{S15}}{30}$$

where $\tau_{stand} = \dfrac{1000}{30}$ is the standard insulator resource.

As an assessment of the insulation state, along with the self-discharge voltage, the return voltage is also used (Fig.2). It is measured after disconnecting the charged insulation from the power supply and a short-term discharge to the ground. Herewith, the insulator struck capacity becomes equal to zero, but the internal absorption charge does not have time to discharge in a short time. Due to this fact, a return voltage U_R is formed at the insulator terminals. Therefore, the return voltage curve has its maximum. In [12], it is recommended to use the value of the return voltage U_{R30} measured at the 30th second after the start of the testing to assess the aging and state of insulation.

For assessing it is recommended to use indicator P, determined according to Fig. 2 by the following formula [8]: $P = U_{max} \cdot t_{max}$.

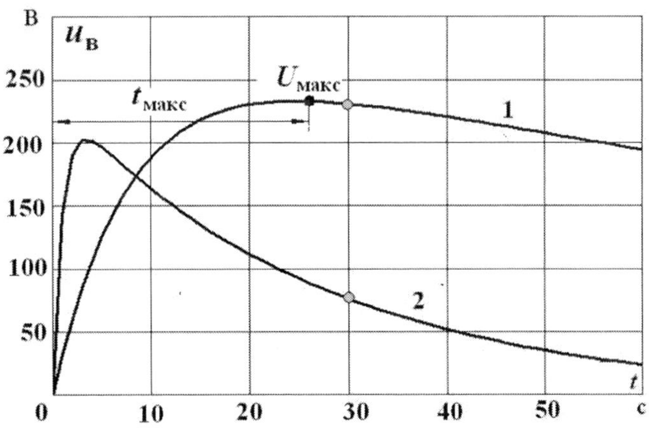

Figure 2. Time dependence of return voltage for a distribution transformer with different operational life: 1 – new transformer at installation; 2 – after 10 years of operation; 3-after 28 years of operation.

As practice shows, as the insulation ages, maximum value of the return voltage U_{max} as well as the time of its forming t_{max} both change. For distribution transformers, the P index is proportional to the remaining resource. It equals 6000 V·s for new insulator in distribution transformers. For transformers that have run out of life, it tends to zero, decreasing by an average of 200 V·s per year.

The remaining resource τ_0 is measured with the formula [12]:

$$\tau_0 = \frac{P}{200} = \frac{U_{max} \cdot t_{max}}{200}.$$

Transformers which have $P = U_{max} \cdot t_{max}$ less than 300 V·s are considered worn out of insulator more than for 90% and are needed to be replaced by new ones.

Within the international practice, to assess the state of paper-oil insulation of cables according to the return voltage, the following ratio of essential parameters is used. It is called *p*-indicator (Fig.3) [24]:

$$p = \frac{U_{R\cdot max}}{s \cdot t_{max}} = \frac{t'}{t_{max}},$$

where $U_{R\cdot max}$ is the maximum value of the return voltage, t_{max} is time at which the maximum return voltage is observed, *s* is initial front of the return voltage curve; *t'* is the time at which a straight line drawn at the angle of the initial front *s* of the return voltage curve reaches the maximum value of the return voltage $U_{R\cdot max}$. The *p* coefficient increases with humidification and aging of the insulator. This trend is observed for paper-oil insulated cables and power oil transformers.

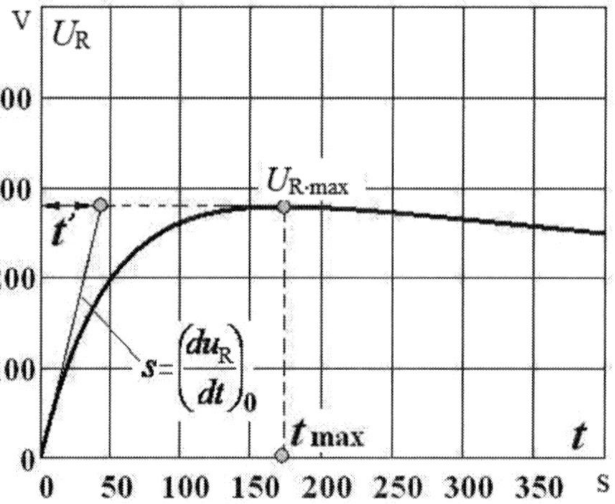

Figure 3. Time dependence of the return voltage for a paper-oil insulated cable.

3. Results.

So, the aging of insulation without its destruction, as shown by research, can be assessed by the nature of polarization processes [13, 14]. Based on the measurement of the parameters described above, a conclusion is made about the state of the insulation and its remaining resource [12].

The block diagram of the proposed device for quality control of the electrical insulation is shown in Fig.4 [15]. The device comprises a test voltage source 1, reference resistor 2, charge key 3, discharge key 4, discharge resistor 5, first voltage constant multiplier 6, test object 7, break blocking contact of the charge key 8, closing blocking contact of the charge key 9, closing blocking contact of the discharge key 10, two-channel digital meter with the storage device 11, display 12, differentiating element 13, zero-comparator 14, light indicator 15, first peak detector 16, second peak detector 17, time counter 18, first voltage multiplier unit 19, second constant multiplier 20, first digital indicator 21, control unit with outputs “Charge”, “Discharge”, “Start” and “Zero setting” 22, clock pulse generator

23, control unit 24, second voltage multiplier unit 25, voltage division unit 26, second digital indicator 27, first 28 and second 29 output terminals of the device.

The device works as follows. In the initial state, at the output of the control unit 22 the signals are formed, according to which the key 4 is open, and the electrical capacities of the test object 7 are discharged through the discharge resistor 5 having a small resistance. After the condensers of the test object 7 are discharged, according to the rules of the device, within one minute the control unit 22 send the signal for opening key 4 and then – key 3.

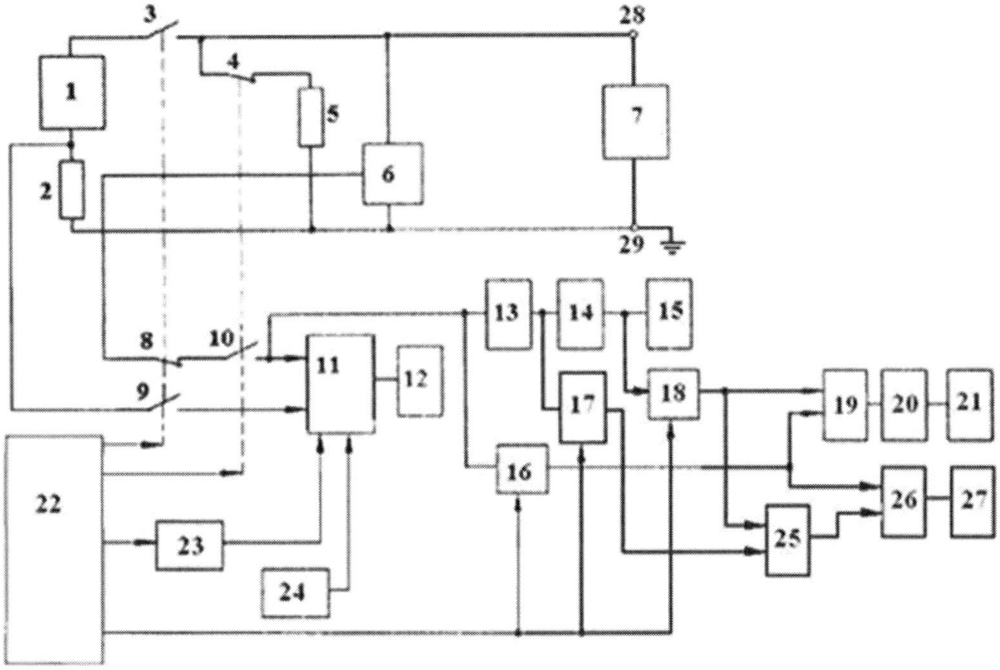

Figure 4. Block diagram of the device for quality control of electrical insulation.

At the specified position of the keys 3 and 4 starts the charge of the test object 7 insulator. Blocking contact 9 is open and the voltage proportional to the leakage current of the test object 7 is applied at the second information input of the two-channel digital meter 11. Output signal "Start" of the control unit 22 starts clock pulse generator 23. The signals of 1s frequency go form the clock pulse generator output to the first control input of the two-channel digital meter 11 and the voltage values proportional to the leakage current are stored in the memory of the meter 11 every second (or optionally after 5 seconds).

After one minute of the insulation additional charge and the disconnection of the charge key 3, the discharge key 4 closes for a short time Δt equal to 5 seconds, during which the struck capacity of the test object 7 is completely discharged through resistor 5 up to zero, and the capacity due to the internal absorbed charge, it remains practically undischarged, since the internal absorption charge cannot change "instantly", in this case for a short period of time Δt = 5 s. After the time Δt discharge key 4 opens. Charge key 3 remains open. The process of measuring the return voltage, which is formed due to the charge of the struck capacity by the internal absorption charge, begins. Voltage at the output of the first voltage constant multiplier 6 is proportional to the return voltage. It is fed through the closed blocking contact 8 and 10 to the second information output of the two-channel digital meter with the storage device 11, which records the measured return voltage every second for one minute. At the be-

IOP Conf. Series: Materials Science and Engineering 1089 (2021) 012015 doi:10.1088/1757-899X/1089/1/012015

ginning of the measurement, the control unit 22 gives the signal "Zero setting" which zeroes the peak detectors 16 and 17; the time counter 18 start counting from zero, too.

The value of the maximum return voltage is recorded by the first peak detector 16. The initial front s of the return voltage (maximum value of the return voltage derivative at the initial time at the output of the differentiating element 13) is recorded by the second peak detector 17. When the return voltage reaches its maximum, its first derivative equals to zero as well as the signal at the output of the differentiating element 13; zero-comparator 14 stops the time counter 18 which is shown by the light indicator 15. The signal is stored at the output of the time counter 18. It is proportional to the changed time t_{max} indicating the maximum return voltage U_{max}.

The first voltage multiplier unit 19 multiplies U_{max} and t_{max} values and gives the results on the digital indicator 21 display. This value corresponds to the remaining resource P of the insulation operational life. In practice, it is convenient to use the measuring of the relative remaining operational life of the insulation, estimating it in comparison with the operational life of the new insulation, measured when installing high-voltage electrical equipment.

The second voltage multiplier unit 25 multiplies the values of s and t_{max}, the signals from 16 and 25 units' outputs are fed to the input of the voltage division unit 26. It is then fed to the second digital indicator's input 27, which displays the parameter value:

$$p = \frac{U_{max}}{s \cdot t_{max}}.$$

After that, the device turns off.

The technical and economic effect of the proposed device is determined by increasing the operational reliability of the tested high-voltage electrical equipment due to a more objective assessment of the electrical insulation state and its remaining operational life.

4. Conclusion.

1. Different physical values relied to the absorption phenomenon are considered as well as their use as the means of assessment of the state of heterogeneous high-voltage insulators.

2. The device is developed to objectively monitor of the state of heterogeneous high-voltage insulators as well as to determine its remaining resource using estimates of the insulation state based on the absorption phenomenon.

References

[1] Gayfutdinova E R, Bulatova V M and Ilkevich A R 2019 The method of power transformers and autotransformers windings condition determination *IOP Conf. Ser.: Mater. Sci. Eng.* **791** 012028

[2] Gonzalez A A, Jacobo J C and Ruiz S S 2012 Incipient failure detection in power transformers through Frequency Sweep Response *Lecture Notes in Engineering and Computer Science* **2** 1010-4

[3] Wang M, Vandermaar A J and Srivastava K D 2004 Transformer winding movement monitoring in service – Key factors affecting FRA measurements *IEEE Electrical Insulation Magazine* **20(5)** 5-12

[4] Islam A, Khan S I and Hoque A 2011 Detection of mechanical deformation in old aged power transformer using cross correlation coefficient analysis method *Energy and Power Eng.* **3(4)** 585-91

[5] Wang M and Vandermaar A J 2002 Review of condition assessment of power transformers in service *IEEE Electrical Insulation Magazine* **18(6)** 12-25

[6] Irungu G K, Akumu A O and Munda J L 2015 Transformer condition assessment using dissolved gas analysis, oil testing and evidential reasoning approach *IEEE Electrical Insulation Conf.* (Seattle: IEEE) pp 145-9

[7] Jurcik J, Gutten M and Korenciak D 2011 Analysis of Transient Actions Influence in Power

Transformer *Advances in Electrical and Electronic Engineering* **9(2)** 65–9
[8] Menzel J, Patsch R and Kamenka D 2009 Measurement of the dielectric properties of paper-oil cables and the cellulose-oil insulation system of power transformers *IEEE 9th Int. Conf. on the Properties and Applications of Dielectric Materials* (Harbin: IEEE) pp 1020-5
[9] Sidorova A, Semenov D, Cheremukhin A and Astakhova T 2019 Algorithm for determining the state of impregnated paper insulation of high-voltage cables *Int. J. of Emerging Electric Power Systems* **20(4)** 20190002
[10] Song Z, Yang F, Wu G and Gao B. 2016 Aging effects of transformer oil and insulation paper on extended Debye model parameters *High Voltage Apparatus* **8(52)** 101-7
[11] Martin D and Saha T 2017 A review of the techniques used by utilities to measure the water content of transformer insulation paper *IEEE Electrical Insulation Magazine* **3(33)** 8-16
[12] Semenov D, Sidorova A, Romanov P and Kuvshinov A 2018 Examination of State of the Cable Insulation by the Return Voltage *Int. J. of Emerging Electric Power Systems* **19(6)** 20180167
[13] Al-Amar E, Karady G G and Hevia O P 2007 Improved technique for fault detection sensitivity in transformer maintenance test *IEEE Power Eng. Society, General Meeting* (Tampa: IEEE) 505-8
[14] Weindl C et al 2010 Development of the p-factor in an accelerated ageing experiment of the MV PILC cables *10th IEEE Int. Conf. on Solid Dielectrics* (Potsdam: IEEE) 5568071 pp 1-4
[15] Serebryakov A S, Semenov D A and Osokin V L 2020 *Patent RU 2730535C1 Ustrojstvo kontrolya kachestva elektricheskoj izolyacii* [Patent RU 2730535C1 Device for quality control of electrical insulation] (Moscow: Rospatent) [In Russian]

Non-stationary modes of electric networks

D P Andrianov, N P Badalyan and G P Kolesnik

Department of power engineering and electrical engineering, Vladimir State University named after A.G. and N.G. Stoletovs, Gorky St., 87, Vladimir, 600000, Russia

E-mail: andrianov@vlsu.ru

Abstract. Ensuring energy efficiency and solving problems of energy saving of power supply systems is impossible without forecasting the state of energy sources and elements of the electric network. Distribution networks consist of a set of elements whose parameters follow stochastic regularities and collectively organize a nonlinear dynamic system with parameters that depend on and do not depend on the mode. Description of networks requires the involvement of higher mathematics in conjunction with modern hardware and software of computer technology. The article is devoted to modeling of non-stationary modes of operation of multi-node electric networks with a large number of degrees of freedom and a high order of differential equations describing the processes under study in the model under consideration. The influence of such factors on transients as abrupt changes in the load of consumers, disconnection-connection of one of the voltage sources, short circuits on the consumer side, stability of the oscillatory process is considered. On the example of a 6-node circuit with 3 alternating voltage sources and 6 consumers, the nature of changes in the generated electrical power of the sources is analyzed. The system of differential equations corresponding to the scheme under study is solved by numerical methods. The proposed approach allows us to evaluate the efficiency of the power supply network while maintaining the mode parameters and predict the possibility of undesirable overloads during short circuits and conditions for exceeding variables that determine the practical stability of the system.

1. Introduction

The modern energy market covers power supply systems that ensure the production, transmission and consumption of electric power and requires flexible integrated solutions to a number of complex problems. Constant growth of electricity consumption in conditions of limited global fuel resources requires optimization of processes at all levels of the energy market [1].

The nearest future of the energy complex is the development and implementation of new technologies based on digitalization [2], which cannot be implemented without an objective assessment of the state of power systems and a clear definition of operating parameters.

One of the key problems of power supply at the present time is dispatching the flows of transmitted power [3].

In branched power supply schemes that have a complex topology, an unfavorable combination of the required modes on the part of consumers can lead not only to overloads, but also to instability of the entire network [4].

Multi-purpose optimization of power losses in power supply systems is possible if a mathematical apparatus is developed that combines adaptive control algorithms [5] and methods of evolutionary calculations [6], which can be implemented in voltage and reactive power management [7].

Hardware for reactive power compensation is possible using thyristor compensators [8], SAPF compensators [9], and GRID systems [10].

Modeling of power supply systems is becoming increasingly important. There are three major problems that determine the quality of solving the problem of preliminary calculations of technical objects:

- description of objects: the number and location of elements of the power supply system, their characteristics, topology and connection parameters [11];
- operating conditions of the power system, the nature and intensity of the load, the presence of random factors [12, 13];
- mathematical methods [14], hardware and software, and computer system resources.

A promising direction is the use of Micro-Grid systems that solve the problems of mode management based on monitoring the parameters of the state of the electric network using renewable energy sources [15], modeling is carried out on the basis of MATLAB SIMULINK [16].

2. Materials and Methods

A 6-node circuit with 3 AC voltage sources and 6 electric power consumers is studied (Figure 1).

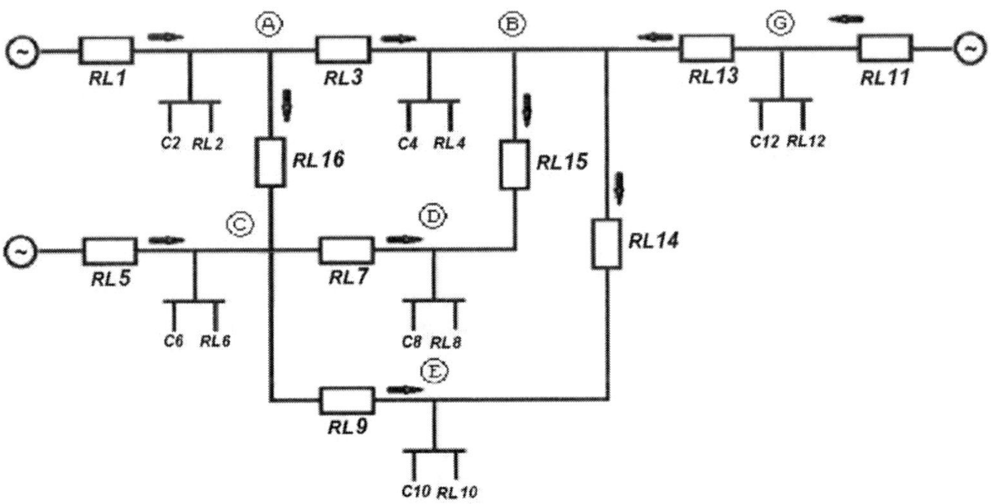

Figure 1. Diagram of the simulated power supply network.

The following assumptions were made when drawing up the calculation model:

- AC voltage sources generate a common-mode sinusoidal voltage.
- Network branches connecting nodes, sources, and consumers are represented as active-inductive load R-L.
- The branch of the electric power consumer is represented as a step-down transformer with a transformation coefficient K, feeding the active-inductive load R-L and the capacitive load C (Figure 2).
- Magnetic hysteresis, mutual influence of windings, and transformer current leakage are not taken into account.

The presented electrical network diagram is considered as a linear dynamic model consisting of 21 1-order differential equations in concentrated parameters.

The research method is numerical integration of a system of differential equations taking into account time – varying parameters of the simulated system.

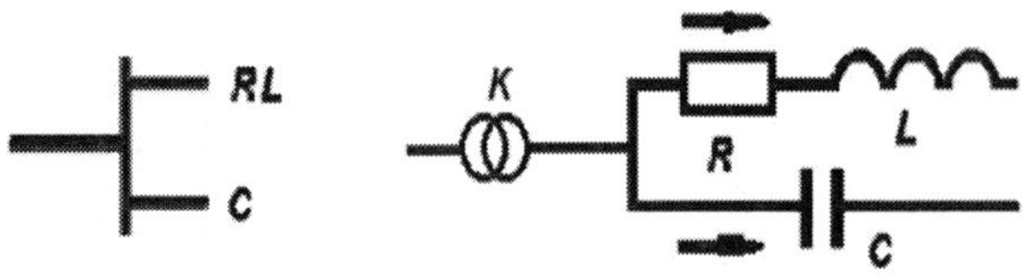

Figure 2. Schematic designation of the electric power consumer.

The system of equations is reduced to a typical Cauchy form with the derivations of the unknowns highlighted on the left side. The system of equations is represented in matrix form for using subroutines of the mathematical package. Standard functions of the mathematical package implement numerical algorithms for solving differential equations (4-order Runge-Kutta method with automatic selection of the integration step).

The following possible changes in the state of the network under study are simulated:

- disconnection-connection of one of the AC voltage sources,
- an abrupt change in the load of the consumer,
- short circuits on the consumer side,
- stability of the oscillatory process depending on the combination of network parameters.

The analysis is based on the resulting calculations of oscillograms of changes in the strength of currents, voltages, and electrical capacities.

3. Results

Simulation of disconnection-connection to the network of one of the AC voltage sources showed a significant impact on the nature of the transition process of its phase matching.

As an example, the transition process was calculated when the 3rd source was disconnected from the network at a time of 0.0505 s (corresponds to the mark 252 of the abscissa axis of the waveform in Figure 3) and connected to the network at a time of 0.1075 s (mark 537 of the waveform in Figure 3).

Disconnection-connecting the 3rd source caused a corresponding change in the load on the remaining sources, and the connection was accompanied by a power jump caused by a phase mismatch between the connected source and the network. By adjusting the oscillation phase of the 3rd source, you can eliminate an undesirable jump in the power generated by it.

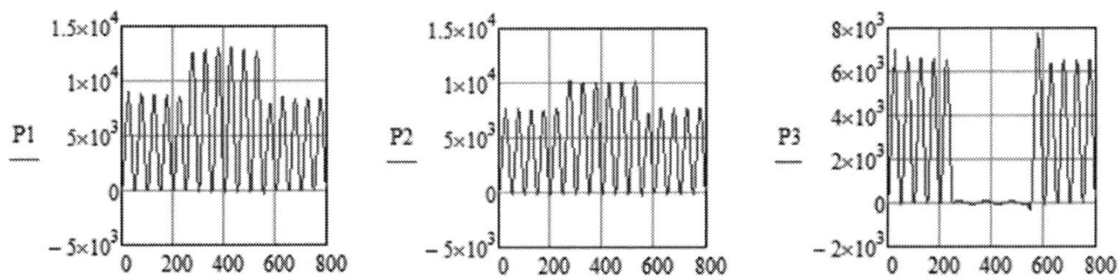

Figure 3. Waveforms of power sources when the 3rd source is turned off or on.

Modeling of the effect of changes in a load of consumers was carried out with a jump change in the load resistance (coefficient of change of 0.02) for a certain period of time, after which the load value was abruptly restored.

So, when the load of the consumer R12 changes, the power waveforms from the network sources (Figure 4) showed that the main load falls on the 3 voltage source closest to the consumer.

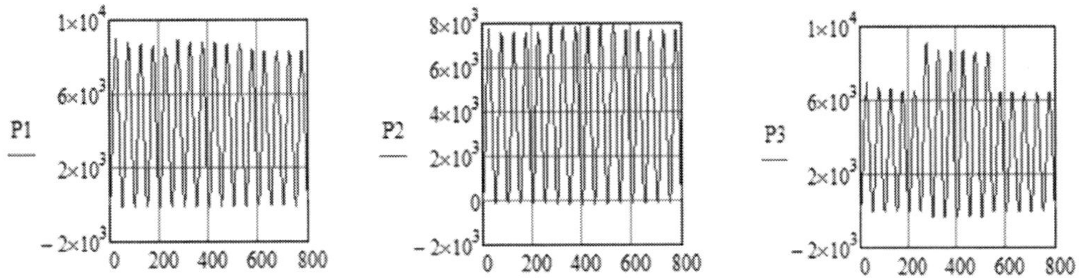

Figure 4. Waveforms of power sources when changing the load of the consumer R12.

The current waveforms (figure 5) in the branches connecting the consumer to the network confirm the above – the current jump takes place in the RL11 branch. Differences in current values are caused by the influence of the transformer.

The study of the effect of short circuits on the network was carried out on the example of a consumer circuit closure.

Short circuits were modeled by an abrupt change in the resistance of the selected consumer (coefficient of change of 0.0002) at a given time. After a certain period of time, the consumer's resistance was restored.

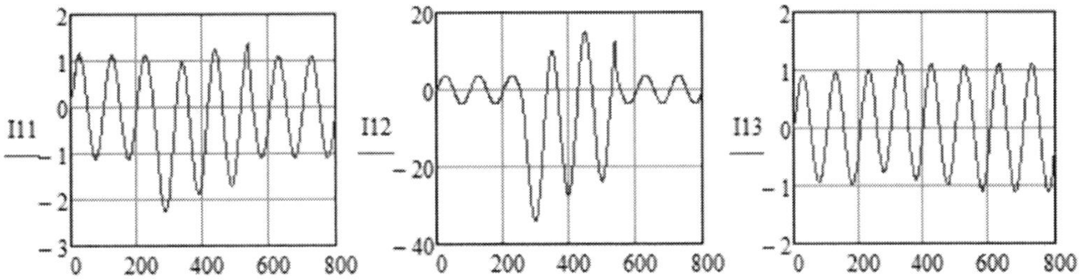

Figure 5. Waveforms of the current strength in the RL11, RL12 and RL13 branches when the load of the R12 consumer changes.

As an example, the reaction of power supply sources to a short circuit of the consumer R12 is shown in figure 6. The main load falls on 3 voltage sources, as the closest to the place of the short circuit.

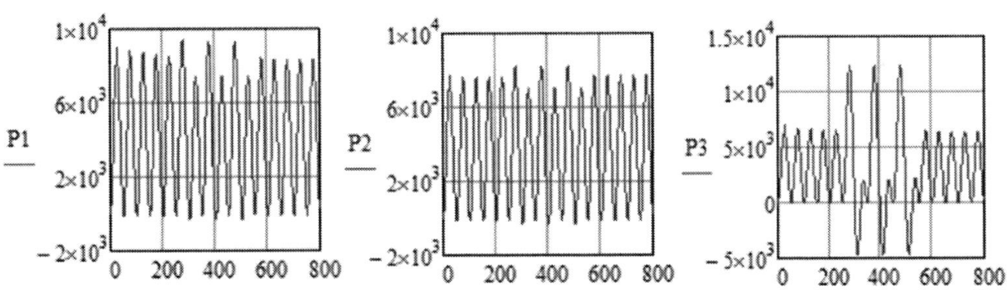

Figure 6. Waveforms of power sources with a short circuit R12.

A comparison of the current waveforms (figure 7) in the RL11 and RL12 branches shows the softening effect of the transformer located between the power lines and the RL12 consumer.

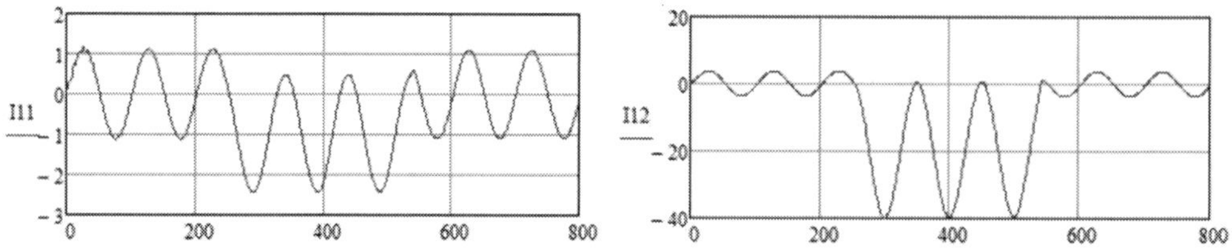

Figure 7. Waveforms of the current strength in the RL11 and RL12 branches when R12 is short-circuited.

Issues of ensuring the stability of the power supply network appeared when trying to model when the voltage level increases. Thus, an increase in the voltage from 6 kV to 10 kV showed an increase in current fluctuations in the connecting branches between the nodes (figure 8).

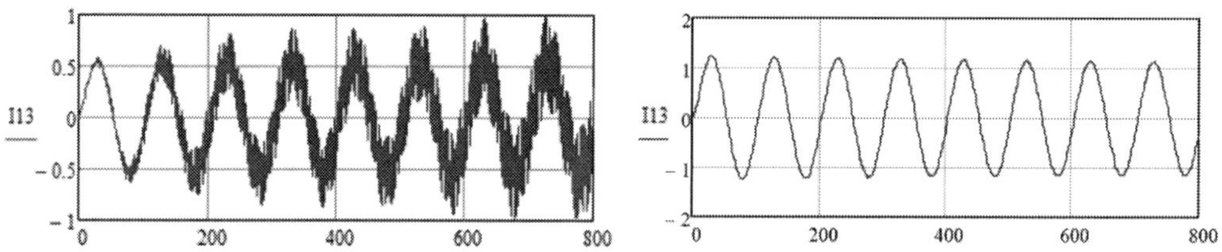

Figure 8. Waveforms of the current in the RL11 branch at a voltage of 10 kV and 6 kV.

4. Discussion

Modeling of power supply networks from the perspective of calculating transients makes it possible to predict the dynamics of the system behavior under random influences, which can include the current load change, short circuits, and disconnection-connection of sources. Of particular interest is to obtain load characteristics of sources that can be used to predict the consumption of fuel resources required for generating electricity. Issues of dispatching electricity flows, as well as related economic calculations, can be partially solved at the level of designing power supply systems.

Taking into account the possibility of short circuits in the power supply system provides a justification for choosing protective equipment that ensures reliable operation of the entire system as a whole.

Changing the load in electric networks can lead to undesirable power overflows, which increases the financial costs of consumers.

By the nature of the power transmitted in electric networks, it is possible to judge the operating mode of electrical equipment, and in case of its overload, to initiate measures to improve its reliability and durability.

Simulation of transients by numerical methods requires appropriate hardware and software and qualification of calculators. Consideration of multi-node power supply schemes with a number of nodes in several tens leads to the consideration of high-order calculation matrices that can be formed only by formal methods. As presented in the article example, the formation of the design matrix was conducted manually, without formal methods (matrix, topological relations, etc.). The number of nodes and links of the considered example (sparse matrix) allows using the interface of the mathematical package that allows you to directly control the process of generating payroll results (Figure 9).

Ensuring the stability of a dynamic system requires separate consideration involving methods of automatic control theory and automatic control theory. A simple calculation of the transition process gives a direct answer to the question – whether the system is stable under these parameters or not.

$$DD(t,xx) := \begin{pmatrix}
\frac{-R1}{L1} & 0 & 0 & 0 & 0 & 0 & 0 & 0 & 0 & 0 & 0 & 0 & 0 & 0 & 0 & 0 & \frac{-K}{L1} & 0 & 0 & 0 & 0 & 0 \\
0 & \frac{-R2}{L2} & 0 & 0 & 0 & 0 & 0 & 0 & 0 & 0 & 0 & 0 & 0 & 0 & 0 & 0 & \frac{1}{L2} & 0 & 0 & 0 & 0 & 0 \\
0 & 0 & \frac{-R3}{L3} & 0 & 0 & 0 & 0 & 0 & 0 & 0 & 0 & 0 & 0 & 0 & 0 & 0 & \frac{K}{L3} & \frac{-K}{L3} & 0 & 0 & 0 & 0 \\
0 & 0 & 0 & \frac{-R4}{L4} & 0 & 0 & 0 & 0 & 0 & 0 & 0 & 0 & 0 & 0 & 0 & 0 & 0 & \frac{1}{L4} & 0 & 0 & 0 & 0 \\
0 & 0 & 0 & 0 & \frac{-R5}{L5} & 0 & 0 & 0 & 0 & 0 & 0 & 0 & 0 & 0 & 0 & 0 & 0 & 0 & \frac{-K}{L5} & 0 & 0 & 0 \\
0 & 0 & 0 & 0 & 0 & \frac{-R6}{L6} & 0 & 0 & 0 & 0 & 0 & 0 & 0 & 0 & 0 & 0 & 0 & 0 & \frac{1}{L6} & 0 & 0 & 0 \\
0 & 0 & 0 & 0 & 0 & 0 & \frac{-R7}{L7} & 0 & 0 & 0 & 0 & 0 & 0 & 0 & 0 & 0 & 0 & 0 & \frac{K}{L7} & \frac{-K}{L7} & 0 & 0 \\
0 & 0 & 0 & 0 & 0 & 0 & 0 & \frac{-R8}{L8} & 0 & 0 & 0 & 0 & 0 & 0 & 0 & 0 & 0 & 0 & 0 & \frac{1}{L8} & 0 & 0 \\
0 & 0 & 0 & 0 & 0 & 0 & 0 & 0 & \frac{-R9}{L9} & 0 & 0 & 0 & 0 & 0 & 0 & 0 & 0 & 0 & \frac{K}{L9} & 0 & \frac{-K}{L9} & 0 \\
0 & 0 & 0 & 0 & 0 & 0 & 0 & 0 & 0 & \frac{-R10}{L10} & 0 & 0 & 0 & 0 & 0 & 0 & 0 & 0 & 0 & 0 & \frac{1}{L10} & 0 \\
0 & 0 & 0 & 0 & 0 & 0 & 0 & 0 & 0 & 0 & \frac{-R11}{L11} & 0 & 0 & 0 & 0 & 0 & 0 & 0 & 0 & 0 & 0 & \frac{-K}{L11} \\
0 & 0 & 0 & 0 & 0 & 0 & 0 & 0 & 0 & 0 & 0 & \frac{-R12 \cdot F1(t)}{L12} & 0 & 0 & 0 & 0 & 0 & 0 & 0 & 0 & 0 & \frac{1}{L12} \\
0 & 0 & 0 & 0 & 0 & 0 & 0 & 0 & 0 & 0 & 0 & 0 & \frac{-R13}{L13} & 0 & 0 & 0 & 0 & \frac{-K}{L13} & 0 & 0 & 0 & \frac{K}{L13} \\
0 & 0 & 0 & 0 & 0 & 0 & 0 & 0 & 0 & 0 & 0 & 0 & 0 & \frac{-R14}{L14} & 0 & 0 & 0 & \frac{K}{L14} & 0 & 0 & \frac{-K}{L14} & 0 \\
0 & 0 & 0 & 0 & 0 & 0 & 0 & 0 & 0 & 0 & 0 & 0 & 0 & 0 & \frac{-R15}{L15} & 0 & 0 & \frac{K}{L15} & 0 & \frac{-K}{L15} & 0 & 0 \\
0 & 0 & 0 & 0 & 0 & 0 & 0 & 0 & 0 & 0 & 0 & 0 & 0 & 0 & 0 & \frac{-R16}{L16} & \frac{K}{L16} & 0 & \frac{-K}{L16} & 0 & 0 & 0 \\
\frac{K}{C2} & \frac{-1}{C2} & \frac{-K}{C2} & 0 & 0 & 0 & 0 & 0 & 0 & 0 & 0 & 0 & 0 & 0 & 0 & \frac{-K}{C2} & 0 & 0 & 0 & 0 & 0 & 0 \\
0 & 0 & \frac{K}{C4} & \frac{-1}{C4} & 0 & 0 & 0 & 0 & 0 & 0 & 0 & 0 & \frac{K}{C4} & \frac{-K}{C4} & \frac{-K}{C4} & 0 & 0 & 0 & 0 & 0 & 0 & 0 \\
0 & 0 & 0 & 0 & \frac{K}{C6} & \frac{-1}{C6} & \frac{-K}{C6} & 0 & \frac{-K}{C6} & 0 & 0 & 0 & 0 & 0 & 0 & \frac{K}{C6} & 0 & 0 & 0 & 0 & 0 & 0 \\
0 & 0 & 0 & 0 & 0 & 0 & \frac{K}{C8} & \frac{-1}{C8} & 0 & 0 & 0 & 0 & 0 & 0 & \frac{K}{C8} & 0 & 0 & 0 & 0 & 0 & 0 & 0 \\
0 & 0 & 0 & 0 & 0 & 0 & 0 & 0 & \frac{K}{C10} & \frac{-1}{C10} & 0 & 0 & 0 & \frac{K}{C10} & 0 & 0 & 0 & 0 & 0 & 0 & 0 & 0 \\
0 & 0 & 0 & 0 & 0 & 0 & 0 & 0 & 0 & 0 & \frac{K}{C12} & \frac{-1}{C12} & \frac{-K}{C12} & 0 & 0 & 0 & 0 & 0 & 0 & 0 & 0 & 0
\end{pmatrix} \cdot xx + \begin{pmatrix}
\frac{U1}{L1} \cdot \sin(\omega \cdot t + \varphi 1) \\ 0 \\ 0 \\ 0 \\
\frac{U2}{L5} \cdot \sin(\omega \cdot t + \varphi 2) \\ 0 \\ 0 \\ 0 \\ 0 \\ 0 \\
\frac{U3}{L11} \cdot \sin(\omega \cdot t + \varphi 3) \\ 0 \\ 0 \\ 0 \\ 0 \\ 0 \\ 0 \\ 0 \\ 0 \\ 0 \\ 0 \\ 0
\end{pmatrix}$$

Figure 9. Example of the calculation matrix of the considered electrical network scheme.

It is of interest to consider the influence of nonlinearities introduced by transformers on the mode of electric networks, as well as unbalanced load modes in the case of three-phase electric circuits. Issues related to the input-output interface of mathematical packages, which determine the complexity of solving the problem at the preparation stage, are of particular importance.

5. Conclusion

Modeling of non-stationary modes of multi-node electric networks based on solving systems of differential equations makes it possible to take into account the entire set of specific features of the functioning of electric networks. The proposed approach makes it possible to predict the state of the electric network with a changing number of sources and a variable load of consumers, which makes it possible to avoid possible overloads of electric energy sources.

References

[1] Kumar S and Goyal S K 2014 A Particle Swarm Optimization for Reactive Power Optimization *Int. J. of Computational Eng. Research* **4(11)** 40-6
[2] Khovalova T V 2019 Innovacii v e`lektroe`nergetike: vidy`, klassifikaciya i e`ffekty` vnedreniya [Innovations in the electric power industry: types, classification and effects of implementation] *Strategicheskie Resheniya i Risk-Menedzhment* **10(3)** 274-83 [In Russian]
[3] Patel V and Patel R 2019 Reactive Power Management through Dispatch Analysis using Evolutionary Algorithms International *J. of Innovative Technology and Exploring Eng.* **8(12)** 759-65
[4] Songhao Y, Masahide H and Baohui Z 2018 A phase-plane trajectory vector-based method for real-time identification of critical machines *Transactions on Electrical and Electronic Eng.* **13(11)** 1578-85
[5] Mandal B 2013 Optimal reactive power dispatch using quasi-oppositional teaching learning based optimization *Electrical Power and Energy Systems* **53** 123-34
[6] Sahu G, Swarnkar K 2014 Review of Reactive Power Optimization Using Evolutionary Computation *Techniques Advance in Electronic and Electric Eng.* **4(1)** 73-82
[7] Wu P and Zhang J 2013 Power system reactive power optimization based on global differential evolution algorithm *25th Chinese Control and Decision Conf.* (Guiyang: IEEE) 6561073
[8] Dixon J, Moran L, Rodríguez J and Domke R 2005 Reactive power compensation technologies, state-of-the-art review *Proceedings of the IEEE* **93(12)** 2144-64
[9] Mujawar I, Patil D and Mujawar I 2013 Reactive Power Compensation and Harmonic Mitigation of Distribution System using SAPF Compensator *Int. J. of Eng. Sci. and Innovative Technology* **2(3)** 494-502
[10] Patidar R D and Singh S P 2009 Active and reactive power control and quality management in dg-grid interfaced systems *J. of Eng. and Applied Sci.* **4(3)** 81-9
[11] Sharma P and Batish N 2015 Computational Analysis of IEEE 57 Bus System Using N-R Method *Int. J. of Advanced Research in Electrical, Electronics and Inst. Eng.* **4(11)** 8859-69
[12] Bouffard F, Galiana F and Conejo A 2005 Market clearing with stochastic security - part I: formulation *IEEE Transactions on Power Systems* **20** 1818-26
[13] Bouffard F, Galiana F and Conejo A 2005 Market clearing with stochastic security - part II: Case Studio *IEEE Transactions on Power Systems* **20** 1827-35
[14] Mishra S and Das D 2008 Distribution System Load Flow Methods: A Review *The Icfai University J. of Electrical & Electronics Eng.* **1(2)** 7-25
[15] Ahshan R, Iqbal M T, Mann G K I and Quaicoe J E 2013 Modeling and Analysis of a Micro-Grid System Powered by Renewable Energy Sources *The Open Renewable Energy J.* **6(1)** 7-22
[16] Fouad M A, Badr M A and Ibrahim M M 2017 Modeling of micro-grid system components using MATLAB/SIMULINK *Global Scientific J.* **5(5)** 163-77

IOP Conf. Series: Materials Science and Engineering 1089 (2021) 012017 doi:10.1088/1757-899X/1089/1/012017

Optimization of impacts parameters on the equipment of electrical networks during operation according to the technical condition

V M Levin[1], N P Guzhov[2], N A Chernenko[3] and N F Cheganova[3]

[1]Department of Automated Electrical Power Systems, Novosibirsk State Technical University, Karl Marx ave., 20, Novosibirsk, 630073, Russia
[2]Department of Power Supply Systems Enterprises, Novosibirsk State Technical University, Karl Marx ave., 20, Novosibirsk, 630073, Russia
[3]Department of High Voltage Engineering and Electrophysics, Novosibirsk State Technical University, Karl Marx ave., 20, Novosibirsk, 630073, Russia

E-mail: levin@corp.nstu.ru

Abstract. Scheduled operation of power grid facilities is regulated by the relevant regulations based on a system of scheduled preventive maintenance. Maintenance and repair (M&R) of power grid facilities due to their technical condition increases demands for planning methods and models. In these conditions, the goals of the M&R management are aimed at improving the efficiency of the operation of electrical networks. They differ depending on the purpose and form of ownership of the latter. The article suggests a combined approach to planning maintenance of electrical networks using strategy depending on their technical condition based on optimization of impact parameters such as frequency and labor intensity under specified constraints and variability of management goals. To implement the proposed planning procedure, the authors developed and verified models of process maps. Each of them represents a local database including the information about the composition and complexity of operations, depending on the type of equipment and the type of corrective actions. In addition, a set of operational algorithms, that solve the issues of process automation in the maintenance system of electrical network equipment, has been developed. The analysis and discussion of the developed models and algorithms are carried out on a specific example illustrating the possibilities of improving the efficiency of equipment operation in the electrical networks on the oil fields.

1. Introduction

Long-lasting safe operation of electrical network equipment is ensured by compliance with a set of statutory and technical requirements, according to which the main corrective actions are maintenance and repair. Scheduled maintenance and repair (M&R) of electrical equipment (EE) imply advance scheduling for the introduction of corrective actions with the registration of their type, volume, and cost in the planning interval ΔT (year). Depending on the type of EE, an M&R strategy is identified. It determines the planning approaches (Rules for organization of electric power facilities maintenance and repair. Approved by order No. 1013 dated 25.10.2017 of the Ministry of Energy of the Russia).

Until recently, the strategy "by periodicity, operating time", which regulates planning and implementing corrective actions for M&R of EE at certain precisely fixed time intervals, was recognized as practically the only one [1]. This strategy, along with its advantages, has a significant disadvantage. It assigns to each unit of EE the implementation of a full set of operations as part of a certain corrective action as it reaches its planned period, regardless of whether there is an actual need for it. This determines the unjustified redundancy of the strategy "by periodicity, operating time" with respect to the total time spent, labor and material resources involved in technical M&R of EE. In addition, increasing the frequency and duration of planned outages of EE reduces the reliability of power supply to consumers, and thus the efficiency of electrical network operation [2, 3].

With the introduction of methods for condition index (CI) calculation for units and/or groups of single-type EE in operating practice of electrical networks the scope of M&R application with the strategy "by technical condition" expanded (Methodology for assessing the technical condition of the main technological equipment and power lines of power plants and electrical networks. Approved by Order of the Ministry of Energy of Russia dated July 26, 2017 No. 676) [4]. This strategy prescribes the implementation of corrective actions with an arbitrary frequency, precisely as necessary, and the necessity is determined as a result of monitoring and diagnosing the status parameters of EE with a quantitative assessment of the CI.

The mass switch of EE to the strategy "by technical condition" opens up new opportunities for improving the operating efficiency of power distribution networks for various purposes. This will require formalization of the description and solution of a set of tasks to justify the performance criteria in accordance with the priority control goals of the M&R, developing models, methods and scheduling algorithms, updating the parameters of technical impacts on the main EE, taking the influence of key risk factors into account, and so on. Many of these tasks have been brought to acceptable practical results, promising methods and approaches to solving them are discussed, but the process of finding the best scientific-based solution is still far from complete [5-12].

Electrical networks of oil-production industries, in contrast, for example, to the facilities of JSC Rosseti, supply power only to technological consumers of oil-field facilities (OFF). This determines the difference in the target priorities and performance criteria of such facilities operation. Thus, for general-purpose electrical networks a long-term strategic priority is to ensure the maximum level of reliability of power supply to consumers, which is regulated by reliability indexes such as SAIFI (System Average Interruption Frequency Index) - the index, which shows the average frequency of interruptions in the system, SAIDI (System Average Interruption Duration Index) - the index of the average duration of interruptions in the system, etc. [13]. For electrical networks, providing electricity for oil fields, the strategic target priority is the continuity of technological processes and minimum total damage from underproduction (loss of oil production). Thus, there are obvious differences in optimization approaches to planning M&R and choosing parameters of corrective actions on EE in frequency and volume. It is necessary to point out, that the solution of the optimization problem is usually performed under various restrictions, for example, on available material, labor, and financial resources (costs for M&R of EE), as well as taking the influence of numerous risk factors into account.

For the electrical networks of oil fields, a key risk event is the failure of the EE that supplies energy to the OFF. Failure of EE, considered as an accident, is characterized by parameters such as probability Q and frequency ω, depending on a number of factors. There are gradual and sudden failures of EE [14]. Gradual failures can usually be predicted. They are caused by aging (worsening of the material properties and construction elements) of EE. Sudden failures cannot be predicted. They occur due to the influence of internal and external occasional factors in relation to electrical equipment, such as environmental influences, operating errors, etc. Thus, the probability (frequency) of EE failure is directly related to its technical condition. CI is its integral characteristic. Is determined on the basis of technical documentation data and diagnostic results. The relationship for the i-th unit of electrical equipment is taken into account in the following formula:

$$Q_i = 1 - \exp(-\omega_i) \cdot J_i, \quad (1)$$

where ω_i (year^{-1}) and J_i (relative units) - the values of the average failure rate and the CI of the i-th unit of EE.

In addition to the probability, an accident is characterized by the severity of the consequences (damage from technology violations, equipment failures, threats to the environment and personnel safety, product underperformance) [15]. The components of total damage (Dam) can be ranked and quantified for each typical OFF [16]. For these purposes, the author of the article enumerated in the references uses the term "criticality", which serves as an integral estimative probability and severity of the consequences of a key accident for each type of OFF (GOST 27.310-95 Reliability in technology. Analysis of types, consequences, and criticality of failures). Thus, the criticality or risk function of EE failure can be shown as follows:

$$R = Q \cdot M[Dam], \tag{2}$$

where:
$M[Dam] = P_{load} \cdot \Delta t \cdot D \cdot c$ - damage expectation value;
P_{load} (kW) - load power;
Δt (hour) - power supply interruption;
D (tons/kWh) - specific capacity of the OFF consumer;
c (USD/ton) - unit cost of losses.

The total number and scope of corrective actions for M&R of EE serve as optimization parameters on the planning interval ΔT since they ultimately determine one of the main components of the integrated production resources ($C_{M\&R}$) of the network enterprise, which are always limited [17].

Taking this into account, the minimum of the risk function *Min (R)* should be taken as the optimization target function, and the following relations should be taken as restrictions:

$$C_{M\&R} \leq C_{lim} \quad CI_i \geq CI_{lim} \text{ при } i = 1,N, \tag{3}$$

where:
C_{lim} - an available integrated resource of a network enterprise for M&R of EE for the planning year (for example, in man-hours or rubles);
CI_{lim} - CI acceptable limit values for each equipment unit of a certain type;
N - number of units EE to be M&R for the planning year (number of units).

The parameters for corrective action optimization will be the following:

1) frequency of M&R activities for each unit of EE in the planned interval $T_{M\&R}$;
2) the total labor intensity of a set of operations to eliminate defects $(\sum L I_i^j)$ as part of each i-th corrective action of a certain type for each j-th unit of N units of EE to be scheduled M&R.

Considering the information mentioned above, it is obvious that there is a need for a decision-making methodology that would take into account and combine the scenario approach, the variety of influencing factors, and the variability of management goals. This article presents and discusses the core components of the developed methodology.

2. Materials and Methods

2.1. Planning model for M&R of EE

EE operation with the "by technical condition" strategy implies the initial formation of the scheduled plan of the work execution plan (WEP) for a calendar year. The principles of its formation are based on the concept of a repair cycle – the time interval between two consecutive overhauls (OH), which are designed to restore partially lost functional capabilities of EE. During the repair cycle interval, T_{rc} of EE with a fixed standard rate, other types of corrective actions are scheduled. They include:

- inspection (I) - visual testing of EE;
- detailed inspection (DI) - instrumentation control with the use of facilities of technical diagnostics and nondestructive testing (TD&NT) of EE;
- maintenance (M) - complex of operations to maintain the EE working condition during operation without disconnecting it;
- routine maintenance (RM) - a set of operations to guarantee the operability of EE with its withdrawal from the operation.

When the scheduled plan of the WEP for the current calendar year is being prepared, the year of the previous OH and/or RM is taken into account (T_{rWEP}). The plan drawn up in a similar way is fixed and represents a "basic grid" for obtaining a scheduled plan for M&R of EE "by technical condition". The M&R scheduled plan "by technical condition" is not static. It is periodically adjusted by shifting the timing of repair works in accordance with the actual need, which is determined by the technical condition of the EE. Also CI am an integral characteristic of the technical condition.

Thus, obtaining information about the value of the CI of an EE unit is a necessary and sufficient condition for making a decision on adjusting the overhaul period in the scheduled interval. The practice of operating EE in the distribution networks of oil-fields has proved the validity of the following rules:

1) If for an EE unit 25% ≤ CI <50%, the repair is scheduled for the next calendar month with the date fixed in the M&R schedule. This makes it possible to exclude possible failures of electrical equipment, including those with serious consequences.

2) if the value is CI ≥ 50%, then the adjustment of the overhaul period of the RM is made according to the mathematical expression (4):

$$T_{op} = k \cdot T_{stand}, \tag{4}$$

where:
T_{op} - estimated duration of the overhaul period;
T_{stand} - standard duration of the overhaul period for the WEP strategy;
k - correction factor, that takes values within the range $1 \le k \le 1{,}5$.

The specified range is set by an expert method based on an analysis of the functioning of EE with a "by technical condition" strategy. After receiving the adjusted value of the overhaul period, the date of repair is determined:

$$T_r = T_{rWER} + T_{op}, \tag{5}$$

where: T_{rWER} - date of the previous repair from the scheduled plan of the WEP.

Table 1 shows the values of standard overhaul periods for some types of EE 6(10) kV by type of main corrective action.

Table 1. Standard overhaul period values for EE 6(10) kV.

Name of equipment	Periodicity, months		
	M	RM	OH
Overhead line on metal and reinforced concrete supports 6(10) kV	6	24	120
Cable line 6(10) kV	6(3)	26	60
Power transformer 6(10) kV	3	12	144
Oil breaker 6(10) kV	1	12	96
Vacuum breaker 6(10) kV	6	24	120
Disconnecting switch 6(10) kV	1	12	96

The value of the correction factor k is determined using Table 2 in accordance with the values of the EE CI calculated on the basis of data obtained as a result of I, IDI and diagnostic examinations.

Table 2. Selection of the correction index based on the value of the EE CI.

CI, %	50 – 70	70 – 85	85 – 100
k, relative units	1.0	1.25	1.5

2.2. Adaptive model of the EE process map

The process map (PM) serves as the main organizational and technological document in the management of M&R EE. It contains a set of measures for human engineering using the most effective modern means of mechanization, processing equipment, tools, and devices. As a rule, the PM includes the most advanced and rational methods and technologies of M&R that help to increase labor productivity, improve the quality of work, and reduce their cost. The standard model and methodology for creating an MP WEP of electric power facilities are approved by the regulation (Organization standard STO 56947007-29.240.55.168-2014 "Methodological guidelines for the development of flow charts and projects for the maintenance and repair of overhead lines". Approved by the Order of JSC FGC UES dated 02.04.2014 No 165). In accordance with requirements, MP is divided by types of impacts on the object (I, DI, M, RM, OH) and contain the following information:

1) characteristics of the facility;
2) the standard list of required operations;
3) the labor intensity of each operation in man-hours;
4) the number of employees involved in accordance with the functions (brigade staff);
5) processed materials, equipment, tools, special mechanisms, including means of transport.

Management of M&R of EE with "by technical condition" strategy based on the minimum risk criterion relies on making decisions under the influence of random factors, such as damage to EE, power failure to OFF, the severity of the consequences of failure, the availability of the necessary resource, etc. Under these conditions, the standard model of the PM is unsuitable for planning EE M&R due to its redundancy and determinism. We need an adaptive model of the PM that allows us to adjust the total labor intensity of EE repair work based on the minimum set of operations necessary and sufficient for high-quality restoration and verification of the serviceability of the maintained facility (MF) in compliance with standard measures and safety rules.

The adaptive model of the PM of MF repair is based on the typical division of the complete list of operations into two blocks: organizational and technological ones. The organizational block cannot be corrected and it includes operations connected with logistics, workplace preparation, compliance with health standards, access to the MF, reporting documents, and it cannot be corrected. The technological block includes operations aimed at MF fault repair, adjustment and functionality test. It allows the necessary adjustment. The formation of an adaptive process map model should follow the following key principles:

1) full identification of possible malfunctions (defects) of the EE functional units and determination of appropriate corrective operations;

2) objective assessment of each corrective operation labor intensity under the conditions of its execution on the MF;

3) ranking of defects in terms of criticality (from the point of view of extending the operation of the MF), determined by TD&NT, for example, as a non-critical defect (NCD) or a critical defect (CD).

Compliance with these principles ensures that the resulting labor intensity is ($\sum LI_i^j$) determined not as the sum of labor costs for the full list of operations contained in the standard technological map for the repair of an object, but by a reduced list of corrective operations requiring actual performance in specific conditions. Table 3 shows a fragment of the corrected PM for the repair of an overhead line 6(10) kV, derived from the developed adaptive model.

2.3. The algorithm of formation the discrepancy list for MF

A discrepancy list is one of the most popular tools for planning repairs of electrical equipment with the "by technical condition" strategy. A discrepancy list is formed for each MF and contains information from the inspection sheets [18]. In fact, this is a list of defects identified on MF during and I or DI, but not eliminated at the current time, indicating the criticality extent for each defect and assigning a certain type of corrective action to it. As a rule, such types of impacts as M and RM are assigned as corrective actions of M&R. If it is necessary to monitor the extent of EE failure and clarify the scope of repairs, and DI is scheduled, identifying a specific method of TD&NT. Figure 1 shows the structure of the discrepancy list of the generalized MF.

Table 3. Corrected process map of repair overhead power lines 6(10) kV (fragment).

Equipment			Overhead power line 6(10) kV			
List of operations per 1 km of overhead power line length			Repair			
Sr. No	Block of operation	Working sequence	Number, man	Time of operation, minute	Labor costs, man-hours	NOTATION (Defect code)
1	Organizational / start of the operation	MF testing for the previously detected defects and the amount of operating time. If there are defects, determining the necessary mechanisms, components, materials, etc.	1	30	0.50	Analysis of the of the previous DI and M results, discrepancy list, inspection sheets, operating hours, thermovision inspection, ontrolled parameters, etc.
2		Implementation of administrative procedures ensuring work safety in electrical installations, including the implementation of additional measures when working at height, conducting fire-hazard work, etc.	2	30	1.00	Registration of works by order, instruction, list of works. Getting permission to prepare the workplace. Getting work permission, supervision
i *i+1* … *n*	Technological					
1	Organizational / operation completion	Cleaning of the workplace, removal of grounding clusters of the brigade, posters, etc.	3	60	3.00	
2		Registration of operation completion.	1	10	0.17	
3		Filling in the passport of an overhead power line	1	40	0.67	
Total labor costs taking into account correction factors, man-hours			Σ	Σ	Σ	

Depending on the type of EE, each defect is assigned its personal identity code, which is used to define a specific method of TD&NT to the defect. The defect code is also used when forming an adaptive PM of MF repairs to identify the detected defect and the related list of corrective operations. To illustrate this, Table 4 shows a fragment of the discrepancy list for the overhead power line 6(10) kV.

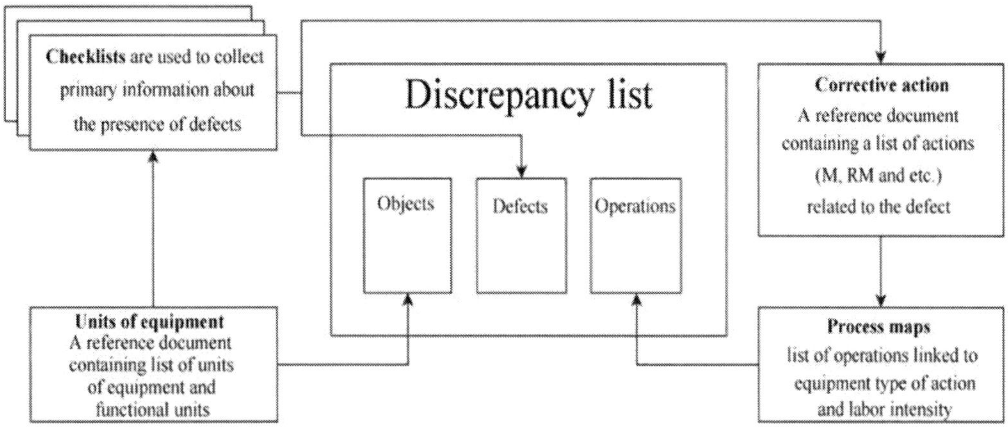

Figure 1. Structure of the discrepancy list.

Table 4. Discrepancy list for overhead power lines 6(10) kV (fragment).

Functi-onal unit	Defect code	Name (characteristic) of the defect	Controlled parameter	Level of criticality		Corrective action	
				CD	NCD	CD	NCD
Right-of-way	T01	The presence of objects not accounted for in the project	$L < 10$ metre	+	–	M	–
	..	..	..	..	..	..	..
	T03	The presence of trees on the edge of the forest corridor threatening to fall on the overhead power line wires		+	–	M	–
Support		Metal support					
	C01	Deformation of the metal grating elements of support structure		+	+	RM	IDI
	C02	Break of elements construction supports		+	+	RM	M
	..	..	..	..	..	..	..
	C08	Absence or loosening of mounting nuts on the anchor bolts		+	–	M	–
Span	П07	Changing the slacks and distances from the overhead line wires to the ground, to intersecting objects, between phases to values other than acceptable		+	–	RM	–

3. Results

As an example illustrating the developed methodology for managing EE repairs in oilfield distribution networks, let's consider two identical objects operating in similar conditions. Both network objects have 2 and 3 km long double-circuit overhead transmission line (OTL), made on metal supports with AS-95 wire. The first object (OTL-1) provides power supply to process load of a booster pipeline pumping station with a device for initial reservoir water separation (BPPS with DIRWS) with a power $P_{load}=1000\ kW$, a supply interruption time $\Delta t=1\ hour$ and a specific throughput $D=0{,}01\ tons/kWh$. The second object (OTL-2) supplies energy to consumers of the oil multiple-well platform with a total power $P_{load}=1000$ *kW*, $\Delta t=0{,}015$ *h*, $D=10$ *tons/kWh*. Let the condition of OTL-1 and OTL-2 be characterized by the presence of defects in the discrepancy list (Table 5).

Table 5. Structure of defects OTL-1 and OTL-2.

Functional unit	Defect code	Name (characteristic) of the defect	Controlled parameter	Level of criticality CD	Level of criticality NCD	Corrective action CD	Corrective action NCD
		OTL-**1**					
Right-of-way	T09	Presence of object not accounted for in the project	$L < 10$ metre	+	–	M	–
Support		Metal support					
	C02	Break of elements construction supports		+	+	RM	M
Span	П07	Changing the slacks and distances from the overhead line wires to the ground, to intersecting objects, between phases to values other than acceptable		+	–	RM	–
		OTL-**2**					
Support	C08	Lack of perpendicularity		+	–	RM	M
Span	П01	Wire repair, bandaging with lowering the wire		+	–	RM	–

Table 6 provides a list of processing steps of the adaptive PM for the repair of each MF (OTL-1 and OTL-2), and shows differences in the total labor costs for their implementation, considering the correction factors that take into account natural and climatic conditions and transport accessibility.

Table 6. List of processing steps of the adaptive PM for the repair of OTL-1 and OTL-2.

OTL-1			OTL-2		
Defect code	Corrective operations	$(\sum LI_i^j)$, Man / hour	Defect code	Corrective operations	$(\sum LI_i^j)$, Man / hour
T09	Clearing the right-of-way from scrubs, eliminating foreign objects, cutting down trees that threaten to fall.	10.0	C08	Checking the condition of the supports and their installation (lack of perpendicularity, skewing elements), if necessary, supports alignment	9.3
C02	Anchor bolt tightening, restoration of corners of the cross brace				
П07	Pulling separate segments of wires, tightening and adjusting the wire sag.		П01	Wire repair, bandaging with lowering the wire	

The calculations of the CI OTL using method [4] are shown in Table 7. The calculations used scorecard groups of separate functional units status parameters of the item within the range from 10 (best) to 1 (worst). Based on the results of the calculations for CI functional units, the CI score for each object as a whole is calculated:

$$J_{OTL-1} = (0{,}91 \cdot 0{,}334 + 0{,}1 \cdot 0{,}666) \cdot 100\% = 37\%;$$

$$J_{OTL-2} = (1 \cdot 0{,}334 + 0{,}37 \cdot 0{,}666) \cdot 100\% = 58\%.$$

As follows from the calculations, the condition of the OTL-1 by the CI value falls into the category ≥ 25% and < 50% (unsatisfactory), in which the repair must be planned directly for the next calendar month after the calculated one, with the date fixed in the Maintenance schedule M&R. For OTL-2 the correction factor k of the repair frequency adjustment takes value 1.0 (Table 2), i.e. the overhaul period is not adjusted in the M&R schedule. Thus, based on the assessment of the technical condition, the repair of OTL-1 is of a higher priority than the repair of OTL-2 and should be planned according to the M&R schedule at an earlier time.

Table 7. Calculations results of CI OTL-1 and OTL-2.

Defect code	Parameters group	Scorecard of parameters group OTL-1 / OTL-2	"Weight" of a parameters group in the node estimation	CI of functional units
Support	Fixing support, stand, traverse, hitch , insulator on the support stand, wire	10	0.3	OTL-1 (0.3·10 + 0.1·1 + 0.2·10+ 0.1·10 + 0.3·10)/10 = 0.91
	Attachment, stand, strut	1/10	0.1	
	Traverse, hitch, insulator on the traverse	10	0.2	OTL-2 (0.3·10+0.1·10+ 0.2·10+ 0.1·10+0.3·10)/10=1
	Grounding connection	10	0.1	
	Switching units, surge arrestor	10	0.3	
Span	Right-of-way OTL	1/10	0.3	OTL-1 (0.3·1 + 0.7·1)/10 = 0.1
	Wire, intermediate cable	1/1	0.7	OTL-2 (0.3·10 + 0.7·1)/10 = 0.37

When calculating the probability of an MF failure, taking into account their actual technical condition, formula (1) uses the value of the average failure rate for an OTL 6 kV with a length of 2-3 km, equal to ω=0,202 year^{-1} [19, 20]:

$$Q_{OTL-1} = 1 - \exp(-0.202) \cdot 0.37 = 0.698;$$

$$Q_{OTL-2} = 1 - \exp(-0.202) \cdot 0.58 = 0.526.$$

Calculations of the risk function (optimization target function) for each of the OTL are performed using formula (2) when the value of the incremental cost of oil production losses c = 200 USD/ton.

Obtained results:

$$R_{OTL-1} = 1000 \cdot 1 \cdot 0.01 \cdot 0.698 \cdot 200 = 1396 \text{ USD};$$

$$R_{OTL-2} = 1000 \cdot 0.015 \cdot 10 \cdot 0.526 \cdot 200 = 15780 \text{ USD},$$

indicate that the risk of interruption of consumer power supply of OFF in the OTL-2 is well above than the risk of OTL-1 failure.

Thus, according to the minimum risk criterion, the highest priority in planning is given to the repair of OTL-2, the failure of which leads to the most severe consequences. From the point of view of checking the fulfillment of conditions (3), we will limit ourselves to general considerations. As the most likely situation, we assume that the total cost of equipment maintenance does not exceed the specified value (C_{lim}), and CI_{lim}=25%. This situation will allow us to recognize the optimal solution to the problem of planning the repair of OTL-2 within a calendar month from the estimated date, and to shift the repair of OTL-1 a month after OTL-2. This, in turn, will increase the index of its technical condition and reduce the probability of failure, and therefore the amount of total risk.

4. Conclusion

1. Improving the methodology for managing M&R of EE with a "by technical condition" strategy on risk-based models and decision-making algorithms contributes to improving the efficiency of electrical networks and industrial process complexes.

2. The original methodology proposed by the authors is focused on the application of OFF in distribution networks and power supply systems. The methodology allows for a formal description and solution of a set of M&R planning tasks for selecting and adjusting the type, volume, and frequency of technical impacts on the EE with a "by technical condition" strategy. The target function of optimizing the parameters of technical impacts is to minimize the total risk from interruptions of power supply to consumers of OFF. The main restrictions on the scope of acceptable solutions are the available resources of the enterprise (total M&R costs) and reliability (the maximum allowable value of the EE CI). The calculation of the total risk involves the determination of oil production loss from each process consumer, taking into account the individual characteristics of its power supply, as well as determining the probability of EE failure, paying attention to its actual technical condition.

3. One of the main options in the methodology is an adaptive model of the PM for the repair of EE with the strategy "by technical condition". The basis of its modeling is featured by a number of key pillars, such as complete identification of possible defects of EE functional units and relevant corrective operations, an objective assessment of the labor intensity of each correcting operation in the conditions of its implementation on the MF, ranking defects by the extent of criticality, defined by the TD&NT. This ensures that the amount of corrective action on electrical equipment is determined only by a reduced list of corrective operations requiring actual performance under certain conditions.

4. Certain example illustrates the logical and computational-analytical advantages of the developed methodology. Unfortunately, the scope of the article limits the ability to demonstrate the proposed set of solutions to the full, but the example clearly shows an accurate sequence of calculations, the reliability of the results, provided by the correctness of the models' application, verified by practice, the logic of excluding contradictions when choosing priorities and evaluating the effectiveness of the results obtained. It should be noted that the presented components of the developed methodology have found practical application in the terms of real operation of EE OFF.

References

[1] Sinyagin N N 1984 *Sistema planovo*-predupreditel'nogo *remonta elektrooborudovaniya promyshlennykh predpriyatiy* [System for preventive maintenance of electrical equipment of industrial enterprises] (Moscow: Energiya) [In Russian]

[2] Korotkevich M A 2003 *Osnovnye napravleniya sovershenstvovaniya ekspluatacii elektricheskih setej* [Main directions for improving the operation of electric networks] (Minsk: Tekhnoperspektiva)

[3] Grabchak E P 2017 Otsenka tekhnicheskogo sostoyaniya energeticheskogo oborudovaniya v usloviyakh tsifrovoy ekonomiki [Assessment of the technical condition of power equipment in the digital economy] *Nadezhnost' i bezopasnost' energetiki* [Reliability and Safety of Energy] **10(4)** 268-74 [In Russian]

[4] Nazarychev A N 2002 *Modeli i metody optimizatsii remonta elektrooborudovaniya ob"yektov energetiki s uchetom tekhnicheskogo sostoyaniya* [Models and methods for optimizing the repair of electrical equipment for energy facilities, taking into account the technical condition] (Ivanovo: ISEU) [In Russian]
[5] Wang M, Vandermaar A J and Srivastava K D 2002 Review of condition assessment of power transformers in service *IEEE Electrical Insulation Magazine* **18(6)** 12-25
[6] Gavrilovs G 2011 Technical condition asset management of power transformers *2nd IEEE PES Int. Conf. and Exhibition on Innovative Smart Grid Technologies* (Manchester: IEEE) 6162762
[7] Aladon M B 2018 *RCM3: Risk-Based Reliability Centered Maintenance* (NY: Industrial Press)
[8] Levin V M 2018 Methodological Aspects of Assessing State of HPP Transformers in Monitoring Mode *XIV Int. Scientific-Technical Conf. on Actual Problems of Electronics Instrument Eng.* (Novosibirsk: NGTU) pp 238-43
[9] Petrosenko A V and Tulsky V N 2019 Primeneniye mnogokriterial'nogo podkhoda i kombinirovannogo analiza pri formirovanii proizvodstvennoy programmy organizatsiy [Application of the multicriteria approach and combined analysis in the formation of the production program of organizations] *E`lektroe`nergiya. Peredacha i Raspredelenie* **56(5)** 38-49 [In Russian]
[10] Grabchak E P, Medvedeva E A et al. 2019 O metodologii rascheta tekhnicheskogo riska na osnove veroyatnosti i posledstvij otkaza funkcional'nogo uzla i edinicy osnovnogo tekhnologicheskogo oborudovaniya [On the methodology for calculating technical risk based on the probability and consequences of failure of a functional unit and a unit of main technological equipment] *E`lektroe`nergiya. Peredacha i Raspredelenie* **52(1)** 22-9 [In Russian]
[11] Antonenko I N 2019 Metodologiya RCM: retrospektiva i perspektiva nadezhnostno-orientirovannogo tekhnicheskogo obsluzhivaniya [RCM methodology: a retrospective and perspective of reliability-oriented maintenance] *E`nergiya Edinoj Seti* **1(43)** 34-46 [In Russian]
[12] Rychagova E A and Levin V M 2019 Improving the efficiency of electric network equipment operational service *E3S Web of Conf.* **114** 03001
[13] Saharova I V 2013 Ob uchete kachestva uslug v tarifnom regulirovanii raspredelitel'nyh elektrosetevyh kompanij v rossijskoj i zarubezhnoj praktike [About accounting of quality of services in tariff regulation of distribution electric grid companies in the Russian and foreign practice] *Sovremennaya E`konomika: Problemy` i Resheniya* **5** 43-51 [In Russian]
[14] Levin V M and Sekretarev Yu A 2019 Ocenka vliyaniya na nadezhnost' sistemy elektrosnabzheniya razlichnogo roda defektov ee osnovnyh elementov [Assessment of the impact on the reliability of the power supply system of various types of defects in its main elements] *Vestnik KGEU* **44(4)** 55-63 [In Russian]
[15] Nordgärd D E, Welte T M and Heggset J 2010 Using life curves as input to quantitative risk analysis in electricity distribution system asset management *Proceedings of the Institution of Mechanical Engineers, Part O: J. of Risk and Reliability* **224(2)** 63-74
[16] Lesnyh V V, Timofeeva T B and Petrov V S 2017 Problemy ocenki ekonomicheskogo ushcherba, vyzvannogo pereryvami v elektrosnabzhenii [Problems of assessment of economic damage caused by interruptions in the power supply] *Regional'naya Ekonomika* **3** 847-58 [In Russian]
[17] Antonenko I N 2020 Risk-orientirovannyj podhod k upravleniyu proizvodstvennymi aktivami energetiki [Risk-based approach to managing energy production assets] *Energoexpert* **1** 26-33 [In Russian]
[18] Voronin V V and Davydov O A 2019 Relative Equivalence of Defects *2019 Int. Multi-Conf. on Ind. Eng. and Modern Technologies* (Vladivostok: IEEE) 8933857

[19] Napoleone A, Roda L and Macchi M 2016 The implications of condition monitoring on asset-related decision-making in the Italian power distribution sector *IFAC-PapersOnLine* **49(28)** 108-13
[20] Levin V M 2020 Innovacionnye resheniya v upravlenii remontami energeticheskogo oborudovaniya neftedobyvayushchego kompleksa [Innovative solutions in the management of repairs of power equipment of the oil production complex] *Glavny`j E`nergetik* **1** 30-9 [In Russian]

Rectangular current pulse generator to test varistors with pulsed electrical load

N A Chernenko[1], A V Morozov[2], A A Myulbaer[1] N V Sheglov[1], S S Shevchenko[1] and I V Shmakov[1]

[1] Department of High Voltage Engineering and Electrophysics, Novosibirsk State Technical University, Karl Marx ave., 20, Novosibirsk, 630073, Russia
[2]Laboratory of Experimental Technique, Novosibirsk State Technical University, Karl Marx ave., 20, Novosibirsk, 630073, Russia

E-mail: N.Chernenko@corp.nstu.ru

Abstract. A brand-new circuit of a rectangular pulse current generator is presented. As a rule, generators consisting of 6-10 LC cells connected in series are used to obtain rectangular current pulses. Such units are very cumbersome and complicated in the selection of element parameters and adjustment, so a new scheme of pulse current generator was proposed. The main difference from the classical circuit is the method of generating a rectangular current pulse, i.e. by adding the currents from two parallel circuits connected to the load. The first circuit at discharge generates an aperiodic current pulse, the second generates an oscillatory current pulse. The total current pulse flowing through varistor has constant amplitude for some time. In a specified period of time, while the current pulse has a constant amplitude, there is an interruption of current with the help of a switching device. The given circuit solution is possible to implement when using fundamentally new switching devices of the IGBT module. A connection diagram of a pulse current generator for testing varistors with a classification voltage to 1200 V with adjustable pulse duration of 50, 1000, 2000 μs is proposed. The parameters of the generator circuit elements are determined. The measures aimed at suppression of overvoltages arising at break of current in circuits with high inductance as well as the tests confirming the operability of the designed circuit are carried out.

1. Introduction

The production of varistors requires constant monitoring of their main parameters, one of which is the ability to withstand the effects of a pulsed electrical load. Varistors are tested using two types of pulse current generators (hereinafter referred to as just "generator"). The first type includes generators that create aperiodic current pulse with a rise time of 8 μs, a pulse length of 20 μs and amplitude of up to several tens of thousands of amperes, the second is generators, generates rectangular current pulses with on-time of 50, 1000 or 2000 μs and amplitude of up to 1000 amperes. The circuit design of pulse current generators was developed in the middle of the last century and is described in the key issues of high voltage engineering [1, 2].

2. Materials and Methods

One can generate a rectangular current pulse of a long on-time using a (classical) circuit made up of series-connected *LC* cells (Figure 1). This circuit simulates a long uniform line that allows to get close to ideal rectangular current pulses when the load resistance R_n is equal to the characteristic resistance of the circuit Z [3]:

$$R_n = Z = \sqrt{\frac{L}{C}}, \quad (1)$$

where: $L = L_1 = L_2 = \ldots = L_n$, and $C = C_1 = C_2 = \ldots = C_n$ – parameters of LC cells; the load resistance R_n includes the damping resistance R and the dynamic resistance of the test varistor R_0 (see Fig. 1).

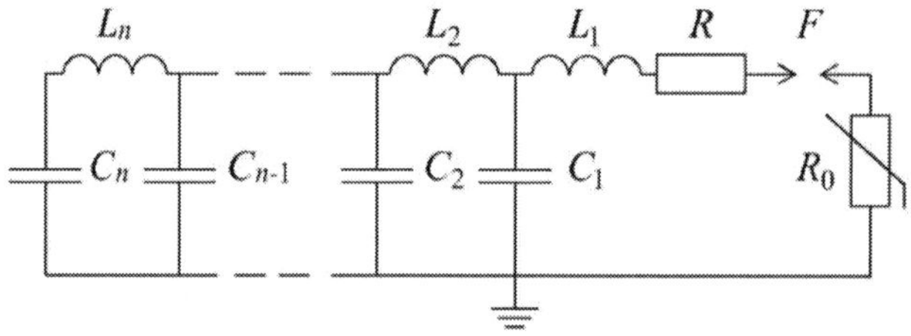

Figure 1. Classical circuit of a rectangular pulse current generator. F – spark gap.

The total values of capacitance and inductance in the circuit are determined by the formulas (2, 3):

$$C_\Sigma \approx \frac{n \cdot T_p}{2R \cdot (n-1)}, \quad (2)$$

$$L_\Sigma \approx C \cdot R^2, \quad (3)$$

where:
C_Σ is the total capacity of the generator, n is the number of cells;
T_p is the time during which the current exceeds 0.9 of the amplitude value;
R_n is the load resistance, and L_Σ is the total inductance of the generator.

However, even with a large number of cells, the shape of the current pulse will deviate from rectangular and approach the trapezoidal shape. To solve this problem, one needs to use different parameters of LC cells. Moreover, as a rule, the capacitors C_1, C_2, ... C_n take the same capacitance, and the inductance of the coils is varied so as to obtain the desired pulse shape. Experience shows that testing varistors using the classical generator circuit requires from 8 to 12 LC cells. In this case, the inductors L_1, L_2 should have the largest value, and L_{n-1}, L_n – the smallest of all available [3].

The following significant drawbacks are typical for the classical generator circuit:

- the complexity of the circuit. In addition to requiring many LC cells, the magnetic coupling between them must be minimized so that its effect on the shape of the test pulse is negligible. Therefore, large distances are required between the cells, which increases the unit area;
- the shape of the current pulse strongly depends on the test object. The load resistance must match the characteristic resistance of the LC circuits. This does not allow to standardize the unit;
- there is no possibility to control the on-time of the current pulse, namely, a separate generator is required for each pulse on-time;
- the load resistance strongly affects the shape of the current pulse;
- complex configuration of the unit parameters to obtain a current pulse of the required shape due to the large number of elements;
- the charging voltage of the generator must be at least twice the remaining voltage of the tested varistors;
- high cost.

To eliminate these shortcomings in the classical circuit of a rectangular current pulse generator, it was proposed to develop a unit using a component base of power electronics.

In the last few years, new electronic power equipment has emerged that can break a circuit of hundreds of amperes and switch voltages of several kilovolts [4, 5]. Such switching devices can be easily controlled, unlike classic spark gaps, which depend on external conditions. Some of these devices operate on the basis of an IGBT transistor and are called an IGBT module, i.e. a bipolar transistor with a sealed gate. An IGBT transistor is a device that is a combination of a field-effect and bipolar transistor. Such combination has resulted in it having the positive qualities of both types of these transistors, with the field-effect transistor driving a powerful bipolar transistor.

The most significant and used in the development of the generator is the ability of IGBT modules to break the current pulse without passing through the zero value and the rated voltage up to 6.5 kV [6-8]. Taking this into account, a fundamentally new circuit of the pulse current generator was developed and further modeled in the OrCAD software (Figure 2).

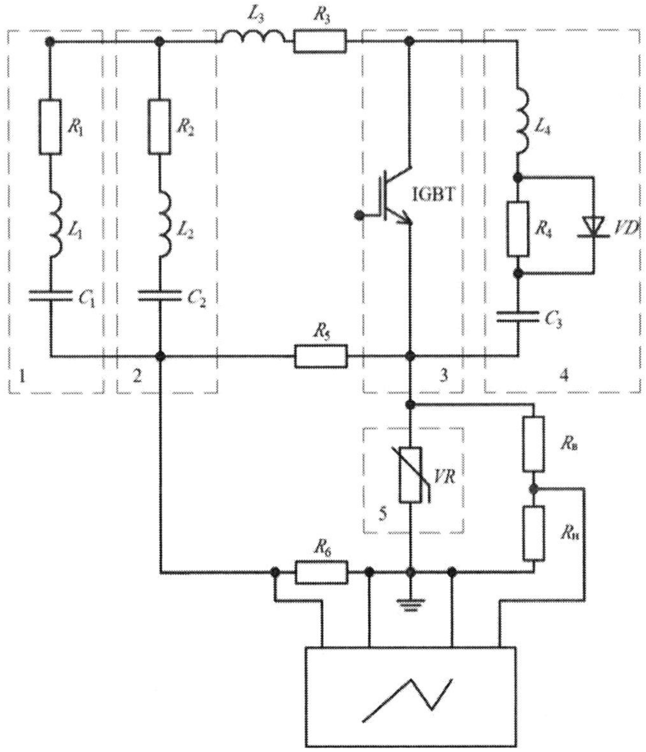

Figure 2. Connection diagram of the designed generator. VD – high-voltage diode; R_{vn} – upper arm of the voltage divider; R_{nn} – lower arm of the voltage divider.

The generator circuit consists of two discharge circuits (1, 2), a switching device – an IGBT module (3), a snubber circuit (4) of an IGBT module (a protective RCD circuit), and a varistor as a load (5).

The discharge circuits of the circuit are series-connected capacitive, inductive, and active elements. The accumulated charge in the capacitor during switching is discharged to the load for a specified time (50, 1000, 2000 microseconds). This process can be either aperiodic or oscillatory. The parameters of the discharge circuits must be chosen so that the current in the first one changes according to the aperiodic law, and in the second one – according to the oscillatory law, respectively. Figure 3 illustrates the generation of a rectangular current pulse: the current of the first circuit changes in time according

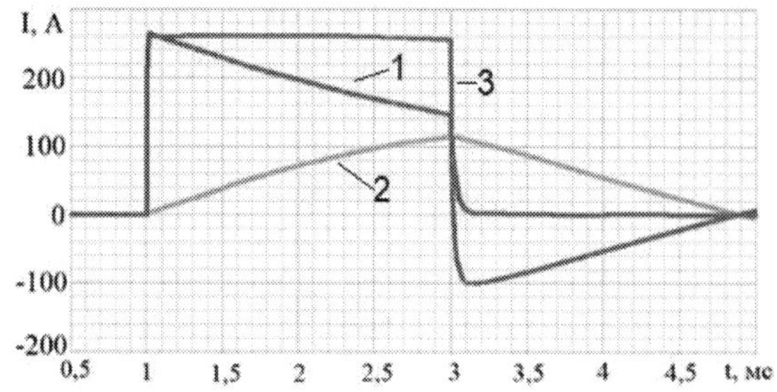

Figure 3. Current waveform (at t_1=1 ms – switching on the IGBT module, at t_2=3 ms – switching off the IGBT module). 1 – current of the first discharge circuit, 2 – current of the second discharge circuit, 3 – total current.

to the aperiodic law (1), the current of the second circuit has an oscillatory feature (2). The addition of two current pulses gives the desired rectangular current pulse with the specified parameters (3). It should be noted that in the absence of a measured varistor in short-circuit mode, the generator circuit can withstand a maximum current of 1000 A.

When the IGBT module is disconnected, the current flowing through the inductive elements of the circuit is cut off, as a result of which significant overvoltages occur on the module terminals. To prevent this, an *RCD* circuit is connected in parallel to the IGBT module, which is also called a protective or snubber circuit [9, 10]. A snubber circuit consists of a series-connected resistance and capacitance. The capacitor contained in the *RCD* circuit must absorb the energy that has been accumulated in the inductance in kinetic form, and, as a result, protect the switching device from the effects of overvoltage. A high-voltage pulse diode, which is designed to regulate the charge and discharge time of the capacitor [11], is connected in parallel to the snubber circuit resistor. To avoid overvoltage, the resistor and capacitor of the snubber circuit must also have the lowest possible inductance. It is worth noting that the resistor and capacitor always have their own inductance, but the simulation results showed that the inductance of 0.1 μh is acceptable in this case [12, 13]. Figure 4 shows the voltage change at the module terminals in the absence and presence of a snubber circuit. The voltage on the IGBT module during switching in the absence of a snubber circuit (curve 1) reaches 9 kV, which will certainly lead to its destruction. If a snubber circuit is installed parallel to the IGBT module, the overvoltage will be completely suppressed (curve 2).

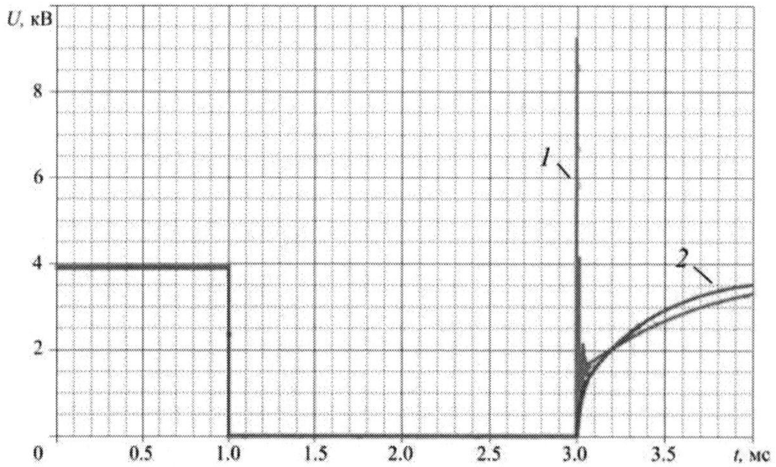

Figure 4. Voltage waveform on the IGBT module during switching (at t_1=1 ms – switching on the IGBT module, at t_2=3 ms – switching off the IGBT module).

Table 1 shows the parameters of the generator circuit elements that allow creating a rectangular current pulse in accordance with GOST (all-Union State Standard) [14, 15].

Table 1. The parameters and purpose of circuit elements of the generator.

Element	Rating value	The purpose of the element
C_1	1.1 mF	Energy storage
C_2	0.3 mF	Energy storage
C_3	4 μF	Absorption of energy stored in the inductance
L_1	10 μH	The formation of the pulse shape
L_2	10 mH	The formation of the pulse shape
L_3	2 μH	Self-inductance of the generator busbar
L_4	0.1 μH	Self-inductance of the snubber circuit
R_1	3 Ohms	The formation of the pulse shape
R_2	1 Ohms	The formation of the pulse shape
R_3	1 Ohms	The formation of the pulse shape
R_4	2 Ohms	Limiting the discharge current of the snubber capacitor
R_5	100 Ohms	The discharge of the varistor capacitance. Charging the snubber capacitor in the absence of the test object
R_6	50 mOhms	Current measuring shunt
VD	6000 V, 30 A	Regulation of the charge and discharge rate of the snubber capacitor
IGBT module	6500 V, 500 A	Switching device

3. Results

Figure 5 shows a photo of the IGBT module mounted on the radiator, along with the snubber chain. The correct choice of parameters of the snubber circuit is proved by oscillography of the voltage at the terminals of the IGBT module during switching. From the waveform shown in Fig. 6, it can be seen that there are no overvoltages when the current is turned off.

Figure 5. Photo of an IGBT module with a snubber chain.

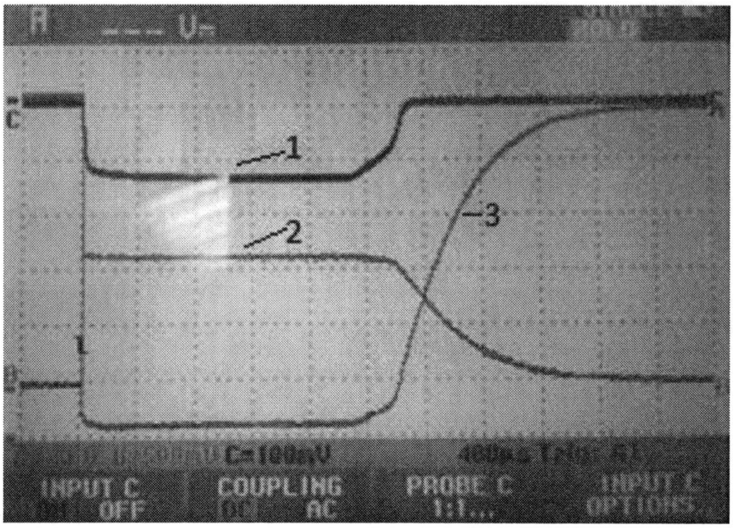

Figure 6. Waveforms: 1 – current pulse (the signal is inverted), 2 – remaining voltage on the varistor, 3 – voltage on the IGBT module terminals.

4. Discussion

The proposed circuit of the pulse current generator makes the device much more compact and mobile. In addition, due to the small number of elements, compared to the classic circuit, setting the unit parameters has been much simplified. New oscillator circuit is designed using the latest achievements in the field of power electronics. It should be noted that due to the use of the IGBT module, it is now possible to control the on-time of the current pulse. The charging voltage of the developed generator is slightly higher than the remaining voltage of the tested varistors. The test varistor does not affect the shape of the pulse, but only affects the current amplitude, which can be adjusted using the charging voltage to the desired value.

5. Conclusion

In the developed circuit of the rectangular pulse current generator, which offers a fundamentally new circuit solution, a high-voltage IGBT module is used as a switching device, which makes it possible to generate test current pulses with an on-time of 50, 1000 and 2000 microseconds on one and the same unit. The developed unit can be used for various testing and technological circuits.

6. Acknowledgments

The research has been carried out under the research contract No. AAAA-Б19-219012990011-9.

References

[1] Glasoe G N and Lebacqz J V 1948 *Pulse Generator* (N.Y.: McGraw-Hill)
[2] Stefanov K S 1967 *Tekhnika vysokikh napryazhenii* [High-voltage engineering] (Leningrad: Energiya) [In Russian]
[3] Beiler M, Beck V and Zaengl V 1989 *Tekhnika vysokikh napyazhenii: teoreticheskie i prakticheskie osnovy* [High-voltage engineering: theoretical and practical application] (Moscow: Energoatomizdat) [In Russian]
[4] Giesselmann M, Palmer B, Neuber A and Donlon J 2005 High voltage impulse generator using HV-IGBTs *2005 IEEE Pulsed Power Conf.* (Monterey: IEEE) 763-6
[5] Tanaka N, Ota K, Iura S, Kusakabe Ya, Nakamura K, Wiesner E and Thal E 2015 Robust

HVIGBT module design against high humidity *PCIM Europe* (Nuremberg: VDE VERLAG GMBH) 368-73

[6] Auerbach F et al. 1999 6.5 kV IGBT-modules *IEEE Industry Applications Conf. Thirty-Forth IAS Annual Meeting* (Phoenix: IEEE) 1770-4

[7] Werber A D et al. 2014 6.5kV RCDC: For increased power density in IGBT-modules *IEEE 26th Int. Symposium on Power Semiconductor Devices & IC's ISPSD* (Waikoloa: IEEE) 35-8

[8] Pourzaki A and Mirzaee H 2009 New high voltage pulse generators *Recent Patents on Electrical & Electronic Engineering* **2(1)** 65-76

[9] Linkov A O 2014 Issledovanie tsepei zashity ot kommutatsionnykh perenapryazhenii vypryamitelnoi ustanovki vozbuzhdeniya electovoza na IBGT-transistorakh [Research of protection circuits against switching overvoltages of a rectifier electric locomotive excitation system on IGBT transistors] *Vestnik IrSTU* **5(88)** 119-26 [In Russian]

[10] Beukes H J, Enslin J H R and Spee R 1996 Active snubber for high power IGBT modules *Proceedings of IEEE. AFRICON* **1** 456-61

[11] Zhang Y, Sobhani S and Chokhawaia R 2000 Snubber considerations for IGBT applications URL: https://www.infineon.com/dgdl/Infineon-Snubber_Considerations_for_IGBT_Applications-Article-v01_00-EN.pdf?fileId=5546d462533600a401535748b5103fe8

[12] Naayagi R T, Shuttleworth R and Forsyth A J 2011 Investigating the effect of snubber capacitor on high power IGBT turn-off *1st Int. Conf. on Electrical Energy Systems* (Newport Beach: IEEE) pp 50-5

[13] Wang C, Zhang Q H and Streaker C 2000 A 12 kV solid state high voltage pulse generator for a bench top PEF machine *Third Int. Power Electronics and Motion Control Conf.* (Beijing: IEEE) pp 1347-52

[14] Moller R, Vocke J, Muhlbeier A and Puffer R Development of a test bench for the investigation of the breakdown voltage of insulation materials at medium frequency rectangular voltages *IEEE Transactions on Power Electronics* **36(3)** 2621-8

[15] Muhammad H and Rashid M H 2017 *Power electronics handbook: devices, circuits, and applications handbook* (Canada: Academic Pressl)

Transformation Justification of types of filter compensating devices in 6 kV mine networks with powerful nonlinear electric receivers

E V Zhilin, D A Prasol and N Yu Savvin

Department of Electric Power Engineering and Automation, Belgorod State Technological University named after V.G. Shukhov, Kostyukov St., 46, Belgorod, 308012, Russia

E-mail: zhilinevg@mail.ru

Abstract. In this paper presents the features and disadvantages of networks of mining enterprises with a voltage of 6 kV with high-power thyristor DC electric drives of hoisting installations, made according to the TP-D system. In most of these networks, negative phenomena are observed due to the presence of powerful nonlinear electrical receivers. Various power filters are widely used to eliminate negative phenomena and reduce additional power losses in mine networks. Moreover, the choice of a specific type of filter and its design is a complex multi-criteria task. The paper presents an in-depth analysis of various types of power filters, and considers factors that influence the choice of FCD. The disadvantages of various existing methods of selecting FCD are revealed. Justification of the types of filter-compensating devices is based on the application of a multi-criteria approach and optimization of the structure of the FCD using the fuzzy set apparatus. The optimal design of FCD for use in 6 kV mine networks with powerful thyristor DC electric drives of hoisting installations, made according to the TP-D system, has been determined. At the same time, the use of the developed FCD design in 6 kV mine networks will allow to compensate for the most pronounced HH currents and voltages, significantly reduce additional power losses, improve the quality of electrical energy and improve the overall electromagnetic environment.

1. Introduction

A feature of power supply systems and electric networks of industrial enterprises is the presence of powerful nonlinear electric receivers, which significantly disrupt the electromagnetic environment and negatively affect many electrical parameters.

For example, distribution networks of medium voltage 6-10 kV are problematic for almost all mining enterprises. This is due to the peculiarities of the technological process of ore extraction by open or closed methods. When mining iron ore in an underground (closed) way, the mine field is opened with skip and crate vertical shafts equipped with hoisting installation (HI). At many operating mining enterprises, electric drives made according to the system of thyristor Converter – DC motor with independent excitation (TP-D) are used as HI drives [1-3], which are powerful nonlinear electric receivers (figure 1). Powerful thyristor electric drives are the reason for generating higher harmonics (HH) of currents and voltages in the high-voltage mine network and create non-sinusoidal dynamic modes of operation. Such an impact on the mine supply network is the cause of the disturbance of the general electromagnetic environment and deterioration of the quality of electrical energy.

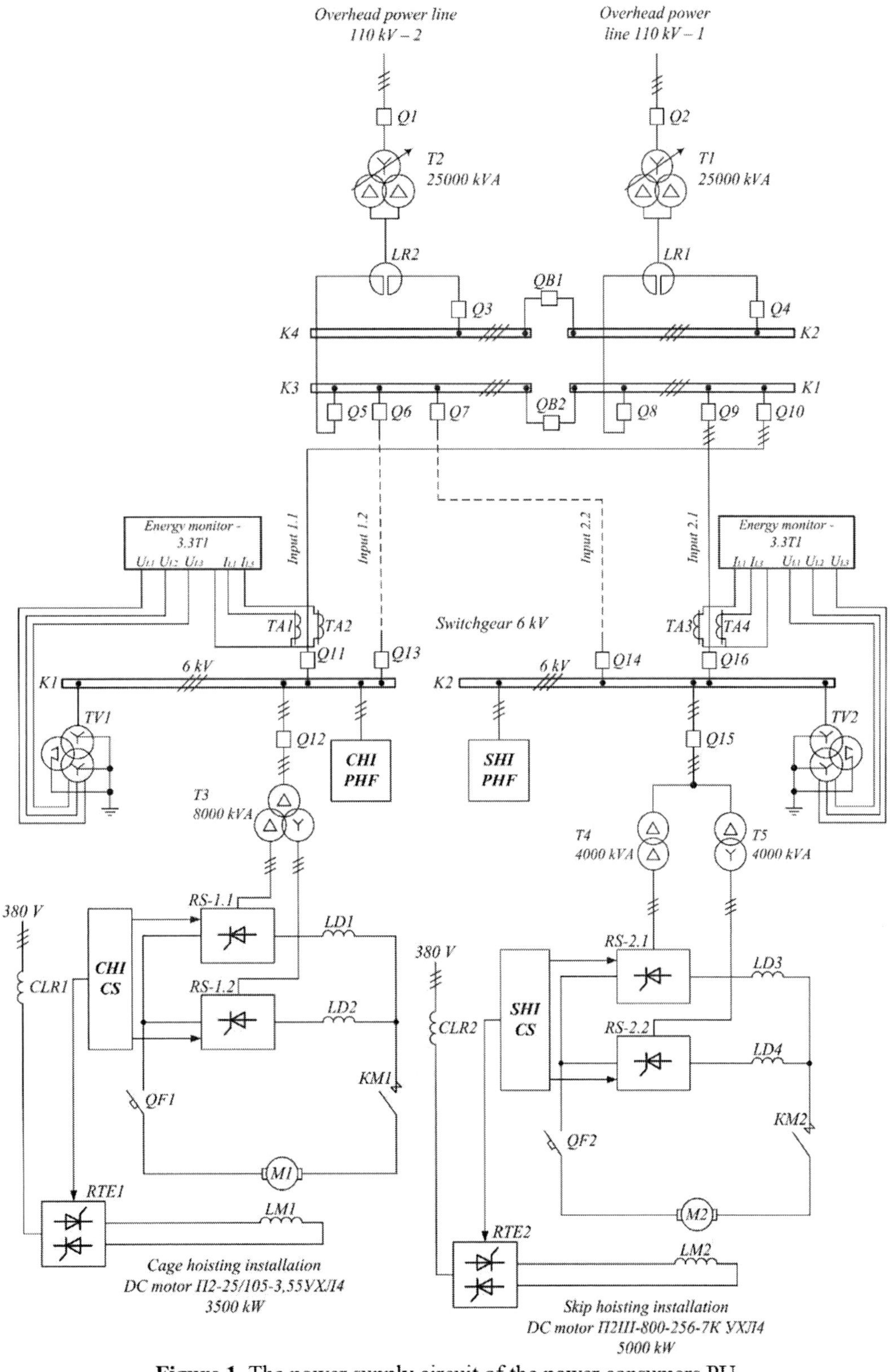

Figure 1. The power supply circuit of the power consumers PU.

2. Power loss study

For the power supply system of the Yakovlev's mine in the Belgorod region, experimental studies, simulation modeling and analytical calculations were carried out, which showed that the most pronounced are the 11, 13, 23, 25, 35 and 37 harmonics of currents and voltages (figure 4). At the same time, the total coefficients of harmonic components of stresses in the 6 kV mine network reach an average of KU=10-17% per hoisting cycle and exceed the normalized values. At the same time, the reliability of the data obtained is confirmed by a high correlation of measurement results, simulation modeling, and analytical calculations, the degree of discrepancy of which does not exceed 10-12 % [1-4].

The presence of higher harmonics of current and voltage in the mine network leads to many negative consequences, one of which is additional loss of power and electricity in the elements of the power supply system.

Additional losses of power and electrical energy in the elements of the power supply system are defined as the sum of additional losses in cable and overhead lines, reactors and transformers from a single HH source. Additional power losses in power transmission lines, reactors, and power transformers of the power supply scheme were determined using the methods presented in [5-7].

In a result of the assessment, the additional power losses caused by the action of HH currents and voltages in the power supply system of the ore mining enterprise amounted to the values shown in figure 2. At the same time, additional power losses in non-sinusoidal modes in comparison with the total losses, respectively, were: for skip HI δadd = 16.99%; for cage HI δadd = 44.3%.

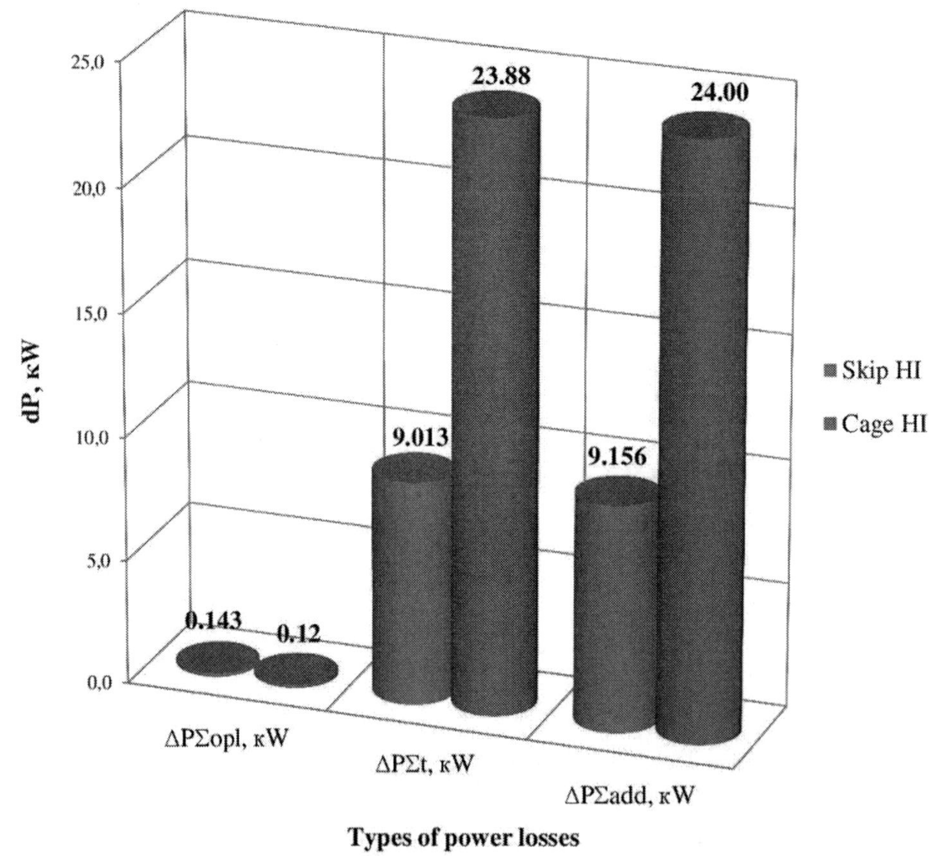

Figure 2. Histograms of additional power losses in the mine power supply system at non-sinusoidal modes.

3. Technical devices for reducing power losses

Now, to compensate for HH currents and voltages, the main distribution is organizational, technical measures and measures to improve the systems of calculation and technical accounting of electricity, as well as various types of technical means.

Various types of power filters are widely used to compensate for HH currents and voltages and reduce additional power losses [8-13]:

- passive harmonic filters (PHF);
- active harmonic filters (AHF);
- hybrid filters (GF or active-passive filters);
- passive filters with special settings (PFSN).

In addition to attenuating the higher harmonics of currents and voltages, power filters can perform the functions of reactive power compensation and voltage regulation at the connection point.

To compensate for the HH in the 6 kV mine network with thyristor converters made according to the 12-pulse rectification scheme, the installation of a FCD consisting of:

- PHF tuned to 11, 13, 23 and 25 harmonics can be provided;
- AHF tuned to the entire harmonic series;
- combinations of PHF tuned to 11, 13 harmonics,

and AHF tuned to other harmonics (HF – hybrid filter).

An important feature is that due to the technical features of mine power supply systems and the low quality of electricity, the installation of FCD is necessary in the 6 kV network directly near non-linear loads. This leads to a significant cost of the FCD design.

Based on the analysis of numerous works, the specific average cost of PHF is about 380.9 rubles/kvar, AHF – 7162.6 rubles/kvar, HF – 6994.2 rubles/kvar [14 – 18].

Taking into account the significant power of electric drives of HI, as well as the specific average cost of power filters, the most appropriate and cost-effective option for compensating HH currents and voltages in the 6 kV mine network is to install filter-compensating devices consisting of passive harmonic filters.

In the electric power industry, the following PHF configurations are used (figure 3): first, second, and third order filters and C-type filters. A comparative analysis of passive harmonic filters, their main characteristics, scope and purpose, advantages and disadvantages are presented in table 1.

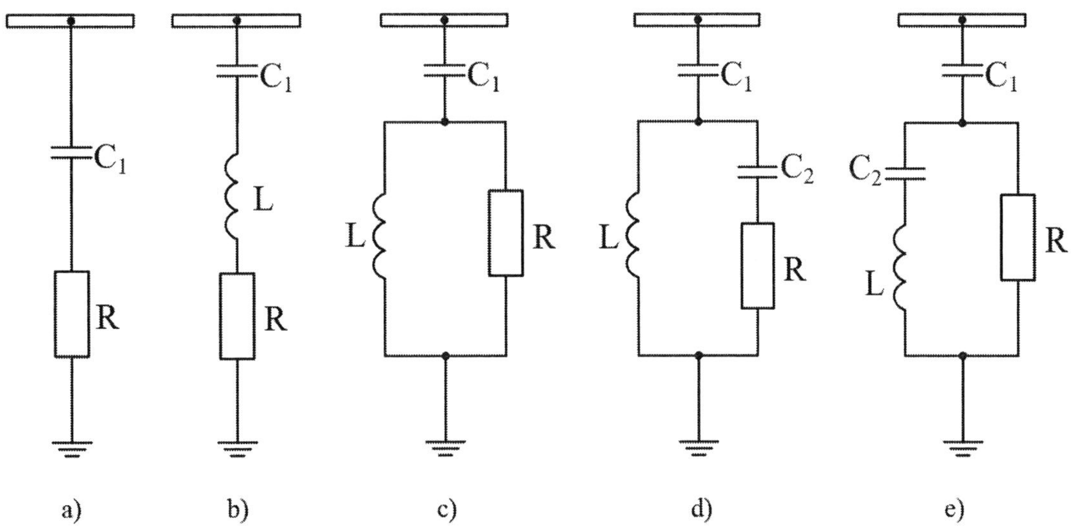

Figure 3. Configurations of passive harmonic filters: a – first order PHF; b – resonant or narrowband PHF; c – second order PHF; d – third order PHF; e – C-type PHF.

To compensate for HH currents and voltages in the 6 kV mine network, almost equivalent versions of FCD can be used, consisting of various configurations of PHF in their composition. To select the best version of the FCD, a number of important criteria were determined, which each proposed type of FCD must meet to some extent. The choice of a specific variant of the FCD was determined by solving a multi-criteria optimization problem, which means choosing the best option simultaneously according to many criteria.

4. Selection of the optimal configuration of technical devices

Many existing methods for selecting the location and parameters of the FCD in power supply systems are based on minimizing active power losses and reduced costs [3, 19-21]. The proposed solutions have a number of disadvantages: not taking into account the properties of the frequency characteristics of the "filter – external network"; power losses in the FCD. The calculations use deterministic values of parameters of power supply systems and empirical formulas for accounting for changes in network resistance depending on the HH numbers. a more accurate justification for the use of a specific type of FCD is provided by a multi-criteria approach and optimization of the FCD structure.

The following important criteria are applied to each specific type of FCD installed in a 6 kV mine network with high-power thyristor electric drives for compensation of HH currents and voltages:

- capital expenditures, including installation and installation of FCD and operating costs;
- properties of the "filter – external network" frequency characteristics in terms of compensation of canonical harmonics and amplification of interharmonics and subharmonics;
- minimization of additional power losses in the network;
- power losses in the elements of FCD;
- ability to compensate for reactive power on the main harmonic.

Capital expenditures depend on the number and configuration of PHFs in the FCD, on the capacity and total capacity of capacitors, and on the manufacturer.

Properties of the frequency characteristics of the "filter – external network" significantly expands the idea of its operating modes with different composition and variations of the load at higher harmonics. This criterion allows us to evaluate not only the efficiency of compensation of HH currents and voltages, but also warns about possible resonances of current and voltage, amplification of interharmonics and subharmonics [4, 6, 9, 22, 23].

All other things being equal, it is important to consider the power loss in the FCD, consisting of various configurations of the PHF. Power loss estimation at the fundamental and harmonic frequencies for the SHPF can be performed similarly to a resonant PHF tuned to a single frequency.

The ability to compensate for the reactive power of the FCD on the main harmonic is taken into account at the design stage. The parameters of the PHF configurations in the FCD were calculated taking into account reactive power compensation in accordance with the required power factor.

To compensate for HH currents and voltages in the 6 kV mine network, four alternative equivalent types of FCS were proposed and calculated, consisting of various PF configurations:

- resonant PHFs tuned to the 11th and 13th harmonics, and second-order SPF tuned to compensate for the HH, starting from the 23rd;
- Dual tuning PHF for the 11th and 13th harmonics and second-order SPF tuned for HH compensation, starting from the 23rd;
- A C-type PHF tuned to the 11th and 13th harmonics, and a second-order SPF tuned to compensate for the HH, starting from the 23rd;
- resonant PHFs tuned to 11, 13, 23, 25, 35, and 37 harmonics.

The problem of choosing the optimal type of FCD is solved using the fuzzy set apparatus with the same and different importance of the proposed criteria. The results of solving the problem showed that the most optimal type of FCD is the first alternative (resonant PHFs tuned to the 11th and 13th harmonics, and a second-order PHF tuned to compensate for the SH, starting from the 23rd).

Table 1. Comparative analysis of passive harmonic filter configurations.

Types of filter	PFG First load	Resonant or narrowband PFG	PFG Second load	PFG Third load	C-type PFG
Brief description	A parallel resonant circuit with the inductance of the external network.	Adjusts to the frequency of a specific harmonic.	Simultaneous suppression of several harmonics.	Simultaneous suppression of several harmonics. It is a device consisting of two capacitors C1, C2, a reactor L and an active resistance R.	Inclusion in the cross branch of the second-order filter of the capacitor C2. The capacitance of the capacitor C2 must be such that the resonant frequency of the oscillating circuit LC2 coincides with the frequency of the main harmonic.
Quality factor	–	30–50 30–60	0.5-5 Higher values are used when it is necessary to attenuate closely spaced harmonics. The quality factor is chosen such as to ensure that a given characteristic is obtained in a certain frequency range.	0.5-5 Higher values are used when it is necessary to attenuate closely spaced harmonics. The quality factor is chosen such as to ensure that a given characteristic is obtained in a certain frequency range.	3-5 The quality factor is chosen such as to ensure that a given characteristic is obtained in a certain frequency range. The frequency response of the filter can be varied by changing the resistance of the damping resistor R.
Advantages	Simplicity	Effectively weakens harmonic components whose frequencies are close to the resonant frequency. Simplicity, economy and reliability. It is widely used in networks with nonlinear loads.	It is convenient in operation, in comparison with the first order filter, it has lower losses at the fundamental frequency. Provides a reduction in the values of harmonic voltage components in electrical networks in a wider frequency range.	Small (lower) losses at the fundamental frequency in comparison with a second-order filter, provides a decrease in the values of harmonic components of voltage in electrical networks in a wider frequency range.	Lower power loss at the fundamental harmonic frequency compared to narrow-band filters of the second and third orders. Providing reduction of values of harmonic components of voltage in electric networks in a wider frequency range. Provide reactive power compensation at the frequency of the main harmonic and attenuation of higher harmonics.
Disadvantages	It has high losses at the fundamental frequency due to the use of a high power capacitor, which limits its use.	Efficiency decreases when the harmonic composition of currents and voltages and network parameters change. The possibility of resonance in a parallel oscillatory circuit formed by the filter and the inductance of the supply network, at frequencies close to the frequencies of higher harmonics.	It is widely used in industry. The losses at the fundamental frequency are large compared to a narrow-band filter.	Reducing the possibility of obtaining the required frequency response in the attenuation band. It is rarely used in industry due to its high cost and complexity.	Fine tuning to the fundamental frequency, a wide range of capacitor ratings and, accordingly, a large total capacitance. High sensitivity to changes in fundamental frequency and deviations of parameters of elements. The degree of attenuation of harmonics, the frequencies of which exceed the resonant frequency, are small.

The parameters of all resonant PHFS were determined on the basis of canonical schemes of reactive bipolar circuits using a matrix of hybrid circuit parameters through the system resistance and filter [9]. For the obtained optimal type of FCD, the parameters of resonant PHFS were calculated for the most pronounced canonical 11th and 13th higher harmonics. The SPFS were tuned to the 23rd harmonic component, the most pronounced harmonic in the high-frequency region after the 13th harmonic. The results of calculations of the parameters of resonant and broadband PHF are presented in table 2.

Table 2. Parameters of resonant and broadband PHF.

HI type	P, kW	QK, kvar	H	HH number	ki	Li, mGн	Ci, mkF	Qi, kvar	Q	R, Om
Resonant PF										
Skip HI	1381	807.89	0.274	11	1.752	1.817	46.09	193.1	60	0.1047
				13	1.904	1.672	35.87	150.0		0.1138
Cage HI	2246	788.35	0.312	11	1.537	2.071	40.43	169.5		0.1193
				13	1.674	1.905	31.47	131.6		0.1297
Broadband PF 2nd order										
Skip HI	–	807.89	–	23	–	0.268	71.43	–	1	1.937
Cage HI	–	788.35	–	23	–	0.275	69.71	–	1	1.985

Note: H – normalization coefficient; Q – q factor

Installation of the FCD is supposed to be in the SG-6 kV of the building of hoisting machines, on the side of the higher voltage of matching transformers in the form of separate additional cells (figure 1).

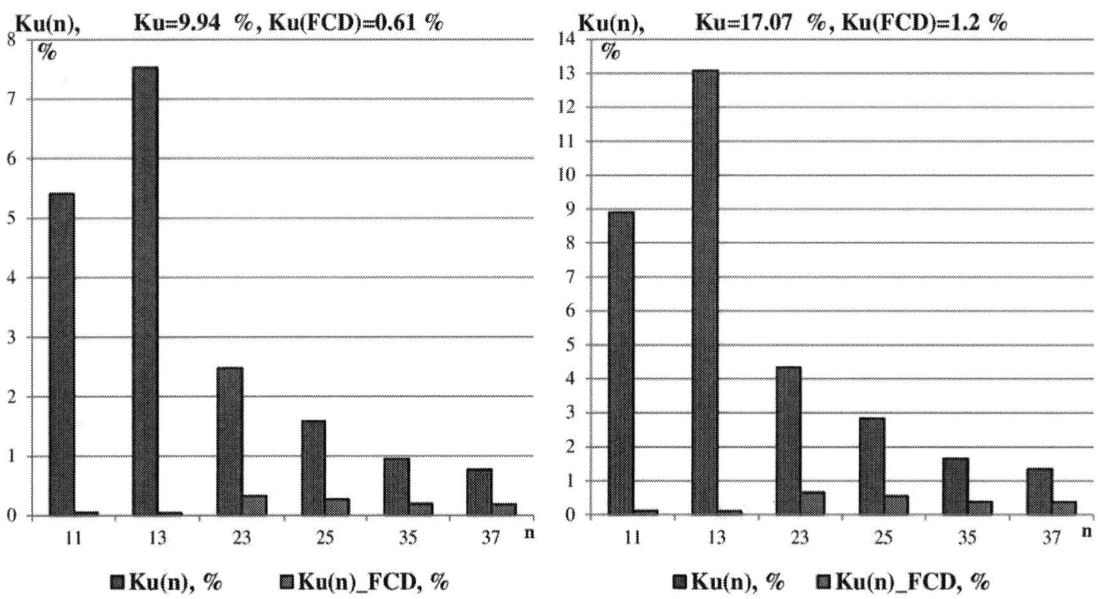

a b

Figure 4. Spectra of harmonic voltages in the underground network of 6 kV: a – Skip HI; b – Cage HI.

As follows from the simulation results, the currents and voltages in the phases of power supply systems of HI take forms close to sinusoidal, and the total coefficients of the harmonic components of the voltage are equal: for skip HI: $K_U = 0.61$ %; for cage HI: $K_U = 1.2$ %.

The histogram of additional power losses in the elements of the power supply systems of skip and cage HI when installing the FCD is shown in figure 5 a – skip HI; b – cage HI.

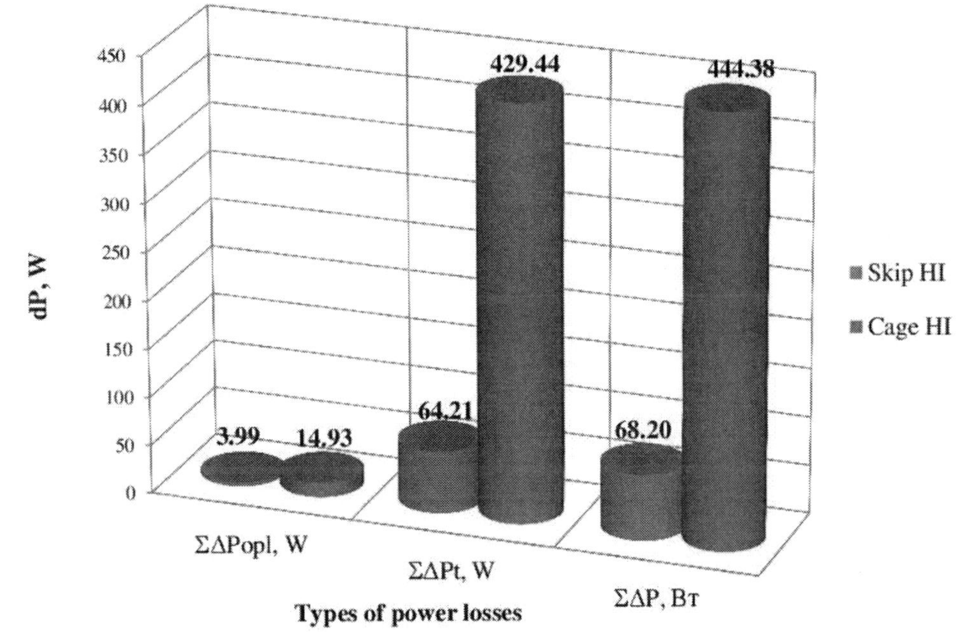

Figure 5. Histogram of additional power losses in the power supply systems of skip and cage HI during installation of FCD.

As follows from the simulation results, additional power losses caused by non-sinusoidal operating modes during the installation of the FCD decreased by more than 90% and amounted to ΔP = 0.51 kW. Thus, compensation of HH currents and voltages helps to reduce additional power losses in the elements of power supply systems for skip and cage HI under non-sinusoidal operating modes.

5. Conclusions

Most networks of mining enterprises with a voltage of 6 kV are characterized by a violation of the General electromagnetic environment, low quality of electric energy in terms of non-sinusoidal currents and voltages, and significant additional power losses in non-sinusoidal modes. The reason for these phenomena are powerful thyristor DC electric drives of hoisting installations, made according to the TP-D system.

Various power filters are widely used to compensate for HH currents and voltages and reduce additional power losses in mine networks. Moreover, the choice of a specific type of filter and its design is a complex multi-criteria task.

The choice and justification of FCD types in 6 kV mine networks is based on a deep analysis of various types of power filters, taking into account a large number of factors and studying the methods of selecting FCD. In addition, a multi-criteria approach was applied to optimize the structure of the FCD, using the fuzzy set apparatus. As a result, the optimal design of the FCD was determined for use in 6 kV mine networks with powerful thyristor DC electric drives of hoisting installations made using the TP-D system (resonant PHFS tuned to the 11th and 13th harmonics, and second-order SHPF tuned to compensate for the HH, starting from the 23rd).

The use of the optimal design of the FCD in 6 kV mine networks with powerful thyristor DC electric drives of hoisting installations will allow you to compensate for the most pronounced HH currents and voltages, significantly reduce additional power losses, improve the quality of electrical energy and improve the overall electromagnetic environment.

6. Acknowledgments
The article was prepared within development program of the Flagship Regional University on the basis of Belgorod State Technological University named after V.G. Shukhov References.

References
[1] Averbux M A, Prasol D A and Xvorostenko S V 2017 E`ksperimental`naya ocenka parametrov rezhimov v vy`sokovol`tny`x rudnichny`x setyax s moshhny`mi nelinejny`mi e`lektropriemnikami [Experimental evaluation of the parameters of modes in high-voltage mine networks with powerful nonlinear electrical receivers] *Vestnik IrGTU* **21(2)** 75-84 [In Russian]
[2] Averbukh M A and Prasol D A 2018 Influence of high-power nonlinear consumers on electric energy losses in mining high-voltage power line *IOP Conf. Ser.: Mater. Sci. Eng.* **327** 052028
[3] Averbukh M A, Pogorelov A V and Prasol D A 2018 Minimizaciya poter` moshhnosti pri nesinusoidal`ny`x rezhimax v vy`sokovol`tny`x rudnichny`x setyax [Minimization of power losses during non-sinusoidal modes in high-voltage mine networks] *Promy`shlennaya E`nergetika* **7** 38-46 [In Russian]
[4] Averbukh M A, Prasol D A and Zhilin E V 2019 Special aspects of application of filter-compensating devices in power supply systems with non-linear electrical receivers *IOP Conf. Ser.: Mater. Sci. Eng.* **643** 012087
[5] Xu Y, Dong Z Y, Wong K P, Liu E and Yue B 2013 Optimal capacitor placement to distribution transformers for power loss reduction in radial distribution systems *IEEE Transactions on Power Systems* **28(4)** 4072-9
[6] Temerbaev S A and Dovgun V P 2014 Improvement of power quality in distributed generation systems using hybrid power filters *16th Int. Conf. on Harmonics and Quality of Power* (Bucharest: IEEE) pp 694-8
[7] Zhilin E V and Prasol DA 2020 Optimization approach to reduce the negative impact of high harmonic components of currents and voltages in high-voltage mine networks *IOP Conf. Ser.: Mater. Sci. Eng.* **791** 012038
[8] Hu H, He Z and Gao S 2015 Passive filter design for china high-speed railway with considering harmonic resonance and characteristic harmonics *IEEE Transactions on Power Delivery.* **30(1)** 505-14
[9] Sreenivasarao D, Agarwal P and Das B 2012 Neutral current compensation in three-phase, four-wire systems: a review *Electric Power Systems Research* **86** 170-80
[10] Das J C 2015 *Power System Harmonics and Passive Filter Designs* (Hoboken: John Wiley & Sons)
[11] Hammad A 2008 Taming harmonics in Switzerland *Transmission & Distribution World* **60(10)** 42-7
[12] Agrawal A and Agrawal P 2014 Comparison of various configurations of hybrid active filter with three different control strategies *Int. J. of Eng. Research & Technology* **3(5)** 1672-8
[13] Fujita H and Akagi H 1990 A practical approach to harmonic compensation in power systems series connection of passive and active filters *Conf. Record of the 1990 IEEE Industry Appl. Society Annual Meeting Transactions on Industry Applications.* vol 2 (Seattle: IEEE) pp 1107-12
[14] Aleem S H E A, Zobaa A F and Balci M E 2017 Optimal resonance-free third-order high-pass filters based on minimization of the total cost of the filters using Crow Search Algorithm *Electric Power Systems Research* **151** 381-94
[15] Comsys Active Dynamic Harmonic Filters URL https://comsys.se/ADF/en/

[16] Zamyatin E O, Voytyuk I N and Zamyatina E N 2019 Filter Compensating devices connection technique *2019 IEEE Conf. of Russian Young Researchers in Electrical and Electronic Eng.* (Saint Petersburg and Moscow: IEEE) pp 1114-6
[17] Active harmonic filters URL https://activeharmonicfilter.net/
[18] Shandrigin D A, Egorov D E, Novikov V V and Dovgun V P 2019 Analiz effektivnosti passivnykh fil'trokompensiruyushchikh ustroistv dlya sistem tyagovogo elektrosnabzheniya [Analysis of the effectiveness of passive filter-devices for traction power supply]. Doklady Akademii nauk vysshei shkoly Rossiiskoi Federatsii [Proceedings of the Russian higher school Academy of sciences] **1(42)** 91-103 [In Russian]
[19] Averbukh M A, Zhilin E V, Sizganova E Yu and Abdulvaxxab M V 2020 Selection of nodes for connecting distributed generation in the networks of the Republic of Iraq *Electrical Systems and Complexes* **2(47)** 4-10 [In Russian]
[20] Averbukh M A, Zhilin E V and Prokopishin D I 2018 Electric Power Loss Minimization in Power Supply Systems for the Suburban Houses *Problemele Energeticii Regionale* **2(37)** 31-8
[21] Balci M E and Sakar S 2014 Optimal design of singletuned passive filters to minimize harmonic loss factor *Middle East J. of Scientific Research* **21(11)** 2149-55
[22] Kartashev I I, Tul`skij V N and Shamonov R G 2017 *Upravlenie kachestvom e`lektroe`nergii* [Power quality management] (Moscow: ME`I) [In Russian]
[23] Dakhno A V, Smirnov Yu P, Mashkov A V and Potapenko O V (2018) To an assessment of size of diffusive losses of coolant through a rubber camera – cover *Chemical Bulletin* **1(1)** 18-26

IOP Conf. Series: Materials Science and Engineering 1089 (2021) 012020 doi:10.1088/1757-899X/1089/1/012020

Installation of absorption refrigerating machines for the needs of conditioning of the Armada Mall

V Y Sokolov and T Y Sokolova

Department of Electrical and Heat Power Engineering , Orenburg state University, Shevchenko St., 28, Orenburg, 460005, Russia

E-mail: teploosu@mail.ru, sokolova_25@mail.ru

Abstract. The energy strategy of Russia for the period up to 2020 noted that in order to ensure energy security it is necessary to solve two priority issues: firstly, the modernization of the obsolete and physically worn out technological base of the fuel and energy complex using the best domestic and relevant foreign technologies for our conditions; secondly, it requires a change in the consumption pattern and distribution of production of fuel and energy resources. The work analyzed the state and considered the characteristics of the energy source. The concept of increasing the efficiency of the thermal power plant has been developed. As a result of analysis of existing plants for additional generation of cold from a more heated power generator, a scheme and theoretical cycle of absorption bromide of a lithium refrigerating machine is made. The results were obtained on the expected annual savings of the Armada Mall operating costs at the application of lithium refrigeration machine absorb bromide for the annual cold demand of 3,4 million kWh.

1. Introduction

Currently, steam compression chillers (steam compression chillers) are used in the air conditioning system of the Armada Mall, which cool water directed to the fan coil system [1, 2].

As a systemic source of electricity, CHP plants play a significant role in the structure of generating capacities and electricity production in most of the country's regional and combined energy systems. Of course, having played a large positive role in the previous development of power systems, the existing thermal power plants are seriously outdated, and now a significant part of their equipment needs to be dismantled, modernized or replaced with more modern equipment [3, 7].

The economic efficiency of further use of the accumulated capacity of existing thermal power plants strongly depends on such «systemic» factors as the growth rate of electricity and heat consumption in the respective regions and the rate of aging of the capacity of existing thermal power plants in these regions. At the same time, the efficiency of heating both in the local context and at the system level is highly dependent on external factors, such as fuel prices, technical and economic indicators of all types of existing and new sources of heat and electricity production and transportation, as well as the amount of CO_2 emissions charges [4].

All researchers agree that heating systems should be seriously modernized. The changes being developed today in the model of heat and electricity markets and in the methodology for planning energy development should create the right incentives for market participants to create reliable, balanced energy resources for all types of flexible and economical systems.

Steam compressors use electric energy to produce cold. In contrast, absorption refrigeration machines use heat energy (in hot water or steam) to produce cold, and electric energy primarily for the operation of circulation pumps. According to the estimates of the manufacturers of refrigerating machines with the same refrigerating capacity, the electric energy consumption of absorption refrigerating machines is 21% of the electric power consumption of vapor compression refrigerating machines [5, 6, 8]. In the middle years of the last century, the introduction of a high-quality method of regulation was not facilitated by low prices for fuel and energy resources and the lack of an urgent need for energy conservation in the energy industry. In addition, the implementation of quantitative and qualitative-quantitative methods for regulating thermal load was hindered by the absence or imperfection of automatic control devices.

Therefore, these methods of regulating the thermal load of energy sources are not widely used in domestic heat supply. At present, due to radically changed economic conditions and new technical capabilities, the introduction of these methods of regulation in heating systems allows to achieve a significant energy saving effect and improve the quality of energy supply [6-8]. Thus, the efficiency of most domestic heating systems continues to remain at a rather low level, however, CHPs allow the city to produce electricity and other energy carriers with specific fuel costs that are not achievable outside the heating cycles, and in the amount necessary to meet all urban needs. It is only necessary to develop a set of measures for the modernization of urban heating systems and correctly select their structure, composition of the CHP equipment corresponding to thermal and electrical loads. This problem has become particularly urgent in recent decades due to the transition to a non-volatile economy and the constant increase in energy prices. The analysis of the state of domestic urban thermal systems and experience in their improvement make it possible to formulate the following main conclusions on the concept of increasing their efficiency due to structural and technological modernization.

2. Materials and Methods

When replacing the chillers of steam compression chillers with absorptive chillers, the win-win principle between Armada Mall and PJSC «T Plus» is achieved by:[8-10]

- increase in sales of heat energy in the summer season from the Sakmarskaya CHPP (increase in branch revenue, increase in the base load of the Sakmarskaya CHPP and, as a result, improvement in the equipment economic indicators) - PROFIT for PJSC «T Plus»;
- reduction of the annual operating costs of the air conditioning system - PROFIT for Armada Group of Companies.

Armada Mall is connected to the water heating network from the Sakmarskaya CHPP (M1 highway with a jumper to M3), and currently consumes heat energy only for heating needs in the winter season (the hot water supply of the complex is provided by electric water heaters). In order to reduce capital expenditures, this leads to the use of hot-water absorption chillers with the supply of thermal energy in hot water in the summer season through the existing heating network. Since the calculated temperature of hot water supplied to absorption refrigeration machines (120 ^{0}C) differs from the water temperature in summer in the supply lines from the Sakmarskaya CHPP (75 ^{0}C), it is planned to build a one-pipe section of the heating network from collector power station to rack № 126 (branch to the Armada Mall) with a length of 1 500 m and a nominal diameter of 150 mm along the existing corridors, supports and racks to the former Silk weaving factory. In the part of the return pipeline, it is supposed to use the existing line [9-12].

The circuit diagram of the connection is shown in Figure 1.

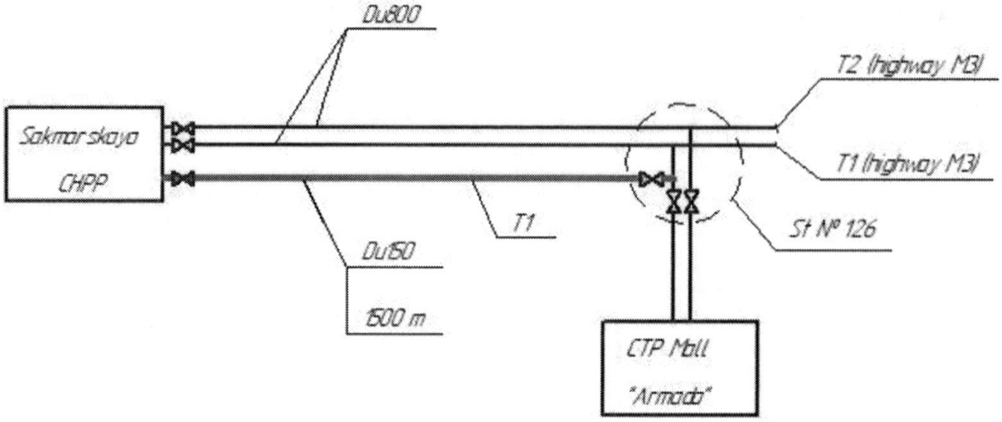

Figure 1.Connection diagram of the Armada Mall to the Sakmarskaya CHPP.

Monthly estimated heat consumption for the generation of cold absorption hot-water refrigeration machines for the needs of the Armada Mall air conditioning and ventilation systems (calculated based on the actual energy consumption provided by the mall for air conditioning and ventilation for 2018) are given in table 1.

Information about the chillers used in the Armada Mall are shown in table 2.

Table 1. Heat consumption for the production of cold Armada Mall.

Indicator	May	June	July	August	September	TOTAL for the summer period:
Heat consumption for cold production, Armada Mall, Gcal	240.96	696.31	1024.63	971.41	316.23	3249.54
Estimated average hourly heat load , Armada Mall, Gcal / h						0.98
Estimated maximum achieved hourly heat load of hard currency of the Armada Mall, Gcal / h						4.77

Table 2. Information from refrigeration machines used in the Armada Mall.

Type, brand of refrigeration machine (chiller)	The number of chillers, pieces.	Cooling power, kW	Total refrigerating power, kW	Serviced air-conditioned area, m^2
MBC-07182	12	554	6648.0	102183.40
EWLDI-SS 600	1	595	596.0	46926.00
EWLDI-SS C60	3	1033	3099.0	46926.00
Total:			10343.0	149109.40

The estimated capital costs for replacing the chillers of steam compression chillers with absorption chillers with the supply of thermal energy in hot water from the Sakmarskaya CHPP are given in table 3 and table 4.

Table 3. Activities on heating networks (costs of PJSC «T Plus»).

Name	Amount	Unit of measurement	Cost, thousand rubles without VAT		Note
			one unit of measure, at initial prices	total volume, in prices of the 1st quarter of 2017	
Construction of a heating network section 1Du150, overhead laying in mineral wool insulation with a cover layer of galvanized steel using existing supports	1.5	km	5500.00	9817.50	According to the estimated cost of the construction (re-construction) of heating networks of BE Engineering PJSC «T Plus», taking into account the conversion coefficient from the prices of 2015 to the prices of the 1st quarter of 2018
Other and unaccounted expenses	–	–	–	981.75	Accepted at a rate of 10% of the cost of building a heat network section
TOTAL				10799.25	

3. Results

Armada Mall here is considered as a collector consumer with the sale of thermal energy in hot water at the collector rate. In order to assess the economic efficiency of connecting the refrigeration load of the Armada Mall to the Sakmarskaya CHPP, the following options were calculated:

1. The entire heat load of the hard drive of the Armada Mall is connected to the Sakmarskaya CHPP at a time. In this case, the annual net supply of heat energy in hot water increases by the amount indicated in table 1.
2. The thermal load of the hard drive of the Armada Mall is connected to the Sakmarskaya CHPP 2 in equal turns for 2 years. At the same time, during the connection period, the useful supply of thermal energy in hot water from the Sakmarskaya CHPP annually increases by 1/2 the value indicated in Table 1.
3. The heat load of the hard drive of the Armada Mall is connected to the Sakmarskaya CHPP in 3 equal stages for 3 years. At the same time, during the connection period, the net supply of thermal energy in hot water from the Sakmarskaya CHPP increases annually by 1/3 of the value indicated in Table 1.
4. The heat load of the hard drive of the Armada Mall is connected to the Sakmarskaya CHPP 4 in equal bursts for 4 years. At the same time, during the connection period, the useful supply of thermal energy in hot water from the Sakmarskaya CHPP annually increases by 1/4 of the value indicated in table 1.

At the same time, the capital costs of PJSC «T Plus» for all options are shown in section 1 of tables 3 and 4.

Indicators of economic efficiency options are shown in table 5.

The results of the analysis of the annual savings in operating costs of Armada Mall when using absorption refrigerating machines of vapor compression machines for the annual need for cold in the amount of 3.4 million kW/h are shown in table 6.

Table 4. Activities on the client side (costs Armada Mall).

Name	Cost, thousand rubles without VAT: one unit of measure, at initial prices	Cost, thousand rubles without VAT: total volume, in prices of the 1st quarter of 2017	Note
Acquisition of absorption bromistolytic refrigeration machine in hot wate(5000 kW)	16.00	80010.00	Based on the data of ARM suppliers, the specific cost of ARM on hot water was accepted in the amount of 254 $ per 1 kW of cold production, the dollar exchange rate in the first quarter of 2018 was assumed to be 63 rubles.
Purchase of a cooling tower for absorption bromistolytic refrigeration machine	2660.00	2793.00	According to ARM suppliers at prices in 2019, taking into account the conversion coefficient from prices in 2018 to prices in the first quarter of 2018
Delivery and storage costs absorption bromistolytic refrigeration machine	321.12	321.12	Accepted at a rate of 12% of the cost of equipment
The cost of design survey works absorption bromistolytic refrigeration machine	299.71	299.71	Accepted at a rate of 10% of the cost of equipment with delivery
Construction installation costs absorption bromistolytic refrigeration machine	899.14	899.14	Accepted at a rate of 30% of the cost of equipment with delivery
The cost of commissioning absorption bromistolytic refrigeration machine	630.00	630.00	Accepted in the amount of 10 000 $ according to suppliers, the dollar exchange rate in the first quarter of 2018 is assumed to be 63 rubles.
Other and unaccounted expenses on the client side	–	8495.30	Accepted in the amount of 10% of the total cost of equipment, delivery, the cost of design survey works, construction installation costs ,the cost of commissioning absorption bromistolytic refrigeration machine
TOTAL		93448.27	

Table 5. Indicators of economic efficiency.

Index	Option 1	Option 2	Option 3	Option 4
Capital expenditures of PJSC «T Plus», mil.rub.	10.80	10.80	10.80	10.80
Net present value, NPV, mil.rub.	2.38	0.89	0.09	–0.64
Internal rate of return, IRR, %	21.09	17.44	15.62	14.03
Payback period is simple, PP, years	2.37	3.14	3.63	4.13
Discount payback period, DPP, years	5.66	7.99	9.76	>14

Table 6. Analysis of annual savings when using ARM instead of VRM.

Index	VRM	ARM
Annual energy costs, thousand rubles[1]	14985.00	3146.85
Annual costs of thermal energy, thousand rubles[2]	0.00	3058.50
Annual savings from the use of ARM, million rubles		8779.65
Capital investments, million rubles [3]		93448.27
Simple payback period due to savings, years		10.6

Note
[1] With an estimated electricity tariff of 4.5 rubles / kW-h without VAT.
[2] With an estimated collector tariff for thermal energy in hot water from Sakmarskaya CHPP, 941.077 rubles/Gcal, excluding VAT.
[3] According to section 2 of tables 3 and 4.

4. Conclusion

The greatest effect for PJSC «T Plus» is achieved by connecting the entire heat load of the hard drive of Armada Mall at a time, while the costs of activities on heating networks are borne by T Plus and the costs of acquisition, installation and commissioning of absorption refrigeration machines - at Armada Group of Companies.

The payback period of absorption chillers for Armada Group due to energy savings is 10 years, and this is subject to the installation of 1 absorption chiller with an installed capacity of 5 MW. By installing several absorption refrigeration machines (and putting the refrigeration load in stages), the costs will increase, and accordingly, the payback period will increase, which can frighten off a potential client.

All researchers agree that heating systems should be seriously modernized. Changes in the model of heat and electricity markets and in the energy planning methodology being developed today should provide market participants with the right incentives to create reliable, balanced energy resources for all types, maneuverable and economical systems.

A significant package of measures to increase the efficiency of the existing CHPP is proposed:

- switching to the CHPP, loads of own and municipal boiler houses, including combining most of the city load of the WAN with jumpers of small diameters (reduction of specific fuel consumption for heat and electricity, increase in the utilization factor of installed capacity);
- formation of favorable proposals on the cost of steam and hot water for enterprises that have their own production boilers, for example, for the price of the fuel component in the production of heat in their own boilers;
- attracting, in cooperation with municipalities, investors for the construction of energy-intensive means (steam production, refrigerators with absorption machines, production with drying processes), primarily to increase the demand for heat in the summer;
- legal division of CHPP with low power generation into the CHP itself and boiler house for accounting in tariffs for heat of real fuel costs;
- transfer of CHPP to operation according to thermal schedule, up to removal of blades in LPC of steam turbines and liquidation of cooling towers (elimination of loss-making modes of equipment operation);
- reduction of the heating network temperature schedule, switching to quantitative control;
- decommissioning or preservation of excessive worst power equipment, including those involved in heating modes, but used less than 1000-2000 hours per year, with compensation of thermal peaks by peak boilers or boilers put into peak mode (significant reduction of operating costs for unloaded equipment);
- superstructure of CHPP with gas turbines (reduction of specific fuel consumption, significant increase in the profitability of summer regimes, additional revenues from the sale of electricity and participation in the coverage of consumption peaks) [13-15].

Most residential urban thermal systems use central heat load quality control on heat sources, which involves changing the temperature of the refrigerant depending on the temperature of the outside air with a constant flow of refrigerant [16].

The main advantage of quality control of heat load is stable hydrocyclic mode of heat supply system. Thus, the efficiency of most domestic heating systems continues to remain at a rather low level, however, CHPs allow the city to produce electricity and other energy carriers with specific fuel costs that are not achievable outside the heating cycles, and in the amount necessary to meet all urban needs. It is only necessary to develop a set of measures for the modernization of urban heating systems and correctly select their structure, composition of the CHPH equipment corresponding to thermal and electrical loads.

References

[1] Sartor K, Quoilin S and Dewallef P 2014 Simulation and optimization of a CHP biomass plant and district heating network *Applied Energy* **130** 474-83
[2] Ahmadi G, Toghraie D and Akbari O A 2018 Technical and environmental analysis of repowering the existing CHP system in a petrochemical plant: A Case Study *Energy* **159** 937-49
[3] Keynia F 2018 An optimal design to provide combined cooling, heating, and power of residential buildings *Int. J. Model. Simulation* **38(4)** 216-31
[4] Guelpa E, Barbero G, Sciacovelli A and Verda V 2017 Peak-shaving in district heating systems through optimal management of the thermal request of buildings *Energy* **137** 706-14
[5] Esen M and Yuksel T 2013 Experimental evaluation of using various renewable energy sources for heating a greenhouse *Energy Build* **65** 340-51
[6] Li Q, Amibe D A and Müller N 2011 Using water vapor as refrigerant in multistage variable speed turbo compressor to improve seasonal energy efficiency ratio of air conditioning *Int. J. of Air-Conditioning and Refrigeration* **19(02)** 131-40
[7] Chamoun M, Rulliere R, Haberschill P and Peureux J L 2014 Experimental and numerical investigations of a new high temperature heat pump for industrial heat recovery using water as refrigerant *Int. J. of Refrigeration* **44** 177-88
[8] Monné C, Alonsoa S, Palacínb F and Serra L 2011 Monitoring and simulation of an existing solar powered absorption cooling system in Zaragoza (Spain) *Applied Thermal Engineering* **31(1)** 28-5
[9] Mann J 1994 The spatial structure of neutral atmospheric surface-layer turbulence *J. Fluid Mechanics* **273** 141-68
[10] Pan Q W, Wang R Z and Wang L W 2015 Comparaison de différentes techniques de récupérations de chaleur appliquées dans un système frigorifique à adsorption *Int. J. of Refrigeration* **55** 37-48
[11] Cers A, Turlays D and Zeltinsh N 2013 Recovery of the waste heat by large capacity heat pumps for Riga city district heating system *Modern Science: Researches, Ideas, Results, Technologies* **2(13)** 38-43
[12] Cao X-Q, Yang W-W, Zhou F and He Y-L 2014 Performance analysis of different high-temperature heat pump systems for low-grade waste heat 125 recovery *Applied Thermal Engineering* **71(1)** 291-300
[13] Havelský V 1999 Energetic efficiency of cogeneration systems for combined heat, cold and power production *Int. J. Refrigeration* **22(6)** 479-85
[14] Pan L, Wang H and Chen Q 2011 Theoretical and experimental study on several refrigerants of moderately high temperature heat pump *Appl. Thermal Engineering* **31(11-12)** 1886-93
[15] Sukhikh A A, Antanenkova I S, Kuznetsov V N and Mereutsa E V 2016 Experimental and numerical investigations of heat regeneration process efficiency in a heat pump with a mixture of refrigerants *Int. J. of Energy for a Clean Environment* **17(2-4)** 209-22
[16] Lukas K 2001 Efficient energy systems on the basis of cogeneration and heat pump technology *Int. J. Thermal Sci.* **40(4)** 338-43

IOP Conf. Series: Materials Science and Engineering 1089 (2021) 012021 doi:10.1088/1757-899X/1089/1/012021

Addressing environmental issues faced by chemical companies when cleaning gas emissions from sulfur dioxide

L A Nikolaeva and E M Khusnutdinova

Department of Technology of Water and Fuel, Kazan State Power Engineering University, 51 Krasnoselskaya street, Kazan 420061, Russia

E-mail: rr-088@mail.ru

Abstract. At the enterprises of the chemical complex, significant harmful gas emissions are generated that pollute the atmospheric air. In this paper, it is proposed to purify gas emissions from industrial enterprises by the adsorption method. The adsorption method is one of the more common methods of gas purification in the industry, which ensures the content of harmful substances in gases at the level of specific air emissions standards. As a sorption material, energy waste is used – chemical water treatment sludge of Naberezhnye Chelny combined heat and power plant, which is formed during coagulation and liming of natural water. The paper presents a laboratory setup and the results of a study of the sorption of sulfurous anhydride by sorption material based on granulated sludge of chemical water treatment. The calculation of the economic and environmental effect of upgrading the technological scheme for cleaning gas emissions from sulfurous anhydride of the sodium bisulfite production line at the L. Y. Karpov Chemical plant JSC was made.

1. Introduction

In the last decade, the amount of gas emissions that pollute the air has been decreasing due to the measures taken. However, the amount of gas emissions to be cleaned off is still a factor. One of the reasons is the lack of protection of the environment from industrial waste.

The Republic of Tatarstan is one of the industrial regions of Russia, in which the main sources of gas emissions that pollute the air are the following industries: fuel, chemical and petrochemical, heat and power, construction, engineering, and agriculture. The chemical and petrochemical industry unites more than 16 companies, the largest of which are located in three cities of the Republic, namely Kazan, Nizhnekamsk, and Mendeleyevsk. The largest contributors to air pollution are Nizhnekamskneftekhim PJSC, TANECO JSC, Nizhnekamsktekhuglerod JSC, L. Ya. Karpov Chemical Plant JSC, Kazanorgsintez PJSC and others. According to the office of Rospotrebnadzor for the Republic of Tatarstan, the leading air pollutants 2 to 5 times exceeding the Threshold Limit Value (TLV) in 2017-2019 in the Republic of Tatarstan were such harmful substances as hydrogen sulfide, ammonia, carbon oxide, nitrogen dioxide, sulfur dioxide, dimethylbenzene, carbon (soot), phenol and its derivatives.

2. Results and discussion

One of the environmental issues that the chemical and petrochemical industry faces is the cleaning of gas emissions from sulfur dioxide (SO_2). It is toxic and is included in the list of pollutants subject to state accounting and rating (by order №. 579 issued by the Ministry of Natural Resources of the Russian Federation on December 31, 2010).

Conventional methods of cleaning industrial gas emissions from SO_2 are split into absorption, catalytic, biochemical and adsorption [1]. The absorption is characterized by the bulkiness of equipment, complexity in operation, high capital expenditures, the formation of liquid runoff, solid precipitation, which makes it difficult to operate the equipment. Catalytic gas treatment is based on heterogeneous catalysis and serves to convert impurities either into harmless compounds or into compounds that can be easily removed from the gas mixture. Biochemical treatment methods are based on the ability of microorganisms to break down and transform various compounds.

One of the most effective ways to clean gas emissions is their adsorption using microporous sorbents [2]. This technology ensures the content of harmful substances in the exhaust gases at the level of standards for specific emissions into the atmosphere and eliminates the formation of liquid effluents.

The authors suggest using a modified carbonate sludge-based Pelletized Sorption Material (PSM) produced by the chemical water treatment plant at Naberezhnye Chelny combined heat and power plant (CHPP) as an adsorbent of gas emissions in the adsorber bed installed on the sodium bisulfite production line of L. Ya. Karpov Chemical Plant JSC in Mendeleyevsk. Every year, tens of thousands of tons of sludge is formed during the pre-treatment of make-up water in thermal power plants and heating plants.

The technologies for cleaning gas emissions produced by plants and factories use pellet adsorber beds. Therefore, to reduce the hydraulic resistance in the bed, through which gas emissions are passed, pelletized adsorbents are used.

To produce pellets, fine sludge with a particle size ranged 0.01 to 0.09 mm is mixed with sodium silicate at a mass and volume ratio of 2:1, respectively [3]. This ratio was picked experimentally. At a lower ratio, the sludge is partially impregnated with sodium silicate; during subsequent firing, the pellets crumble; at a higher ratio, the binder is overspent. The resulting pellets are soaked in the oven at a temperature of 450°C for 3 hours [4-6]. Then it is cooled to room temperature in a desiccator.

Pelletized sludge (PSM) exhibits a high porosity, which is especially important when using its adsorption properties in the gas transmission mode. The low cost of the adsorption material based on chemical water treatment sludge, its availability, and the possibility to be recovered make it possible to use it for cleaning gas emissions with minimal costs and maximum performance.

3. Description of the pilot installation

A dummy installation with a fixed PSM bed made at the Department of Water and Fuel Technology of the Kazan State Power Engineering University was used for studying the sorption properties of PSM in a gas atmosphere that is close in composition to the outgoing gases of the sodium bisulfite production line at L. Ya. Karpov Chemical Plant JSC (figure 1).

The reaction column is made of stainless steel, which ensures its resistance to corrosive environments. The gas mixture was supplied and discharged through pipes 3 and 13. The pipe 13 is threaded to the sprayer 10. This ensures a uniform distribution of gases over the PSM bed 8 on the steel mesh 9. The steel mesh 9 above the PSM bed prevents particles from carrying over. A thermostat with an electric heater 6 was used to heat up the column. The gas temperature was 50 °C, which was controlled by a chromel-drop type thermocouple [7,8].

The gas composition throughout the trial was as follows: O_2 – 4 to 5.7 %; N_2 – 75 to 78 %; H_2O to 3 %. The remaining volume of the gas mixture was CO_2. In the process of sorption, the concentration of sulfur dioxide varied in the range of 0–3000 mg/m^3. The gas flow rate reduced to normal conditions was 4×10^{-4} m^3/s [8]. Sulfur dioxide was produced in the laboratory by reaction:

$$Na_2SO_3 + 2HCl \rightarrow 2NaCl + H_2O + SO_2\uparrow$$

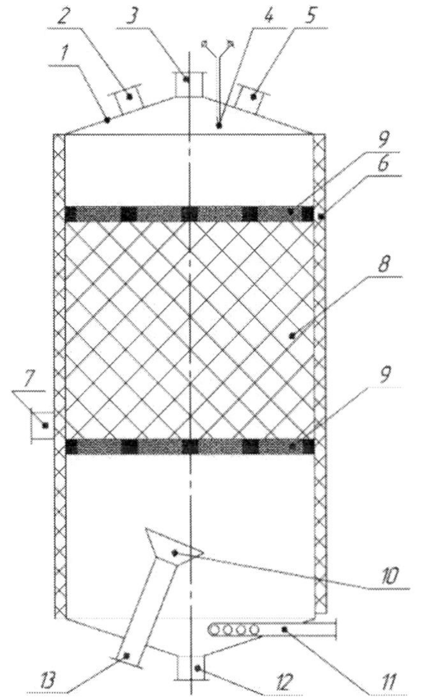

Figure 1. Pelletized sorption material (PSM) fixed bed reaction column.
1 – housing; 2 – adsorbent discharge pipe; 3 – purified gas discharge pipe; 4 – thermocouple; 5 – hatch for loading adsorbent; 6 – electric heater; 7 – hatch for unloading adsorbent; 8 – PSM bed; 9 – steel mesh with holes no more than 1 mm in diameter; 10 – sprayer, 11 – steam supply pipe; 12 – condensate discharge pipe; 13 – gas mixture supply pipe.

The results of the testing the PSM sorption properties relative to SO_2 showed that the adsorption capacity of pelletized sludge works during the first minutes of contact and after 16 minutes the process is almost completed, which indicates the onset of sorption equilibrium [9]. The adsorption capacity reaches 140 mg/g of SO_2. Figure 2 shows the PSM sorption capacity vs. time.

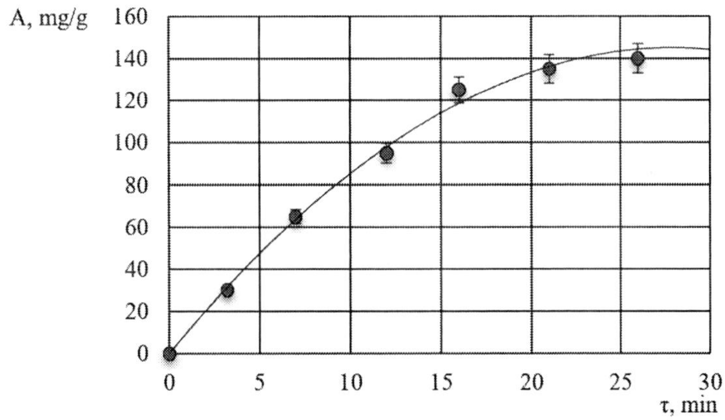

Figure 2. The PSM sulfur dioxide adsorption kinetic curve.

The main results of testing pelletized sorption material, as well as the optimal parameters of industrial treatment plants can be obtained using the sorption isotherm – the sorption capacity *A* vs. changes in the SO_2 concentration, which is depicted in figure 3. The convex shape of the adsorption isotherm belongs to type I according to the Brunauer-Deming-Deming-Teller (BDDT) classification corresponds to the Langmuir isotherm of the L-type group according to the Smith classification, and indicates the presence of micropores in the adsorbent [10, 11].

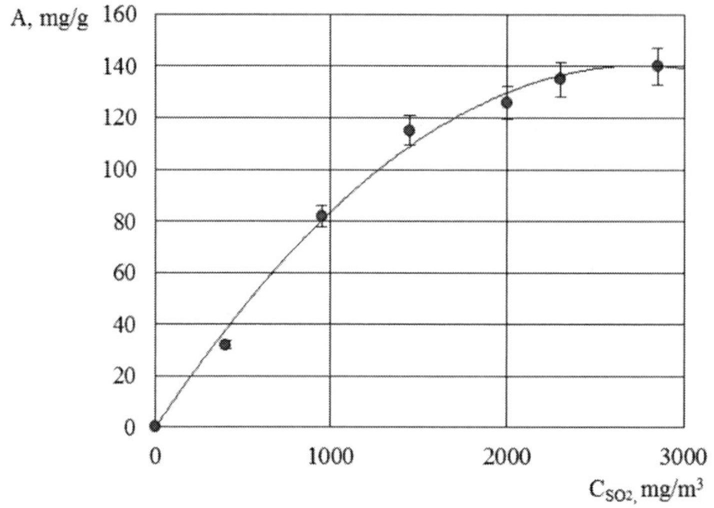

Figure 3. The PSM sorption capacity vs. the concentration of sulfur dioxide.

For studying the mechanism of how the SO_2 is adsorbed by PSM, the process was studied under static conditions at different temperatures (293K, 313K, 323K, 333K). The resulting isotherm and isoster of adsorption on sulfur dioxide is shown in figure 4 (during sorption of sulfur dioxide, the concentration varied in the range of 0–3000 mg/m^3) [12, 13].

Isosters show the relationship between equilibrium temperatures and concentrations at a constant capacity of the sorbent. An increase in temperature leads to a decrease in the SO_2 adsorption capacity, which is characteristic of the exothermic process and indicates the physical nature of the forces. The differential heat of adsorption was determined by the Clausius–Clapeyron relation (1):

$$\frac{\Delta lnC}{\Delta(1/T)} = -Q/R \tag{1}$$

where:
C is the equilibrium concentration of sulfur dioxide in the gas, mol/m^3;
T is the temperature, K; Q is the differential heat of the adsorption isoster, J/mol;
R is the molar gas constant 8.341 J/mol·K.

The differential heat of the adsorption isoster is determined by the equation:

$$Q = -R\frac{\Delta lnC}{\Delta(1/T)}, \tag{2}$$

according to equation (2), the differential heat of adsorption of sulfur dioxide by PSM pellets was calculated according to the isoster angles of inclination. The calculation results are shown in Table 1.

Table 1. Differential heat of adsorption of dissolved sulfur dioxide by PSM.

Adsorption capacity, $A \cdot 10^{-4}$, mol/g	Differential heat of adsorption Q, kJ/mol
17.19	9.31
21.88	12.17
22.97	13.64
23.44	14.22

a

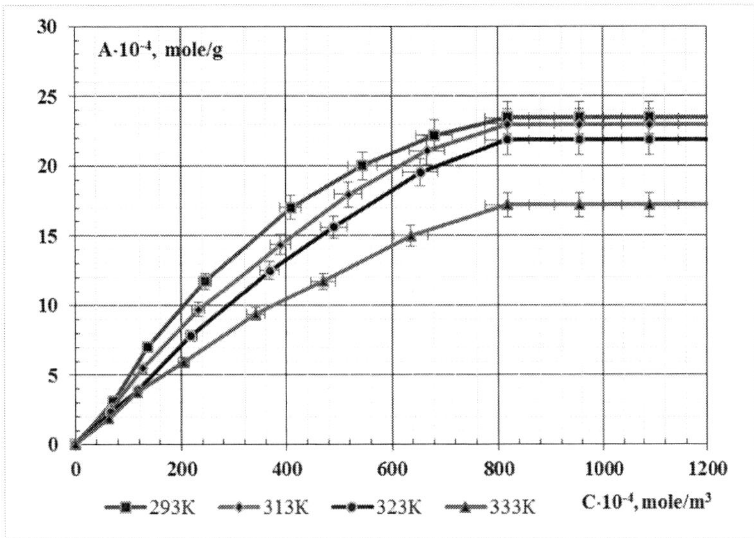

b

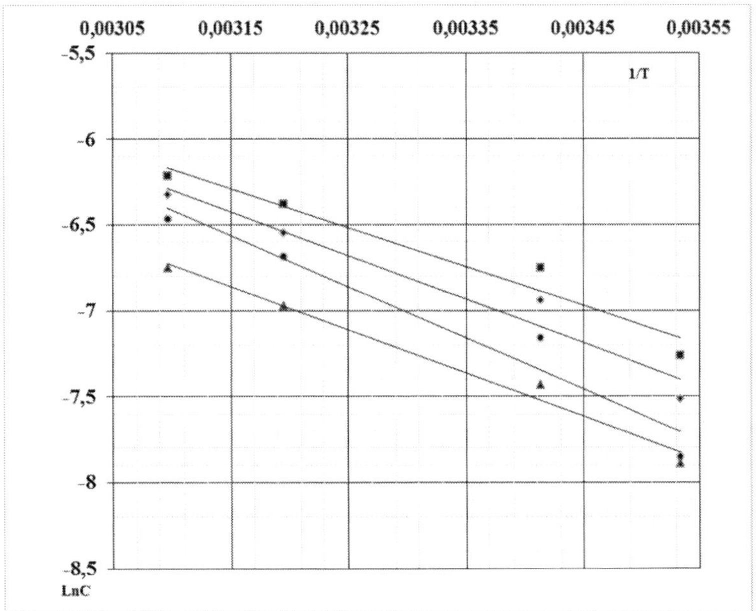

Figure 4. Isotherms (a) and isosters (b) of SO_2 adsorption of PSM at different temperatures.

Analysis of Table 1 shows that with the increase in adsorption of sulfur dioxide, the value of the differential heat of adsorption decreases.

The Gibbs energy of the adsorption process is determined by the equation (3):

$$\Delta G = -R\,T\ln K_1, \tag{3}$$

where:
ΔG is the Gibbs energy, J/mol;
T is the temperature, K;
K_1 is the adsorption equilibrium constant.

The resulting Gibbs free energy values change slightly with increasing temperature and confirm spontaneous SO_2 adsorption by PSM [14].

The kinetics of the SO_2 adsorption by PSM was studied. Kinetic dependences were obtained at different temperatures. Kinetic equations were used to determine the adsorption rate constants. It is shown that with the increase in temperature, the adsorption rate constants decrease, which is typical of non-activated adsorption (Table 2).

Table 2. Influence of the temperature on the adsorption rate constant of sulfur dioxide by PSM.

Value	Temperature, K			
	293	313	323	333
$K\cdot 10^3$, s^{-1}	9.54	8.93	8.17	5.74

The test results were then used to enhance the existing gas emission cleaning process at L. Ya. Karpov Chemical Plant JSC. The sodium bisulfite ($NaHSO_3$) production process has an adverse impact on the environment, since it releases about 18.5 tons of SO_2 emissions into the atmosphere per annum.

To reduce the content of sulfur dioxide in the gas emissions of the sodium bisulfite production, it is proposed to convert the spray trap into an adsorber bed charged with PSM. For continuous operation of the gas emission treatment system, the sodium bisulfite production line at L. Ya. Karpov Chemical Plant JSC is recommended to include PSM with charging it into the adsorber bed (figure 5).

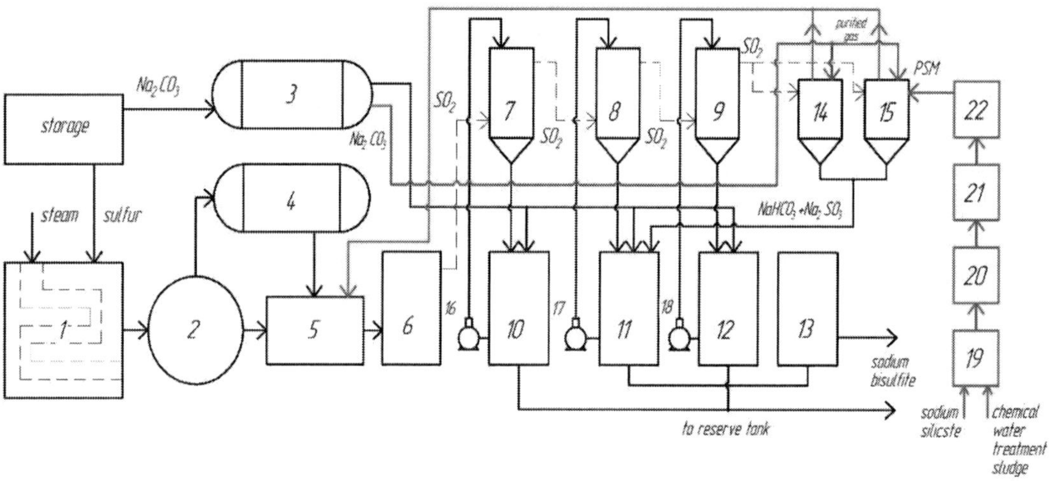

Figure 5. The upgraded sodium bisulfite production train: 1– sulfur smelter; 2 – settling tank; 3 – soda solution head tank; 4 – sulfur dioxide head tank; 5 – cyclone furnace; 6 – cooling tower; 7 – absorber of first stage; 8 – absorber of second stage; 9 – absorber of third stage; 10 – circulation tank; 11 – bisulfite circulation tank; 12 – sanitary tank; 13 – finished product tank; 14, 15 – spray trap; 16, 17, 18 – centrifugal pump; 19 – poppet pelletizer; 20 – muffle furnace; 21 – hopper for cooling and storing the finished sorbent; 22 – sorbent feed hopper.

For the PSM production, standard technological equipment has been selected, which consists of the following stages: pelletizing, calcination, cooling and feeding of PSM to the adsorber bed [15].

Recovery begins with the supply of soda ash (Na_2CO_3) to the adsorber bed from the pressure vessel of the soda solution (3). The concentration of Na_2CO_3 is 250-280 g/dm^3. The washing time is 0.5 hours. Soda ash reacts with sulfur dioxide to form bicarbonate and sodium sulfite by reaction:

$$2Na_2CO_3 + SO_2 + H_2O = 2NaHCO_3 + Na_2SO_3.$$

The resulting solution is sent to the circulating bisulfite circulation tank (11), where the subsequent formation of sodium bisulfite takes place. PSM is dried with fresh gas supplied to the adsorber bed.

The adsorbent is recovered in a closed cycle with the return of SO_2 as sodium sulfate to the bisulfite circulation tank, where the formation of sodium bisulfite takes place. The purified gases are returned to the cyclone furnace (5) as compressed air through the ejectors. While in the furnace, the molten sulfur is fed by gravity to a pneumatic sulfur nozzle, where the molten sulfur is sprayed with compressed air. Thus, during the PSM recovery, sulfur dioxide from the adsorber bed returns to the sodium bisulfite production cycle. The amount of sulfur dioxide returned to production is about 16.65 t/y, which can cause savings in material resources in the production of sodium bisulfite.

4. Conclusion

Upgrading the off-gas treatment of the sodium bisulfite production line at L. Ya. Karpov Chemical Plant JSC makes it possible to bring the cleaning of gas emissions from SO_2 to 99.9% and solves, in part, the issue of utilization of many tons of carbonate sludge produced by chemical water treatment of Naberezhnye Chelny CHPP and saving materials.

The paper provides the calculation of the prevented economic effect and prevented environmental damage as a result of environmental protection activities from soil and land degradation and reduction of SO_2 emissions into the atmosphere in the area under discussion. The total prevented environmental damage will amount to 248.407 kilorubles per annum. The economic effect will be 162.4 kilorubles per annum, the payback period is 4.5 years.

References

[1] Khairullin S R, Kerzhentsev M A, Yasnik S A, Salnikov A V, Teryaeva T N, Ilukhin I V, Sadykov R R and Ismagilov Z R 2015 Sulfur dioxide purification processes for gas emissions from non-ferrous metallurgy enterprises *Chemistry for Sustainable Development* **23(4)** 469-89

[2] Alekseev A M, Klinova L L, Komova Z V, Sergeev S P, Krasnushkina N V and Korabel'nikova L M 2003 Technological aspects of adsorption catalytic removal of organic impurities from industrial gas emissions *Theoretical Foundations of Chemical Engineering* **37** 510-3

[3] Shikina N V, Khairullin S R, Kuznetsov V V, Rudina N A and Ismagilov Z R 2015 Development and investigation of adsorbents based on ore materials for the purification of flue gas from sulphur dioxide *Chemistry for Sustainable Development* **23(2)** 199-208

[4] Gromoglasov A A, Subbotina N P and Kvrivishvili I R 1982 *Patent RU 947067 Sposob obessolivaniya i obeskremnivaniya vody`* [Patent RU 947067 Method of desalting and desilicizing water] (Moscow: Rospatent) [In Russian]

[5] Krylov I O, Lugovskaya I G, Korovushkin V V and Makeev D 2008 Vliyanie termicheskoj obrabotki na sorbcionny`e i kataliticheskie svojstva shungitovy`x porod [Influence of heat treatment on sorption and catalytic properties of shungite rocks] *E`kologiya i Promy`shlennost` Rossii* **10** 24-7 [In Russian]

[6] Terekhova M V and Rusakova S M 2013 Issledovanie adsorbcionny`x zakonomernostej anionov na poverxnosti krasnogo shlama [Investigation of adsorption patterns of anions on the surface of red slime] *Izvestiya MGTU* **3(17)** 147-51 [In Russian]

[7] Nikolaeva L A 2013 Research of sorption processes using chemical water purification sludge for nitrogen and sulfur oxides contained in smoke gases emitted from a thermal power station *Thermal Engineering* **60(4)** 244-7

[8] Shlekova I and Knysh A 2018 Intensifikaciya biologicheskoj ochistki proizvodstvenny`x stochny`x vod v ae`rotenkax s pomoshh`yu adsorbentov [Intensification of Biological Treatment of Industrial Wastewater in Aerotanks Using Adsorbents] *E`kologiya i Promy`shlennost` Rossii* **22(6)** 20-5 [In Russian]

[9] Nikolaeva L A and Khusnutdinov A N 2019 Reshenie e`kologicheskix problem promy`shlenny`x predpriyatij pri ochistke gazovy`x vy`brosov granulirovanny`m shlamom

[Solving environmental problems of industrial enterprises when cleaning gas emissions with granulated sludge] *Voda: Ximiya i E`kologiya* **1-2(118)** 83-9 [In Russian]

[10] Khusnutdinova E M and Nikolaeva L A 2020 Novy`j sorbcionny`j material na osnove otxoda e`nergetiki v texnologiyax ochistki gazovy`x vy`brosov ot dioksida sery` [New sorption material based on energy waste in technologies for cleaning gas emissions from sulfur dioxide] *Ratsional'noye ispol'zovaniye prirodnykh resursov i pererabotka tekhnogennogo syr'ya: fundamental'nyye problemy nauki, materialovedeniye, khimiya i biotekhnologiya* [ational use of natural resources and processing of technogenic raw materials: fundamental problems of science, materials science, chemistry and biotechnology] (Belgorod: BGTU) pp 224-30 [In Russian]

[11] Eshmukhamedov M A, Ponamaryova E G and Raevskaya E G 2020 Ochistka gazovy`x vy`brosov ot dioksida sery` s pomoshh`yu aktivirovannogo granulirovannogo sorbenta [Purification of gas emissions from sulfur dioxide by means of activated granulated sorbent] *Ximicheskaya Bezopasnost`* **4(1)** 170-82 [In Russian]

[12] Nosyrev M, Komlyashev R, Ilyina S and Kulikov M 2018 K voprosu modelirovaniyatexnologicheskoj sxemy`ochistki biogaza [Modeling Technological Scheme of Biogas Treatment] *E`kologiya i Promy`shlennost` Rossii* **22(1)** 10-3 [In Russian]

[13] Nikolaeva L A and Khusnutdinov A N 2019 Adsorbcionnaya ochistka gazovy`x vy`brosov ot serovodoroda granulirovanny`m karbonatny`m shlamom [Adsorption treatment of gas emissions from hydrogen sulfide with granular carbonate sludge issues of modern science and practice] *Voprosy` Sovremennoj Nauki i Praktiki. Universitet V. I. Vernadsky* **3(73)** 37-47 [In Russian]

[14] Zavizion Yu V, Slyusar' N N, Glushankova I S, Zagorskaya Yu M and Korotaev V N 2015 Ocenka sostoyaniya poligonov zaxoroneniya TBO po izmeneniyu organicheskoj sostavlyayushhej [Evaluation of the State of Disposal Polygons of Solid Domestic Waste by the Variation in the Organic Component] *E`kologiya i Promy`shlennost` Rossii* **19(7)** 26-31 [In Russian]

[15] Nikolaeva L A and Khusnutdinov A N 2018 Ochistka gazovy`x vy`brosov predpriyatij ximicheskoj promy`shlennosti karbonatny`m shlamom [Purification of gas emissions of chemical industry enterprises by carbonaceous cutting] *E`kologiya i Promy`shlennost` Rossii* **22(8)** 14-8 [In Russian]

Adsorption Purification of Phenol-Containing Wastewater from Oil Refineries

L A Nikolaeva[1] and N E Aikenova[1,2]

[1] Department of Water and Fuel Technology, Kazan State Power Engineering University, Krasnoselskaya St., 51, Kazan, 420066, Russia
[2]Department of Chemistry and Chemical Technology, Aktobe Regional University named after K.Zhubanov, A.Moldagulova Av., 34, Aktobe, 030000, Kazakhstan

E-mail: larisanik16@mail

Abstract. In this paper, the possibility of adsorptive purification of industrial wastewater from oil refineries from phenols by industrial wastes is considered. It is proposed to use carbonate sludge for water treatment of natural water as one of the most accessible materials, which is a large-tonnage waste of energy, for sewage treatment. Phenols in wastewater are hazardous toxic substances. In legal order, sanitary and hygienic standards establish the maximum permissible concentration (MPC) of phenol in water, which is 0.001 mg / dm^3. Industrial wastewater containing phenols has a number of difficulties faced by technologists when selecting a technological scheme for cleaning these pollutants, or their disposal. The adsorption isotherm, the kinetic curve of phenol adsorption, the output curve of phenol adsorption under dynamic conditions were constructed, the dynamic sorption capacity of the "GrSM-1" sorption material was determined, which prove the effectiveness of using carbonate sludge as an adsorbent. A flowchart for the purification of wastewater from phenols is proposed, the calculation of the adsorption filter with the loading of carbonate sludge in the scheme for the purification of wastewater from phenols is performed.

1. Introduction

Enterprises of the petrochemical and refining industries are large consumers of natural water. Water is used for technological operations, the preparation of solutions necessary for the production of products, heating and cooling, washing process equipment. The waste waters of these enterprises contain a wide range of organic substances dissolved in water, including petroleum products and phenols. In the absence of wastewater treatment plants, the anthropogenic pressure on the environment increases, due to the nature of the industrial activity of the enterprise [1-3].

Traditionally used technological schemes of wastewater treatment are obsolete and physically obsolete by now and do not provide high requirements for water quality, therefore, the improvement of wastewater purification technologies from phenol using effective integrated solutions to achieve normal requirements for the quality of treated water with all the associated Phenol contamination is relevant and has a scientific and practical value [4].

The article proposes an adsorption technology for purification of waste oil-containing waters of «Aktobe Oil Refining» LLP with a large-tonnage waste of energy - sludge from chemical water treatment generated during liming and coagulation at the stage of preliminary purification of raw water.

2. Materials and Methods

For the purification of wastewater from phenols with a sorption material developed on the basis of chemical water treatment sludge, an adsorption isotherm and adsorption kinetics are constructed, a purification process flow diagram is considered, and an adsorption filter is calculated. The technical characteristics and adsorption capacity of the material for phenol have been determined.

The work uses chemical water treatment sludge (ChWP) - waste that is generated at the water treatment plant of thermal power plants at the stage of preliminary water purification. In experimental studies, we used dried fine powder sludge [5].

To determine the sorption capacity of granules of the GRSM-1 material, the adsorption isotherm with respect to phenol was obtained, the kinetics of the adsorption process, the adsorption of phenol GRSM-1 was studied under dynamic conditions.

For the purification of wastewater of LLP "Aktobe Oil Refining" from oil products and phenols with a capacity of 60 m^3/h, a basic technological scheme is proposed.

In this technological scheme, it is proposed to load the developed hydrophobic GRSM-1 into the adsorption column. The calculation of the adsorption filter FSU 3.0–0.6 with the loading of the developed GRSM-1 was performed.

3. Result and Discussion

Sludge ChWP - waste, which is formed at the water treatment plant of thermal power plants. Experimental studies were conducted with carbonate slurry of Aktobe TPP-1 (humidity-3%). The dried sludge is a fine powder from light yellow to brown color. During the study, a fraction with a particle size of 0.09-0.05 mm was used. X-ray qualitative and phase analysis of sludge on a Bruker P8 ADVANCE diffractometer showed the following chemical composition: calcium $CaCO_3$-73%, brucite $Mg(OH)_2$ - 8%, Portland $Ca(OH)_2$ - 1%, quartz SiO_2 - 0.4%, and other substances - 17.6%.

In [6], carbonate sludge is considered as a sorption material for wastewater treatment of industrial enterprises from ITM.

The article discusses the fundamental possibility of purifying wastewater from phenols by sorption material developed on the basis of a ChWP sludge. In earlier works [7], the technical characteristics and the adsorption capacity of the material were determined for emulsified and dissolved petroleum products of the oil from the Shiya field.

he moisture capacity of the ChWP sludge is -57%, which confirms its high hydrophilicity and poor wettability by non-polar compounds. Therefore, to increase the adsorption capacity and wettability by non-polar compounds, a hydrophobic granular sorption material has been developed and studied on the basis of fine sludge.

The “Silor” liquid was chosen as a hydrophobic agent - a product of chemical decomposition of organosiliconvulcanizates in tetraethoxylan in the presence of an alkali solution, which in its chemical composition is similar to organosilicon hydrophobic liquids.

Liquid sodium glass was used as a binder. After heat treatment, granules with a particle diameter of 0.5-2.5 mm are formed by rolling. To select the optimal conditions for the manufacture of material granules, studies have been carried out of the dependence of the total pore volume and specific surface area on the processing temperature of the sludge granules. The maximum value of the total pore volume and specific surface is achieved at a temperature of - 700°C. Determination of the specific surface of the material granules was carried out using a "Sorbi-M" by comparing the volumes of gas - nitrogen (adsorbate) sorbet by the granules with a standard sorbent - silica gel.

Thus, a granulated hydrophobic sorption material was obtained: heat-treated at 700°C for 60 minutes, granules with a diameter of 0.5 to 2.5 mm, at a ratio of 1: 2 with a binder with liquid sodium glass, impregnated with 5% - Sylor water emulsion, dried to constant weight. The technological characteristics of the obtained granules of the material were determined (Table 1).

Table 1. Technological characteristics of "GrSM-1".

Characteristic	Value
Particle size, mm	0.5-2.5
Bulk density, ρ_n, kg/m^3	670
Humidity, %	2.5
Specificsurface, m^2/g	64.9
Total pore volume, cm^3/g	0.84
Waterabsorption, %	1.2
Adsorption capacity for iodine, %	7
Methylene blue adsorption capacity, %	20

To determine the sorption capacity of material granules, an adsorption isotherm with respect to phenol was constructed. The adsorption isotherm is of type V according to the classification of Brunauer, Deming, Deming and Teller (BDDT), a similar S-type isotherm is usually found in the presence of micro and mesopores, and is described by the Freundlich equation: A=0,28$C^{0,75}$.

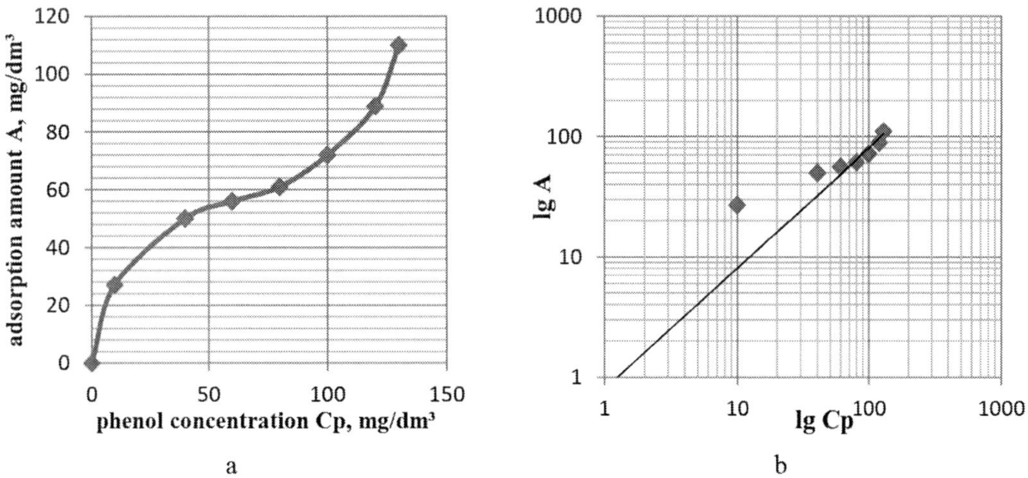

Figure 1. Isotherm of phenol adsorption of the GrSM-1 (a) and its appearance in logarithmic coordinates (b).

To determine the time to reach the adsorption equilibrium, the kinetics of the process of phenol adsorption by material granules from model solutions with a certain concentration was studied (C_{init} = 100 mg/dm^3) [8-10]. The contact time of the granules with a model solution of 0.33; 0.66; 1; 2; 4; 5; 7 o'clock. It is established that the adsorption equilibrium occurs after 3 hours of contact of the adsorbent with the adsorbate.

For production processes, the adsorption of organic impurities in dynamic conditions is of the greatest importance. The process of phenol adsorption was investigated using granulated GrSM-1 (fractions 0.5-2.5 mm) in a laboratory setup, which is a filter column 2.5 cm in diameter. The concentration of phenol in the model solution is 1.5 mg/dm^3, is medium at the entrance to the adsorption filter in the wastewater treatment system. The height of the loading layer is - 20 cm, weight - 54.38 g, filtration speed - 3.5 m/h. The breakthrough is fixed at a concentration of 0.001 mg / dm3. Figure 3 shows the output curve of phenol adsorption under dynamic conditions. During the experiment, the dynamic sorption capacity (DSC), the total sorption capacity (TSC) "GrSM-1" was determined. The result is presented in table 2.

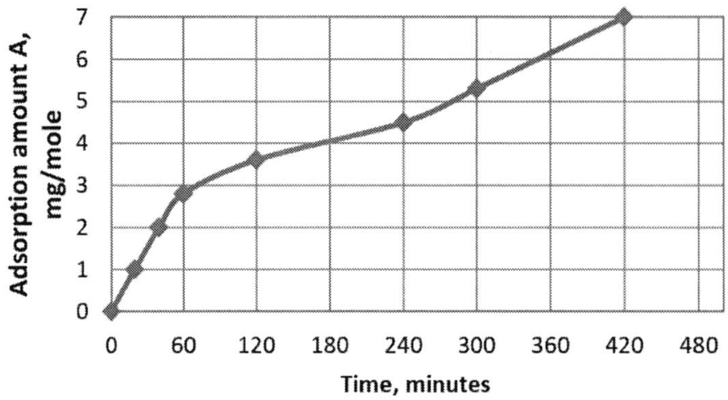

Figure 2. Kinetic curve of phenol adsorption with GrSM-1.

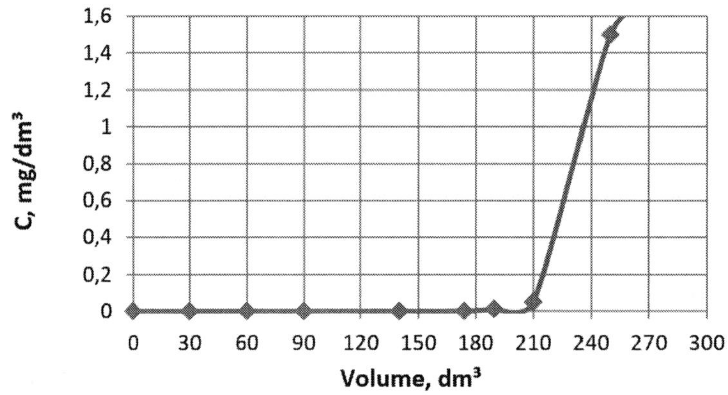

Figure 3. The output curve of the adsorption of phenol GrSM-1 in dynamic conditions.

Table 2. The value of the dynamic sorption capacity and full sorption capacity GrSM-1.

Indicator mg/g	Value	Volume of the passed water, dm^3
DSC	4.8	174.5
TSC	6.9	250.4

According to the Shilov equation, the time τ and the coefficient K of the protective action of the GrSM-1 layer are calculated: τ = 95,2h; K = 612,6 h/m.

For the calculation of technical and economic indicators, the use of a sorbent, GrSM-1 in the wastewater treatment system of "Aktobe Oil Refining" LLP is proposed.

The greatest consumption of water in industrial installations is noted at the stage of oil preparation, in the process of its dehydration and desalting. Oil is pumped into the pump room and comes in the form of a mixture together with the washing water into the electrohydrators, where desalination and dehydration by the action of electric current takes place. The water separated on the installations of electrical desalting plant is discharged into a special sewage network.

During primary oil refining in atmospheric vacuum tubulars (AVT) to obtain light distillates and oil fractions. Oil after electrical desalting plant passes heat exchangers, where it contacts with water - it cools, condenses in heat exchangers and condensers.

Vacuum in barometric condensers for mixing AVT vacuum columns is created due to the direct contact of water with oil product vapors and gases. As a result, the waste water is polluted with vapors of petroleum products and hydrogen sulfide.

Due to the aggressiveness of sulfur compounds to the metal processing equipment, their presence in commercial petroleum products is not allowed. Oil products are purified from sulfur compounds by washing with an aqueous solution of alkali (caustic soda). In this case, hydrogen sulfide, mercaptans and other sulfur compounds, as well as phenols, are transferred from petroleum products to alkaline solution. After repeated use of alkaline solution containing a large amount of sulfur compounds, as well as other pollutants, is discharged into a special sewer network.

At refineries, industrial storm sewage is generated, which is discharged into the sewage system and then fed to wastewater treatment plants. Sources of pollution include process wastewater from condensation equipment, as well as the recovery tanks of the MEA solution. Table 3 shows the chemical composition of the description of the wastewater of "Aktobe Oil Refining" LLP.

Table 3. Description of wastewater of "Aktobe Oil Refining" LLP.

The name of indicators	Unit of measuremen	Concentration	MPC
Hydrogen ions	pH units	7.34	6.5-8.5
Dry residue	mg/dm^3	2310	-
Total hardness	$mg\text{-}eq/dm^3$	4.6	-
Oil products	mg/dm^3	96.8	0.1-5
SAS	mg/dm^3	0.490	0.5
Phenols	mg/dm^3	0.1	0.001
Ammonia nitrogen	mg/dm^3	5.53	-
Common iron	mg/dm^3	2.92	2-20
Cadmium	mg/dm^3	0.0001	0.01-0.6
Manganese	mg/dm^3	0.254	-

For wastewater treatment of "Aktobe Oil Refining" LLP from petroleum products and phenols with a capacity of 60 m^3/h, the following flow chart [11] is presented in Figure 4.

In the shop for processing by-products and wastes, wastewater is formed during the decomposition of sodium phenolate with sulfuric acid. Wastewater contains mostly phenol and acetone. Together with the wastewater from the decomposition shop of isopropyl benzene hydroperoxide, containing up to 30 g/l of phenol, they fall into local cleaning, which consists in the extraction of phenol with diisopropyl ether or acetophenone. Pre-waste water is acidified with sulfuric acid in tank 3 to pH = 1. From here, they are sent through an average 4 for irrigation into column 6 for adsorption of volatile substances from the gases. Extraction of phenol is carried out in pulsation columns 7 with intermittent flow of the extractant (75 pulsations per 1 minute). The ratio of ether and wastewater is 1:3, the degree of extraction of phenol in wastewater when using diisopropyl ether is 99.3%, while using acetophenone is 99.6%.

The extract saturated with phenol is fed to the rectification for the regeneration of the extractant. Rectification is carried out in a packed column 8 with a reflux number of 0.5-1 to a residual content of phenol in the extractant 7-1 mg/l. Pairs of ether from the column (at a temperature of 68-69^{0}C) enter the reflux condenser 9 and the condenser 10; the condensed ether is collected in tank 14 and returned to extraction. The consumption of fresh air to compensate for losses in one cycle is 1% of its quantity circulating in the system. The regenerated phenol enters the tank 15 and returns to production. Water is subjected to Stripping from the ether in column 12 and after cooling in the refrigerator 13 is discharged into the sewer.

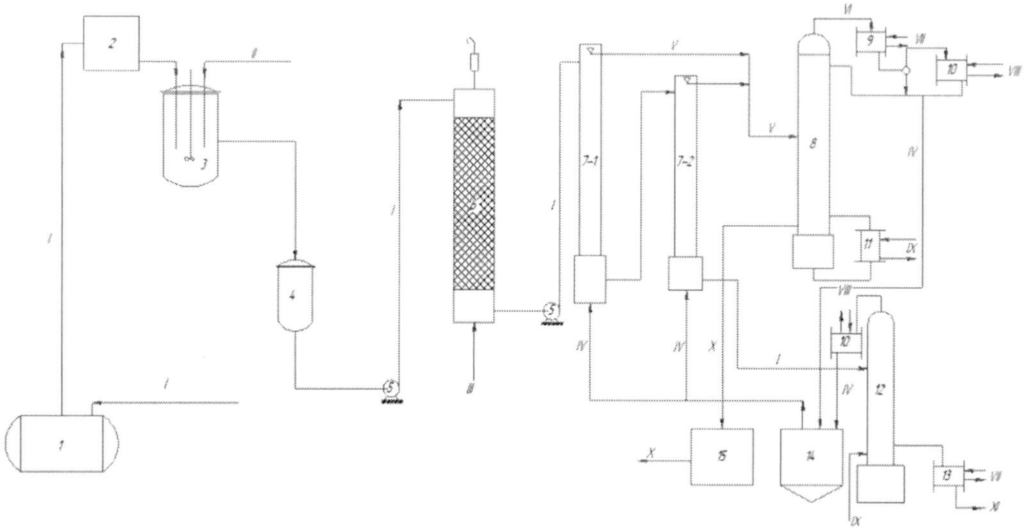

Figure 4. Schematic diagram of the dephenolation of wastewater "Aktobe Oil Refining" LLP. 1 – collective wastewater tank; 2 – pressure tank; 3 – tank for acidification of wastewater; 4 – averaging tank; 5 –pumps; 6 – adsorption column; 7-I and 7-II - extraction columns; 8 - distillation columns; 9 – reflux condenser; 10 – capacitor; 11 – boiler; 12 – Stripping column; 13 – fridge; 14 – capacity for ether; 15 – capacity for phenol; I – phenolic wastewater; II – sulfuric acid; III – exhaust gases; IV – extractant (ether); V – extract; VI – extractant (ether) vapors; VII – cooling water; VIII – cooling brine; IX – pairs; X- phenol; XI – drain to sewer.

In this technological scheme, it is proposed to load the developed hydrophobic granulated material "GrSM-1" into the adsorption column 6. The calculation of this adsorption filter FSU 3.0–0.6 with the loading of the developed sorption material "GrSM-1" was made. Table 4 presents the initial data for the calculation.

Table 4. Original data for the calculation of the adsorption filter with loading "GrSM-1".

Parameter	Value
Specific free volume (porosity), ε	0.407
The density of wastewater $\rho_{w.w}$, kg/m^3	1000
The coefficient taking into account the shape of particles, F	0.9
Loading height H, m	2,5
Kinematic viscosity coefficient of water (at 20 ^{0}C), μ	$1.004 \cdot 10^{-3}$
Number of filters, n	1
Filter performance Q_{per}, m^3/h (m^3/s)	25 (0.0069)
Filter area S_{ads}, m^2	7.1
Bulk density ρ_b, kg/m^3	670
Filtering speed, m/h	3.5

Fictitious flow rate ω_{fic}, m/s, is calculated by the formula

$$\omega_{\text{fic}} = \frac{Q_{per}}{S_{ads}} = \frac{0.069}{7.1} = 0.00097\ \text{m/s} \tag{1}$$

The true flow rate ω_{tr}, m/s, is calculated by the formula

$$\omega_{\text{tr}} = \frac{\omega_{fic}}{\varepsilon} = \frac{0.00097}{0.407} = 0.0024\ \text{m}/s \tag{2}$$

The apparent density of the adsorbent ρ_{app} local, kg/m^3, is calculated by the formula

$$\rho_{\text{app}} = \frac{\rho_{sor}}{1-\varepsilon} = \frac{560}{1-0.407} = 944.35\ \text{kg/m}^3 \tag{3}$$

The particle size of the adsorbent: $d_p = 0.002$ m.
The specific surface a_v, m^2/m^3, is calculated by the formula

$$a_v = \frac{6(1-\varepsilon)}{d_p} = \frac{6(1-0{,}407)}{0.002} = 1779\ m^2/m^3 \tag{4}$$

Determine the flow of flow in the layer.
Reynolds number is calculated by the formula

$$Re = \frac{\omega_{\text{tr}} d_p \rho_{w.w}}{\mu} = \frac{0.0024 \cdot 0.002 \cdot 1000}{1.004 \cdot 10^{-3}} = 4.78 \tag{5}$$

The coefficient of friction is found depending on the nature of the movement [12]:

1) $Re < 50\ \lambda = \frac{220}{Re}$;

2) $Re = 50 - 7200\ \lambda = \frac{11{,}6}{Re^{0.25}}$;

3) $Re > 7200\ \lambda = 1.26$

The coefficient in this case is equal to $\lambda = \frac{220}{4.78} = 46.025$

The pressure drop of the bulk layer ΔP, Pa, is determined by the formula

$$\Delta P = \lambda \cdot \frac{H}{d_p} \cdot \frac{\rho w.w.\cdot \omega_{\text{tr}}^2}{2} = 46.025 \cdot \frac{2.5}{0.002} \cdot \frac{1000 \cdot 0.0024^2}{2} = 331.8\ Pa \tag{6}$$

The mass of GrSM-1 for loading one adsorption filter m_c, kg, is calculated by the formula

$$m_c = \rho_b \cdot S_{ads} \cdot \text{H} = 670 \cdot 7.1 \cdot 2.5 = 11\ 892\ \text{kg} \tag{7}$$

The penetration of the adsorbate into the adsorbent grain is a diffusion process and, therefore, is characterized by a diffusion coefficient. The diffusion coefficient D is the amount of a substance diffusing into a unit of time through a square centimeter of surface with a concentration gradient equal to one [13].

The exact determination of the diffusion coefficient is made by complex formulas. In technical adsorption, it should be limited to its definition D, m^2/s, according to the approximate dependence [6]

$$D = \frac{K \cdot r^2}{\pi^2 \cdot \tau_{0.5}}, \tag{8}$$

where:
r – is the radius of the granules;
K – coefficient depending on the shape of the granules;
$\tau_{0.5}$ – is the half-time of the adsorption capacity, i.e. period of time elapsed from the beginning of the experiment to the moment when the amount of the adsorbed substance reaches 50% of the equilibrium adsorption capacity [13].

For particles of a spherical shape, the coefficient K is equal to 0.308.

Time $\tau_{0.5}$= 120 s (the half-time of the sorption tank "GrSM-1" for the averaged oil product is gasoline, oil).

The diffusion coefficient of phenol molecules inside the adsorbent grain is determined by the equation

$$D = \frac{K \cdot r_p^2}{\pi^2 \cdot \tau_{0,5}} = \frac{0.308 \cdot 0.001^2}{3.14^2 \cdot 120} = 5.201 \cdot 10^{-10} \mathrm{m^2/s}$$

The kinematic viscosity coefficient ν, m^2/s, is determined by the formula

$$\nu = \frac{\mu}{\rho_{ww}} = \frac{1.004 \cdot 10^{-3}}{1000} = 1.004 \cdot 10^{-6}\ \mathrm{m^2/s} \tag{9}$$

To determine the mass transfer coefficient, it is necessary to determine the Prandtl diffusion criterion [14-15] using the formula

$$Pr_{\text{dif}} = \frac{\nu}{D} = \frac{1.004 \cdot 10^{-6}}{5.201 \cdot 10^{-10}} = 1930 \tag{10}$$

The reduced diameter d_{red}, m, is calculated by the formula d_{red}

$$d_{red} = \sqrt[3]{\frac{6\left(\frac{\pi \cdot d_ч^2}{4}\right) \cdot l_ч}{\pi}} = \sqrt[3]{\frac{6\left(\frac{3.14 \cdot 0.002^2}{4}\right) 0,002}{3.14}} = 0.0023\, m \tag{11}$$

$$L = \frac{\pi \cdot d_{\text{red}}}{2} = \frac{3.14 \cdot 0.0023}{2} = 0.0036\ \mathrm{m} \tag{12}$$

The mass transfer coefficient β, m/s, is calculated by the formula

$$\beta = 0.62\left(\frac{\Delta P \cdot \varepsilon \cdot \nu}{a_v \cdot H \cdot L \cdot \rho_{\text{с.в.}}}\right)^{\frac{1}{3}} \cdot 1930^{-\left(\frac{2}{3}\right)} = 0.62\left(\frac{331.38 \cdot 0.407 \cdot 1.004 \cdot 10^{-6}}{1779 \cdot 2.5 \cdot 0.0036 \cdot 1000}\right)^{\frac{1}{3}} \cdot 1930^{-\left(\frac{2}{3}\right)} = 7.3 \cdot 10^{-6}\ \mathrm{m/s} \tag{13}$$

The volume coefficient of mass transfer β_y, s^{-1}, is calculated by the formula

$$\beta_y = \beta \cdot a_v = 7.9 \cdot 10^{-6} \cdot 1779 = 0.014\ s^{-1} \tag{14}$$

4. Conclusion

In this work, the sorption capacity of granules of the material is determined, the adsorption isotherm with respect to phenol is constructed, the adsorption curve for the study of phenol adsorption kinetic processes is constructed, and the adsorption curve of phenol GrSM-1 is constructed under dynamic conditions. Also carried out experimental studies on the adsorption purification of phenol-containing wastewater of "Aktobe Oil Refining" LLP using granulated modified carbonate sludge (GrSM-1). Purification efficiency E> 90%. The calculation of the adsorption filter with the loading of the lubricant in the scheme of purification of wastewater from phenols was made.

5. Acknowledgments

The work was performed in the framework of the basic part of the state task in the field of scientific activity (No. 13.6384.2017 / BP).

References

[1] Sun X, Wang C, Li Y, Wang W and We J 2015 Treatment of phenolic wastewater by combined UF and NF/RO processes *Desalination* **355** 68-74
[2] Lakshmi S, Harshitha M, Vaishali G, Keerthana S and Rhea M 2016 Studies on different methods for removal of phenol in waste water- Review *Int. J. of Science, Eng. and Technology Research* **5(7)** 2488-96
[3] Cordova Villegas L G, Mashhadi N, Chen M, Mukherjee D, Taylor K E and Biswas N 2016 A Short Review of Techniques for Phenol Removal from Wastewater *Current Pollution Reports* **2** 157-67
[4] Franz M, Arafat H A and Pirto N G 2000 Effect of chemical surface heterogeneity on the adsorption mechanism of dissolved aromatics on activated carbon *Carbon* **38(13)**
[5] Nikolaeva L A 2016 Adsorbczionnaya ochistka stochny'h vod modificzirovanny'm karbonatny'm shlamom [Adsorption treatment of industrial wastewater with modified carbonate sludge: DSc thesis] (Kazan: KGEU)
[6] Sklavos S, Gatidou G, Stasinakis A S and Haralambopoulos D 2015 Use of solar distillation for olive mill wastewater drying and recovery of polyphenolic compounds *J. Environ. Manag.* **162** 46-52
[7] Tran V S, Ngo H H, GuoW, Zhang J, Liang S, Ton-That C and Zhang X 2015 Typical low cost biosorbents for adsorptive removal of specific organic pollutants from water *Bioresource Technology* **182** 353-63
[8] Kulkarni S J and Kaware J P 2013 Review on Research for Removal of Phenol from Wastewater *IJSRP* **3(4)** 161089
[9] Mukherjee S, Basak B, Bhunia B, Dey A and Mondal B 2013 Potential use of polyphenol oxidases (PPO) in the bioremediation of phenolic contaminants containing industrial wastewater *Environ. Sci. Biotechnol.* **12** 61-73
[10] Masomi M and Ghoreyshi A A 2014 Adsorption of Phenolic Compounds onto the Activated Carbon Synthesized from Pulp and Paper Mill Sludge: Equilibrium Isotherm, Kinetics, Thermodynamics and Mechanism Studies *Int. J. Eng.* **27(10)** 1485-94
[11] Mukherjee R and De S 2014 Adsorptive removal of phenolic compounds using cellulose acetate phthalate–alumina nanoparticle mixed matrix membrane *J. Hazard. Materials* **265** 8-19

[12] Mohammadi S, Kargari A, Sanaeepur H, Abbassian K, Najafi A and Mofarrah E 2015 Phenol removal from industrial wastewaters: a short review *Desalination and Water Treatment* **53(8)** 2215-34
[13] Sciban M, Klasnja M and Skrbic B 2006 Modified Hardwood Sawdust as Adsorbent of Heavy Metal Ions from Water *Wood Science and Technology* **40(3)** 217-27
[14] Carvajal-Bernal A M, Gómez F, Giraldo L and Moreno Pirajá JC 2015 Chemical modification of activated carbons and its effect on the adsorption of phenolic compounds *Ingeniería Y Competitividad* **17(1)** 109-19
[15] Zwain H M, Vakili M and Dahlan I 2014 Waste Material Adsorbents for Zinc Removal from Wastewater: A Comprehensive Review *Int. J. of Chemical Eng.* **2014** 347912

IOP Conf. Series: Materials Science and Engineering 1089 (2021) 012023 doi:10.1088/1757-899X/1089/1/012023

Application of data envelopment analysis for multi-criteria evaluation of system for technogenic waste recycling in oil refining industry

M Yu Derevyanov[1], Yu E Pleshivtseva[1], A A Afinogentov[2], S A Kirpichenkov[1], A G Mandra[1] and A A Pimenov[3]

[1] Department of Control and Systems Analysis of Thermal Power and Socio-Technical Complexes, Samara State Technical University, Molodogvardeyskaya Str., 244, Samara, 443100, Russia
[2] Department of Pipeline Transportation, Samara State Technical University, Molodogvardeyskaya Str., 244, Samara, 443100, Russia
[3] Department of Gas Processing, Hydrogen and Special Technologies, Samara State Technical University, Molodogvardeyskaya Str., 244, Samara, 443100, Russia

E-mail: mder2007@mail.ru

Abstract. The paper proposes an approach based on the DEA method (Data Envelopment Analysis) to a multi-criteria evaluation of a complex structured system for the processing of technogenic waste in the oil refining industry. The special algorithm is considered for assessing environmental safety and energy efficiency of operating waste processing system facilities, taking into account their resource value and reuse potential. According to the developed algorithm, five interconnected mathematical programming problems are solved. These problems formulated using CCR and Super-efficiency models of DEA method allow to obtain the assessments of the disposal system regarding the chosen efficiency criteria. A new approach to the analysis of the results on the basis of total effectiveness estimates using the Super-efficiency model of the DEA method is proposed. The results of testing the developed algorithm on an oil-containing waste processing system in the Samara region make it possible to form appropriate control decisions in the region's oil refining industry.

1. Introduction

The oil refining industry is the key sector of the economics of the Russian Federation, the development of which currently represents one of the most important strategic objectives. The reduction of energy costs and the harmful environmental impact of the processes of oily waste refining are of particular importance due to the recognition of the importance of energy-saving and environmental safety of production in modern society. The traditional methods of increasing the technical and economic indicators of oil refining processes are focused on the specific nature of the industries in which they were used in the period before the fast development of modern information technologies and intelligent systems. According to statistics, up to 10 million tons of technogenic oil-contaminated waste (OCW) of the oil refining industry (oil and drilled sludge, oil waste, used oils, etc.) are accumulated in Russia annually, and the volume of their reuse is no more than 10% [1, 2]. At the same time, in western countries OCWs are considered as sources of secondary resources of high value [3-7], and the proportion of their secondary processing reaches 50% [8].

The term «value of secondary resources» includes not only a quantitative assessment of their physical and chemical composition and properties, which determines the degree of suitability of the waste for use as material resources in recycling technologies but also the creation of a wide range of measurable benefits in the environmental, economic, social and technical fields evaluated by groups of relevant indica-tors [9-13].

A significant part of the production cycle of complex OCW processing is represented by energy- and resource-intensive technological processes that negatively affect the environment. Given the total costs of storage, transportation, treatment, and disposal of TW, it can be argued that due to losses associated with both the fatal features of the oil refining processes and the removable causes of inefficient use of energy resources and their negative impact on the environmental situation in the present and In the future, the effect expected from the optimization of the integrated processing system may be more significant than the effect from the optimization of primary processing processes.

The problem of a comprehensive assessment of the efficiency of OCW utilization by groups of heterogeneous indicators is an urgent task in modern societies. One approach to solving such a problem is based on the DEA method (Data Envelopment Analysis) [14-15]. The concept of measuring the environmental efficiency of companies' management by aggregating various indicators without explicit values of the coefficients of weight functions using the "input-output" model of the DEA method was proposed in [16].

The environmental performance of a waste recycling system can be measured using the ratio of GDP per capita to greenhouse gas emissions per capita derived from the waste treatment sector [17]. The analysis of the improvement of the environment per unit of investment costs for waste utilization is given in [18], taking into account the separation coefficient of waste during collection and the ratio of the methods used for waste processing.

In [19], on the basis of the DEA method, multifactor analysis of the efficiency of industrial systems at the stage of energy use and the stage of purification from pollution is carried out in order to accurately assess the overall energy efficiency. Three-stage DEA model is used in [20] to measure environmental efficiency and the marginal cost for CO_2 emission decreasing in 37 industries in China during 2005-2014.

In [21], the basic model of the DEA method was used to assess the ecological efficiency of regions by identifying those that are on the boundary of efficiency, and then the Super-efficiency model of the DEA method was used to rank the most effective regions.

To solve the problem of reducing energy consumption and the environmental im-pact of waste treatment processes, it is necessary to make a comparative assessment of the environmental safety and energy efficiency of the operating of the optimized waste recycling system. Toward this goal, let us assume that the system for OCW recycling includes N storages of wastes and M recycling technologies. According to DEA method terminology, all these objects under consideration will be called further as DMUs (Decision Making Units). The article proposes an approach based on the DEA method to assess heterogeneous indicators characterizing DMUs of a complex structured system for the processing of technogenic oil-contaminated wastes. Based on the proposed approach, an algorithm has been developed for a multi-criteria comparative assessment of environmental safety and energy efficiency of functioning of waste processing system facilities, taking into account assessments of resource value and reuse potential.

In the analyzed system, the waste storage facilities and technologies for waste processing are considered as interrelated objects (DMUs) for comparative evaluation.

2. Multi-criteria assessment based on DEA-method

The paper proposes a unified approach based on the DEA method to solve the problem of comparative evaluation of heterogeneous indicators (criteria) characterizing DMUs of a complex structured system for the processing of technogenic waste from the oil and gas industry.

One of the most important quantitative indicators for estimating the efficiency of the secondary use of the waste is the resource value. The resource value of wastes is a quantitative assessment of their physicochemical composition and properties, which determines the degree of suitability of waste for use as material resources in recycling technologies associated with their recycling, recovery and regeneration. The original approach to estimating the resource value proposed in [22] is based on the DEA method. In [23], it is suggested to consider the reuse potential as the complex quantitative criterion of the possibility of effective waste recycling, assessed by the combination of heterogeneous indicators (technological, energetic, environmental, economical, social, technical, etc.) and the resource value of OCW. The estimation of the reuse potential is also based on the DEA method.

Similar to assessments of the resource value and the reuse potential on the basis of the DEA method, one can obtain comparative estimates of DMUs of a complex structured system for the recycling of technogenic oil-contaminated wastes according to heterogeneous indicators (criteria) characterizing the overall efficiency of the system operation. It is proposed to consider the waste storage facilities and the technologies for waste processing as interrelated DMUs in the analyzed system for comparative evaluation.

Problem of mathematical programming (MPP) of searching for the maximum of objective function F_z [14], formulated on the basis of CCR model of DEA method that allows one to obtain evaluations of Z DMUs, can be written in the following form:

$$F_z(X_z, Y_z) = \frac{\sum_{j=1}^{J} u_{jz} y_{jz}}{\sum_{i=1}^{H} v_{iz} x_{iz}} \rightarrow \max_{U,V \in G} \quad (1)$$

subject to the following constraint:

$$\frac{\sum_{j=1}^{J} u_{jz} y_{jz}}{\sum_{i=1}^{H} v_{iz} x_{iz}} \leq 1, \ \forall \begin{pmatrix} z = \overline{1,Z};\ j = \overline{1,J}; i = \overline{1,H}; \\ u_{jz} > 0; v_{iz} > 0 \end{pmatrix} \quad (2)$$

providing that the numerical values of the relative assessments should belong to the interval [0; 1].

Here the objective function F_z represents the ratio of the weighted sum of the output parameters $y_{jz}, j = \overline{1,J}$ that positively affect the estimated indicator, to the weighted sum of the input parameters $x_{iz}, i = \overline{1,H}$ that negatively affect the estimated indicator; $u_{jz}, j = \overline{1,J}, z = \overline{1,Z}$ and $v_{iz}, i = \overline{1,H}, z = \overline{1,Z}$ are input and output "weights". The number H of input parameters and number J of output parameters depend on the physical meaning of the evaluated function F_z, as well as on the possibility of determining the numerical values of inputs and outputs or the characteristics of the compared DMUs used to calculate values of inputs or outputs in particular cases.

Solution of the problem (1) under constraint (2) gives the relative values F_z, $z = \overline{1,Z}$ of efficiency estimation for Z DMUs and appropriate weight coefficients $u_{jz} \in G, j = \overline{1,J}, z = \overline{1,Z}$ и $v_{iz} \in G, i = \overline{1,H}, z = \overline{1,Z}$, maximizing the functional (1). Formulated problem (1)-(2) is related to the problems of mathematical programming (MPP) that can be solved by standard optimization methods [24].

As a result of the solution of Z MPPs, several DMUs in the analyzed group can at once have an assessment $F_z = 1$, meaning that these DMUs have maximum efficiency in the system. The remaining DMUs with a rating of $0 < F_z < 1$ are considered as less efficient ones regarding criterion (1). When several DMUs have maximum possible efficiency ($F_z = 1$), the results of MPP (1)-(2) solution can be not representative from point of view of ranking DMUs in the analyzed group regarding criterion (1).

Super-efficiency model of DEA method is used for comparative evaluation in the cases when it is important to determine which DMU demonstrates the highest efficiency according to the chosen criterion [15].

Then, similar to (1)-(2), the following mathematical programming problem based on Super-efficiency model should be formulated for the comparative evaluation of the DMUs:

$$S_z(X_z,Y_z)=\frac{\sum_{j=1}^{J}\bar{u}_{jz}y_{jz}}{\sum_{i=1}^{H}\bar{v}_{iz}x_{iz}}\rightarrow\max_{\bar{U},\bar{V}\in\bar{G}} \tag{3}$$

$$\frac{\sum_{j=1}^{J}\bar{u}_{jz}y_{jz}}{\sum_{i=1}^{H}\bar{v}_{iz}x_{iz}}\leq 1\ \forall\begin{pmatrix} z=\overline{1,Z};\ z\neq k;\ j=\overline{1,J};\\ i=\overline{1,H};\bar{u}_{jz}>0;\bar{v}_{iz}>0 \end{pmatrix} \tag{4}$$

where $\bar{U}=(\bar{u}_{jz}),j=\overline{1,J};\bar{V}=(\bar{v}_{iz}),\ i=\overline{1,H}$ are vectors of unknown weight coefficients for each z-th DMU, $z=\overline{1,Z}$. It is important to underline that output weights $\bar{u}_{jk},j=\overline{1,J}$ and input weights $\bar{v}_{ik},i=\overline{1,H}$ for k-th DMU with efficiency estimation $F_k=1$ are eliminated from the solution set in problem (1)-(2), where k is the sequential number of DMU having assessment $F_k=1$.

Solution of the problem (3)-(4) gives the relative value $S_z,z=\overline{1,Z}$ of efficiency estimation belonging to the interval [0,∞). As a result, the DMU having maximum estimation S_z can be considered as the most effective one in the analyzed system.

3. Algorithm for evaluation of environmental safety and energy efficiency of the system for waste recycling

Based on the approach described above, the special algorithm has been developed for a multi-criteria comparative evaluation of environmental safety and energy efficiency of waste processing system operating taking into account assessments of resource value and reuse potential of DMUs. This algorithm represented in Figure 1 comprises 5 consecutive stages. On each stage the above-described MPPs (MPPs 1-5 in Figure 1), formulated in the form (1) - (2) on the basis of the CCR model and in the form (3) - (4) on the basis of Super-efficiency model of the DEA method, are solved to obtain comparative assessments of N waste storage facilities and M technologies for waste recycling considered as compared DMUs in the analyzed system.

On the first stage in order to determine the comparative significance of N storages (DMUs) as potential sources of the secondary resources the problem of comparative evaluation of the resource value (Problem 1 in Figure 1) is solved [23].

As DMU's inputs, those smaller amounts are preferable for the higher relative re-source value, the following waste components have been selected in MPP 1: average-weighted content of water, asphaltenes and resins, mechanical and mineral impurities, and sulfur (Figure 1, inputs 1-4). As DMU's outputs, those bigger amounts result in the higher relative resource value of the storage, the following OCW components have been selected in Problem 1: average-weighted content of light hydrocarbons and the ratio of the mass of light carbon-hydrogens to the total mass of harmful impurities and water (Figure 1, outputs 5-6).

Solutions of the MPPs formulated in the forms (1)-(2) and (3)-(4) based on CCR and Super-efficiency models of the DEA method provide relative estimations F_n^1 and S_n^1, $n=\overline{1,N}$ of resource value of n-th DMU (OCW' storage), respectively. These estimations can be obtained for all N storages of the analyzed group. As a result, the obtained assessments S_n^1, $n=\overline{1,N}$ can be compared and the storage with the biggest resource value of OCW can be determined in the analyzed group.

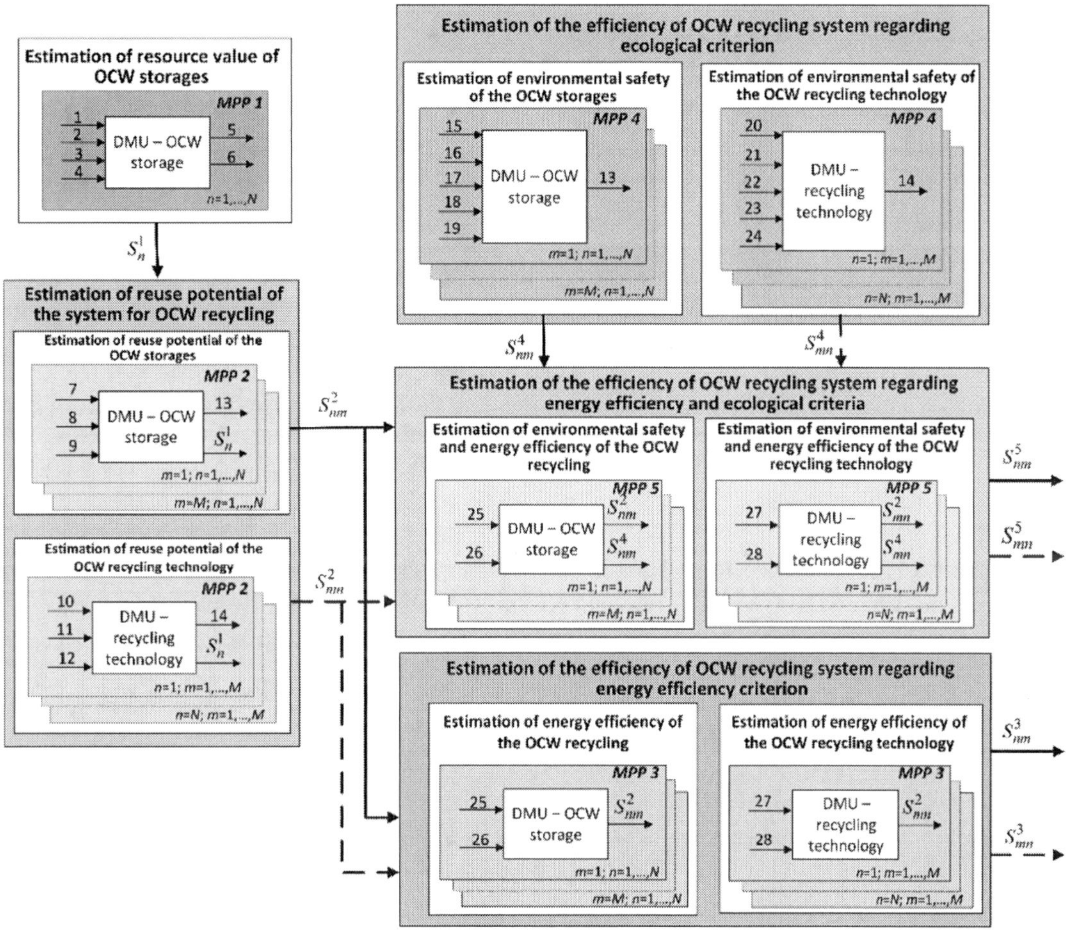

Figure 1. Algorithm for evaluation of environmental safety and energy efficiency of the system for waste recycling.

On the second stage the resource potential is proposed to be evaluated separately for two types of the compared DMUs: storages and technologies. The resource potential of the n-th storage characterizes the cost of useful products obtained as a result of recycling the stored wastes taking into account the cost of recycling process. The resource potential of the *m*-th processing technology characterizes the cost of the useful products obtained as a result of this technology application for the waste utilization in specific storage taking into account the cost of recycling process.

To evaluate the reuse potential, Problem 2 (Figure 1) formulated in the forms (1)-(2) and (3)-(4) is solved. As DMU' inputs, those smaller amounts are preferable for the higher relative reuse potential value of the DMUs, the following parameters have been selected: duration of the recycling process, reagents mass and energy (fuel) consumption (Figure 1, inputs 7-9 and 10-12). As outputs, those bigger amounts are preferable for the higher relative reuse potential value of the DMUs, the following parameters have been selected in the Problem 2: mass of valuable recycling products and estimation S_n^1, $n=\overline{1,N}$ of resource values of *n*-th DMU, obtained as a result of solution of Problem 1 in the form (3)-(4) (Figure 1, outputs 13 and 14).

Solutions of the Problems 2 in the forms (1)-(2) and (3)-(4) based on the CCR and Super-efficiency models of DEA method, respectively, provide the relative values F_{nm}^2 and $S_{nm}^{2.1}$, $n=\overline{1,N}$, $m=\overline{1,M}$ of reuse potential, estimating possibility of effective recycling of OCW in *n*-th storage (DMU) using *m*-th recycling technology. Analysis of the estimations $S^2_{nm}, m=\overline{1,M}, n=\overline{1,N}$ obtained for all N storages allows to determine the storage with the biggest reuse potential in the in the analyzed system.

Similarly, solutions of the Problem 2 give the relative values F^2_{mn} and S^2_{mn} $n=\overline{1,N}$, $m=\overline{1,M}$ of reuse potential, estimating the ability of m-th recycling technology (DMU) to use the waste in n-th storage as a resource. These estimations can be obtained for all M technologies in analyzed system. Analysis of estimations S^2_{mn}, $n=\overline{1,N}$, $m=\overline{1,M}$ allows to determine the technology with the biggest reuse potential in the in the analyzed system.

On the third stage a multi-criterion assessment of a system for waste recycling according to the criterion of energy consumption is proposed taking into account the energy costs of waste transporting from storage facilities to the locations of stationary recycling plants or the cost of transporting mobile recycling installations to the storage facilities. To determine comparative assessments of the DMUs efficiency regarding the criterion of energy consumption, Problem 3 is solved for two types of compared DMUs (storages and technologies, see Figure 1). The input parameters are the indicated specific energy costs for transportation and its duration (inputs 25, 26 and 27, 28 in Figure 1). The output parameters are the assessments S^2_{nm} and S^2_{mn} of resource potential obtained as a result of the decision of the Problem 2.

Solutions of the Problem 3 formulated in the forms (1)-(2) and (3)-(4) based on CCR and Super-efficiency models of DEA method, respectively, provide the relative assessments F^3_{nm} and S^3_{nm}, $m=\overline{1,M}, n=\overline{1,N}$ for energy efficiency of waste recycling in the compared storages. Analysis of the assessments S^3_{nm}, $m=\overline{1,M}, n=\overline{1,N}$ allows determining the storage in which the recycling process using all M technologies requires minimum energy consumption.

Similarly, solutions Problem 3 provide the relative assessments F^3_{mn} and S^3_{mn}, $m=\overline{1,M}, n=\overline{1,N}$ for energy efficiency of the n-th technology application for recycling waste in the n-th storage. Analysis of the assessments S^3_{mn}, $m=\overline{1,M}$, $n=\overline{1,N}$ allows determining the most energy saving technology for the recycling waste in all N storages of the analyzed system.

To estimate the environmental safety of the recycling system operation, Problem 4 is formulated and solved on the fourth stage of algorithm for two types of comparison DMUs: storages and technologies (Figure 1). As DMU' inputs, those smaller values are preferable for the higher environmental safety of DMU, the following parameters have been selected in the Problem 4: the overall emissions of greenhouse gases, sulphur dioxide and mineralex, mass of not processed part of mineral and mechanical impurities, and mass of polluted water formed in the recycling process (inputs 15-19 and 20-24, Figure 1). As DMU' output, those bigger value is preferable for the higher environmental safety, the mass of useful recycling products is considered (outputs 13-14, Figure 1).

Solutions of the Problem 4 in the form (1)-(2) and (3)-(4) based on the CCR and Super-efficiency models of the DEA method give, respectively, the relative values F^4_{nm} and $S^{4.1}_{nm}$, $n=\overline{1,N}, m=\overline{1,M}$, estimating the comparative environmental safety of waste recycling in n-th storage using m-th recycling technology. These estimations can be obtained for all N storages. The analysis of assessments S^4_{nm}, $n=\overline{1,N}, m=\overline{1,M}$, defines which storage in the analyzed system has the highest environmental safety with respect to all M recycling technologies

Similarly, solutions of the Problem 4 in the form (1)-(2) and (3)-(4) based on the CCR and Super-efficiency models of DEA method provide, respectively, the relative values F^4_{mn} and $S^{3.2}_{mn}$, $m=\overline{1,M}$, $n=\overline{1,N}$, estimating the environmental safety of usage of m-th technology for recycling the waste in n-th storage. These estimations can be obtained for all M recycling technologies. The analysis of assessments S^4_{mn}, $m=\overline{1,M}$ shows which technology has the highest environmental safety for waste recycling in all N storages of the analyzed system.

IOP Conf. Series: Materials Science and Engineering 1089 (2021) 012023 doi:10.1088/1757-899X/1089/1/012023

Finally, at the fifth stage of algorithm Problem 5 is formulated and solved to estimate the system for waste recycling regarding either energy efficiency or environmental safety. As DMU' inputs the same input parameters as in the Problem 3 have been selected in the Problems 5 (inputs 25, 26 and 27, 28, Figure 1). As DMU' outputs the following parameters have been selected in the Problems 5: the estimations S^2_{nm} and S^2_{mn} of reuse potential obtained as solutions of the Problems 2 in the form (3)-(4), respectively; the estimations S^4_{nm} and S^4_{mn} of the environmental safety obtained as solutions of Problems 4 in the form (3)-(4), respectively.

Considering the storages as DMUs, solutions of the Problem 5 in the forms (1)-(2) and (3)-(4) based on the CCR and Super-efficiency models of DEA method provide, respectively, the relative values F^5_{nm} и S^5_{nm}, $m=\overline{1,M}$, $n=\overline{1,N}$ estimating energy efficiency and environmental safety of waste recycling in n-th storage using m-th recycling technology. These estimations can be obtained for all N storages of OCW in the analyzed system. Analysis of estimations S^5_{nm}, $m=\overline{1,M}$, $n=\overline{1,N}$, defines the best storage in the analyzed system from the point of view of the energy saving and environmental safety.

Similarly, solutions of the Problem 5 for technologies as DMUs provide the relative values F^5_{mn} and S^5_{mn}, $m=\overline{1,M}$, $n=\overline{1,N}$ estimating the energy efficiency and environmental safety of usage of m-th technology for recycling of waste in n-th storage. These estimations can be obtained for all M recycling technologies of the analyzed system. Analysis of estimations S^5_{mn}, $m=\overline{1,M}$, $n=\overline{1,N}$ defines the most energy saving and environmentally safe technology in the analyzed system.

4. Implementation of the developed algorithm and results analysis

The developed algorithm has been tested for the evaluation of the waste recycling system consisting of N=15 storages of waste and M=8 recycling technologies located in Samara region of Russian Federation. As the results of Problems 1-5 solutions the relative values of assessments F^γ_{nm}, F^ψ_{mn}, S^γ_{nm}, S^ψ_{mn}, $n=\overline{1,N}$, $m=\overline{1,M}$, $\gamma=\overline{1,5}$, $\psi=\overline{2,5}$ have been obtained (Figure 1).

Generalized analysis of the obtained results can be performed using cumulative assessments calculated by formulas:

$$D1^\gamma_n = \sum_{m=1}^{M} S^\gamma_{nm}, \quad n=\overline{1,N}, \gamma=\overline{1,5}; \tag{5}$$

$$D2^\psi_m = \sum_{n=1}^{N} S^\psi_{mn}, m=\overline{1,M}, \psi=\overline{2,5}. \tag{6}$$

Cumulative estimations $D1^\gamma_n$, $\gamma=\overline{1,5}$, obtained by expression (5), for γ-th MPP characterize generalized assessments for M technologies applied for recycling waste in n-th storage, $n=\overline{1,N}$ (Table. 1). Cumulative estimations $D2^\psi_m, \psi=\overline{2,5}$, obtained by expression (6) for ψ-th MPP characterize generalized assessments for the case of application of m-th technology ($m=\overline{1,M}$) for recycling waste in N storages (Table 2).

To perform the final analysis, it is proposed to formulate MPP on the basis of CCR model of DEA method in the form similar to (1)-(2), that allows to obtain a comparative assessment of Z DMUs of the recycling system regarding to the efficiency indicators:

$$Q_z(Y_z) = \sum\nolimits_{\varphi=1}^{\Phi} u_{\varphi z} y_{\varphi z} \to \max_{U\in\Omega}. \tag{7}$$

The following constraint

$$\sum_{\varphi=1}^{\Phi} u_{\varphi z} y_{\varphi z} \leq 1, \ \forall \left(z = \overline{1,Z}; \ \varphi = \overline{1,\Phi}; \ u_{\varphi z} > 0 \right) \tag{8}$$

provides belonging the numerical value of the relative assessments Q_z to the interval [0; 1]. In (7)-(8) $u_{\varphi z}, z = \overline{1,Z}; \ \varphi = \overline{1,\Phi}$ are coefficients that characterize "weights" of outputs $y_{\varphi z}, \varphi = \overline{1,\Phi}$; $y_{\varphi z}, \varphi = \overline{1,\Phi}$ are output parameters that positively affect the assessment Q_z and represent cumulative assessments $D1_n^{\gamma}$, $n = \overline{1,N}, \gamma = \overline{1,5}$ obtained by (5) or $D2_m^{\psi}, m = \overline{1,M}, \psi = \overline{2,5}$ obtained by (6). Number of output parameters $\Phi = 5$ used for cumulative assessment Q_z is determined by number of formulated and solved MPP according to algorithm of evaluation of environmental safety and energy efficiency of the waste recycling system, including *N* storages and *M* recycling technologies (Fig. 1).

If several DMUs have maximum values of assessments $Q_z = 1$, then the results of MPP (7)-(8) solution can be not representative from point of view of ranking DMUs in the analyzed system, and therefore, the Super-efficiency model of DEA method is again should be applied for comparative evaluation.

Then, similar to (7)-(8), the following mathematical programming problem based on Super-efficiency model is formulated for the comparative evaluation of the DMUs as follows:

$$R_z(Y_z) = \sum_{\varphi=1}^{\Phi} \bar{u}_{\varphi z} y_{\varphi z} \to \max_{\bar{U} \in \bar{\Omega}}, \tag{9}$$

$$\sum_{\varphi=1}^{\Phi} \bar{u}_{\varphi z} y_{\varphi z} \leq 1, \ \forall \left(z = \overline{1,Z}; \ \varphi = \overline{1,\Phi}; \ \bar{u}_{\varphi z} > 0 \right) \tag{10}$$

where $\bar{U} = (\bar{u}_{\varphi z}), \varphi = \overline{1,\Phi}$ is vector of weight coefficients for *z*-th DMU, $z = \overline{1,Z}$, that represent unknown parameters excluding coefficients $\bar{u}_{\varphi b}$. It is important to underline that "weights" $\bar{u}_{\varphi b}$, $\varphi = \overline{1,\Phi}$ for *b*-th DMU with estimation $Q_b = 1$ are eliminated from the solution set in problem (9)-(10), where *b* is sequential number of DMU having assessment $Q_b = 1$.

Solution of the problem (9)-(10) gives the relative values R_z, $z = \overline{1,Z}$ of the assessments belonging to the interval $[0,\infty)$. As a result, the DMU having maximum estimation S_z can be considered as the most effective one in the analyzed system.

In Table 1, the assessments $R1_n$, $n = \overline{1,15}$, obtained as a result of solution of MPP (9)-(10) formulated using Super-efficiency model of DEA method characterize relative efficiency of *n*-th waste storage in comparison with other storages of the system taking into account all criteria considered as objective functions in Problems 1-5 (Figure 1).

The results in Table 1 show that it is most effective to utilize the waste in the storage №12, which has maximum evaluation $R1_{12} = 1.907$, using all recycling technologies of the analyzed system. At the same time, analyzing the cumulative assessments obtained for Problems 1-5, one can see that the storage №12 can be considered as the best one only regarding resource value ($D1_{12}^{1} = 2.381$) and environmental safety ($D1_{12}^{4} = 8.301$). It is conditionally not effective to recycle waste from the storage №15, having minimum value of assessment $R1_{15} = 0.553$.

In Table 2, the assessments $R2_m$, $m = \overline{1,8}$, obtained as a result of solution of MPP (9)-(10) formulated using Super-efficiency model of DEA method characterize relative efficiency of *m*-th recycling technology in comparison with other technologies of the system taking into account all criteria considered as objective functions in Problems 1-5 (Figure 1).

IOP Conf. Series: Materials Science and Engineering 1089 (2021) 012023 doi:10.1088/1757-899X/1089/1/012023

Table 1. Cumulative assessments $D1_n^{\gamma}$ of environmental safety and energy efficiency of waste storages.

Number of waste storage ($n=\overline{1,15}$)	Assessments $D1_n^{\gamma}$					Assessment $R1_n$
	$\gamma=1$	$\gamma=2$	$\gamma=3$	$\gamma=4$	$\gamma=5$	
1	1.249	5.789	2.708	6.435	3.097	0.909
2	0.850	4.550	2.692	6.059	3.264	0.839
3	0.682	3.299	1.857	4.828	2.182	0.623
4	0.806	5.035	1.113	5.417	1.245	0.764
5	0.440	3.710	1.346	3.878	1.541	0.555
6	0.837	3.811	1.776	5.181	2.024	0.650
7	0.687	4.367	2.182	5.477	2.439	0.724
8	0.629	7.380	5.271	4.347	5.484	1.655
9	0.527	3.794	0.775	4.061	0.856	0.574
10	0.860	4.448	3.186	6.393	4.791	1.068
11	0.653	4.106	1.717	4.873	1.993	0.654
12	2.381	5.631	2.745	8.301	3.078	**1.907**
13	1.063	7.792	2.719	5.578	2.778	1.106
14	0.548	4.015	1.073	4.897	1.233	0.647
15	0.440	3.047	1.804	3.994	2.132	0.553

Table 2. Cumulative assessments $D2_m^{\psi}$ of environmental safety and energy efficiency of recycling technologies.

Number of recycling technology ($m=\overline{1,8}$)	Assessments $D2_m^{\psi}$				Assessment $R2_m$
	$\psi=2$	$\psi=3$	$\psi=4$	$\psi=5$	
1	11.290	3.106	6.924	3.312	0.892
2	10.684	2.784	12.371	3.405	0.998
3	11.290	3.053	12.371	3.681	1.022
4	5.259	1.228	10.259	2.085	0.815
5	3.663	2.815	12.708	4.240	1.038
6	10.678	9.420	2.529	9.513	1.608
7	5.257	4.662	11.642	5.990	1.047
8	12.653	5.896	10.913	5.912	1.167

The results in Table 2 show that it is most effective to apply the technology №6, which has maximum evaluation $R2_6=1.608$, for recycling waste in all storages of the analyzed system. At the same time, analyzing the cumulative assessments obtained for Problems 2-5, one can see that the technology №6 can be considered as the most energy saving ($D2_6^3=9.42$) and environmentally safe ($D2_6^5=9.513$). It is conditionally not effective to recycle waste using technology №4, having minimum value of assessment $R2_4=0.815$.

Obtained results (Tables 1 and 2) can be applied to make the decisions for the optimization recycling system in Samara region in order to improve its operating efficiency. Developed algorithm of multi-criteria evaluation can be considered as a basis for optimization of complex structured recycling system regarding environmental safety and energy efficiency taking into account the resource value and reuse potential of the system elements.

5. Acknowledgments
The reported study was funded by RFBR, project number 20-08-00353.

References

[1] Kosulina T P and Kononenko E A 2012 *Povyshenie ecologicheskoy bezopasnosti utilizacii neftyanyh shlamov* [Increasing environmental safety of product recycling oil slud] *Nauchnyj Zhurnal KubGAU* **78** 84-93 [In Russian]

[2] Garabadzhiu A B et al 2012 *Klaster tehnologicheskih ustanovok pererabotki mnogotonnazhnyh nakopleniy kislyh gudronov i nefteshlamov* [Cluster of technological units for processing large-tonnage accumulations of acid sludge and oil sludge] *Neftepererabotka i Neftekhimiya. Nauchno-Tekhnicheskiye Dostizheniya i Peredovoy Opyt* **9** 37-48 [In Russian]

[3] Ashby M F 2016 Vision: a circular materials economy *Materials and Sustainable Development* (Boston: Butterworth-Heinemann) pp 211-39

[4] Ghisellini P, Cialani C and Ulgiati S 2016 A review on circular economy: the expected transition to a balanced interplay of environmental and economic systems *J. Clean. Prod.* **114** 11-32

[5] Haas W, Krausmann F, Wiedenhofer D and Heinz M 2015 How circular is the global Economy?: an assessment of material flows, waste production, and recycling in the european union and the world in 2005 *J. Ind. Ecol.* **19(15)** 765-77

[6] Murray A, Skene K and Haynes K 2017 The circular economy: an interdisciplinary exploration of the concept and application in a global context *J. Bus. Ethics.* **140(3)** 369–80

[7] Iacovidou E et al 2017 Metrics for optimising the multi-dimensional value of resources recovered from waste in a circular economy: A critical review *J. Clean. Prod.* **166** 910-38

[8] Mazlova E A and Mesheryakov S V 2001 *Problemy utilizacii nefteshlamov i sposoby ih pererabotki* [Problems of oil sludge utilization and methods of their processing] (Moskow: Noosfera) [In Russian]

[9] Christensen T H, Gentil E, Boldrin A, Larsen A W, Weidema B P and Hauschild M 2009 C balance, carbon dioxide emissions and global warming potentials in LCA-modelling of waste management systems *Waste Manag. Res.* **27** 707-15

[10] Allegrini E, Vadenbo C, Boldrin A and Astrup T F 2015 Life cycle assessment of resource recovery from municipal solid waste incineration bottom ash *J. Environ. Manag.* **151** 132-43

[11] Arvidsson R, Fransson K, Froling M, Svanstrom M and Molander S 2012 Energy use indicators in energy and life cycle assessments of biofuels: review and recommendations *J. Clean. Prod.* **31** 54-61

[12] Puig R, Fullana-i-Palmer P, Baquero G, Riba J-R and Bala A 2013 Cumulative Energy Demand indicator (CED), life cycle based, for industrial waste management decision making *Waste Manag.* **33(12)** 2789-97

[13] Fang K, Heijungs R and De Snoo G R 2015 Understanding the complementary linkages between environmental footprints and planetary boundaries in a footprint–boundary environmental sustainability assessment framework *Ecol. Econ.* **114** 218-26

[14] Charnes A, Cooper W W and Rhodes E 1978 Measuring the efficiency of decision-making units *Eur. J. Oper. Res.* **6(2)** 429-44

[15] Chen Y and Du J 2015 Super-Efficiency in Data Envelopment *Analysis Data Envelopment Analysis: A Handbook of Models and Methods* (Boston: Springer) pp 381-414

[16] Dyckhoff H and Allen K 2001 Measuring ecological efficiency with data envelopment analysis (DEA) *Eur. J. Oper. Res.* **132** 312-25

[17] Pais-Magalhães V, Moutinho V and Marques A C 2020 Scoring method of eco-efficiency using the DEA approach: evidence from European waste sectors *Environment, Development and Sustainability: A Multidisciplinary Approach to the Theory and Practice of Sustainable Development* (Berlin: Springer) pp 1-23

[18] Yang Z, Zhou X and Xu L 2015 Eco-efficiency optimization for municipal solid waste

management *J. Clean. Prod.* **104** 242–9
[19] Wu J, Xiong B, An Q, Sun J and Wu H 2017 Total-factor energy efficiency evaluation of Chinese industry by using two-stage DEA model with shared inputs *Ann. Oper. Res.* **255** 257-76
[20] Xie B, Duan N and Wang Y 2016 Environmental efficiency and abatement cost of China's industrial sectors based on a three-stage data envelopment analysis *J. Clean. Prod.* **153** 626-36
[21] Yang Q, Wan X and Ma H 2015 Assessing Green Development Efficiency of Municipalities and Provinces in China Integrating Models of Super-Efficiency DEA and Malmquist Index *Sustainability* **7(4)** 4492-510
[22] Pleshivtseva Yu, Derevyanov M, Kashirskikh D, Pimenov A, Kerov A and Tyan V 2018 Comparative evaluation of the reuse value of storage for oil-contaminated waste based on DEA method *Oil Industry* **11** 139-44
[23] Derevyanov M, Pleshivtseva Y, Afinogentov A, Mandra A and Pimenov A 2019 Simulation and Multi-Objective Evaluation of Reuse Potential of Waste Recycling System for Oil And Gas Industry *XXI Int. Conf. Complex Systems: Control and Modeling Problems (CSCMP)*(Samara: IEEE) 429-34
[24] Michael J Todd 2002 The many facets of linear programming *Mathematical Programming* **91(3)** 417–36

Calculation of electric energy savings and simulation of tank operation with variable extraction of steam and liquid phases of propane-butane mixtures

A V Rulev, S S Kuznezov and V O Sevostiyanov

Department of Heat and Gas Supply, Ventilation, Water Supply and Applied Hydro-Gas Dynamics, Yuri Gagarin State Technical University of Saratov, 77 Polytechnicheskaya street, Saratov, 410054, Russian Federation

E-mail: nautech@inbox.ru

Abstract. Currently, when consumers use various categories of propane-butane mixtures of liquefied petroleum gas as the main or backup energy carrier, natural and artificial regasification systems based on electric or direct fire evaporators with intermediate heat transfer agent or solid-state intermediate heat transfer agent of various evaporation capacity, which operate depending on the design in various evaporation modes, are widely used. An important element for reducing energy consumption in the artificial liquefied petroleum gas regasification systems is the switching of tank to the combined regasification mode. The combined cycle provides two main periods of tank operation: the initial period with natural regasification of liquefied gas in the day tank, ground tube evaporator, and the subsequent period with artificial regasification of the propane-butane mixture in the evaporator with an intermediate heat transfer agent. For the purpose of theoretical justification of the proposed cycle, a mathematical model of combined regasification of liquefied petroleum gases was developed, which comprehensively takes into account the main periods of tank operation: with natural regasification mode in the day tank, ground heat exchanger and artificial regasification mode in the evaporator. The results of the conducted research show that the implementation of the combined regasification cycle in the liquefied gas supply can significantly reduce the energy costs of its evaporation, increases the energy efficiency of liquefied petroleum gas supply systems and provides an average annual savings of 33.7% in electric energy for regasification.

1. Introduction

In modern domestic and foreign experience of gas power supply to houses and industrial facilities located remotely from the main power station, decentralized gas power supply systems fed with propane-butane mixtures (PBM) of liquefied petroleum gases (LPG) from tanks are increasingly used [1-6]. When using liquefied petroleum gases as the main energy carrier in gas tank systems, they are evaporated artificially in evaporators with an intermediate solid-state or liquid heat transfer agent, under conditions of its natural convection [7-10]. An important element for reducing energy consumption in systems of artificial liquefied petroleum gas regasification is the switching of tanks to the combined regasification mode. The combined cycle provides two main periods of tank operation: the initial period with natural regasification of liquefied gas in the day tank, ground tube evaporator and the subsequent period with artificial LPG regasification in the evaporator with an intermediate heat transfer agent. The natural regasification mode is possible due to the natural heat of the surrounding soil, when propane-butane mixtures, even with a high content of butane fractions, can be partially evaporated in the ground heat exchangers, but currently the heat of the soil mass is not used for the regasification of liquefied petroleum gas with a high concentration of butane fractions [11]. At the same time, electric energy is a more valuable energy carrier and is characterized by a higher cost compared to LPG used as fuel in direct fire evaporators [12].

In order to save heat for regasification, the principle and technical solution was justified, consisting in the evaporation of part of the liquid phase directly in the underground tank 2 (Figure 1), enclosed in a polymer case 3 filled with an inert low-boiling liquid and a flow ground evaporation pipeline 1 by using the natural heat of the soil with alternate supply of the steam and liquid phases to the evaporator using the phase switch valve 7.

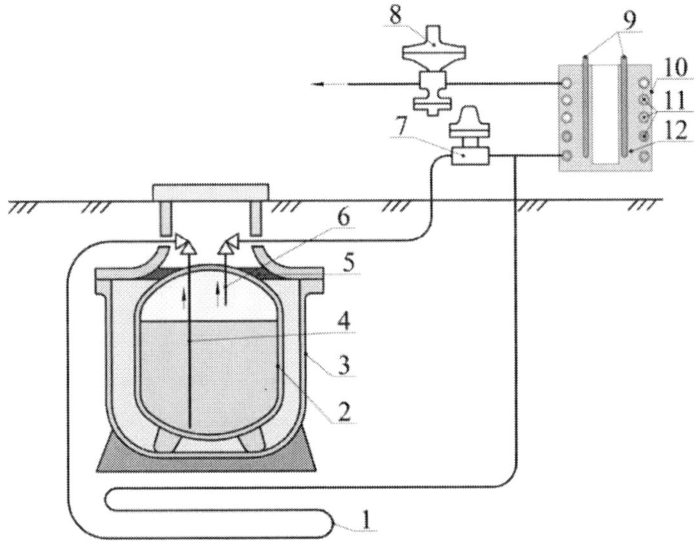

Figure 1. Energy saving diagram of an electric LPG evaporator: 1 – ground tube evaporator; 2 – LPG steel tank; 3 – polymer case filled with antifreeze; 4,6 – liquid and steam phase tubes; 5 – sealed gasket; 7 – steam and liquid phase switching valve; 8 – low pressure control; 9 – tubular electric heaters; 10 – electric evaporator; 11 – evaporation coil; 12 – aluminum intermediate heat transfer medium.

According to the proposed method, during the initial period of operation, the steam phase from underground tank 2 (Figure 1) is fed through the steam stand pipe 6, then through steam and liquid phases selector valve 7 to evaporation coil 11, overheats in it, and then goes to the consumer through the output pipeline and low-pressure regulator 8.

When the pressure and temperature of the saturated steam phase in tank 2 decreases to the calculated value, valve 7 closes the steam phase supply through stand pipe 6 and opens the liquid phase supply to ground tube evaporator 1. The resulting pressure drop between tank 2 and evaporator 1 provides a rise of the liquid phase through the supply pipeline to ground tube evaporator 1, where it partially evaporates due to the heat of the soil to the dryness factor X_s until the boiling point of the propane-butane mixture $t_{e.s}$ rises to a temperature close to the minimum temperature of the soil on the axis of its location $t_{s.min}$. The liquid phase not evaporated in ground evaporator 1 with dryness factor X_s enters evaporation coil 9, where it is completely evaporated due to the heat supplied from the tubular electric heaters 9 through a layer of solid-state heat transfer agent 12. Due to the heat leak from the ground, the steam pressure in tank 2 increases, and valve 7 opens again. The gas is supplied to the consumers again by supplying the vapor phase from tank 2. The cycle is repeated. The proposed solutions provide for the use of a continuous-flow ground evaporator pipeline 2 located at the bottom of the pit around tank 1 at low gas flow rates (Option A, Figure 2). In cases of higher evaporation capacities, the principle of radial horizontal location of pipeline 2 is proposed, which is implemented on the basis of directional drilling of horizontal wells, followed by the installation of U-shaped pipe elements 3 in them (Option B, Figure 2).

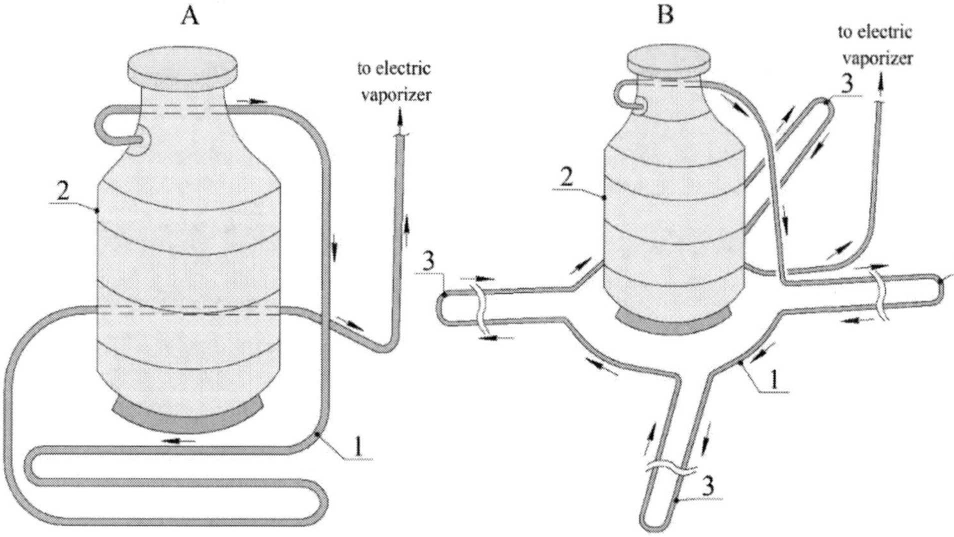

Figure 2. Options for continuous-flow ground evaporator pipeline.

2. Materials and Methods

For the purpose of theoretical justification of the proposed cycle, a mathematical model of combined liquefied gas regasification was developed.

For the convenience of research, we divide the complex problem of combined regasification into two interrelated subtasks in accordance with the tank operation modes.

We divide the total duration of tank operation in the mode of natural regasification τ into S calculated time intervals of duration $\Delta\tau$. b means the beginning of the calculated time interval, and e means the end of the calculated time interval.

Assuming that the tank technical characteristics change very slightly at the point of a separate time interval $\Delta\tau$, the heat balance equation for the i-th time interval is as follows:

$$K_i F_{mix,i}\left(t_s - \frac{t_{l,i}^b + t_{l,i}^e}{2}\right)\Delta\tau + \left(c_m M_{m,i} + c_m M_{l,i}\right)\left(t_{l,i}^b + t_{l,i}^e\right) = rG\Delta\tau, \quad (1)$$

where:
K_i is the tank heat transfer coefficient, kJ/(h·m^2·K);
$F_{\mathrm{mix},i}$ is the tank wetted surface, m^2;
$t_{l,i}$ is the temperature of the LPG liquid phase in the tank, °C;
t_s is temperature of the natural ground soil on the axis of tank location, °C;
c_m, c_l is mass heat capacity of the metal shell and liquefied natural gas, kJ/kg·K;
$M_{m,i}$ is the mass of the metal shell corresponding to the tank wetted surface, kg;
$M_{l,i}$ is mass of liquid in the tank, kg;
r is heat of vaporization of liquefied gas, kJ/kg;
G is gas flow rate from the tank, kg/h.

After the transformations, equation (1) will be as follows:

$$t^{e}_{l,i}=\frac{\Delta\tau\left(K_i F_{mix,i}\left(t_s-\frac{t^{b}_{l,i}}{2}\right)-rG\right)+t^{b}_{l,i}\left(c_m M_{m,i}+c_m M_{l,i}\right)}{0,5K_i F_{mix,i}\Delta\tau+c_m M_{m,i}+c_m M_{l,i}}, \tag{2}$$

Equation (2) is applied by iteration method sequentially for each time interval in accordance with the algorithm presented in [13].

The boiling point of PBM at the ground evaporation pipeline outlet is determined depending on the minimum temperature of the soil on the axis of its location $t_{gr.\min}$ (in March) by the formula:

$$t_{e.s}=t_{s.\min}-\Delta t_{\min}, \tag{3}$$

$\Delta t_{\min}$ – minimum temperature pressure in the ground-liquefied gas system, ^{0}C.

The proportion of the liquid phase numerically equal to the dryness factor that can be evaporated in a ground evaporation pipeline is determined by the known value of the boiling point of PBM at its outlet $t_{e.s}$.

$$X_s=P_{mix}\cdot\left(\frac{\psi}{P_{mix}-10^{A_b-\frac{B_b}{C_b+t_{e.s}}}}+\frac{1-\psi}{P_{mix}-10^{A_p-\frac{B_p}{C_p+t_{e.s}}}}\right); \tag{4}$$

where:

P_{mix} – the LPG pressure in the evaporator, Pa;

ψ – propane content in the liquid phase of the initial PBM, mol;

A_b, B_b, C_b, A_p, B_p, C_p – the weighted average coefficients for butane fraction and the propane within a certain range of temperatures.

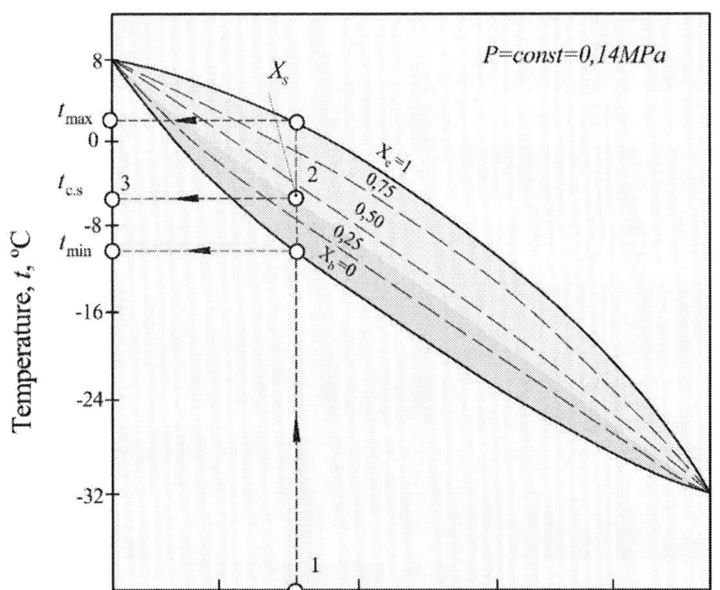

Figure 3. Chart for determining the fractional amount of liquid phase X_s that can be vaporized in a ground evaporator pipeline.

The subsequent period of tank operation provides a combined regasification mode: the G_v evaporator generates a part of the steam phase, the day tank and the continuous-flow ground evaporator pipeline due to the heat leak from the surrounding soil generate the other part of G_p.

The extraction of the steam and liquid phases of liquefied petroleum gas determines the corresponding dynamics of the mass of liquefied gas in the tank M_l, which, in turn, causes changes in other technical parameters: the level of filling the tank with liquefied gas φ, the tank wetted surface F_{mix}, the mass of the tank M_m, the heat transfer coefficient of the tank K.

During the extraction of steams from the tank, the fractional composition of the LPG liquid phase changes, and the content of the butane fraction increases in it. Since the pressure of the steam blanket in tank P_{comb} is constant, the change in the fractional composition of the liquid phase is accompanied by an increase in its temperature t_l.

We divide the total duration of tank operation in the combined regasification mode τ_{comb} into T calculated time intervals of $\Delta\tau$ duration.

Assuming that the technical parameters of the tank change very slightly during certain time intervals, the heat balance equation for the k-th time interval of tank operation in the combined regasification mode will be similar to equation (1).

The total amount of gas obtained due to the natural evaporation efficiency of the tank when the tank is operating in the natural and combined regasification modes is determined by the formula:

$$M = G\tau + \sum_{1}^{S} G_p \Delta\tau + \sum_{1}^{T} G_p \Delta\tau . \tag{5}$$

The contribution of the natural evaporation efficiency of the tank to the total evaporation capacity of the tank is

$$Z = \frac{M}{M_l^0 \left(\varphi_b - \varphi_{ost}\right)}, \tag{6}$$

where:
M_l^0 is a mass of gas in the tank at 100% filling, kg;
φ_b, φ_{ost} is the initial and residual levels of filling the tank,%.

Saving of electric energy for the LPG evaporation is defined as the ratio of the amount of heat obtained due to the natural evaporative capacity of an underground tank Q_n^t and a flow ground evaporator Q_n^{gr} to the total amount of heat energy required for evaporation of the calculated mass flow rate of gas G.

$$\Delta Q = \frac{Q_n^t + X_{gr}(Q_s - Q_n^t)}{Q_s}. \tag{7}$$

X_{gr} - the percentage of the liquid phase (dryness factor) that can be evaporated in the ground evaporation pipeline, unit fraction

The expanded formula for determining the amount of heat obtained by artificial heat supply is as follows:

$$\begin{aligned} Q_a = {} & (1 - X_{gr})[(\xi_b - \xi_r) - (G_p\tau_{cool} + \xi_{al})][r + C_l\,(t_b(P_b) - t_e(P_p)) + C_v\,(t_h - t_e)] + \\ & + C_v\,\{[G_p\tau_{cool} \cdot (t_h - \frac{t_{b.cool}(P_b) + t_{e.cool}(P_p)}{2})] + \xi_{al}(t_h - t_{e.cool}(P_p))\}, \end{aligned} \tag{8}$$

where:
τ_{cool} – duration of the LPG cooling period in the underground tank, h;

ξ_b, ξ_r, ξ_{al} – accordingly, the mass of the liquid phase at the beginning of the cooling period, before the next refueling and during the periods of alternate sampling of liquid and steam phases of LPG from the tank, kg;
C_l, C_v – accordingly, the averaged specific heat capacities of the liquid and steam phases of LPG, kJ/kg*K;
$t_{b.cool}(P_b), t_{e.cool}(P_p)$ – accordingly, the temperature of the liquid phase of LPG at the beginning and end of the cooling period, °C;
P_b, P_p – pressure of saturated LPG vapors at the beginning and end of the cooling period, MPa;
t_b, t_e, t_h – respectively, the temperature of the beginning and end of the boiling of LPG and overheating of its vapors, °C.

To verify the reliability of the proposed mathematical model, field observation was conducted on an experimental unit. An underground tank with a volume of 2.5m^3 was used as a liquefied petroleum gas source. Studies were conducted with a constant extraction of LPG at 3.0 m/h. The initial gas composition in terms of a binary mixture: propane - 64.2%, n-butane - 35.8%. Thermal characteristics of the soil (loam): density p=1793kg/m, humidity ω=12.22 %, thermal conductivity coefficient $\lambda = 1.3$ W/(m*°C) [14].

3. Results

The results of experimental studies of operational parameters in an underground tank with combined extraction of the liquid and steam phases are presented in Figure 4 as data points. Figure 4 shows the results of numerical implementation of the mathematical model as solid lines.

Experimental values of parameters of the tank at the beginning of its operation were used as initial parameters for the calculation. The design parameters of the underground tank with a volume of 2.5m^3 and its technical characteristics, as well as the heat transfer coefficient, taking into account the thermal characteristics of the soil mass, were adopted from open data.

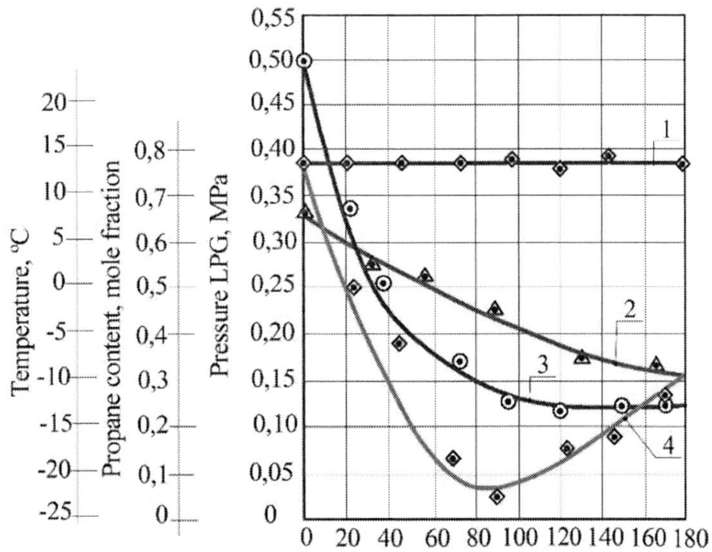

Figure 4. Dependences of the temperature, component composition, and pressure of liquefied petroleum gas on the duration of the experiment: 1 - soil temperature, °C; 2 - component composition of LPG, mole fraction; 3 - LPG pressure, MPa; 4 - LPG temperature, °C.

As we see on Figure 4, the theoretical values of the studied parameters agree quite well with the experimental data. The maximum discrepancy between the theoretical and experimental values with a confidence probability of 0.95 is:

- by the liquefied gas temperature 16.4%;
- by liquefied gas pressure 9.6%;
- by the component composition of liquefied gas 9.1%.

Calculations performed using formulas (7) - (8) for the underground tank with a volume of 10 m^3 and an evaporator with a capacity of 32 kg/h located in a moderately cold climate area, for gas with a butane content of 75 mol.%, show that the application of the proposed scheme provides for at least 33,7% of the average annual savings in electrical energy for the LPG evaporation.

4. Conclusion

Methodological provisions were developed for the use of natural soil heat for regasification by evaporation of part of the liquid phase in a continuous-flow ground heat exchanger and an underground storage tank, which provides an average annual savings of 33.7% in electric energy for regasification. Experimental studies confirm the reliability of the proposed mathematical model. The results of theoretical and experimental studies show that the natural evaporation capacity of LPG day tanks makes a significant contribution to the overall evaporation capacity of tank. In this regard, the switching of tank with artificial regasification to the combined LPG regasification mode provides an adequate reduction in energy consumption for the liquefied gas evaporation. At the same time, the additional costs of installing the liquid and steam phase selection valve are compensated within 1 year.

References

[1] Treloar R D 2010 *Gas Installation Technology 2nd edition* (New York: John Wiley & Sons Ltd.)

[2] Saxon F 2006 *Tolley's Basic Science and Practice of Gas Service* (Amsterdam: Elsevier Science)

[3] Synáka F, Čulíka K, Rievaja V and Gaňaa J 2019 Liquefied petroleum gas as an alternative fuel *Transportation Research Procedia* **40** 527-34

[4] Rachevskij B S 2009 *Szhizhennye uglevodorodnye gazy* [Liquefied petroleum gas] (Moscow: Neft' i gaz) [In Russian]

[5] Wang X and Economides M 2009 *Advanced Natural Gas Engineering* (Gulf Publishing Company)

[6] Shi G-H, Aye L A, Liu Y-C and Du X-J 2018 Dynamic simulation of liquefied petroleum gas vaporisation for burners *Applied Thermal Engineering* **137** 575-83

[7] Fedyaev A V and Lachkov G G 2017 Effektivnost' gazifikacii kotel'nyh udalennogo regiona szhizhennym uglevodorodnym gazom mestnogo mestorozhdeniya [Efficiency of gasification of boiler houses in a remote region with liquefied petroleum gas from a local field] *Vestnik IGTU* **10(129)** 150-9 [In Russian]

[8] Shevtsov S A, Kargashilov D V and Shutkin A N 2018 Fire and Explosion Safe Technology of Storage and Regasification of Liquefied Petroleum Gas *Chemical and Petroleum Engineering* **54** 38-40

[9] Shevcov S A, Kargashilov D V and Habibov M A 2016 Osobennosti proektirovaniya rezervuarnyh ustanovok szhizhennyh uglevodorodnyh gazov v sistemah avtonomnogo gazoenergosnabzheniya s uchetom ocenki pozharnogo riska [Design features of liquefied petroleum gas tank installations in Autonomous gas power supply systems with consideration for fire risk assessment] *Pozharnaya bezopasnost'* **3** 150-5 [In Russian]

[10] Ajeysurya S 2016 Hazop study in LPG installation in process industry *Advances in Natural and Applied Sciences* **10** 149-53

[11] Kuritsyn B N, Osipova N N and Kuznetsov S S 2016 On the definition of operational

parameters of tank installations in combined regasification of liquefied petroleum gas *Scientific Herald of the Voronezh State University of Architecture and Civil Engineering. Construction and Architecture* **1(29)** 7-12

[12] Shurayts A L, Rulev A V and Usacheva T A. 2010 *Sistemnye issledovaniya po povysheniyu intensivnosti teploobmena regazifikatorov snizhennogo uglevodorodnogo gaza* [System studies on increasing the heat exchange intensity of reduced hydrocarbon gas regasifiers] (Saratov: SSTU) [In Russian]

[13] Kuritsyn B N, Osipova N N and Kuznetsov S S 2011 *Razrabotka matematicheskoj modeli kombinirovannoj regazifikacii szhizhennogo uglevodorodnogo gaza* [Development of a mathematical model of combined regasification of liquefied petroleum gas] *Vestik SGTU* **4 (59)** 218-24

[14] Osipova N N, Kuznezov S S and Rulev A V 2019 Modeling and studying basic parameters of tank sets in combined regasification *IOP Conf. Ser.: Mater. Sci. Eng.* **687** 044020

Ecology and STEM

E V Esina[1], M A Velichko[2], Yu P Gladkikh[2], O N Satler[2] and R A Nortsova[3]

[1]Department of Foreign Languages and Professional Communication, Belgorod State University, 85, Pobeda St., Belgorod, 308015, Russia
[2]Department of Computer Science, Natural Sciences and Teaching Methods, Belgorod State University, 85, Pobeda St., Belgorod, 308015, Russia
[3]Gymnasium #12, 4, Khikhlushki St., Belgorod, 308014, Russia

E-mail: velichko@bsu.edu.ru

Abstract. This paper is devoted to combining ecology challenges and STEM technologies. We consider ecology as an integral part of STEM process. The including of ecology in the STEM allows not only to develop mathematician calculations and engineering skills but also forms students' ecological consciousness. The attempt to combine ecology and STEM is aimed to make the knowledge in ecological sphere more practical. STEM education can help in solving the main ecological problems and challenges. STEM lessons, projects, conferences, exchanges could explain and give a living example of damages caused to our planet by irrational using of natural resources. It can improve public awareness and show the careful attitude to our planet.
Key words: ecology challenges, STEM, STEM technologies, ecological consciousness.

1. Introduction

Ecology is a science about environment. The most important purpose of ecological science is the study of environment and the relationship in it. Ecology is closely associated with conservation and environmentalism today, but it is not only this. Ecology can also touch human gut flora, the processes of the urban environment functioning and the ecology of the nutrient cycles of the soil. The word "ecology" has come from the Greek language and today it means "house study or living relations study". That mostly defines its true destination - the studying of relationships between those who occupy a house. Ecology touches upon the problems of organic life and examines such elements as spatial distribution (local or general) abundance and their relationship with our environment. This includes the interaction of different elements with other organisms within that environment - essentially their "interrelatedness" as a functioning network [1]. It is considered to be a form of environmentalism and it is usually associated with ecology sciences. Besides ecology also includes the aspects of such sciences as biology, botany, zoology, genetics, bacteriology, chemistry and physics. Ecology is a science about biodiversity in a given environment. It has much in common with many other sciences. For example, it has common features with physiology, behavioral sciences and the evolutionary sciences as well as it has with environmental sciences [2] in concerning time and space such as physics. In this paper we'll try to consider ecology like a part of STEM education.

2. Materials and methods

The are many areas of ecology, they include adaptation, distribution, abundance, spatial concentration and biodiversity. These processes that constitute biological life. But what are the reasons that make ecosystems be changed when subject to external pressures? They change to ones and move of the others, energy and materials are constantly being renovated through an ecosystem. Some ecosystems can change quickly, some of them remain more constant and change slowly. This means an ecological study can include almost everything. It can be something from bacterial cells [3] to gut flora. It can be the fact how many generations of aphids can populate a plant. Or all the ways how deserts form and maintain a constant balance. And the impact of rainforests and ice fields on the Earth's natural processes. Or how all these processes are impacted by planetary fluctuations [4].

Ecology and its various subdisciplines are entering both a new time full of challenges and excitings. Some challenges are accelerating the ecology development. From one hand new technologies could help practitioners to streamline some processes. On the other hand, decision-making politicians could improve public awareness.

Ecology now faces various challenges. The first and biggest challenge is in conservation ecology. One of the most important problems is increasing species extinction because of human-induced climate change. Actually, we can see the influence of average temperature rises on the global ecology [5]. International agreements continue to be made and acted on to reduce carbon emissions targets. We know that ocean acidification leads to damaging aquatic ecologies, damaging coral and affecting biological life that depends on those ecosystems. Ecology and ecologists are expected to remain at the forefront of understanding these serious problems and possible solutions.

Biodiversity is also an important challenge. As land clearance in ecologically sensitive areas makes way for farmland or urban landscapes, this affects the species that comprise this landscape. Loss of habitat without provision for conservation can create serious imbalances between predators and their prey [6], remove vital food sources for herbivorous and carnivorous species. Urban environments are constantly searching how to improve biodiversity within their own environments [7]. Now the urban designers and architects are continuing to promote biodiversity and find survival opportunities for the species that are being displaced or those that have already made a home in our urban landscapes.

The developing ecological damage is another challenge. As previously poor countries improve their economies, livelihoods, lifestyles and global economic power [8], developers are cutting down natural landscapes to make way for urban centers, factories, road network and infrastructure. Balancing of these economics needs against the global ecology is expected to be a major discussion point in both international trade and climate agreements.

Access to water, water security and scarcity together with all of the above is likely to underpin many of our ecological problems.

As we can see ecology is a science itself with many serious global problems to be solved. And it seems to be impossible to solve all these problems by itself. Ecology is not just a science in an abstract sense, it is the part of our everyday life because we all live here and it also can be the part of new stream in an educational and scientific space called STEM. The term STEM (Science (biology, ecology, etc.), Technology, Engineering and Mathematics) is an acronym used by those relevant to the educational method concerning the fields of Science, Technology, Engineering and Mathematics. This term was first used in 2011 by the biologist Judith A. Ramaley, who as head of the Natural Science Institution of the U.S.A., was in charge of developing new educational programs. STEM is an educational approach designed to combine technology and engineering together with science and mathematics, and this combination will be more helpful in order to understand more the laws of the universe.

In our attempt to combine ecology and STEM the basic question to be answered is “How can we give such an understanding to ecology so that children would want to do their best for our nature and planet, improve their knowledge in this sphere and perform actions in order to start academic education and enter working market?” [9, 10].

The answer to this question is simple: Students would have to be trained properly so that they can understand that their own prosperity depends on the quality of the whole world they live in, in other words they should be educated in matters like the climatic change, renewable energy sources, natural environment etc. [11]. A nation deeply educated is able to meet the everyday challenges and innovate. With STEM children get the opportunity to develop their capabilities in an ideal environment. STEM education is now an economic must in the country, as more than 30 prospering occupations will require STEM knowledge in the near future [12]. STEM education is an attempt to evolve from the tutor-centered approach into a teaching method that involves problem solving skills, creativeness and initiatives, as well as research and hands-on activities. STEM offers a chance for children to develop their abilities, to have new ideas, to form new thinking. In STEM education children are encouraged to answer questions and are involved with fun activities based on science, mathematics, engineering and technology. It is really impressive to see the children's reaction to this method, as they seem to find it more interesting and appealing. By applying STEM through various projects, students learn not only to process facts and focus on solving problems but also gain abilities fitting to global education. In such way of education they learn to develop critical thinking and work in a team, while it has also been reported that the knowledge gap between children from different countries has significantly been decreased. In this philosophy, a new project was born, called STEM. From one hand STEM is aiming to initiate social awareness and ensure everyone's involvement in achieving real increase in children's performance in science, technology, engineering and mathematics. From the other hand STEM is based on the knowledge of science, technology, engineering and mathematics.

There's no doubt in the effectiveness of STEM education and STEM methods in education. There is no doubt in the special role of ecology in the science [13, 14]. But where is the ecology place in the hierarchy of STEM processes? We are just trying to answer this question and to find its place.

Ecology studies the relationship and problems in our environment and STEM helps us to understand the functioning and problems of our world - environment.

STEM education can help in solving of the main ecological problems and challenges. STEM lessons, projects, conferences, exchanges could explain and give a living example of damages caused to our planet by irrational using of natural resources [15]. It can improve public awareness and show how the careful attitude to our planet - our home can save it and us.

3. Discussion

Ecology is increasingly penetrating into various strata of society. Solutions to many important issues are now evaluated based on their impact on the environment. Such an assessment is becoming increasingly difficult, given the need to control and monitor a large number of factors. Environmental mistakes will have a greater influence on our children. To be able to avoid new mistakes and preserve the environment, the younger generation must be very competent in environmental matters. Even now, children are actively involved in various environmental activities, such as those designed to draw public attention to climate change issues (the most famous activist is Greta Thunberg even spoke at the UN). In different countries, environmental lessons are included in the mandatory course of the school curriculum, and the importance of environmental protection is emphasized in the lessons of the "Surrounding world". In high school, students have the opportunity to assess the impact of human actions on nature through the prism of their knowledge of physics, chemistry and biology, and learn more about environmental disasters that have occurred. Thus, environmental issues are intertwined in various aspects of the school curriculum, in fact, being a link of interdisciplinary interaction. This is especially important in countries where STEM is not yet a mandatory curriculum. The topic of ecology is also among the top most popular topics of project activity of schoolchildren, and project activity and STEM are a single whole. The activities of large and medium-sized enterprises cannot do without addressing environmental safety issues. Often entire departments are created for this purpose. Automation systems for managing such complexes are one of the possible tasks of future STEM specialists.

The faculty of mathematics and science education of Belgorod State University with the support of the European Union (Erasmus+ project) this year has opened a new program for training future masters in STEM disciplines. In this program, as many as 3 disciplines are devoted to ecology: Fundamentals of ecology, Environmental Monitoring and Greening of educational programs. STEM teachers who have defended their master's thesis in this area will be able to train STEM specialists who will be well versed, in particular, in the field of environmental safety.

IOP Conf. Series: Materials Science and Engineering 1089 (2021) 012025 doi:10.1088/1757-899X/1089/1/012025

One of the main ecological problems is the environment pollution - water pollution, air pollution, plastic pollution. It is a well-known fact that a plastic bottle is used for 15 minutes on average and takes about 200 years to decompose. About one million plastic bottles are sold in the world every minute. And only 9% of used plastic is recycled, 12% gets burned, and the rest ends up in garbage dumps or the ocean. Because of this we find it in our drinking water and in our bodies (and in the bodies of other animals too). And this is just one of the many types of pollution that is currently taking place on Earth. It seems it's high time for each of us to think about how we can take part in saving our planet.

One of the easiest examples of the STEM practice organizing may be the problem solving task which is offered to students at the beginning of the lesson.

For example, we can offer to look at the photo showing the problem of plastic pollution. On the left we can see a huge heap of plastic. On the right - the plastic has been recycled and transformed into shoes. So, we see plastic pollution and plastic recycling. And we give to the students the opportunity to choose, to think, to make conclusions by answering them a simple question:

Where is the plastic bottle you've used today – on the left or on the right?

Where can this plastic be tomorrow?

Figure 1. The problem of plastic recycling.

The visualization of the problem during STEM lessons "turns on" the thinking processes and starts the mechanism of creation. After looking at this picture and thinking the students could unlikely remain indifferent to the ecological problems in general and to the problem of pollution in particularly. Minimum they can do - pick up the plastic bottle and put into special container for recycling. But some of the students can do their maximum - create ecological projects, which may help or even save our planet in the future. In the sheet bellow we can observe the steps of the STEM activity in ecology.

The students can stop at any of these steps and they can go to the end and even further. And that is the main purpose of STEM education in ecology - through visualization and thinking to creation and invention.

Table 1. STEM staircase.

Step 1	Look at the photo, watch film, etc.
Step 2	Make choice and decision
Step 3	Find out the ways of problem solving
Step 4	Chose the methods and techniques
Step 5	Calculate the expenses and cost
Step 6	Create the model
Step 7	Find the practical appliance
Step 8	Analyze the results
Step 9	Make conclusions

4. Results

We are going to list successful ecological projects that have been created by people who care about the pollution problem. These Tremendous Ecological Projects that really make our planet healthier and can be realized during STEM practice.

1. Footwear made from recycled plastic bottles

Figure 2. Footwear made from recycled plastic bottles https://m.facebook.com/rothys.

The Rothy's shoe brand creates shoes from recycled plastic bottles. The most popular collection consists of colorful and washable flats which Meghan Markle put on during her recent visit to Australia. The company has already recycled 25 million plastic bottles.

Table 2. Ideas for students STEM practice #1.

1. Study the problem of plastic pollution (Science)
2. Count the cost of one shoe pair production (Mathematics)
3. Create your own shoe pair design (Engineering)
4. Make one shoe pair on 3D printer (Technology)

2. Kids furniture made of broken plastic toys

Figure 3. Kids furniture made of broken plastic toys.

A Dutch project called EcoBirdy creates furniture from broken and discarded kids toys. Apart from making furniture, the company has also launched an educational project about the harm done by plastic and has installed special containers in kindergartens where kids can leave their broken toys. The company has already recycled 25 tons of plastic toys.

Table 3. Ideas for students STEM practice #2.

1. Collect all thr old toys at home and study the problem of plastic pollution (Science)
2. Count the cost of all your aold toys plastic recycling (Mathematics)
3. Create your own model of furniture for kids (Engineering)
4. Make your piece of furniture on 3D printer (Technology)

3. Newspaper wood

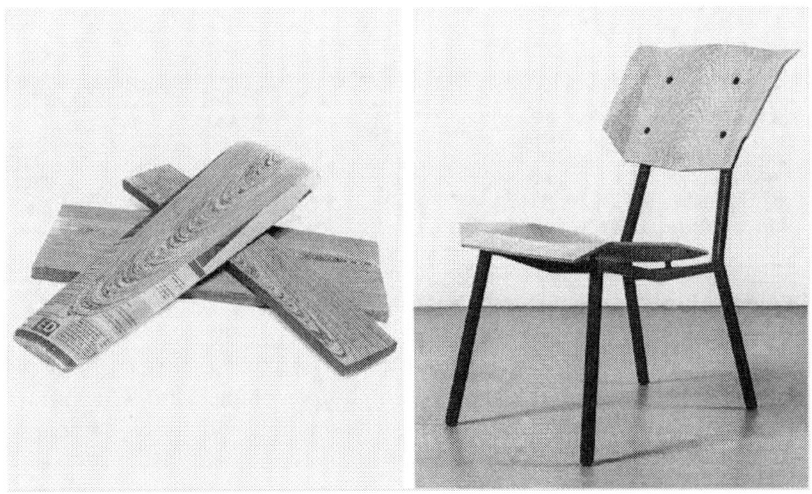

Figure 4. Newspaper wood project. https://newspaperwood.com/projects/.

Dutch designer Mieke Meijer has always been perturbed by unsold newspapers because a lot of trees are cut down to produce them. That's how and why he came up with the 'Newspaper wood' idea — a material made from extruded old newspapers and organic glue that resembles wood in its appearance and strength. It can be sawed, polished, and drilled. Today, furniture and interior decorations are made using this brilliant material.

Table 4. Ideas for students STEM practice #3

1. Collect all old newspapers at home and study the problem of the wood cutting (Science)
2. Count the cost of one newspaper chair (Mathematics)
3. Create your own model of a newspaper chair (Engineering)
4. Make your newspaper chair on 3D printer (Technology)

4. The ocean cleanup

The plastic floating in the ocean weighs as much as 1,000 Eiffel Towers and it could take as long as 79,000 years to collect it. The good news is that a 24-year-old Dutchman named Boyan Slat has created a litter trap. It is a floating U-shaped system that is able to catch plastic floating on the surface of the ocean. Slat hopes to catch all the ocean trash within 20 years with the help of these traps. You can watch the principle of this system work here. The project also has famous investors like the founder of PayPal, Peter Thiel. The first trap was set on the water in September 2018. It is planned that a special garbage collecting ship will take out the collected plastic once every several months and send it to be recycled. The bonus is a floating off-shore city made of plastic trash.

Figure 5. A floating off-shore city made of plastic trash https://www.freischaerler.archi/united-plastic-nation/.

An Australian company called 'Freischärler' came up with another method of fighting ocean pollution — a project to build a floating city out of ocean plastic with the help of 3D printers and drones. 'United plastic nation,' which resembles an iceberg with a visible underwater section, will become an independent off-shore zone where plants will be grown with the help of the aquaponic method and where all trash will be recycled.

Table 5. Ideas for students STEM practice #4.

1. Collect all old plastic at home and study the problem of water pollution (Science)
2. Count the cost of your old plastic recycling (Mathematics)
3. Create your own model of what you can do from this plastic (Engineering)
4. Make the computer model (Technology)

5. Conclusion

These projects are being realized now by different international companies. But they can be also realized by students during STEM lessons. Drones, 3D printers, microscopes connected with computers can be used on the ecology STEM lessons. The main advantage of such lessons in fact that they allow not only to develop mathematician calculations and engineering skills but also they form sdudents' ecological consciousness. The realizing of these projects by students themselves forms the new level of ecological personality.

This approach to learning and combining ecology and STEM is certainly not an easy task, but the benefits to students and the entire school community are tremendous. Students and teachers engaged in ecology via STEM make more real-life connections. So that our planet is not a place where you live but instead becomes the entire experience of life and learning itself. We are always learning, always growing, always experimenting and make our world better.

References

[1] Kostina I B, Gladkikh Y P, Velichko M A, Krasovskaya LV and Satler O N 2019 Assessment of ecological consciousness formation among adolescent girls in the learning process by means of specialized computer systems *Int. J. of Recent Technology and Eng.* **8 (2S10)** 665-8

[2] Chernyavskikh S D, Velichko M A, Kostina I B, Krasovskaya L V and Satler O N 2018 Improving educational process quality in the lessons of natural and mathematical cycle by means of stem-training *Cypriot J. of Educational Sciences* **13(4)** 501-10

[3] Shtyleva L V 2018 Genderny`j komponent pedagogicheskoj kul`tury` i problema orientacii devochek na stem- obrazovanie i stem-professii [Gender component of pedagogical culture and the problem of girls' orientation to stem education and steam professions] *Gendernoe Izmerenie Cifrovoj E`Konomiki: ot Strategii k Dejstviyu* (Ivanovo: IGU) pp 98-102 [In Russian]

[4] Rugh M S, Calabrese J E , Madson M A, Capraro R M, Barroso L R, Capraro M M and Bicer A 2019 Stem language can be the stem of the problem *IEEE Frontiers Education Conf.* (San Jose: IEEE) 8659231

[5] Babarović T and Šimunović M 2019 The role of parental stem-supportive behaviours in formation of children's stem career interest *XVI European Congress of Psychology* (Moscow: MGU) 1074 [in Russian]

[6] Su C H 2019 The effect of users' behavioral intention on gamification augmented reality in stem (gar-stem) education *J of baltic science education* **18(3)** 450-65

[7] Halvorsen H P, Tretjakova R, Timmerberg J, Thiriet J M, Mylvaganam S 2019 Steam for stem-include 'art' in stem (science, technology, engineering and mathematics) *29th Annual Conf. of the European Association for Education in Electrical and Inf. Eng.* (Ruse: IEEE) 9000460

[8] Kuzmina Yu A and Yashina N V 2018 Vozmozhnosti ispol`zovaniya texnologii stem-obrazovaniya pri zakreplenii znanij uchashhixsya na urokax v nachal`noj shkole [Opportunities to use steam education technology to consolidate students ' knowledge in primary school lessons] *Innovacionnoe Razvitie* **4(21)** 220-2 [In Russian]

[9] Ananchenkova P I 2018 Stem-frilans: novy`e vozmozhnosti zhenskoj samozanyatosti [Stem freelancing: new opportunities for women's self-employment] *Gendernoe Izmerenie Cifrovoj E`konomiki: ot Strategii k Dejstviyu (2018-2030)* (Ivanovo: IGU) pp 131-4 [In Russian]

[10] Ganicheva A N, Kaitov A P, Lvova A S, Lyubchenko O A, Marinyuk A A and Serebrennikova Yu A 2018 Ispol`zovanie pedagogami stem-texnologij v obuchenii detej mladshego shkol`nogo vozrasta [The use of stem technologies in teaching young school children] *Materialy` V Vserossijskoj Nauchno-Prakt. Konf.: Rebenok v Sovremennom Obrazovatel`nom Prostranstve Megapolisa* (Moscow: Izvestiya instituta pedagogiki i psixologii obrazovaniya) pp 32-6 [In Russian]

[11] Lyalina V V 2019 Orientaciya doshkol`nikov na inzhenerny`e professii posredstvom stem-obrazovaniya [Orientation of preschoolers to engineering professions through stem education] *Sbornik Mater. VII Vserossijskoj Nauchno-Prakt. Konf.: Pedagogika i Psixologiya V Xxi Veke: Sovremennoe Sostoyanie i Tendencii Issledovaniya* (Kirov: MCITO) pp 415-9 [In Russian]

[12] Barlow S E and O'neill M A 2020 Technological advances in field studies of pollinator ecology and the future of e-ecology *Current opinion in insect science* **38** 15-25

[13] Dodds W K and Whiles M R 2019 *Freshwater ecology: concepts and environmental applications of limnology* (Amsterdam: Elsevier)

[14] Mallet Noa 2020 The digital ecology: the key to a sustainable evolution *VII-I Annual all-Russian Sci. Practical Conf. "Digitalization of Society: State, Problems, Prospects"* (Moscow: FGBOU VO RE`U im. G. V. Plexanova) pp 16-20

[15] Korneeva R V 2020 Informacionny`e texnologii v e`kologii [Information technologies in ecology] *Mater. Mezhregional`ny`x Nauchno-Prakt. Konf. Realizaciya celej ustojchivogo razvitiya oon v sfere e`kologii na primere Kaluzhskoj oblasti* (Kaluga: AKF Politop) pp 140-51 [In Russian]

IOP Conf. Series: Materials Science and Engineering 1089 (2021) 012026 doi:10.1088/1757-899X/1089/1/012026

Nature-like economy as a sphere of energy-saving behavior of households

R A Burganov, E A Dolonina and L R Urazbakhtina

Department of Economics and Organization of Production, Kazan State Power Engineering University, Krasnoselskaya Str., 51, Kazan, 420066, Russia.

E-mail: burraabr@gmail.com

Abstract. This article discusses the author's position on the essence and manifestations of a nature-like economy based on the use of nature-like technologies. The author draws attention to the need to respect the use of energy of all types by households, especially in the context of widespread use of electronic products. The paper notes that the creation of nature-like technologies of various levels should be based on an economic base. Within the framework of logic, the subject and object of research of a nature-like economy are identified as an area of scientific research. In the system of practical activity, energy is identified as the main factor in such an economy, the main participant in the economy is artificially created living systems. And production is based on the use of nature-like technologies. The use of various types of energy is the basis for the functioning of a nature-like economy. Enormous investments are needed to develop nature-like technologies and economies. The article also provides examples of the experience of using nature-like technologies and economies in various countries of the world in order to optimize the use of various types of energy. The article highlights mono and polyfactorial methods of analysis, the scientific basis for studying the actions of households is highlighted. According to the authors, the formation and development of a nature-like economy will affect the change in the thinking style of household members in the field of energy conservation, which will take quite a long time.

1. Introduction

The need to search for new reserves in energy conservation of various institutions of society requires in-depth studies of the advantages of certain types of social and economic devices based on technologies. In particular, at present, there is an impact of the widespread use of electronic products on the growth of energy consumption by households. The so-called "vampire" electronics - devices and devices that spend energy, even when they are not actively used, significantly affect the level of energy saving. For example, a microwave oven consumes energy even when food is not being prepared - it keeps the digital clock running. And even when your computer is "asleep", it is still using energy. Scientists at the Lawrence Berkeley National Laboratory of the US Department of Energy claim that standby energy accounts for 5-10% of household electricity consumption. If you turn off unused devices from the network (TV, VCR, stereo, charger), then this will reduce a significant amount of electricity consumption. According to researchers, cloud-based forms of information storage in the next 10 years can consume about 11 percent of global consumption.

In such an environment, it is necessary to find new reserves in energy saving. It should be noted that in each state a regulatory framework for energy conservation has been created, a large number of organizational and technological measures will be carried out at the level of the country, regions, industries, organizations and firms aimed at creating conditions for optimizing energy consumption. Despite the efforts made by politicians and business executives and a considerable amount of research and the results obtained, the theoretical and methodological basis for studying energy saving of households on the basis of economic theory has not been created. There is a study of individual manifestations of the energy-saving process of a narrow nature, in particular, the problems of energy conservation management in an apartment building, in everyday life. The fact that energy saving in Russia is based on more than 70 regulatory documents indicates the need for a systematic and integrated approach to studying problems in this area and making scientifically sound decisions to resolve them

2. Materials and Methods

Currently, in scientific circles and practical activities, the provisions of the innovative economy, the digital economy, and "Industry 4.0" are actively applied. Each of them has its own characteristics of formation and development, and also relies on certain core technologies. Thus, digital technologies are the backbone of the digital economy, and innovative technologies are the basis of an innovative economy. Industry 4.0 is big data.

Natural technologies have also come to the attention of scientists and statesmen. Russia also pays great attention to using the advantages of such technologies. The Strategy of Scientific and Technological Development of the Russian Federation as a strategic priority: "in the long term, research in the field of ... the development of nature-like technologies ..." (strategy of scientific and technological development of the Russian Federation. Decree of the President of the Russian Federation of 01.12. 2016 No. 642 http://kremlin.ru/acts/bank/41449).

3. Results and Discussion

The creation and development of nature-like technologies should be based on a certain economic base and obtaining economic benefits. It should be noted that there is no fundamental research on the content and features of the implementation of a nature-like economy. In contrast to a cyclical economy, a nature-like economy relies on the use of benefits from operations, where natural and natural factors and resources of human and firm life are the beneficiary. The work of Abderraouf Bouakkaz and others considered a hybrid power system [1]The first paragraph after a heading is not indented (Bodytext style).

For the purpose of a specific and systematic analysis of the manifestations of the features and prospects for the development of a nature-like economy, it is necessary to give it the following scientific definition. Nature-like economy is a field of scientific and economic activity that considers the behavior of subjects of financial and economic activity in the production and consumption of various types and forms of energy, reproduced using nature-like technologies. Nature-like technologies are based on the use of energy. As for the concept of "energy", it should be noted that there is no generally accepted meaning. The word "energy" itself has ancient Greek roots and is used in the meaning of "active force.

A nature-like economy as a theory should have its own subject and object of study (table 1).

Table 1. Characterization of nature-like economy as a field of scientific activity of the theory

No.	Scientific elements	The content of scientific elements
1	Subject of research	Economic and technological phenomena or processes associated with energy consumption generated using nature-like technology
2	Object of study	A set of relationships (economic, financial, managerial, technological, digital, etc.) that affect the state of consumption and production of energy by nature-like technologies.

A nature-like economy as a sphere of economic activity is characterized by the following data (table 2).

Table 2. Characteristic features of implementing a nature-like economy in practice

No.	Elements of a nature-like economy	Content of elements
1	The main factor in the activities of economic entities	Energy in different types and forms, produced by nature-like technologies.
2	Production	Based on the use of nature-like technologies
3	The main participant in the activity	Artificially created living systems
4	Production control and management	Integrated human-machine systems

In general, a nature-like economy should be located at the junction of the combination of "living" and "inanimate" aspects of society's activities, which meets the requirements of the development of civilization, and, accordingly, fundamental research in this area of economic science is needed. At the same time, scientific papers note that "computational models based on economic principles and methods are powerful tools for understanding and analyzing energy and environmental problems, as well as for developing policies to solve them" [2]. Other paragraphs are indented (BodytextIndented style).

Energy conservation is one of the many goals of various actors in society and the economy [3]. The energy-saving behavior of a household in a nature-like economy will differ from its behavior in a traditional economy. In recent years, the energy-saving behavior of households has become an object of research by scientists from around the world. Basic approaches to energy demand modeling have been reviewed by Afeees A by others [4]. Among the significant works in recent years, one can single out the study by Mahmoud Salari and Roxana J. Javid on modeling household energy expenditures in the United States [5], Boudet X. and others on clustering energy-saving behavior in the family [6], Ito and Koichiro on the consumption or the average price of electricity [7].

Significant results on attitudes towards energy conservation were obtained based on the analysis of empirical data by researchers in France [8]. In most cases, energy saving is considered not from the point of view of analyzing the behavior of households, but from the point of view of using energy-saving building technologies. So, the behavior of households when buying can be influenced by the level of heat storage on different floors of a multi-storey building [9]. For example, researchers Marlyne Sahakian and Béatrice Bertho have identified the effect of household emotions on both reducing and improving energy use [10].

The energy-saving behavior of households in a nature-like economy should be based on the use of energy that occurs naturally. At the same time, the analysis of the formation of the behavior of households in the field of energy saving depends on the used scientific research base.

Thus, from the point of view of representatives of the neoclassical direction of economic theory, the behavior of households in the use of energy resources is characterized by such concepts as "utility", "benefit", "profitability", "marginal", etc. Energy elasticity is essential for shaping household behavior. It is known that the most stimulating direction in energy saving is getting benefits from energy saving, that is, there is a market model of consumer behavior. In recent years, household consumer spending on electricity has been growing (table 3).

After all, the less a household spends energy, the less it pays for it. If you take an average family with a minimum income level, then the family will certainly save energy in the house. It should be emphasized that in Russia about 15% of households receive income at or below the subsistence level. Conventionally, if in the previous month the family paid about 3-4% of the total income for energy, then after saving and using energy-saving funds it can already be 1.5-2%. The increase in electricity debt is one of the most pressing problems. Turning off the lights on time, good housekeeping and the use of energy-saving lamps in the household can save a lot of energy.

Table 3. The structure of consumer spending by households in the Russian Federation (as a percentage of the total)

Year	2010	2011	2012	2013	2014	2015	2016	2017	2018
Consumer expenses	100	100	100	100	100	100	100	100	100
including on electricity, gas and others	5,5	5,5	5,1	5,1	5,1	5,4	5,6	5,4	5,5
of them electricity	*1,2*	*1,3*	*1,2*	*1,3*	*1,3*	*1,4*	*1,4*	*1,4*	*1,4*

A source. The website of the Federal State Statistics Service, https://www.gks.ru

The energy-saving behavior of households in terms of institutional theory is defined by such concepts as "institutions", "rules and regulations", "property", "energy contracts", etc. The economic meaning of the very concept of "institution" is multifaceted. Discussions about the institution are conducted on the issues of its origin, role and evolution.

Of particular importance within the framework of institutional theory is the concept based on the use of a system of contracts and agreements. In particular, households in the field of energy supply can conclude such agreements as an agreement on the supply of electricity, on the provision of services for the development of a list of energy service activities; contract agreement for the installation of energy service equipment; an agreement on the provision of services for the analysis and maintenance of energy service activities, etc. The process of concluding each agreement and energy contract has its own characteristics.

The institutional behavior of households in the energy sector is also influenced by informal norms. Among the problems is theft of electricity. It is also necessary to take into account technical losses, unaccounted electricity consumption. It is also worth highlighting the creation and functioning of institutions regulating energy consumption. Thus, laws, by-laws and other governing rules of the game in the energy sector are constantly reviewed, supplemented, etc. This is especially noticeable in the field of providing energy services to the population. Each year, households are faced with changes in electricity tariffs, billing, etc.

According to the moral and educational concept as a part of institutionalism, the energy consumption of households is related to such institutions as customs and traditions, a careful attitude to energy - heat resources.

The technological concept of energy-saving behavior of households involves the use of modern energy-saving technologies. “Smart home”, “smart city”, “smart grids” are already becoming common concepts. Energy efficient technologies hold great promise for lowering household finance costs. For example, it is no longer a secret for anyone that LED lamps are more profitable than incandescent lamps of the old type. Technologies do not stand still, therefore there are a huge number of devices and systems for energy saving and energy efficiencyBut you need a heat pump, that is, a device that transfers heat energy from a heat source to a heat reservoir for subsequent use. An invaluable resource is the thermal energy produced by the human body. So, scientists at the University of Massachut came up with wall paint that could provide a house with energy. In fact, the paint must absorb the energy of living organisms. Assembled the Air-gen device, which extracts electricity from water molecules in the air using bacterial nanowires.

Stockholm Central Station has long learned to use the energy of human heat. With the help of heat exchangers in the station ventilation system, excess heat energy is transformed into hot water, which is subsequently pumped into the heating system of the neighboring building. Thus, the lack of energy resources, as well as their cost, reveal the potential of households to generate their own energy and the possibility of its further distribution for a certain fee.

A nature-like model of energy-saving behavior of households involves the use of modern energy-saving technologies. “Smart home”, “smart city”, “smart grids” are already becoming common concepts. Fang Zhao and others pay attention to the features of the development of e-government and the digital economy [11]. Energy efficient technologies hold great promise for reducing the financial costs and environmental damage associated with energy use. Technologies do not stand still, therefore there are a huge number of devices and systems for energy saving and energy efficiency. In particular, researchers from the University of Gothenburg have found a way to turn ordinary windows into solar-powered heaters that can significantly increase the temperature of the glass, even in freezing weather. The main functional components of the invention are plasmonic nanoantennas. With the help of plasmons, nanoantennas are capable of intensely absorbing light, which then heats up the entire surface.

The experience of foreign countries, where innovative methods of energy-saving behavior of households are used, is also very interesting. So, in Germany there is a so-called "passive housing". The thermos inside keeps the temperature almost constant for a long time without the need for additional heating or cold. Likewise, in a passive house, the desired constant room temperature can be maintained in all weather conditions with minimal operating costs. Unlike an “active” energy efficient home, a “passive” one prioritizes not generating alternative energy, but minimizing costs. Its energy consumption is 40-90 percent less than that of a typical typical building. Such a house heats itself in winter and cools in summer thanks to high-quality insulation.

In Japan, living curtains and green walls. Energy efficient curtain in Japan is the solution for hot summer and cold winter nights. On a hot summer day, the walls of houses in areas exposed to the sun can heat up to temperatures above the air temperature, and then heated objects will radiate heat into the environment, increasing the heat around them. A plant wall provides shade and cooling for windows and balconies, reducing building wall temperatures by 10 degrees by reflecting direct sunlight. In winter, these green walls can insulate a home and reduce heating energy costs.

In South Korea: tents inside the house. Due to high electricity and heat tariffs, some Koreans sleep in tents. If the temperature in the apartment is 19 ° C, then inside the tent the air heats up to 23 ° C. Not entirely comfortable, but this allows them to save on heating appliances.

For a deeper analysis of the behavior of households in the field of energy saving, it is necessary to carry out its strategic modeling. On the basis of an economic-mathematical model, the process of energy-saving behavior of households can be used one-factor or poly-factor representations.

The model with one variable factor is based on the use of one of the variables, in particular, an increase or decrease in household income, which most of all shapes the energy-saving behavior of energy users. In the event of an increase in income, households can transform their behavior towards obtaining more benefits from using an additional unit of energy. And vice versa. A decrease in household income contributes to the formation of its economical behavior, for example, they will refuse certain types of benefits.

In multi-parameter mathematical models, the implementation of the influence of external (main and secondary) and internal (main and auxiliary) functions in the energy-saving behavior of households is taken into account.

As you know, external (general object) functions are performed by the object of analysis as a whole and reflect the functional relationship between the object of analysis and the scope of its application (external environment). They reflect the relationship of the object of analysis as a whole with the scope of its application, i.e. with the system of which it is an element. Also reflect the operating conditions (use of the object).

In turn, external functions, according to the degree of satisfaction of a given need, are divided into main and secondary ones.

The main ones are the functions of appointment, the main purpose of the existence of the object as a whole.

Secondary (enriching, opportunistic) functions reflect the side effects of the object, its additional properties.

Secondary functions do not fundamentally affect the performance of the object and reflect either restrictions that must be observed when the object performs its main (main) function, or additional conveniences and advantages. These are functions that provide functionality and usability/

Each variable can be positively or negatively reflected in the rational or irrational behavior of households in energy conservation.

So, external functions, that is, influencing the energy saving process from outside, include:

- the main function (determining the purpose and meaning of energy saving), for example, changing prices for energy resources or introducing a tax on income received as a result of energy saving in households;
- minor functions (affecting the actions and interconnections of the elements of the energy saving system), for example, the use of outdated house infrastructure of power grids, an increase in the number of used gadgets.

Internal functions of energy saving in households, that is, influencing the process of energy saving from within the energy saving system, include:

- basic (target settings in energy saving of family members), as using the principle "when leaving, turn off the light";
- auxiliary (for example, the level of equipment with energy-saving devices, the presence in the household of energy receivers with the class "A +++".

The factorial approach will allow finding more viable models for the development of energy conservation in household.

4. Conclusion

However, the state should take certain measures to form the optimal energy-saving behavior of households, taking into account their possibilities of using nature-like technologies. Without the purposeful work of state bodies to create institutional conditions, the effectiveness of the decisions made will be questionable. A number of states have such examples. For example, in France, the cost of purchasing energy efficient equipment is deducted from the tax base of citizens. Lower tariffs apply for more efficient equipment in homes. The United States provides government subsidies ranging from $ 50 to $ 200 for the purchase of new, more efficient home appliances. The Chinese government has also taken a number of measures to curb the growth of energy consumption in households, such as energy efficiency labels, discounts on energy efficient appliances, and the introduction of tiered tariffs. [12].

At the same time, it is necessary to pay attention to the results of the economic calculation of using the advantages of using nature-like technology. It is necessary to take into account the cost of all types of energy [13]. Household income is maximized only with an economical attitude to all types of energy, including electrical, thermal, mechanical, human, etc. For example, when preparing food, a large amount of heat energy is accumulated in the kitchen. In fact, they can be accumulated for later use for other purposes.

The difficulties in calculating energy efficiency were noted in a number of works by scientists [14], [15]. The imperfection of methods for calculating energy efficiency indicators should be noted. Since the principle of economic theory "with other explicit conditions" is often applied, that is, not all quantitative indicators of the consumption of all types of energy are taken into the settlement operation. This also includes the problems of accounting for all types of energy in accounting.

The very formation and development of a nature-like economy will depend on a change in the thinking style of household members in the field of energy conservation, which takes a long time horizon. At the same time, it is necessary to segment energy consumers, that is, a part of the population that has minimal restrictions on the purchase and use of the latest technologies based on economical energy consumption will adapt to new conditions faster than other parts. For example, a small part of the population owns smart home technologies. And among low-income strata of the population, the transformation of energy-saving behavior will take quite a long time, and then with the help of state and public structures.

5. Acknowledgments
The reported study was funded by RFBR, project number 20-01-00001-00099 "Theoretical and methodological approaches to the development of models of energy-saving behavior of households in a nature-like economy.

References

[1] Bouakkaz A, Mena A J G, Haddad S and Ferrari M L 2021 Efficient energy scheduling considering cost reduction and energy saving in hybrid energy system with energy storage *J. of Energy Storage* **33** 101887
[2] Sanstsd A H 2018 Energy-economic policy modeling *AIP Conference Proceedings* **1924(1)** 020011
[3] Burganov R A and Yudina N A 2018 To the question of the creation of the energy consumer firm theory *Journal of Entrepreneurship Education* **21(1)** 7000
[4] Salisr A A and Ayende T O 2016 Modeling energy demand: Some emerging issues *Renewable and Sustainable Energy Reviews* **54** 1470-80
[5] Mahmoud S and Roxana J 2017 Modeling household energy expenditure in the United States *Renewable and Sustainable Energy Reviews* **69** 822-32
[6] Boudet H S, Flora J A and Armel K C 2016 Clustering household energy-saving behaviors by behavioral attribute *Energy Policy* **92** 444-54
[7] Koichiro I 2014 Do Consumers Respond to Marginal or Average Price? Evidence from Nonlinear Electricity Pricing *American Economic Review* **104(2)** 537–63
[8] Belaïd F and Joumni H 2020 Behavioral attitudes towards energy saving: Empirical evidence from France *Energy Policy* **140** 111406
[9] Elistratova Y V, Kireev V M and Seminenko A. S. Contribution of heat loss by infiltration to energy saving and microclimate in multi-family residential buildings *IOP Conf. Ser.: Mater. Sci. and Eng.* **791(1)** 012007
[10] Sahakian M and Bertho B 2018 Exploring emotions and norms around Swiss household energy usage: When methods inform understandings of the social *Energy Research & Social Science* **45** 81-90
[11] Zhao F and Mohini J W 2015 E-government development and the digital economy: a reciprocal relationship *Internet Research* **25(5)** 734–66
[12] Andrews-Speed P and Ma G 2016 Household Energy Saving in China: The Challenge of Changing Behavior *China's Energy Efficiency and Conservation* ed B Su and E Thomson (Singapore: Springer) pp 23–40
[13] Burganov R A, Maimakova L V, Urazbakhtina L R and Golitsyna L A 2020 On the need to improve the methodology for calculating energy saving and energy efficiency of enterprises. *IOP Conf. Ser.: Mater. Sci. and Eng.* **791(1)** 012040
[14] Filippini M and Hunt L C 2015 Measurement of energy efficiency based on economic foundations *Energy Economics* **52(1)** S5-S16
[15] Xin-gang Z, Xin M, Ying Z and Pei-ling L 2020 Policy inducement effect in energy efficiency: An empirical analysis of China **211** 118726

Research and development of ice storage in building energy supply system

Y V Yavorovsky[1], I A Sultanguzin[1], B A Khristenko[1], D A Kruglikov[1] and V Y Chaikin[1]

Department of Industrial Energy Systems, National Research University Moscow Power Engineering Institute, Krasnokazarmennaya St., 14, Moscow, 111250, Russia

E-mail: SultanguzinIA@mpei.ru

Abstract. In this work, the most suitable type of heat exchanger for an ice accumulator was determined based on studies of the dynamics of ice growth on heat exchange surfaces of various shapes and its configuration was determined. The calculation of the accumulating capacity of the ice accumulator has been made. The task is set to create a mathematical model in the Ansys Fluent and TRNSYS environment, as well as BIM and BEM models of the main elements of the system. A ready-made design solution for the introduction of an ice storage device into the existing heat supply system of an energy-efficient house with an assessment of the cost of costs and required materials has been developed. An assessment was made of the cost of various solutions for heating a house with a load of 10 kW. A test bench layout has been developed for monitoring and analyzing the operation of the system, as well as the impact on the environment. The functionality of the current system for recording and processing the operating parameters of the equipment allows not only to confirm the correctness of calculations and the operability of the mathematical model, but also to assess the impact of the ice accumulator on the environment.

1. Introduction

In the Russian scientific community, at the moment, more than 68600 articles have been written on the topic "heat pumps" and methods of their application (as requested on Elibrary.ru). However, most of them consider standard solutions for extracting heat from such primary sources as air, ground (vertical and horizontal ground collectors), groundwater, heat from waste water and reservoirs. Each of the previously listed sources has certain advantages and disadvantages. In cases where there are restrictions on the outside air temperature, drilling deep wells, it is possible to use ice accumulators.

In the Russian and foreign scientific community there are already scientific works [1-3] on a similar topic (for example, Daniel Carbonell (h-index according to Scopus = 8), Daniel Philippen (h-index according to Scopus = 6), Martin Granzotto (h- Scopus index = 9), Michel Y. Haller (Scopus h-index = 14) and Elimar Frank. (h-index = 9), Zmeureanu , Radu G. (Scopus h-index = 22)), but they describe only the principle of operation of this installation, proposed mathematical models describing small units of the system and there is no general model for calculating and designing integral systems.

In addition, the considered mathematical models ignore the effect of temperature on the thermal conductivity of ice, which leads to large errors in calculating the dynamics of ice growth on the heat exchange surface.

For the first time, a patent for this technology was developed by Isocal, but in 2012 the German company Viessmann acquired this organization, and along with it a patent for this technology. Since 2012, this organization has implemented more than 1100 standard solutions for the domestic sector.

The official representative office of the Viessmann organization is located in Russia, which after communicating with the members of the project team is ready to provide technical support in an ongoing study, since the further research of this technology will not only expand the scope of heat pumps, but also increase the company's sales.

It is also worth noting that there is not a single completed object on the territory of the Russian Federation. The project provides for the implementation of the object within 2 years.

2. Application of the ice accumulator

The use of ice accumulators as a low-potential heat source for brine-water heat pumps has not been officially recorded in Russia. The main advantage of this system is that they do not require special building permits, unlike geothermal probes [4]. This helps to solve a number of problems arising at the stages: design, construction and further operation. The main difficulty for some regions is the large depth of soil freezing.

The installation scheme consists of one or two ice accumulators (located at a distance of more than 2 meters from each other), the required number of solar collectors selected according to the design calculation, a reversible heat pump and a buffer tank for preparing hot water [2].

The solar-air absorbers installed on the roof remove heat from the environment and in the daytime from insolation and supply it to the heat pump. If the energy is not taken by the heat pump, it is used to regenerate the ice accumulator or to heat the buffer tank. An ice accumulator with a built-in heat exchanger is installed below the freezing depth of the soil, filled with treated water and used as a low-potential heat source. The heat pump must be able to operate in reversible mode. In the summer, when there is no heating demand and the heat load on hot water supply is minimal, solar collectors supply heat energy to the storage tank. When the heating season begins, heat loads increase and the demand for air conditioning decreases, then the heat pump begins to take heat from the ice accumulator, up to controlled icing at 0 ° C. Further there is a phase transition of water from liquid to solid state. In parallel, the process of heat exchange between the ice accumulator and the environment (soil) is carried out in the storage. At the end of the heating period the heating demand of the building decreases and the heat pump switches to reversible operation. In this mode, the ice accumulator is regenerated. The cold is taken from the storage and sent to air conditioning the premises. The ice in the accumulator tank starts to melt and there is a gradual accumulation of heat. Thus, the cycle is closed.

3. Main technical tasks

3.1. Design of heat exchanger coils and concrete tank

The first task is to choose the placement of the coils in the tank. The location of the coils determines the uniformity of heat removal, the correct formation of ice on the heat exchange surfaces, and the efficiency of the ice accumulator. Improper backfilling can lead to intense ice formation and damage to the concrete structure.

Despite the fact that ice accumulators have been around for many years. To date, there is no ready developed methodology for calculating the ice accumulator. The main task is to choose the turn-to-turn distance. In [3], practically experiments were carried out aimed at calculating the rate of ice formation on different surfaces. Based on the data obtained in [3], it can be noted that the rate of ice formation on surfaces decreases when a certain ice thickness is reached. In order to ensure uniform formation of ice on the surface of the coil it is necessary to determine the diameter at which a decrease in growth dynamics is observed. Based on the data received, set the turn-to-turn distance.

The second task is to select the material and shape (dimensions) of the concrete structure. The concrete structure must be strong and ensure reliable and safe operation of the installation.

When designing an ice accumulator, it is worth considering that it makes no sense to heat water in a container over 30 ℃, because there will be significant losses through the enclosing structures of the tank, and this will also lead to a decrease in the moisture content in the soil (decrease in the regenerative function of the soil [5]

3.2. *Installation of a concrete tank*

The third challenge is the correct choice of the location and depth of the ice accumulator. Obviously, in order to regenerate and prevent chilling of the tank, it is necessary to mount it to a depth below the freezing of the soil [6].

There are several methods and technologies for installing and using heat accumulators in the heat supply system [5]. Basically, installation is used near the building and under the building (Viessmann has implemented several objects with storage directly under the building). In the second case (introduction into the existing heat supply system), only the installation of the tank next to the building remains.

Since one of the main goals of this work is the development of a design solution for the implementation of a passive house based on a solar collector and a ground heat pump into the existing heat supply system, it is necessary to install a capacity for heat accumulation at an optimal distance from the house and geothermal wells and other communications and pipelines..

4. Development of calculation methods

4.1. *Selecting the type of heat exchanger for the ice accumulator*

In Figure 1 is presented a graph of the intensity formation of ice on the surface of the heat transfer of different shapes (flat wall, the outer and inner walls of the tube). Analyzing this graph, we can conclude that the rate of ice formation on the inner surfaces of the cylindrical tube is too intense, which can lead to uneven ice formation in the concrete tank (also the concept of an ice accumulator does not allow considering this form of a heat exchanger). The best options are a plate or a coil heat exchanger.

Previous German colleagues [2] has shown that the use of flat coils in this scheme requires compulsory regeneration process, and ice formation rate is high and uneven.

Excluding two types of heat exchangers, only a cylindrical heat exchanger remains (from the outside the coolant is water; through the tubes - brine), which will be immersed in a concrete "barrel".

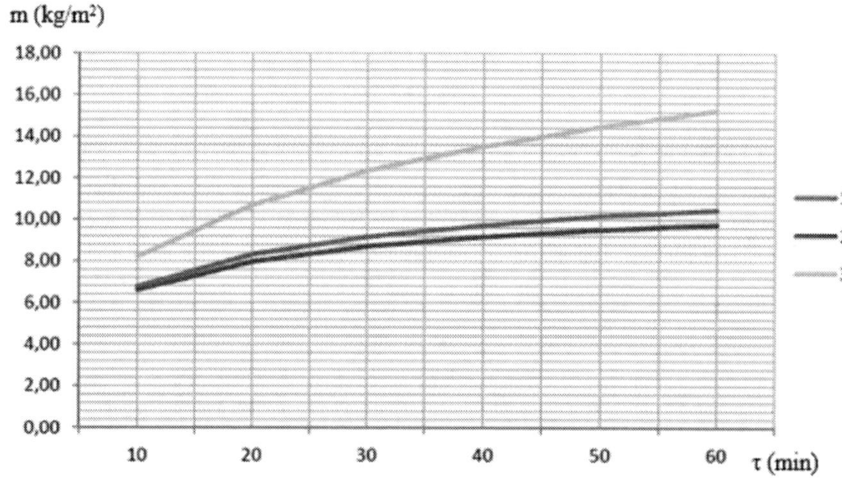

Figure 1. Graph of the dynamics of the growth of ice mass on the surface of heat exchange of various shapes at the same surface temperatures. 1 - flat surface; 2 - outer cylindrical surface of the tube; 3 - inner cylindrical surface of the tube.

In [7], recommendations are given for the design of coil heat exchangers, but they are applicable only to two states of aggregation (gaseous and liquid). Using the model [3], one can estimate the intensity of the increase in the ice layer on the surface of the heat exchanger. On the basis it should determine when the ice thickness is reached, the rate of ice formation decreases, set the inter-turn distance.

At a constant temperature of the coolant, the thickness of ice that forms on tubes of various diameters is practically the same. We add 2ξ to the turn-to-turn distance, which will reduce the intensity of freezing of the heat exchanger and facilitate its regeneration. In "brine water" heat pumps, plastic pipes are used as the material of the soil collector, which are made of low pressure polyethylene (HDPE).

4.2. Requirements for concrete structure

All types of concrete and reinforced concrete structures must meet the following requirements [8]:

- for security;
- for serviceability;
- for durability;
- design requirements.

To meet safety requirements, structures must have characteristics so that under various design influences during construction, installation and operation, there is no threat to the life or health of citizens, property, and the environment.

To meet the requirements for serviceability and durability, the structure must have such characteristics that, under various influences, the formation of cracks and leakage does not occur. It must be taken into account that the concrete structure will:

- installed below the 0.000 m level (it is necessary to provide a foundation);
- filled with water (ensuring the impermeability and tightness of the structure);
- alternate freezing and thawing of water.

The concrete structure of the ice accumulator consists of 2 main parts - a concrete "bowl" and a lid through which the container is inspected and filled.

4.3. Coil heat exchanger design

Problem statement: average water temperature in the ice accumulator during the year $t_B = 5℃$. As pipes, HDPE pipes 32x3 mm are used. From the side of the pipes, a ready-made mixture based on ethylene glycol with corrosion inhibitors down to -20℃ is used as a heat carrier. Co-operation with a Buderus Logatherm WPS 11 heat pump (heating power 10.9 kW) in heating mode is provided.

Let's set the diameter of the battery tank:

$$D_{\text{tank.in}} = 2{,}5\ m - \text{internal diameter of the concrete construction.}$$

In the design of the coil, it is taken into account that the minimum bend radius for pipes DN32 (Uponor GeoPEx , MuoviTech) is 0.32-0.4 m.

Taking into account the data obtained (Figure 3) and recommendations [7], we determine the distance from the pipe to the concrete structure:$\mathrm{a_w} = 0{,}2$ m.

$$\mathrm{a_w} = 0{,}2\ \mathrm{m}$$

$$\mathrm{a_b} = 0{,}2 + 0{,}015 + \frac{(d_\mathrm{n} + 2 \cdot \delta_\mathrm{n})}{2}, \qquad (1)$$

where:
ab – is the distance from the pipe to the bottom of the concrete structure,
aw – is the distance from the pipe to the side walls of the concrete structure,
d_n – is the flow diameter of the pipe, m,
δn – is the wall thickness of the pipe, m.

Because in the upper part of the concrete bowl, in addition to water, collectors will be located, additional space must be provided:

$$a_{\text{collector}} = a_{\text{b}} + 0.1, \tag{2}$$

Determine the diameter of the loop of the first coil:

$$D_{\text{L1}} = D_{\text{tank.in}} - (2 \cdot a_{\text{c}}) = 2.1 \text{ m}, \tag{3}$$

Determine the turn-to-turn distance:

$$s_{\text{coil}} = (d_{\text{n}} + 2 \cdot \delta_n) \cdot 3 + 3 \cdot 0.015. \tag{4}$$

The length of one loop (one complete turn) of the first coil:

$$l_{\text{1loop}} = \sqrt{(\pi \cdot D_{\text{L1}}{}^2 + s_{\text{coil}}{}^2)}, \tag{5}$$

Let's set the number of turns of the first coil:

$$n_{\text{coil1}} = 11 \text{ pcs.}$$

Determine the length of the first coil (taking into account the installation length of 5 m):

$$l_{\text{coil1}} = l_{\text{1loop}} \cdot n_{\text{coil1}} + 5. \tag{6}$$

Let's calculate the surface area of the first coil:

$$F_{\text{coil1}} = l_{\text{coil1}} \cdot d_{\text{тр.н}} \cdot \pi. \tag{7}$$

The coil height:

$$H_{\text{coil1}} = n_{\text{зм1}} \cdot d_{\text{n}} + (n_{\text{coil1}} - 1) \cdot s_{\text{coil}}. \tag{8}$$

Coil arrangement diagram according to Figure 2.3. The distance between the pipes is 20 cm.

By analogy, we make a calculation for all coils (4 pcs.). The calculation results are summarized in Table 1.

Table 1. Summary table of geometrical parameters of ice accumulator coils.

No	Parameter	Units. rev	Coil				Total
			I	II	III	IV	
1	Loop diameter	m	2.100	1.754	1.407	1.061	–
2	Loop length	m	3.728	3.155	2.502	1.891	–
3	Coil length	m	46.03	39.27	32.52	25.81	143.63
4	Coil area	m	5.058	4.318	3.576	2.837	15.789

Figure 2 shows a drawing of a heat exchanger in a concrete structure, as well as the principal location of the supply and return headers.

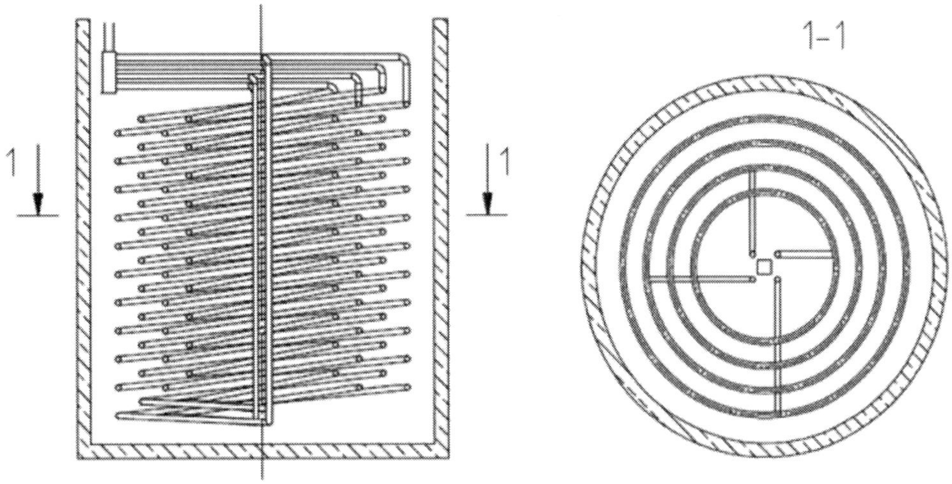

Figure 2. Ice storage drawing.

Knowing the height of the coil, you can determine the height of the water level in the concrete "bowl":

$$\mathrm{H_w} = H_{\mathrm{coil1}} + \mathrm{a_b} + \mathrm{a_{collector}} = 2{,}97\ \text{м}. \tag{9}$$

Let's round the resulting value to 3 m.

The volume of water in the concrete "bowl":

$$V_{\mathrm{w}} = \mathrm{H_w} \cdot \frac{\pi}{4} \cdot D_{\mathrm{tank.in}}{}^{2} - l_{\mathrm{coil.total}} \cdot \frac{\pi}{4} \cdot d_{\mathrm{n}}. \tag{10}$$

Since the density of ice at a temperature of -5℃ is 917.5 kg m^{-3} (the density of water at 0.01℃ is 999.8 kg m^{-3}), when designing the bowl, it is necessary to take into account that the volume will increase to 15.75 m^3 (an increase in volume by 8%, water level rise by 0.16 m). To avoid overflow, it is necessary to increase the height of the concrete structure:

$$H = \mathrm{H_w} + 0.3 \tag{11}$$

4.4. Coil heat exchanger design

According to the research [8] and the requirements [9], to ensure the reliability of a concrete or reinforced concrete structure, it is necessary to make a calculation using the semi - probabilistic method, assessing the acting loads and operating conditions. In addition, it is necessary to consider the possibility of transportation (if necessary), or pouring on site. The concrete structure should consist of 4 elements: a concrete ring, bottom, cover and foundation.

Previously, the main parameters of the concrete structure were set, namely the inner diameter and height of the bowl, which corresponds to D = 2.5 m and H = 3.30 m. The calculation of the parameters and reliability of the ring concrete structure was performed using the Perpendicular.pro software. The cover for this tank can be used the same as for drainage wells.

4.5. Calculation of the maximum storage capacity of the ice accumulator

The calculation will be made by analogy with the article [10].Let's write down the formula for calculating the accumulating capacity of the ice accumulator when it is discharged:

$$Q_{\mathrm{a}} = Q_{I} + Q_{II} + Q_{III} - Q_{\mathrm{loss}} + Q_{\mathrm{reg}} + Q_{\mathrm{sol}}, \tag{12}$$

where:
Q_I - heat extraction from the ice accumulator (from 15℃ to 0.1℃),
Q_{II} - phase transition,
Q_{III} - ice supercooling (0),
Q_{loss} - losses through the concrete structure,
Q_{reg} – the regeneration of the ice accumulator,
Q_{sol} - heat gains from the solar collector (if available).

$$Q_I = V_w \cdot Cp_w \cdot \Delta t_1 \cdot \rho_w = 250 \text{ kWh}, \quad (13)$$

$$Q_{II} = V_w \cdot \rho_w \cdot L = 1338 \text{ kWh}. \quad (14)$$

The project does not provide for ice supercooling below -0.1℃

$$Q_{III} = 0$$

Before proceeding with the calculation of heat losses through the enclosing structures, we will calculate the required time of continuous operation of the heat pump to extract heat from the ice accumulator (cooling capacity according to the passport is 8.6 kW).

$$\tau_I = \frac{Q_I}{Q_c \cdot 3600} = 29 \text{ h } 5 \text{ min}, \quad (15)$$

$$\tau_{II} = \frac{Q_{II}}{Q_c \cdot 3600} = 155 \text{ h } 34 \text{ min}. \quad (16)$$

The total operating time for a complete discharge of the tank (excluding regeneration from the ground and solar collectors) is 184 hours and 39 minutes. Let's calculate the losses and heat inflows through the walls for the given duration of work.

$$Q_{\text{loss}} = 58 \text{ kWh}$$

$$Q_{\text{reg}} = 516 \text{ kWh}$$

Since the sun is often the main source of heat for such ice accumulators [11], the Q_a. value was calculated using the T * Sol Pro 5.5 program. The system of solar collectors was specified, consisted of 15 Vitosol 100-F flat panels, with a total area of 37.77 m^2 (active area 34.94 m^2). Due to the fact that the program is given standard schemes, the heating scheme of the open pool was chosen as the calculated one (the dimensions of the pool correspond to the dimensions of the ice accumulator).

$$Q_a = 2686 \text{ kWh}$$

5. Object of design and implementation of technical solutions

5.1. Design object
The design object is an existing energy efficient house located in the Moscow region. The total area of the building is 205 m^2 (living area 167 m^2, plot area 1 264 m^2). Building dimensions – the building length 12 m, the width 11 m, the height to the ridge of the roof - 10 m. Description of the house is presented in works [12-14].

Design and construction work was carried out using Passive House technologies [15]. That is why the shape of the building is close to a cube, because this solution allows to reduce heat losses of the building through the building envelope. The roof is gable, the roof area on the south side is 130 m^2 (this area is used to accommodate solar collectors). There is a hinged veranda on the south side. It is designed in such way that in summer it limits the flow of heat due to insolation, but in winter, on the contrary, when the sun is low, the sun's rays freely illuminate the window and glass door, thereby providing heating of the internal air in winter.

The main calculation of the thermal loads of this object was carried out in the PHPP software package and is described in [12-14]. Based on the results obtained by the program, the thickness of the layers of the enclosing structures was determined. Later, these results were used in [16] for comparative analysis of various software.

5.2. Existing heat supply system

The phased implementation of equipment is underway at the projected facility. Schematic diagram heating detail is described in the work [12-14]

To date, 2 wells have been drilled and geothermal probes installed in them. The operation of the heat pump shown that at long operation (long cycles described in [17]), there is a drop temperature feedback line below -1 ℃, that indicates the low productivity of the well. This operating mode limits the operating time of the heat pump and does not allow full use of the night electricity tariff. In this connection, it was decided to use the ice accumulator as an additional source of heat. This was done based on research about joint work [18].

To determine the effectiveness of the application of the introduced technology was determined COP for three cases: the heat pump with the existing geothermal probes, a heat pump with ice accumulator`s volume of 14.5 $m^3$ and a heat pump with the ice accumulator`s volume of 50 m^3. The results were determined according to the COP calculation method proposed by the manufacturer of the Buderus Logatherm WPS11 heat pump. The data for calculating the COP of the heat pump in the existing scheme was taken from the existing monitoring system [19]. The heating circuit flow temperature is 35 ° C.

Table 2. COP indicators.

№	Scheme	COP
1	Heat pump with 2 geothermal probes	6.0 – 3.8
2	Heat pump with ice storage volume 14.5 m^3	6.5 – 4.4 (4.7)
3	Heat pump with ice storage volume 53.4 m^3	6.5 – 4.4 (5.5)

The data are presented in Table 2 and listed with the range. For the first scheme, this range means the maximum possible initial COP and the minimum at the end of one cycle of work. For the second and the third schemes, the operating range is presented from the maximum design mode of the ice accumulator (the water temperature in the ice accumulator is 15 ℃ higher) to the moment of complete crystallization (the water temperature is -0.1 ℃, without ice overcooling). The value in the brackets indicates the average operating factor coefficient power.

6. Conclusion

It was shown that in the current practice there is no model and software for calculating the ice accumulator, which could allow for a reliable calculation. All models have a number of assumptions that introduce large errors in the final result.

The paper proposes a simplified method for designing an ice accumulator. Recommendations for design were developed, namely:

- choice of turn-to-turn distance based on the existing recommendations for the design of coil heat exchangers and the maximum rate of ice layer growth on the coil surface;

- the recommended number of turns - 9 - 11 pcs. With an increase in the number of coils, the height of the ice accumulator and the depth of its installation increase. If groundwater is located nearby, it can lead to the impossibility of installing an ice accumulator;
- the walls of the ice accumulator should not be insulated, because this will lead to a decrease in the regeneration of the ice accumulator (regeneration will occur only due to solar collectors and the reversible mode of the heat pump);
- the principle of connecting a heat pump to an icemaker is completely similar to geothermal collectors. Their joint work is allowed;
- it is not recommended to maintain and strive for the temperature in the ice accumulator above 15℃, because this mode will lead to an increase in heat loss and a waste of the received solar energy;
- it is not recommended to raise the temperature of the brine in the wells above 30℃. This can lead to a decrease in the amount of moisture in the soil, which will lead to a decrease in the amount of heat that can be taken away.

The second fulfilled goal is a ready-made design solution for implementation into the existing heat supply system of a passive house. Based on the existing problem with wells chilling during long-term operation, it was necessary to increase the number of wells or to develop an alternative solution to the problem. The solution to this problem will be the introduction of an ice accumulator. It will allow you to avoid chilling wells, as well as accumulate additional energy. In the work, the calculation of the concrete structure was made, the composition of concrete and the section of reinforcement were selected to increase the strength of the structure. The heat exchanger of the ice accumulator was designed and the turn-to-turn distance was chosen, which makes the dynamics of ice growth the same throughout the entire volume of the ice accumulator.

References

[1] Marinyuk B T 2009 *Teploobmenny`e apparaty` TNT: Konstruktivny`e shemy` i raschet* [Heat exchangers TNT: Structural diagrams and calculation] (Moscow: Energoizdat) [In Russian]

[2] Carbonell D, Philippen D, Haller M Y, Frank E 2014 Development and validation of mathematical model for ice storages with heat exchangers that can be de-iced *Energy Procedia* **57** 2342-52

[3] Ugolnikova M A 2017 Modelirovanie processov teploobmena pri namorazhivanii vodnogo l`da na neizolirovanny`x e`lementax nizkotemperaturnogo oborudovaniya: Dissertatsiya na soiskaniye uchenoy stepeni kandidata tekhnicheskikh nauk [Modeling of heat transfer processes during freezing of water ice on non-insulated elements of low-temperature equipment: PhD Thesis] (Moscow: MPI) [In Russian]

[4] Khristenko B A, Kalyakin I D and Govorin A V *2017* Primenenie akkumulyatorov l`da dlya kruglogodichnogo e`nergosnabzheniya e`nergoe`ffektivnogo doma [Using ice accumulators for year-round energy supply of an energy efficient home] *Tezisy` dokladov Dvadczat` tret`ej Mezhdunarodnoj Nauchno-Texnicheskoj Konferencii Studentov i Aspirantov* (Moscow: ME`I) [In Russian]

[5] Pinel P, Cruickshank C A, Beausoleil-Morrison I and Wills A 2011 A review of available methods for seasonal storage of solar thermal energy in residential applications *Renewable and Sustainable Energy Reviews* **15(7)** 3341-59

[6] Hesaraki A, Holmberg S and Haghighat F 2015 Seasonal thermal energy storage with heat pumps and low temperatures in building projects – A comparative review *Renewable and Sustainable Energy Reviews* **43** 1199-213

[7] Zorin V M and Grigoriev V A 1991 *Promy`shlennaya teploe`nergetika i teplotexnika* [Industrial heat power engineering and heat engineering] (Moscow: Energopromizdat) pp 170-1 [In Russian]

[8] Yumrutas R and Unsal M 2012 Energy analysis and modeling of a solar assisted house heating system with a heat pump and an underground energy storage tank *Solar Energy* **86(3)** 983-93

[9] Vinogradov I M 1977 *Matematicheskaya e`nciklopediya v 5 tomax* [Mathematical encyclopedia in 5 volumes] (Moscow: Encyclopedia) [In Russian]

[10] Sultanguzin I A and Khristenko B A 2018 Primenenie l`doakkumulyatorov sovmestno s solnechny`mi kollektorami [Use of ice accumulators in conjunction with solar collectors] *Innovacionny`e Nauchny`e Issledovaniya: Teoriya, Metodologiya, Praktika XII Mezhdunarodnaya Nauchno-Prakticheskaya Konf.: v 2 chastyax* (Penza: Nauka i Prosveshhenie) pp 40-2 [In Russian]

[11] Clarke J, Colclough S, Griffiths P and McLeskey Jr J T 2014 A passive house with seasonal solar energy store: in situ data and numerical modelling *Int. J. of Ambient Energy* **35(1)** 37-50

[12] Sultanguzin I A, Kruglikov D A, Yatsyuk T V, Kalyakin I D, Yavorovsky Y V and Govorin A V 2019 Using of BIM, BEM and CFD technologies for design and construction of energy-efficiant houses *ES3 Web of Conf. SES-2019* **124** 03014

[13] Sultanguzin I, Kalyakin I, Govorin A, Khristenko B and Yavorovsky Yu 2016 Optimization of the energy efficient active house *3 Ingenieurtag 2016* ed S Simon (Cottbus-Senftenberg: Brandenburgische Technische Universitat) 8-12

[14] Sultanguzin I A, Shutz W, Kalyakin I D, Govorin A V, Yavorovskiy Yu V, Zhigulina E V, Khromchenkov V G, Malakhova A A, Lancke B and Damen A 2017 Raschyot, stroitel`stvo i obsledovanie fasada e`nergoe`ffektivnogo doma [Calculation, construction and inspection of the facade of an energy-efficient house] *Stroitel`stvo I Rekonstrukciya* **4(72)** 110-8 [In Russian]

[15] Yavorovskiy Yu V, Sultanguzin I A, Kruglikov D A, Kalyakin I D, Yatsyuk T V 2019 Cravnenie rezul`tatov e`nergeticheskogo modelirovaniya zhilogo doma s pomoshh`yu razny`x programmny`x sredstv [Comparison of the results from modeling the power performance of a residential building using different software tools] *Vestnik MEI* **3(3)** 31-9 [In Russian]

[16] Kruglikov D A 2019 Sovremenny`j kompleksny`j podxod k proektirovaniyu, stroitel`stvu i e`ksplua-tacii e`nergoe`ffektivny`x domov na osnove BIM-, BEM- i CFD- texnologij: Magisterskaya dissertatsiya [A modern integrated approach to the design, construction and operation of energy efficient houses based on BIM, BEM and CFD technologies: Master's dissertation] (Moscow: ME`I)

[17] Kalyakin I D, Kruglikov D A, Xristenko B A, Rudomazin V V, Shuvalov D G and Sultanguzin I A 2018 Vnedrenie e`nergosberegayushhix texnologij v sistemu teplosnabzheniya rekonstruiruemogo administrativnogo zdaniya [Introduction of energy-saving technologies in the heat supply system of the reconstructed administrative building] *E`nergosberezhenie-Teoriya i Praktika trudy` Devyatoj Mezhdunarodnoj shkoly`-seminara molody`x ucheny`x i specialistov* (Moscow: ME`I) pp 181-6 [In Russian]

[18] Navarro L, Gracia A, Colclough S, Browne M, McCormack S J, Griffiths P and Cabeza L F 2016 Thermal energy storage in building integrated thermal systems: A review. Part 1. Active storage systems *Renewable Energy* **88** 526-47

[19] Chaikin V Y, Petrov V A, Sultanguzin I A, Basalaev A A, Nechaev A N and Kapitinov N N 2020 Osobennosti sovmestnogo ispol`zovaniya sistem monitoringa i upravleniya e`nergoe`ffektivnogo doma [Features of joint use of monitoring and control systems of an energy efficient home] *E`nergosberezhenie Teoriya i Praktika trudy` Desyatoj Mezhdunarodnoj shkoly`-seminara molody`x ucheny`x i specialistov* (Kursk: Universitetskaya kniga) pp 291-7 [In Russian]

Resource-saving dialysis technology for electroplating wastewater treatment

S A Loza, N A Romanyuk, A N Korzhov and N O Kovalchuk

Department of Chemistry and High Technology, Kuban State University, Stavropolskaya St., 149, Krasnodar, 350049, Russia

E-mail: s_loza@mail.ru

Abstract. In this work, the separation process of components of electroplating effluent containing sulfuric acid and nickel sulfate by dialysis through various ion-exchange membranes was carried out. Kinetic curves of changes in the concentration of sulfuric acid and nickel sulfate in chambers with permeate and retentate were obtained. The flux densities, nickel losses, the degree of sulfuric acid recovery and the separation factor were calculated. It was shown that the use of cation-exchange membranes does not allow to separate nickel sulfate and sulfuric acid. The use of commercial anion exchange membranes allows 90% sulfuric acid to be recovered with a nickel loss of 8%. The profiling of the anion exchange membranes significantly intensifies the process, but leads to an increase leakage of nickel sulfate through the membranes. The prospects of dialysis wastewater treatment from galvanic production for return of valuable components to production are shown.

1. Introduction

As a result of the operation of mining and metalworking enterprises, wastewater is generated containing acids and heavy metal ions. The most common acid in wastewater is the sulfuric acid, and among the salts is nickel sulfate, which appears in wastewater as a result of electroplating, metalworking, mining and other technological processes.

These substances are toxic to humans and the environment. Nickel sulfate is a highly toxic mutagen and cancerogenic [1,2]. Nickel sulfate in wastewater mixes with water bodies and can spread in nature, including in the soil. $NiSO_4$ has a detrimental effect on plant growth and is a root growth inhibitor. It was shown in [3] that when the content of nickel sulfate in the soil with a concentration of more than 80 μM in plants, point necrosis of leaves is observed. Nickel sulfate is also very toxic to aquatic life with long lasting effects as shown in the International Chemical Safety Card.

Sulfuric acid is also a hazardous substance that can be damaging and corrosive. Thus, these substances must not be allowed to enter the environment. Also, the recovered sulfuric acid and nickel can be recycled back into the process to create resource-saving technologies. In addition, due to the growth in the production of stainless steel, batteries for electric vehicles, nickel cast iron, experts predict an increase in global demand for nickel. Thus, the treatment of wastewater from electroplating enterprises is important not only from an environmental point of view, but can also bring significant economic benefits.

There are various methods of wastewater treatment from heavy metals: sorption [4-7], reagent method [8], ion exchange [9-11], flotation [12,13], and other methods [14]. These technologies have disadvantages, such as high consumption of reagents, loss of valuable substances (acids), high consumption of electricity, re-pollution of waters during the regeneration of the sorbent and ion-exchange resins.

Therefore, it is especially important to improve old and develop new treatment systems that would not only allow cleaning wastewater, but also return valuable components back to the production cycle. Dialysis with ion exchange membranes is promising in this regard. It has proven itself well in water treatment for the removal of nitrates [15, 16], borates [17], arsenates [18] from drinking water, as well as in the creation of resource-saving technologies and purification systems [19, 20]. Therefore, the purpose of this work is to evaluate the effectiveness of dialysis for separating a solution containing nickel sulfate and sulfuric acid and recovering valuable components with the possibility of their reuse.

2. Materials and Methods

Dialysis separation of the solution components was performed on a dialyzer (figure 1), alternately formed by 10 ion exchange membranes with separators with the size of the working area of the membranes 5x20 cm^2, the distance between the membranes is 0.1 cm.

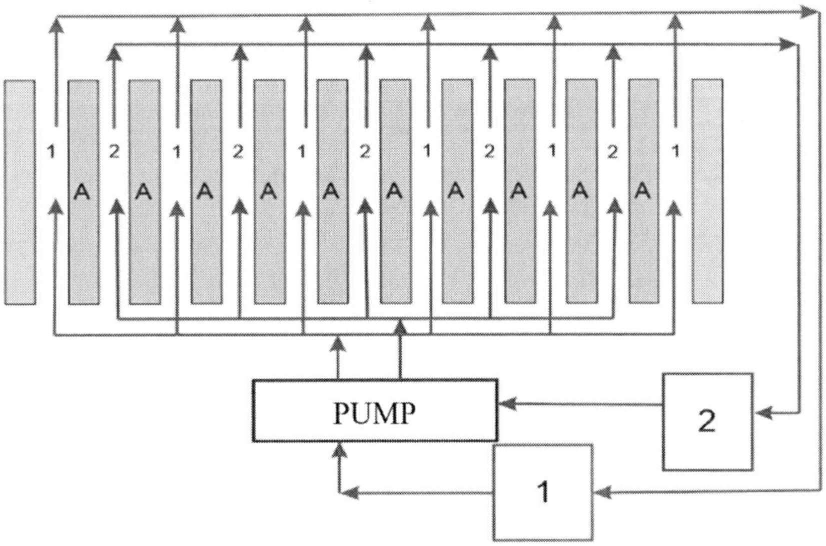

Figure 1. Principal scheme of the dialyzer: 1 – permeate chamber; 2 – retentate chamber; A – anion exchange membrane.

During the experiments, Ralex AM-PP, Ralex CM-PP membranes were used. Ralex AM-PP and Ralex CM-PP membranes made by "MEGA" a.s., Czech Republic are sheets of ion-exchange heterogeneous film with fixed quaternary amino and sulfo groups, respectively. Polyethylene acts as an inert binder. To give the membranes mechanical rigidity, they are reinforced with a polypropylene fabric that is resistant to acids and alkalis. These membranes are intended for use in electrodialysis, deionization and membrane electrolysis.

In the course of the work, the Ralex AM-PP anion-exchange membranes were modified for use in dialysis of the processed solution. Hereinafter, these membranes were referred to as Ralex AM-PP_Prof. The modification was carried out according to the method [21], and consists of pressing pre-swollen membranes by 15 MPa at a temperature of 90 °C and a holding time under a load of 30 seconds.

Modification by profiling increases the microporosity and diffusion permeability of the membrane [22], which should have a positive effect on the dialysis rate of the test solution. Such profiled membranes made it possible to significantly intensify the process of separation of phenylalanine and sodium chloride by diffusion dialysis [23].

For separation (retentate chamber), a 1 liter of electroplating waste solution was used containing approximately 0.3 M nickel sulfate and 2.5 M sulfuric acid. Distilled water was fed into the permeate chamber. The volume of the permeate chamber was 1 liter. The distilled water in the permeate chamber was changed occasionally. Samples for determining the concentration of nickel and sulfuric acid were taken at the inlet to the chambers with permeate and retentate, at fixed intervals. The concentration of sulfuric acid was determined using acid-base titration, nickel – by photometry at a wavelength 394 nm.

3. Results and discussion

To evaluate the applicability of dialysis for separating a solution consisting of sulfuric acid and nickel sulfate, some experiments were carried out on the dialysis processing of a solution of 2.5 M sulfuric acid and 0.3 M nickel sulfate using various membranes: Ralex AM-PP, Ralex CM-PP, Ralex AM-PP_Prof.

Experimental data on dialysis using Ralex AM-PP membranes are presented in figure 2.

Analysis of the data obtained shows that as the dialysis process is carried out, the difference in acid concentration in the retentate and dialyzate decreases and the process slows down (Figure 2 *a*). After changing the dialysate to a fresh portion of distilled water (shown with vertical dashed lines in Figure 2), the process intensifies.

At the initial stage, the nickel concentration slightly decreased, and then remained approximately constant. At the same time, the concentration of nickel sulfate in the dialysate almost always did not exceed 1 g/L (which corresponds to the concentration of an acid solution acceptable for acid reuse in production). Figure 2 *c* is an enlarged fragment of figure 2 *b*.

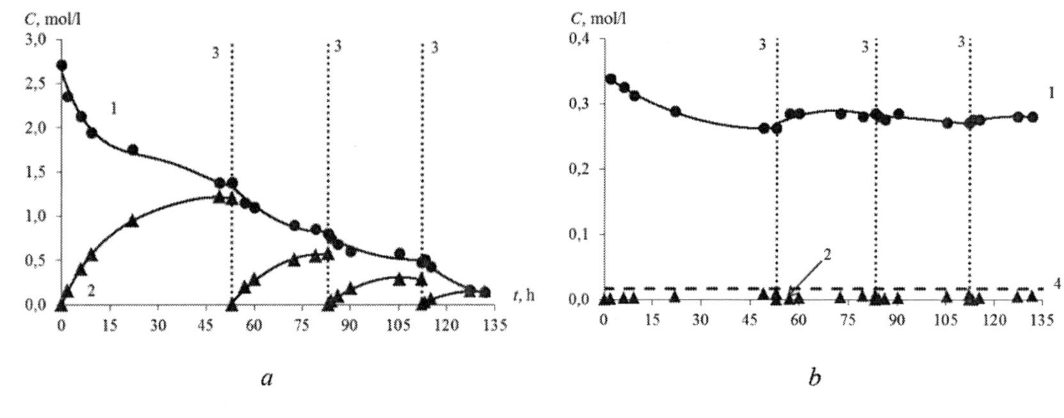

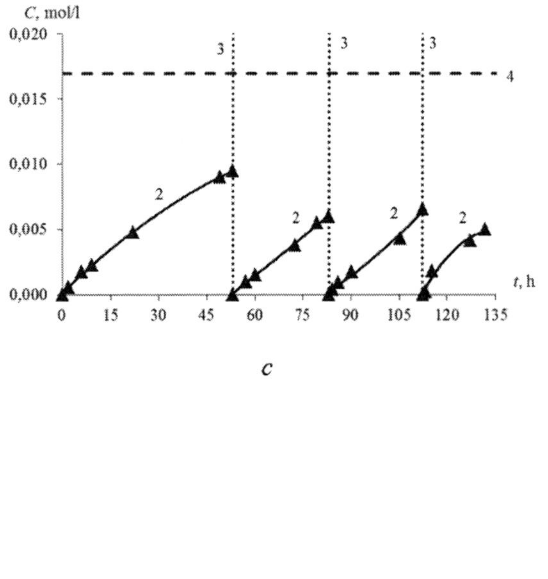

Figure 2. Kinetic curves of changes in the concentration of sulfuric acid and nickel sulfate during dialysis using membranes Ralex AM-PP.
1 - data for the camera with retentate;
2 - data for the chamber with permeate;
3 - lines showing the change of water in the retentate chamber;4 - limit of nickel content (1 g/L)
a – concentration of sulfuric acid;
b, c - concentration nickel sulfate.

In the course of the experiments, kinetic curves of changes in the concentration of sulfuric acid and nickel sulfate in chambers with retentate and permeate were obtained, on the basis of which the average flux density was calculated according to equation (1):

$$J = \frac{n_0 - n}{St}, \tag{1}$$

where:
J is the flux density, (mol/(m^2·h));
n_0 – the amount of substance at the beginning of the experiment, (mol); n is the amount of substance at the end of the experiment, (mol);
S is the area of membranes, (m^2); t - time, (h).

The average specific flux of sulfuric acid through the membranes was 0.195 mol/(m^2 h). In this experiment, despite the high degree of separation of nickel and sulfuric acid, the dialysis rate was periodically slowed down due to a decrease in the concentration gradient between the chambers.

An attempt has also been made to use cation exchange membranes for dialysis. For this, an experiment was carried out on dialysis of the initial solution using Ralex CM-PP membranes. The kinetic curves of changes in the concentration of nickel sulfate and sulfuric acid are shown in Figure 3.

Analysis of the data obtained shows that the concentration of nickel sulfate and sulfuric acid in the retentate chamber decreases, while the concentration in the permeate chamber increases insignificantly. In this case, a high increase in the volume of the solution was observed in the chamber with the retentate. This is due to the osmotic flow of water from the tract with the permeate to the tract with the treated solution, while a decrease in the concentration of the initial substances due to dilution is formally observed, while the total amount of the substance in the chamber remains unchanged. The intense osmotic flow of water through the cation-exchange membranes significantly reduces the diffusion rate, thus dialysis with cation-exchange membranes is not applicable for the separation of nickel sulfate and sulfuric acid.

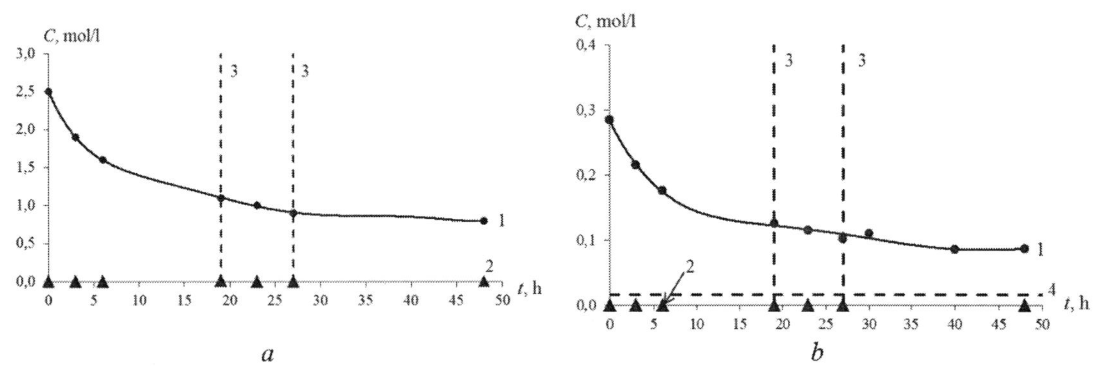

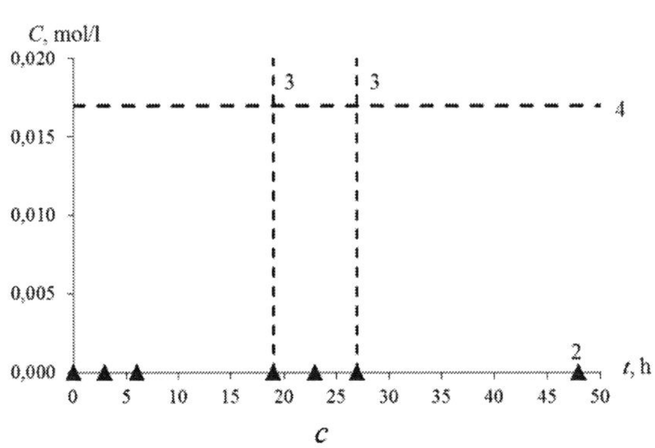

Figure 3. Kinetic curves of changes in the concentration of sulfuric acid and nickel sulfate during dialysis using membranes Ralex CM-PP.
1 - data for the camera with retentate;
2 - data for the chamber with permeate;
3 - lines showing the change of water in the retentate chamber;
4 - limit of nickel content (1 g/L)
a - concentration of sulfuric acid;
b, *c* - concentration nickel sulfate.

To increase the dialysis rate, profiled anion exchange membranes Ralex AM PP_Prof were used (figure 4).

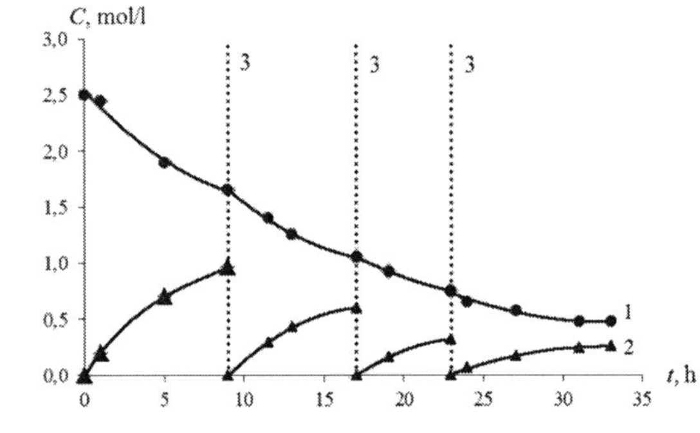

a

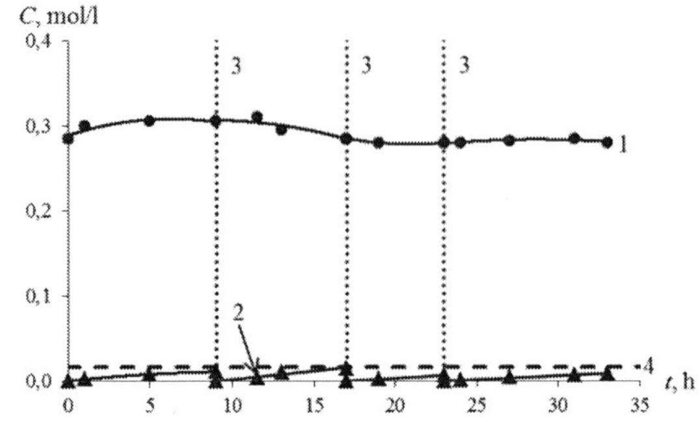

b

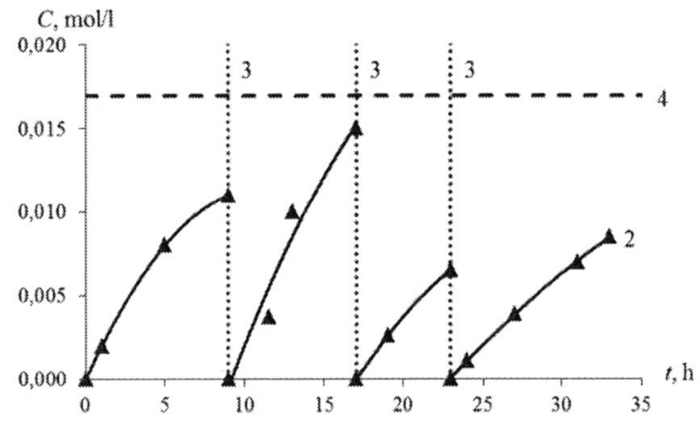

c

Figure 4. Kinetic curves of changes in the concentration of sulfuric acid and nickel sulfate during dialysis using membranes Ralex AM-PP_Prof.
1 - data for the camera with retentate;
2 - data for the chamber with permeate;
3 - lines showing the change of water in the retentate chamber;
4 - limit of nickel content (1 g/L)
a – concentration of sulfuric acid;
b, c - concentration nickel sulfate.

The average ratio of the concentration of sulfuric acid to the concentration of nickel sulfate in the dialysate was 30 times, while in the initial solution the concentration ratio was 3.7, and in the treated solution it decreased to 0.49. In this case, the average flux density of sulfuric acid for profiled membranes Ralex AM-PP_Prof is 0.61 mol/(m^2·h), which is higher than in the experiment with commercially available anion-exchange membranes.

Thus, the profiling of anion-exchange membranes makes it possible to increase the rate of acid transfer through the membrane by about 3 times with a slight decrease in the selectivity for nickel sulfate.

To assess the selectivity of the dialysis process to nickel sulfate, the average losses of nickel sulfate from the initial content per hour were calculated by equation (2):

$$\omega = \left(100 \cdot \frac{\Delta n}{n_0}\right) \cdot \frac{1}{t}, \qquad (2)$$

where:
ω – loss of nickel sulfate per hour, (%/h);
Δn – the amount of nickel sulfate transferred through the membranes during the experiment, (mol);
n_0 – the amount of nickel in the chamber with the initial solution, (mol);
t – time of experiment, (h).

The separation factor (SF) was calculated by (3):

$$SF = \frac{C_{H_2SO_4}}{C_{NiSO_4}} \cdot \frac{C^0_{NiSO_4}}{C^0_{H_2SO_4}}, \quad (3)$$

where:

$C_{H_2SO_4}$, C_{NiSO_4} – concentrations of sulfuric acid and nickel sulfate in diluate (mol/L);

$C^0_{H_2SO_4}$, $C^0_{NiSO_4}$ – concentrations of sulfuric acid and nickel sulfate in initial solution, (mol/L);

Summary characteristics of the dialysis process are presented in table. 1.

Table 1. Summary characteristics of the dialysis process.

Membrane	Average acid flux through the membrane, mol/(m^2·h)	Loss rate of Ni_2SO_4 with acid, %/h	Separation factor (SF) of sulfuric acid and nickel	Nickel loss during 90% sulfuric acid recovery
Ralex AM-PP	0.195	0.06	11.3	8.0
Ralex CM-PP	0.027	–	–	–
Ralex AM-PP_Prof	0.630	0.44	6.3	14.4

Analysis of the data in Table 1 shows that the maximum losses of nickel sulfate correspond to the use of Ralex AM-PP_Prof membranes. However, the use of profiled membranes makes it possible to process the solution 3 times faster than the original membranes.

Thus, dialysis can successfully separate sulfuric acid and nickel sulfate with a high separation factor. At the same time, 85-92% of nickel remains in the initial solution at a sulfuric acid recovery rate of 90%. This solution can be used for nickel recovery, for example, for the manufacture of batteries. The disadvantages of the dialysis method include a large amount of dilute sulfuric acid containing a small amount of nickel sulfate, which must be additionally concentrated using electromembrane methods such as electrodialysis and electrodialysis concentration to return to galvanic production.

4. Conclusion

A study of the dialysis processing of galvanic production waste containing 2.5 mol/L sulfuric acid and 0.3 mol/L nickel sulfate using various ion-exchange membranes was carried out. The use of cation exchange membranes for dialysis does not solve the problem of separating sulfuric acid and nickel sulfate. When carrying out dialysis through commercially available anion-exchange membranes, it is possible to successfully separate sulfuric acid and nickel sulfate with a separation factor of more than 11. The use of profiled membranes reduces the separation factor to 6.3 and increases nickel losses from 8 to 14.4%, but allows to increase the acid flow through the membranes by 3 times.

5. Acknowledgments

This work was funded by the State Assignment Project FZEN–2020–0022.

References

[1] Mahmood A and Malik R N 2014 Human health risk assessment of heavy metals via consumption of contaminated vegetables collected from different irrigation sources in Lahore, Pakistan *Arab. J. Chem.* **7(1)** 91-9

[2] Zambelli B, Uversky V N and Ciurli S 2016 Nickel impact on human health: an intrinsic disorder perspective. *Biochim. Biophys. Acta - Proteins Proteomics* **1864(12)** 1714–31

[3] Bashmakova E B, Pashkovskiy P P, Radyukina N L and Kuznetsov V V 2016 Interactive effects of zinc and nickel on the glutathione system state in Mimulus guttatus plants. *Rus. J. Plant Physiol.* **63** 626-35

[4] Li C 2015 A novel conversion process for waste residue: Synthesis of zeolite from electrolytic manganese residue and its application to the removal of heavy metals *Colloids and Surfaces* **470** 258-67
[5] Zhai Y, Wei X, Zeng G, Zhang D and Chu K 2004 Study of adsorbent derived from sewage sludge for the removal of Cd^{2+}, Ni^{2+} in aqueous solutions *Sep. Purif. Technol.* **38(2)** 191-6
[6] Shilina A S, Bakhtin VD, Burukhin S B and Askhadullin S R 2017 Sorption of cations of heavy metals and radionuclides from the aqueous media by new synthetic zeolite-like sorbent *Nuclear Energy and Technology* **3(4)** 249-54
[7] Duan Ch, Ma T,Wang J and Zhou Y 2020 Removal of heavy metals from aqueous solution using carbon-based adsorbents: A review *J. Wat. Pro. Engin.* **37** 101339
[8] Kurniawan T A, Chan G Y S, Lo W-H and Babel S 2006 Physico-chemical treatment techniques for wastewater laden with heavy metals *Chem. Eng. J.* **118(1-2)** 83-98
[9] Dąbrowski A, Hubicki Z, Podkościelny P and Robens E 2004 Selective removal of the heavy metal ions from waters and industrial wastewaters by ion-exchange method *Chemosphere* **56(2)** 91-106
[10] Lacour S, Bollinger J-C, Serpaud B, Chantron P and Arcos R 2001 Removal of heavy metals in industrial wastewaters by ion-exchanger grafted textiles *Analytica Chimica Acta* **428(1)** 121-32
[11] Kang S Y, Lee J U, Moon S H and Kim K W 2004 Competitive adsorption characteristics of Co^{2+}, Ni^{2+}, and Cr^{3+} by IRN-77 cation exchange resin in synthesized wastewater *Chemosphere* **56(2)** 141-7
[12] Polat H and Erdogan D. 2007 Heavy metal removal from waste waters by ion flotation *J. Hazard. Mater* **148(1-2)** 267-73
[13] Yuan X Z, Meng Y T, Zeng G M, Fang Y Y and Shi J G 2008 Evaluation of tea-derived biosurfactant on removing heavy metal ions from dilute wastewater by ion flotation *Colloid Surf.* **317(1-3)** 256-61
[14] Fu F and Wang Q 2011 Removal of heavy metal ions from wastewaters: a review *J. Environ. Manag.* **92(3)** 407-18
[15] Schaetzel P, Nguyen D N and Nguyen Q T 2004 Batch ion-exchange dialysis to extract nitrate from drinking water: a simplified ion transport model for the best membrane selection *Desalination* **164(3)** 261-8
[16] Altintas O, Tor A, Cengeloglu Y and Ersoz M 2009 Removal of nitrate from the aqueous phase by Donnan dialysis *Desalination* **239(1-3)** 276-82
[17] Bryjak M and Duraj I 2013 Anion-exchange membranes for separation of borates by Donnan dialysis *Desalination* **310** 39-42
[18] Zhao B, Zhao H and Ni J 2010 Arsenate removal by Donnan dialysis: Effects of the accompanying components *Sep. Purif. Technol.* **72(3)** 250-55
[19] Sonoc A C, Jeswiet J, Murayama N and Shibata J 2018 A study of the application of Donnan dialysis to the recycling of lithium ion batteries *Hydrometallurgy* **175** 133-43
[20] Agarwal C, Cattrall R W and Kolev S D 2016 Donnan dialysis based separation of gold (III) from electronic waste solutions using an anion exchange pore-filled membrane *J. Membr. Sci.* **514** 210-6
[21] Zabolotskii V I, Loza S A and Sharafan M V 2005 *Patent RU 2284851 Sposob profilirovaniya geterogennykh ionoobmennykh membran* [Patent RU 2284851 Method for profiling heterogeneous ion exchange membranes] (Moscow: Rospatent) [In Russian]
[22] Zabolotskii V I, Loza S A and Sharafan M V 2005 Fiziko-ximicheskie svojstva profilirovanny`x geterogenny`x ionoobmenny`x membran [Physicochemical properties of profiled heterogeneous ion-exchange membranes] *E`lektroximiya* **41(10)** 1185-92 [In Russian]
[23] Vasil'eva V, Goleva E, Pismenskaya N, Kozmai A and Nikonenko V 2019 Effect of surface profiling of a cation-exchange membrane on the phenylalanine and NaCl separation performances in diffusion dialysis *Sep. Purif. Technol.* **210** 48-59

Study of toxic impact of fly ash bearing construction materials on aquatic organisms

E N Goncharova[1], M S Lebedev[2], M I Vasilenko[1] and N I Kozhukhova[3]

[1]Department of Industrial Ecology, Belgorod State Technological University named after V.G. Shukhov, Kostyukov St., 46, Belgorod, 308012, Russia
[2]Center of High Technologies, Belgorod State Technological University named after V.G. Shukhov, Kostyukov St., 46, Belgorod, 308012, Russia
[3]Department of Material Science and Material Technology, Belgorod State Technological University named after V.G. Shukhov, Kostyukov St., 46, Belgorod, 308012, Russia

E-mail: kozhuhovanata@7andex.ru

Abstract. The article studies the toxicological characteristics and behavior of geopolymer samples GB-1 and GB-2 based on fly ash from thermal power plants. Cement paste also was used to provide comprehensive comparative analysis. Aerated tap water was used as a reference medium. The toxicity of the studied materials was determined by applying standard biotesting methods using such aquatic organisms as *Scenedesmus quadricauda* microalgae and cladocerans *Daphnia magna*. Based on the obtained experimental and analytical data, it was found that the most favorable environment for algae is sample GB-2. Different behavior is observed in crustaceans. In this case, the most favorable medium for acute and chronic toxicity is a cement paste. The highest toxic effect is observed in the GB-1 sample medium vs. the others studied samples for all considered parameters of toxicity, including chronic one: the decrease in the concentration of algae is 30-35 %; the death of daphnia is 50-80 %. Chronic toxicity, which was determined by the value of the total fertility of daphnia in GB-1 medium, is 5-18 times less compared to the Reference sample.

1. Introduction

T In modern conditions of the constantly increasing standard of human life, there is a tendency to tighten the requirements for the quality of the surrounding infrastructure, including, building materials. The eco- and bio-safety of building materials is one of the key quality indicators. According to some scientific studies, minimizing the use of toxic building materials is one of the main principles of the sustainable development of the construction industry. However, despite this, nowadays structures are often built from toxic building materials. This phenomenon can be explained by two reasons:

- poor awareness of builders about the characteristics of the building materials used;
- the economic aspect. Resolving the issue of environmental friendliness, as a rule, requires additional material investments [1].

One of the striking examples of the use of toxic materials can be noted impregnation for wood (fungicides, creosote, copper, chromium, and arsenic, etc.), which are used to protect wooden structures from mold and insects [2]. Another equally common example is the use of toxic mineral industrial waste in structures, for example, road structures, which are liable to leach during operation [3].

In both cases, the negative effect manifests itself when rainwater hits the material, as a result of which a significant proportion of pollutants and hazardous substances enter the environment. Another case of toxicity in building materials is associated with materials that emit toxic fumes during fires. Some studies show that most of the deaths in fires are due to the inhalation of toxic fumes [4]. In addition to chemical pollution, building materials can be characterized by the presence of radioactive elements such as ^{226}Ra, ^{232}Th, ^{40}K, which are also quite dangerous and can be classified as toxic. It is known that the use of wastes with a risk of radiological contamination is dangerous at the level of human health. The danger is that long-term exposure, even at low dosages, can develop cancers. In most cases, building materials are toxic to a much lesser extent than raw materials that are used to synthesize materials and composites [5]. According to the analysis provided by Pacheco-Torgala F. et al. [6], the most toxic types of raw materials are phosphogypsum, some blast furnace slags [7], and some fly ash types [8–10].

One of the most important methods in ecology, allowing to answer many questions about toxicity, is biotesting, which, using certain organisms (test objects), solves a wide range of environmental problems. Most often, this is an assessment of the quality of wastewater treatment, the safety of chemicals for various purposes; an assessment of the composition of bottom soil, and control of the water quality of fishery reservoirs. Biotesting, unlike other, even the most subtle analytical methods, allows the determination of how dangerous the factor under study is for the life of biosystems [11].

Experiments were carried out to study the acute and chronic toxicity of daphnia *Daphnia magna* and *Scenedesmus quadricauda* algae to study the environmental friendliness of fly ash-containing building materials. In toxicological studies, test organisms are represented by planktonic unicellular algae from different zones (*Scenedesmus quardicauda*, *Chlorella vulgaris*, *Phaeodactylum tricornutum*, etc.) [12–14]. Crustaceans, most often – the cladocerans *Daphnia magna*, *D. longispina*, *Ceriodaphnia affinis*, and others, are used as typical representatives of zooplankton [15-17].

An indicator of the toxicity of substances is a whole range of biological criteria, ranging from the death of an organism to subtle molecular rearrangements of its enzymatic systems [18–20]. The most significant indicators of the biological resistance of any organism are its survival rate, fertility, and the quality of the offspring, which determine its safety as a species [11, 13, 15]. Acute toxicity is determined in short-term experiments lasting from several minutes to 4 days, however, based on these data, it is impossible to predict what will happen to the body in the future, especially when exposed to low concentrations of the toxicant. Long-term experiments, which serve to determine chronic toxicity, allow predicting the state of biological processes in the environment, and recognizing latent toxic effects [15, 22]. Determination of water quality using microalgae is necessary stage in the biotesting procedure in all countries of the world [21].

In Russia, the species *Scenedesmus quadricauda* is recommended as a test object for the standard biotesting procedure (ISO 8692:2004 2012), whereas in Europe and the United States, the alga *P. subcapitata* is used more often (Russian Federal Register.1.39.2007.032232007).

S. quadricauda belongs to coenobial chlorococcal algae; 2- and 4-cell coenobia of elongated oval cells with rounded ends (8–36 × 2.1–12 μm) are more common. The marginal cells of one coenobium have two spines bent outward. During reproduction, 2–4 autospores are formed in the mother cell, which, upon exiting, form coenobia [22, 23].

2. Materials and Methods

2.1. Materials

Some text. The studied systems were geopolymers samples GB-1 and GB-2 obtained on the basis of fly ash from thermal power plants of various production. For biotesting, the pure culture of *Scenedesmus quadricauda* algae in an exponential growth stage (3–5 days after reseeding) and genetically homogeneous culture of daphnia *Daphnia magna* were used.

2.2. Methods

Biotesting methods were used in the study to determine toxic characteristics of geopolymer samples. The biotesting method using daphnia is widely used in Russia and abroad both for testing waste and natural waters and in order to identify biological effects and mechanisms of negative influence of anthropogenic factors when earlier ecotoxicological research reults. Biotesting for *Daphnia magna* is realized in the framework of ISO and Russian standards [OCDE-2011; MU 2.1.4.783-99; PND F T 14.1: 2: 4.12-06]. Experimental and control cultures of algae were grown in the medium "Uspensky 1", which is suitable for growing microalgae. A synchronized culture was taken into the experiment.

For synchronization, the test culture was kept in complete darkness at ambient temperature for 2-3 days. As a result, the old cells settled to the bottom, while the young ones remained in the upper layer of the cylinder. The upper layer was transferred without stirring into a sterile nutrient medium. To obtain a double-synchronized culture, the synchronization procedure was repeated after 5-7 days of the exponential growth of the culture. To maintain the exponential stage of algal growth, reseeding was carried out once every 7 days

3. Results

3.1. Algae biotesting

Before the start of the experiment, the initial culture of algae (at the age of 3–5 days) was concentrated until a suspension was formed, and the initial number of cells of the suspension of algae in the Goryaev chamber was determined to calculate the required amount of suspension addition (in control and test vessels), which would provide the required cell density when inoculated at the beginning of the experiment. The number of algae at the beginning of biotesting should be 25–35 thousand cells/ml.

After sowing and thorough mixing, the algae in the control and test vessels were counted in the Goryaev chamber before the start of the experiment. The repeated counting of algae in the Goryaev chamber was carried out after 72 hours. The experimental results are presented in the Table. 1.

Table 1. Results of biotesting test in model media of the studied samples after 72 hours of algae cultivation.

Sample	Number of microalgae (cells / ml) in model media of the studied sample after rotation (days), in brackets – a decrease in the amount of algae relative to the control	
	1 day	**3 day**
GB-1	$3.25\cdot10^6$ (35%)	$3.5\cdot10^6$ (3%)
GB-2	$4.5\cdot10^6$ (10%)	$4.5\cdot10^6$ (10%)
Cement paste	$3.75\cdot10^6$ (25%)	$4.0\cdot10^6$ (20%)
Reference	$5.0\cdot10^6$	$5.0\cdot10^6$

The experimental results are considered reasonable if the number of algae in the control increased by more than 10 times within 72 hours of exposure. In this case, the concentration of algae increased 200 times. In addition, it is necessary that the change in pH at the end of the experiment should not exceed 1.5. This requirement was also met. In accordance with the procedure for conducting an experiment on biotesting on microscopic algae, non-toxic media are those in which the decrease in the concentration of algae does not exceed 20 %. In all considered variants of the experiment, medium of GB-2 samples lead to a decrease in the concentration of algae by only 10–16 %, therefore, they are the safest for microalgae. As a result of experiments carried out on microscopic algae, it was found that the most unfavorable for the environment are samples GB-1, then - Cement paste. Thus, medium of GB-2 samples are more dangerous for algae.

3.2. *Acute toxicity*

During the experiments on the study of acute toxicity on daphnia *Daphnia magna*, the prepared solutions of the extract of building materials and the control solution were poured into 100 ml vessels in triplicate for each extract; 10 individuals were seated, whose age did not exceed 24 hours. The number of survived daphnia was recorded after 3, 24, 48, 72, 96 hours. The dissolved oxygen concentration must be above 3 mg / L at the beginning and throughout the test. The pH should be in the range of 6.5–8.5 and should not change more than 1.5 units during the test. The results of a short-term experiment on daphnia are shown in Table 2 and in Figures 1 and 2.

Table 2. Number of survived dafnias in model media of the studied samples.

Sample	Sample rotation time, days	Number of survived dafnias (units) under different time period of test (hours)				
		3	24	48	72	96
GB-1	1	10	10	10	6	5
	3	10	10	10	2	2
GB-2	1	10	10	10	8	8
	3	10	10	10	8	6
Cement paste	1	10	10	10	10	10
	3	10	10	10	10	10
Reference	-	10	10	10	10	10

As can be seen from the results of an acute toxicity test, the most favorable habitat for daphnia was the cement stone medium. During the observation period of 96 h, no death of the animals was recorded.

In the case of GB-1 samples, the most toxic was the model medium with a long sample rotation time - 3 days. In this case, on the second day of observation, the death of daphnia equal to 80 % was recorded. With the samples that were in contact for a shorter time - 1 and 3 days; the death of animals was observed at a later time - on the 3rd day of observation. At the end of the experiment (after 96 h), all the options considered showed acute toxicity. In the case of GB-2 medium, a long sample rotation time (3 days) was also the most toxic. In this case, on the 3rd day of observation, the death of daphnia was recorded equal to 20 %, and on the next day, the death was 40 %. Thus, at the final stage of the test (after 96 h), these model media did not have such high toxicity as in the case of GB-1 (20 % and 40 %) vs. GB-2 (50 and 80 %).

3.3. *Chronic toxicity*

During the studies of chronic toxicity, the prepared solutions of the toxicant and the control solution were poured into 0.1 L beakers in 3 replicates, 10 individuals up to 24 hours were placed in each one to register the first litter for each daphnia. The appearance of juveniles, which were removed, was monitored daily. The chronic experiment lasted 21 days.

The crustaceans were fed daily chlorella *Chlorella vulgaris Beyer*, which is cultured in the laboratory. Optimal conditions of detention such as temperature 23 ± 0.5 °C and light regime day-night (16-8 hours) were maintained automatically. In long-term experiments, the medium was completely changed after 5 days. For this, immediately at the start of the experiment, reserve solutions of all model solutions were prepared.

The detection and quantification of the content of pollutants in the environment are not enough to answer the question about the sustainability of building material. It is also necessary to determine their danger to living organisms most sensitive to pollution. Thus, research on lower aquatic organisms (protozoa and crustaceans) can solve many important ecological problems. *Cladocera Daphnia magna Straus* is a ubiquitous species with well-studied biology. Daphnia cultivated in laboratory conditions have a short life cycle, which rarely exceeds 2-3 months.

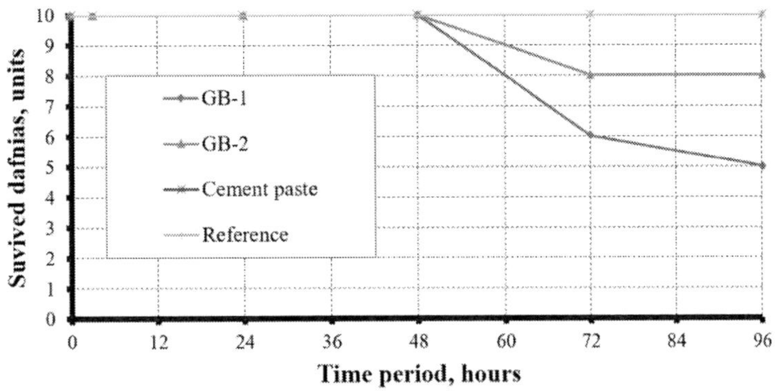

Figure 1. The number of survived daphnia in model media of the studied samples after rotation for a 1 day.

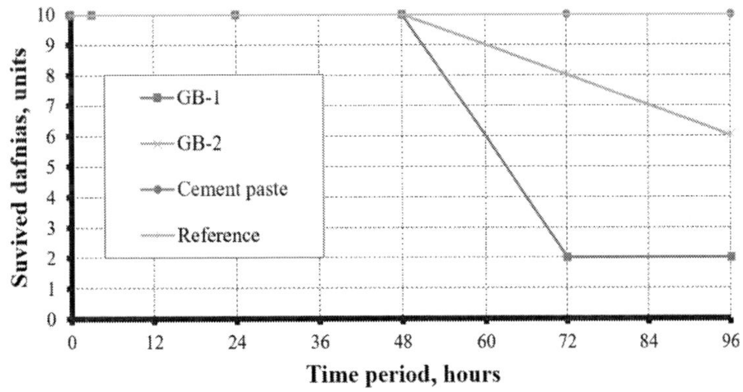

Figure 2. The number of survived daphnia in model media of the studied samples after rotation for 3 days.

An important biological effect of toxic exposure is the fertility and reproductive success of the population because the long-term survival of a species depends on the fertility of individuals and the quality of the offspring. Changes in reproductive processes are classified as one of the most significant effects of pollution. For this purpose, a chronic experiment on daphnia was carried out for 21 days. The results of a chronic experiment on daphnia are given in the Table. 3.

Table 3. Results of a chronic toxicity test on in model medium of the studied samples.

Sample	Sample rotation time, days	Total fertility, units	First litter dates, days	Number of litters for 21 days, units.
GB-1	1	15	10	2
	3	4	18	1
GB-2	1	38	7	4
	3	36	8	4
Cement paste	1	68	7	7
	3	56	6	6
Reference	-	72	6	7

The most favorable medium among those considered ones was the medium formed as a result of the interaction of the cement stone. In the considered variant, the aqueous media had the same chronic effect as in the reference medium. Slightly less (by one) the number of litters in the reference medium with a rotation time of 3 days, the total fertility is less by 22 %. GB-2 medium, with chronic exposure, also inhibits the development of daphnia in all variants of model media. However, this kind of impact is much less than GB-1 medium. In the case of this sample, the total fertility of daphnia is 5-18 times less than in the Reference sample and 2 times less than in GB-2 medium. The same pattern is observed with the onset of the first litter and with the total number of litters.

4. Discussion

The results of experiments on biotesting showed a different degree of toxicity of the studied media of samples, which increases in the following sequence: Cement paste → GB-2 → GB-1. GB-2 reduces algae concentration by only 10 %. It was revealed that from the point of view of acute toxicity the GB-1 medium of the sample is the most dangerous, where the death of *Daphnia* in a short-term experiment was up to 80 %. At the same time, the Cement paste medium is a favorable environment for *Daphnia magna*, where no animal deaths were recorded.

The results of studies of chronic toxicity, which was assessed by the total fertility of daphnia, demonstrated that GB-1 medium is again the most toxic, then in GB-2 medium. According to this indicator, a medium based on cement paste is also favorable for daphnia.

5. Conclusion

Thus, xenobiotics entering water bodies from leached building materials can have a negative effect on various parameters of the vital activity of aquatic organisms, including survival and fertility. From the point of view of the toxic effect, the medium of the GB-1 sample has a greater negative effect, both with short-term and long-term exposure to the aquatic ecosystem.

6. Acknowledgments

This work was realized under the support of the President Scholarship; in the framework of the Program of flagship university development on the base of the Belgorod State Technological University named after V G Shukhov, using equipment of High Technology Center at BSTU named after V G Shukhov.

References

[1] Evdokimov P A 2019 Features of assessment of anthropogenic loads on natural complexes *Construction Materials and Products* **2(5)** 19-24

[2] Morrell J 2002 Wood-based building components: what have we learned *International Biodeterioration* **49** 253-8

[3] Sepelev I I, Eskova E N, Potapova S O, Khizhnyak S V and Bochkov N N 2019 Ecological aspects of technogenic material application in road construction technologies *IOP Conf. Ser.: Earth and Environ. Sci.* **315** 52019

[4] Gann R G, Babrauskas V, Peacock R D and Hall J R 1994 Fire conditions for smoke toxicity measurements *Fire and Materials* **18(3)** 193-9

[5] Papaefthmiou H and Gouseti O 2008 Natural radioactivity and associated radiation hazards in building materials used in Peloponnese, Greece *Radiation Measurements* **43** 1453-7

[6] Pacheco-Torgala F and Jalalib S 2011 Toxicity of building materials: a key issue in sustainable construction *Int. J. of Sustainable Eng.* **4(3)** 281-7

[7] Fomina E V, Voitovich E V, Fomin A E, Lebedev M S and Kozhukhova N I 2015 Ocenka radiacionnogo kachestva shlaka oe`mk dlya primeneniya ego v stroitel`ny`x kompozitax [Assessment of radiation quality slag OEMK for use of its composites in construction] *Vestnik BGTU im V. G. Shukhova* [Bulletin of BSTU named after V G Shukhov] **6** 130-3 [In Russian]

[8] Kozhukhova N I, Lebedev M S, Vasilenko M I and Goncharova E N 2019 Toxic effect of fly ash on biological environment *IOP Conf. Ser.: Earth and Environ. Sci.* **272** 022065
[9] Vasilenko M I, Lebedev M S, Goncharova E N, Kozhukhova N I and Kozhukhova M I 2019 The study of ecological impact of fly ash-based geopolymer binders on soil and aquatic biota *IOP Conf. Ser.: Mater. Sci. Eng.* **791** 012049
[10] Kozhukhova N I, Lebedev M S, Vasilenko, M I and Goncharova E N 2016 Ecology-toxicology study of low-calcium solid wastes from power plants *Int. J. of Pharmacy and Technology* **8(3)** 15349-60
[11] Strokova V, Nelyubova V and Rykunova M 2017 Testing of Binders Toxicological Effects *IOP Conf. Ser.: Mater. Sci. Eng.* **262(1)** 012030
[12] Filenko O F 2007 Biologicheskie metody` v kontrole kachestva okruzhayushhej sredy` [The biological methods in the control of environment's quality] *E`kologicheskie Sistemy` i Pribory`* **6** 18-20 [In Russian]
[13] Barinova S, Klochenko P and Bilous O 2015 Algae as indicators of ecological state of water bodies: methods and perspectives *Hydrobiological J.* **51(4)** 3-23
[14] Filenko O F, Dmitrieva A G and Marushkina E V 2007 Study of the laboratory population of the alga Scenedesmus quadricauda (Turp.) Brèb. microculture method *Hydrobiological J.* **42(5)** 80-8
[15] Moiseenko T I 2009 *Vodnaya E`kotoksikologiya: Teoreticheskie i prikladny`e aspekty`* [Aquatic Ecotoxicology: Theoretical and applied aspects] (Moscow: Nauka) [In Russian]
[16] Olkova A S, Sannikova E A, Budina D V, Kutyavina T I and Zimonina N M 2017 Ocenka toksichnosti prirodny`x i texnogenny`x sred po dvigatel`noj aktivnosti Daphnia Magna [Study of the toxicity of natural and technogenic environments on the motor activity of Daphnia Magna] *Sovremenny`e Problemy` Nauki i Obrazovaniya* **3** [In Russian]
[17] Olkova A S 2014 Biotestirovanie v nauchno-issledovatel`skoj i prirodooxrannoj praktike Rossii [Biotesting in research and environmental practice in Russia] *Uspexi Sovremennoj Biologii* **134(6)** 614-22 [In Russian]
[18] Olkova A S 2018 Aktual`ny`e napravleniya razvitiya metodologii biotestirovaniya vodny`x sred [Current trends in the development of the methodology of bioassay aquatic environments] *Voda i E`kologiya: Problemy` i Resheniya* **2(74)** 40-50 [In Russian]
[19] Kutsenko S A 2004 *Osnovy` Toksikologii* [Fundamentals of Toxicology] (SPb.: Foliant) [In Russian]
[20] Tonkopii V, Zagrebin A and Iofina I 2008 *Bioidentification of xenobiotics as a basis of water management. Sustainable use and development of watersheds* (Berlin: Springer)
[21] Zagrebin A O, Rumyantsev V A and Tonkopii V D 2016 Developing methods for bioidentification of xenobiotics for water quality assessment *Water Resources* **43(1)** 141-4
[22] Alexandrova V V 2009 *Primenenie metoda biotestirovaniya v analize toksichnosti prirodny`x i stochny`x vod* [Application of the biotesting method in the analysis of toxicity of natural and waste waters] (Nizhnevartovsk: NGU) [In Russian]
[23] Anisimova O V and Gololobova M A 2006 *Kratkij opredelitel` rodov vodoroslej* [A short guide to the genera of algae] (Moscow: MGU) [In Russian]

Studying the kinetics mechanism of emulsified petroleum products recovery from waste waters in dynamic conditions

I V Starostina, D O Polovneva, E V Porozhnyuk and R G Shevtsova

Department of of Industrial Ecology, Belgorod State Technological University named after V.G. Shukhov, Kostyukov St., 46, Belgorod, 308012, Russia

E-mail: starostinairinav@yandex.ru

Abstract. Petroleum products belong to the most common man-made pollutants of surface watercourses, and sometimes even underground drinking water sources. They get to the environment at the extraction and transportation of oil and oil products, as well as at industrial disasters. The oil-contaminated sewage waters produce the general-toxic, embryotoxic and carcinogenic effect, and weaken the immune system. To purify wastewaters from petroleum products the most efficient and environmentally friendly is the sorption method. This paper considers using a carbon-containing material, obtained by thermal modification at 450°C of waste kieselguhr sludge from vegetable oil extraction industry, as petroleum sorbent. The designed sorbent is characterized with mesoporous structure, high hydrophobic properties and the presence of oxygen-containing functional groups – carboxyl, hydroxyl and lactone – on the carbon layer surface. Research has been carried out in purification of test water media from emulsified petroleum products in dynamic conditions with the use of the designed sorbent. The number of passed sorption volumes to the «breakthrough» and the full saturation of the sorbent with oil products were determined. The calculated dynamic exchange capacity of the sorbent for petroleum products is 2.9 mg/g, the total exchange capacity – 8.4 mg/g. The limiting stage of petroleum products sorption process kinetics with the carbon-containing sorption material in the studied conditions is diffusion inside the sorbent grains.

1. Introduction

Environmental pollution with petroleum and petroleum products is one of the global environmental problems of the present-day world. At extraction, storage, transportation, processing and utilization of petroleum and petroleum products the contamination of water, air and soil occurs and petroleum spills produce the adverse effect on plant and animal life. Technological accidents, discharge of untreated or undertreated sewage waters, chaotic organization of storm water runoffs from contaminated industrial territories cause the increase of petroleum products content in surface watercourses, and sometimes even in underground drinking water sources. The total amount of petroleum and petroleum products, getting into the global ocean each year makes up from 6 to 12 mln ton.

Sewage waters, containing petroleum products, produce the general-toxic, embryotoxic and carcinogenic effect and weaken the immune system. The petroleum products content in water bodies within the range 0.3 – 0.5 mg/dm^3 inhibits the growth of fish and other aquatic organisms. The influence of such pollutant on the human body causes cardiovascular and nervous systems disorders, can influence the blood values (hemoglobin and erythrocytes decrease), affect liver or skin (dermatitises or eczemas). So, the problem of the efficient purification of oil-contaminated sewage waters is one of the most topical ones.

Getting into water bodies, oil products are mostly contained in them in coarsely-dispersed (drop-wise) state, easily float to the water surface and form a floating film or layer. Another, smaller part of petroleum products is contained in the water in the form of emulsions or solutions. But of all types of petroleum products, which contaminate water bodies, the special attention nowadays is paid to emulsified products, as they are among the most dangerous sources of environment pollution [1, 2]. Emulsified petroleum products are presented with homogeneous disperse systems, consisting of two non-miscible liquids. The dispersed phase of the emulsion is petroleum products (the particle size is over 0.45 nm), and the dispersive medium is water. The emulsion stability depends on the emulsified particles size and concentration, electrokinetic properties of the system, stirring intensity of emulsion's components, surface tension and the presence of stabilizers.

The sewage waters, containing emulsified petroleum products, include waste cooling-lubricating fluids, used in the modern metalworking technological processes, oily wastes of glass fibers and synthetic fibers production, sewage waters of oil-production enterprises and oil refineries [3, 4]. Heavily polluted are the wastewaters, generated at cleaning and washing liquid cargo vessels' tanks, which is done in case of accepting the cargo of higher quality than the previously transported cargo. Due to the usage of cleaning agents and other chemical reagents (surface active agents, alkalies, sodas etc.) these wastewaters are presented with very stable emulsions with petroleum products content up to 80-150 g/dm^3 [5].

To recover emulsified petroleum products from sewage waters the fine purification is performed, with the usage of physical-chemical (flotation, coagulation, flocculation, sorption, coalescence etc.) [5 - 7] and chemical methods (adding reagents - acids, salts, demulsifiers) [8, 9]. At present the adsorption method has got widespread use, which is conditioned by its high purification efficiency, low cost and availability of the used adsorption materials. The main characteristics of sorbents, used for recovering emulsified oil products from water media, include: oil-binding property and petroleum-binding property, hydrophobicity of its surface, possibility of recovery and the subsequent disposal of the used sorption material [3, 4].

At present the area of using industrial and agricultural waste as sorption materials is rapidly developing [10, 11]. The usage of waste products, as they are, is complicated, which is conditioned by their low sorption properties, high hydrophilic properties, as well as the possibility of washing-out of harmful components and recontamination of water media. So, in order to stabilize their structure and improve their sorption properties, the waste products are subjected to various modification methods – chemical (using acids, alkalies and oxidants solutions), physical (ultrasound treatment, UV and IR radiation) and thermal (carbonization and high-temperature treatment) [12].

Carbonization process is one of the basic methods of obtaining carbon-containing sorption materials. As the raw material source, various mineral and plant materials, as well as garden waste, can be used – moss, brown coal, rice husk, sugar cane bagasse, nutshells, sawdust and leaf litter [13, 14].

The earlier studies have demonstrated [15, 16] that as a source material for obtaining carbon-containing sorption materials to recover petroleum products from water media the sludge waste of vegetable oil extraction industry can be used, for example, waste diatomite (kieselguhr) sludge. Waste kieselguhr sludge (WKS) is an organo-mineral slow pasty mass with organic impurities content up to 70%. These organic impurities contain vegetable oil, low-hydratable phospholipids and unsaponifiables - vegetable waxes, hydrocarbons, sterols, carotenoids, alcohols etc.

According to experimental data, presented in [15, 16], the thermal modification of WKS at temperatures 450-500 °C in conditions of lack of oxygen provides the partial oxidation of organic impurities and the formation of carbonic soot-like layer on the surface of mineral (diatomite) particles. This results in the formation of new carbon-containing sorption material – thermally-modified kieselguhr sludge (TKS).

The purpose of this work is to study the process of emulsified petroleum products recovery from water media in dynamic conditions with the use of the carbon-containing sorption material, obtained from waste kieselguhr sludge of vegetable oil extraction industry.

IOP Conf. Series: Materials Science and Engineering 1089 (2021) 012030 doi:10.1088/1757-899X/1089/1/012030

2. Materials and Methods

As a sorption material, the refined oil extraction industry waste kieselguhr sludge from the AO «EFKO» plant, Alekseevka city, Belgorod region, Russia, was considered, thermally modified at temperature 450°C and conditionally designated TKS_{450}. The main characteristics of TKS_{450} are presented in table 1.

Table 1. Characteristics of carbon-containing sorption material TKS_{450}.

Parameter	Value
Color	black
Carbon content, %	10 - 12
Loose density, kg/m^3	310
True density, kg/m^3	2540

As a petroleum product the mineral oil I-20A was taken, the properties of which are presented in table 2.

Table 2. Technological properties of the industrial oil I-20A.

Parameter name	Result
Kinematic viscosity at 40°C, mm^2/s	25-35
Acid number, mg KOH for 1 g of oil, no more than	0.03
Ash content, %, no more than	0.005
Mechanical impurities content	no
Water content	trace
Density at 20°C, kg/m^3, no more than	890
Chilling temperature, °C, no more than	–15
Color on CNT colorimeter, CNT units, no more than	3.0
Open-cup flash point, °C, not lower than	180

As a test solution for this research the oil-water emulsion with oil concentration 15 mg/dm^3 was used. The emulsion was prepared by mixing distilled water and oil with a shaker within 6 hours.

As a result of the petroleum product's contact with water a multicomponent system is formed – which contains both emulsified and dissolved forms of the oil. To describe sorption of a multicomponent mixture, the calculation method of adsorption from two-component solutions was used – a «conventional component» method. As a «conventional component» the total content of oil products in the water – emulsified and dissolved – was considered.

The petroleum products in the test solutions after their purification were analyzed by means of IR-spectrophotometry using a consistency meter KN-3. Petroleum products were extracted with tetrachloride carbon (CCl_4) c.p., the quantitative determination was performed by studying the intensity of C-H bonds in the infra-red range of spectrum [16, 17].

IR spectra of the sorption material TKS_{450} were obtained with a FTIR spectrometer «Bruker Vertex 70». Measuring range is 4000- 400 cm^{-1}; reference sample – KBr.

To perform the purification process of the test emulsion from oil products in dynamic conditions a filtration plant was used, the diagram of which is presented in figure 1.

As a filtration plant a separating column 30 mm in diameter was used; the lower layer of the adsorption material was 40 mm thick (mass 5.83 g), the upper sand layer - 2 mm thick. The necessity of using sand as a weight was preconditioned by the high buoyancy of the sorbent under study.

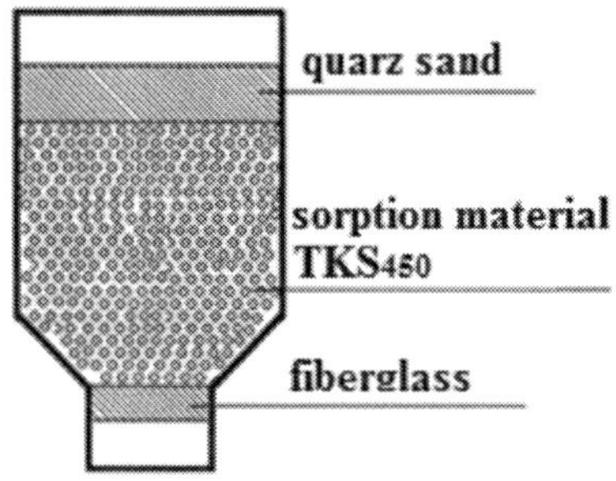

Figure 1. Diagram of the filtration plant.

The sand was preliminarily washed with distilled water and calcinated in a muffle furnace at temperature 500 °C within 3 hours. The motion of test wastewaters in the plant was performed from top to bottom at a rate of 1.15 cm/min.

The dynamic exchange capacity (DEC), mg/g, was determined by the formula

$$DEC = \frac{V \cdot C_e}{m},$$

where:
V – volume of the purified wastewater, flown out of the column from the start of the experiment to the appearance of oil product in the eluate with a «breakthrough» concentration, dm^3;
C_e – equilibrium concentration of the wastewater, mg/dm^3;
m – weight of the sorption material, g.

The total exchange capacity (TEC), mg/g, was determined by the formula

$$TEC = \frac{V_к \cdot C_e}{m},$$

where:
$V_к$ – volume of the purified wastewater, flown out of the column from the start of the experiment to the moment, when the concentration of oil products in the filtrate is equal to its initial concentration in the test emulsion, dm^3;
C_e – equilibrium concentration of the wastewater, mg/ dm^3.

3. Results

Obtaining the carbon-containing sorption material TKS_{450} is based on the thermal modification of an organo-mineral material - waste kieselguhr sludge - at 450°C. At this temperature and lack of oxygen the partial oxidation of organic impurities takes place. The generated organic matter discomposition products and preproducts sedimentate in macropores and large mesopores of kieselguhr particles, «sealing» or blocking them. This is accompanied with the formation of mostly large mesopores over 50 nm and the formation of material with the specific surface 0.9 – 1.5 m^2/g.

The qualitative assessment of the sorption material TKS_{450} surface was preformed by IR spectroscopy and electronic microscopy methods.

The findings of IR spectroscopy of TKS_{450} are presented in figure 2.

At the IR-spectrum of TKS_{450} (figure 3) the intensive bands are registered at 2926 and 2853 cm^{-1}, which correspond to the range (2850 – 3000 cm^{-1}) and belong to the valence vibrations, and the average intensity band at 1470 cm^{-1} - to the deformation vibrations of methylene groups and characterize the presence of C_{sp3}-H bond on the carbon surface of TKS_{450}. The valence vibrations at 1719 cm^{-1} correspond to the carbonyl group (C=O) of compound ethers or lactone, contained in vegetable waxes and waxlike substances, which provides TKS_{450} with high surface hydrophobicity.

The obtained data correlate well with the research findings, presented in figure 3. A drop of water, placed onto the compacted surface of TKS_{450}, forms a sitting drop, the contact wetting angle is 156°.

So, the considered carbon-containing sorption material TKS_{450} is characterized with the mesoporous structure, high hydrophobic properties of the surface and various oxygen-containing functional groups – carboxyl, hydroxyl and lactone – on the carbon layer surface.

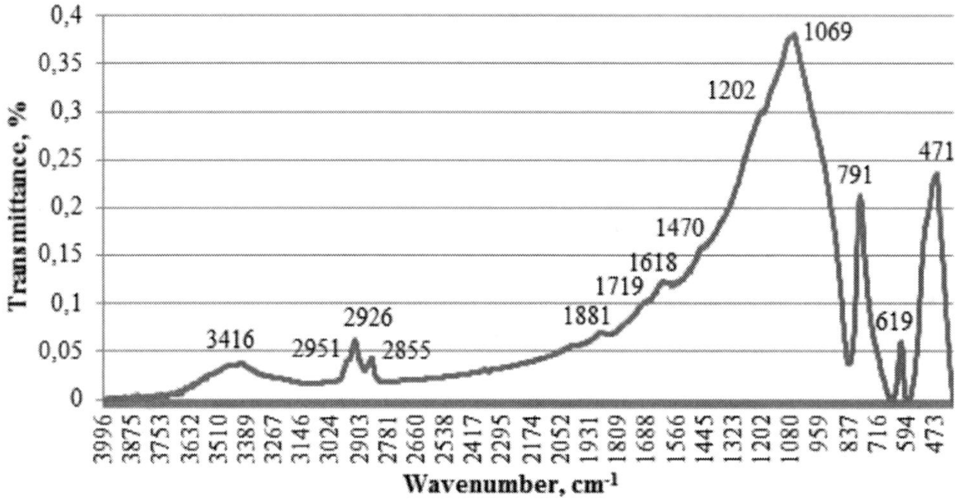

Figure 2. IR-spectrum of the carbon-containing sorption material – TKS_{450}.

The physical and chemical properties of the sorption material TKS450 indicate the promising outlook of its usage in purification of water media from petroleum and petroleum products. The obtained carbon-containing sorption material TKS450 was studied for recovering emulsified petroleum products from test water media in dynamic conditions.

The technological, operational and economic advantages of dynamic conditions in wastewater purification predetermine their wide use at the plant conditions. At developing sewage waters after-purification technologies by the sorption method one of the most important characteristics of sorption materials is their dynamic exchange capacity (DEC).

Figure 3. Water drop on the surface of TKS_{450}.

The dynamic exchange capacity was determined by the «breakthrough» of the absorbed substance at the exit of the purified water from the column. The «breakthrough» concentration is determined with the analytical method's quantification level, but in the real conditions of technological process it is usually correlated with the maximum allowable concentration (MAC), which depends on the requirements of a certain industry, and thus can have different values. In this experiment as a «breakthrough» concentration of oil products the maximum allowable concentration for cultural and general objects of water use – 0.3 mg/dm^3 – was taken. According to the results, presented in figure 4, the volume of the filtrate, flown out of the column from the start of the experiment to the appearance of oil product in the eluate and reaching the concentration 0.3 mg/dm^3, amounted to 45 passed volumes or 1125 ml of the emulsion with the initial concentration 15 mg/dm^3.

The total exchange capacity (TEC) was determined by saturating the sorption material with the absorbed substance until the concentration of oil products in the filtrate is equal to its initial concentration in the test emulsion.

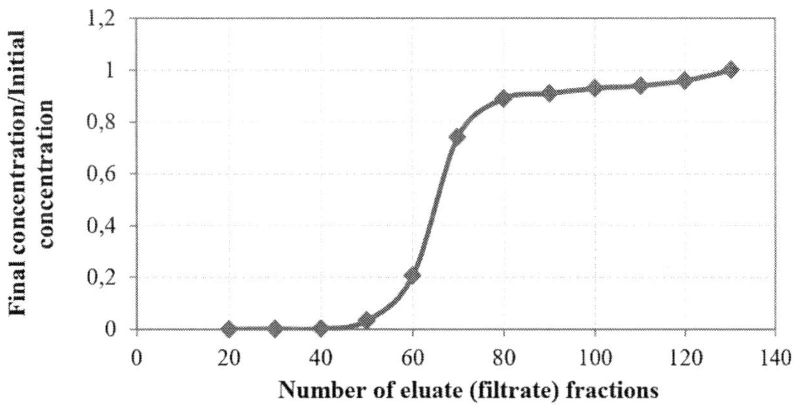

Figure 4. Elution curve of recovering petroleum products by using the carbon-containing sorption material TKS_{450}.

The complete saturation of the TKS_{450} sorption material with petroleum products takes place at 130 passed volumes of the emulsion, which amounts to 3250 ml. The calculation results of DEC and TEC are presented in table 3.

Table 3. DEC and TEC of the TKS_{450} sorption material for petroleum products.

C_{init}, mg/dm^3	DEC, mg/g	TEC, mg/g
15	2.9	8.4

It should be pointed out that the adsorption efficiency in dynamic conditions depends to a great extent on kinetic factors. Determining the kinetics mechanism of a sorption process is a complex problem, which requires taking into account a number of factors, which influence its rate. These are the sorption material's particle size, filtration rate, environment temperature etc.

The adsorption process consists of 3 stages: the approach of the absorbed substance molecules to the outer surface of the sorption material (external diffusion), diffusion inside the particles (internal diffusion) and condensation of molecules on the inner surface of the sorption material (sorption act). So, the rate of adsorption process is determined with diffusion. According to the limiting stage, we can single out 3 kinetic modes of the sorption process: external- diffusional, internal-diffusional and combined.

The limiting mechanism of sorption kinetics was determined by sorption process interrupt method or by «kinetic memory» method in dynamic conditions. For this purpose, a test emulsion with petroleum products concentration 15 mg/dm^3 was passed through the filtration plant, in fractions by 25 ml. The process interrupt amounted to 24 hours. After 24 hours the test emulsion passing through the filtering unit was resumed, and the adsorption process continued. In the filtrate the residual concentration of hydrocarbons was determined.

In figure 5 the elution curves of petroleum products sorption with the carbon-containing sorption material TKS_{450} in conditions of process interrupt for 24 hours are presented.

As we can see from the curves (figure 5), in all cases with sorption process interrupt the reduction of oil products concentration in the solution, flowing out of the column, was observed. The break of continuity on the elution curves of sorption allows us making a conclusion that the limiting stage of the petroleum products sorption process kinetics with the carbon-containing sorption material TKS_{450} in the conditions under study is diffusion inside the sorbent grains. So, the sorption rate is inhibited with the internal-diffusion processes.

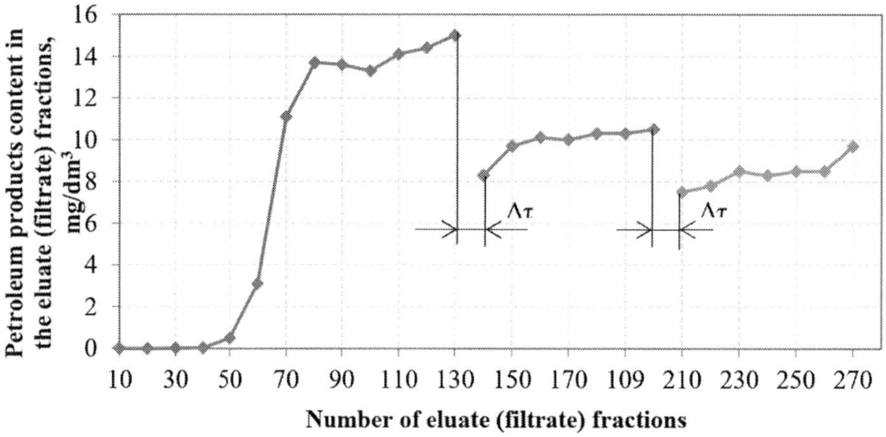

Figure 5. Elution curves of petroleum products recovery by using the carbon-containing sorption material TKS_{450} in conditions of process interrupt for 24 hours, volume of the eluate (filtrate) fraction is 25 ml.

4. Conclusion

The carried-out research of petroleum products sorption process in dynamic conditions with the use of carbon-containing sorption material TKS_{450}, obtained by the thermal modification of waste kieselguhr sludge from vegetable oil extraction industry, allows us making the following conclusions:

1. The carbon-containing sorption material TKS_{450} has the developed, mostly mesoporous structure, high hydrophobic properties and various oxygen-containing functional groups on the carbon layer surface. The obtained data confirm the supposition, that the considered sorbent would be efficient in purification and after-purification of sewage waters from the pollutants, particularly from the emulsified petroleum products, in dynamic conditions.
2. The obtained carbon-containing sorption material TKS_{450} with particle fraction below 0.315 mm was studied for recovering emulsified petroleum products from test emulsions in dynamic conditions. DEC amounted to 2.9 mg/g, TEC – to 8.4 mg/g.

5. Acknowledgments

The paper is prepared in the framework of the Flagship University Development Program on the base of the Belgorod State Technological University named after V.G. Shukhov, using the equipment of High Technology Center at BSTU named after V.G. Shukhov.

References

[1] Ali Almasi, Leila Yavari, Mitra Mohammadi and Seyyed Alireza Mousavi 2020 Optimization of an integrated system for refinery wastewater treatment *Toxin Reviews* **39(4)** 408-21

[2] Diya'Uddeen B H, Daud W M A W and Abdul Aziz A R 2011 Treatment technologies for petroleum refinery effluents: a review *Process Safety and Environmental Protection* **89(2)** 95-105

[3] Okiel K, El-Sayed M and El-Kady M Y 2011 Treatment of oil–water emulsions by adsorption onto activated carbon, bentonite and deposited carbon *Egypt J Petrol* **20(2)** 9-15

[4] Viju S, Thilagavathi G, Vignesh B and Brindha R 2019 Oil sorption behavior of acetylated nettle fiber *The J. of the Textile Institute* **110(10)** 1415-23

[5] Bing Wang, Yiyu Shui, Hongyang Ren and Min He 2017 Research of combined adsorption-coagulation process in treating petroleum refinery effluent *Environmental Technology* **38(4)** 456-66

[6] Abdeen Z and Moustafa Y M M 2016 Treatment of Oily Wastewater by Using Porous PVA Hydrogels as Oil Adsorbent *J. of Dispersion Science and Technology* **37(6)** 799-805

[7] Hor Kosal, Wachirapan Pattanachot and Chaisri Suksaroj 2014 Treatment of oily wastewater from waste glycerol by acidification and the coalescer process *Desalination and Water Treatment* **52(16-18)** 2996-3003

[8] Kulik N, Trapido M, Veressinina Y and Munter R 2007 Treatment of Surfactant Stabilized Oil-In-Water Emulsions by Means of Chemical Oxidation and Coagulation *Environmental Technology* **28(12)** 1345-55

[9] Maha A Tony, Patrick J Purcell and Yaqian Zhao 2012 Oil refinery wastewater treatment using physicochemical, Fenton and Photo-Fenton oxidation processes *J. of Environmental Science and Health, Part A* **47(3)** 435-40

[10] Sabir S 2015 Approach of costeffective adsorbents for oil removal from oily water *Critical Reviews in Environmental Science and Technology* **45(17)** 1916-45

[11] Eremin I S 2017 Zhidkie raketny`e topliva: kontrol` i ocenka e`kologicheskoj opasnosti [Development of Sorbent material Based on Sugar Cane] *E`kologiya i Promy`shlennost` Rossii* **21(2)** 14-7 [In Russsian]

[12] Xu X, Gao B, Jin B and Yue Q 2016 Removal of anionic pollutants from liquids by biomass materials: a review *J. Molecular Liquids* **215** 565-95

[13] El-Gendy N S and Nassar H N 2016 Study on the effectiveness of spent waste sugarcane bagasse for adsorption of different petroleum hydrocarbons water pollutants: kinetic and equilibrium isotherm *Desalination and Water Treatment* **57(12)** 5514-28

[14] Svyatchenko A V Sverguzova S V, Sapronova Zh A and Shaikhiev I G 2020 Ispol`zovanie listovogo kashtanovogo opada v ochistke vodny`x sred ot dizel`nogo topliva [Using Chestnut Leaf Litter in Aqueous media Purification from Diesel Fuel] *E`kologiya i Promy`shlennost` Rossii* **24(8)** 46-50 [In Russsian]

[15] Starostina I V, Stolyarov D V and Kosukhin M M 2017 Carbonic and Mineral Adsorbent, Received out of a Slurry Withdrawal of Oil-Extracting *Production Solid State Phenomena* **265** 319-24

[16] Starostina I V, Cherevatova A V, Stolyarov D V and Anishhenko I V 2020 Issledovanie teksturny`x, strukturny`x i sorbcionny`x svojstv uglerodsoderzhashhix materialov na osnove shlamovogo otxoda proizvodstva rastitel`ny`x masel [Investigating the textural, structural and sorption properties of carbon-containing materials based on sludge waste from the production of vegetable oils] *Trudy` RGU Nefti I Gaza im. I.M. Gubkina* **2(299)** 119-32 [In Russsian]

[17] Tarasova G I 2018 The development of compositions of silicate-based paints thermalizing conveyor-washing sediment - waste of the sugar industry *Construction Materials and Products* **1(1)** 21-31

Technical and economic optimization of fuel facilities – energy complex using elements of fuzzy logic and artificial neural networks

A L Rutskov[1], Ya P Fedorov[2] and L V Volkov[3]

[1]Branch of JSC "UK EFKO" in Voronezh, Taranchenko St., 40, Voronezh, 394018, Russia
[2]Department of corporate Finance and corporate governance Financial University under the government of the Russian Federation, M. Zlatoustinsky lane, 7, Moscow, 101000, Russia
[3]Department of management Financial University under the government of the Russian Federation, M. Zlatoustinsky lane, 7, Moscow, 101000, Russia

E-mail: alex_8_90@mail.ru

Abstract. This article discusses the possibilities of optimizing the fuel and energy complex by improving the accuracy of the forecast of the state in both the operational and medium-term periods: from an hour to several years. The model used to implement the mentioned energy efficiency improvement is based on the use of fuzzy neural networks together with the function of minimizing power losses in the elements of electric power systems. The Mamdani modification is used, which takes into account the database of previous periods of the system state, represented as static load characteristics. The organization of this approach allows you to take into account weakly formalized factors, thereby improving the quality of forecasting and dispatching management. At the same time, it is possible to use models for a wide class of objects, which is achieved due to its scalability. The estimation of forecasting accuracy of the proposed approach exceeds similar indicators of currently used statistical methods and regression implementations. These factors are direct economic drivers of reducing production costs.

1. Introduction

The functioning of modern facilities of the fuel and energy complex (FECF): re-source extraction, supply complexes and end users is associated with the need to optimize the technological processes of using electric energy.

The time variations inherent in the functioning of the FECF are significantly indistinct, which is expressed in the multi-factor dependence of finding the optimum of electric energy losses. To solve this problem, the primary factor is the creation of a model of energy consumption dynamics that has the function of forecasting and developing control actions.

The main disadvantages of most of the developed models are the actual non-accounting (or weak accounting) of weakly formalized factors due to the lack of accuracy in implementing the forecast of energy consumption. At the same time, the corresponding fluctuations in the FECF are represented by linear approximations, which in principle does not allow optimizing the modes with an accuracy exceeding the error of the noted forecasting process.

The functioning of the model considered in this article is organized using combined fuzzy neural networks (FNN), which make it possible to increase the accuracy of indicators for estimating future periods (for solving optimization problems) in comparison with existing classical implementations (regression models).

2. Purpose and objectives of the work

The aim of the work is to increase the competitiveness of the FECF, which is achievable by optimizing the indicator of electric power losses during transmission in local power grid elements (power lines, power transformers, elements of distribution networks).

To achieve this goal, a number of tasks have been solved:

- the basic optimization model is defined for the subsequent development of algorithms for predicting energy resources in the FECF;
- developed variations of algorithms for the FNN -based model;
- evaluation of the effectiveness of the developed algorithms in relation to simulation models of FECF functioning for short-term, medium-term and long-term forecasting is obtained.

3. Materials and methods

There are two basic optimization approaches for the considered class of problems [1 - 5]:

- Lagrange equations;
- gradient models (Newton-Raphson algorithm).

For the purpose of practical use of optimization algorithms, the characteristics of the active and reactive components of FECF are presented as static load characteristics (SLC):

$$\begin{cases} P(U) = P_{ном}(U)[a_0 + a_1 \cdot \left(\dfrac{U}{U_{ном}}\right) + a_2 \cdot \left(\dfrac{U}{U_{ном}}\right)^2 + \ldots + a_n \cdot \left(\dfrac{U}{U_{ном}}\right)^n + \xi_P(U)]; \\ Q(U) = Q_{ном}(U)[b_0 + b_1 \cdot \left(\dfrac{U}{U_{ном}}\right) + b_2 \cdot \left(\dfrac{U}{U_{ном}}\right)^2 + \ldots + b_n \cdot \left(\dfrac{U}{U_{ном}}\right)^n + \xi_Q(U)]; \end{cases} \quad (1)$$

where:

$P(U)$, $Q(U)$ – is the active and reactive power;

$P_{ном}(U)$, $Q_{ном}(U)$, $U_{ном}$ – nominal values of active and reactive power and voltages under normal conditions;

$\xi_P(U)$, $\xi_Q(U)$ – dependences of the distribution of active and reactive power on weakly formalized factors;

$a_0, a_1, a_2, \ldots, a_n, b_0, b_1, b_2, \ldots, b_n$, – parameters of the characteristic technological process of FECF functioning;

U – current measured voltage value.

The optimization model based on the minimum power loss criterion in FECF is implemented as an objective function [2]:

$$F_j = \sum_{i=1}^{k} W_i = \sum_{i=1}^{k} \left[\frac{\left(\sqrt{P_i(U)^2 + Q_i(U)^2}\right) \cdot R_{Эi}}{U_i^2} \cdot \left(\frac{1}{\left(1 + \dfrac{\Delta U_{\%i}}{100}\right)^2} - 1 \right) + \right. \quad (2)$$

$$\left. + \Delta P_{xxi} \cdot T_{pi} \cdot \left(\frac{U_i}{U_{номi}}\right)^2 + \Delta p_{корi} \cdot L_i \cdot k_{Uкорi} \right] \rightarrow \min,$$

where:

W – is the sum of power losses in FECF elements;

$R_Э$, L – equivalent resistance values and the length of the FECF elements;
$\Delta U_{\%}$ – supply voltage as a percentage;
ΔP_{xxi} – idling loss power in FECF elements;
T_{Pi} – number of operating hours of FECF equipment;
$\Delta p_{\kappa opi}$ – the average specific losses to the crown;
$k_{U\kappa opi}$ – loss factor for the crown.
At the same time, it is mandatory to fulfill the balance restrictions:

$$\begin{cases} \left|P_{вні} + P_{ci}\right| > 0; \\ \left|Q_{вні} + Q_{ci}\right| > 0; \\ i = 1, ..., k; \end{cases} \quad (3)$$

where $P_{вн\,i}$, $Q_{вн\,i}$, $P_{c\,i}$, $Q_{c\,I}$ – are the values of internal and external flows of active and reactive power in FECF elements.

Equations (1) - (3) form a universal free-scale model for the redistribution of power flows in the FECF, which will allow optimization by the criterion of mini-mum losses, taking into account weakly formalized and uncertain factors [5].

In [4-7], due to the simplicity of the organization and high accuracy indicators, Mamdani was implemented together with the modified Newton - Raphson algorithm:

$$R^u : If \;\; x(t-1) \;\; is \;\; X_1^u \;\; and, ..., and \;\; x(t-r) \;\; is \;\; X_r^u \, then \;\; y(t) = a^u, u = \overline{1, u} \quad (4)$$

where:
R^U – is a set of expert rules for forming the FNN;
$x(t) = (x_1(t), \ldots , x_r(t))$ – vector of input variables;
$y(t)$– vector of the output value;
M_1^u, M_r^u, Y^u – areas of input and output effects of FNN;
a^u – output constant containing a weighted estimate of the input values and the network operation process.

4. Results

For the developed modification of the forecasting model and further opti-mization of FECF production indicators based on FNN (4), together with criteria (1)-(3), the accuracy indicators of the process of managing the dynamics of pow-er flows in the short, medium and long-term periods were studied. The results were evaluated using Matlab software [4, 5].

When implementing the FNN model based on the Mamdani algorithm, ANFIS, a block of the Matlab simulation modeling software package, was used. The accuracy of optimization of FECF production indicators [4, 5] was studied through the forecasting process as a function of:

- variation of the membership function of input terms;
- implementations of the training set;
- changes to the ins architecture along with training time.

For numerical evaluation of the forecasting process and subsequent

The best results from the sample were shown by the implementation of the FNN with the architecture of input terms ($6 \times 3 \times 3$), as well as the psigmf function [4, 5].

The results of training the model based on the FNN are shown in figures 1, 2.

Figures 3 - 5 show the results of the model functioning in comparison with similar results for the regression model.

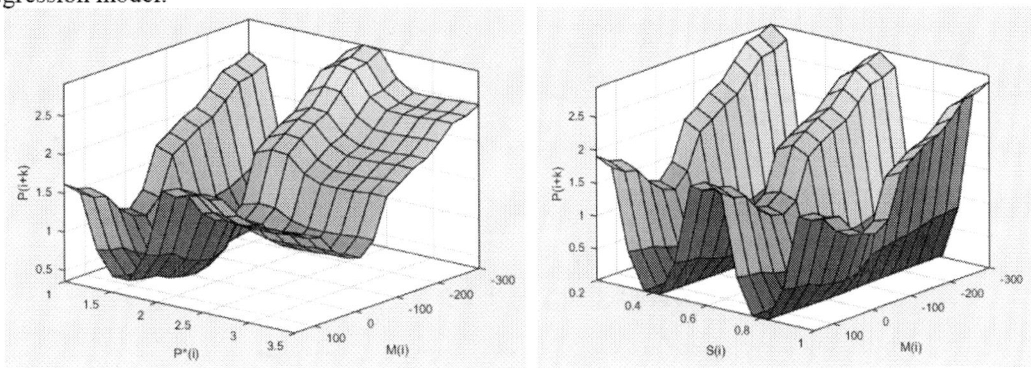

Figure 1. Forecast of FECF production parameters.

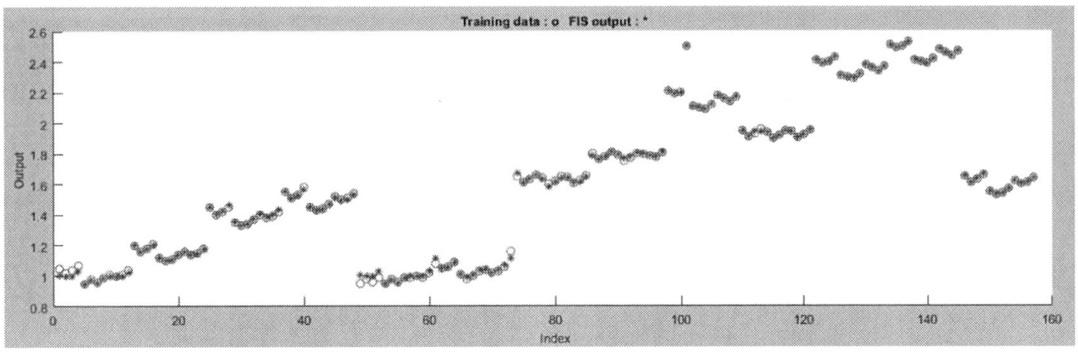

Figure 2. Results of training the FNN model (psigmf: $6 \times 3 \times 3$).

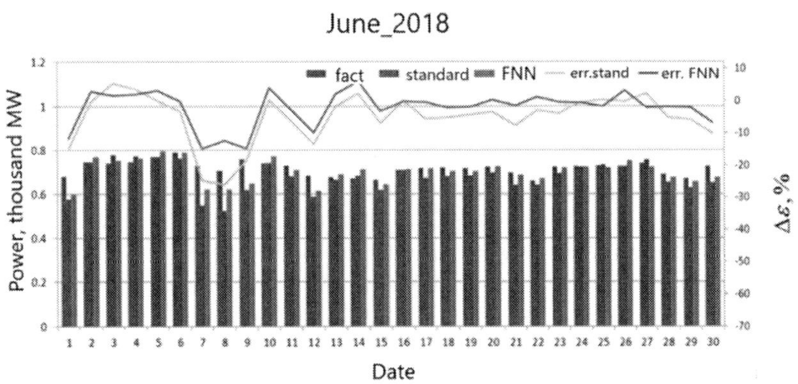

Figure 3. Forecasting the dynamics of energy consumption of the FECF (June 2018): the fact is, the standard (regression), FNN – thousand MW; and an error in the functioning of forecast systems based on the normative method and the application of the FNN - %.

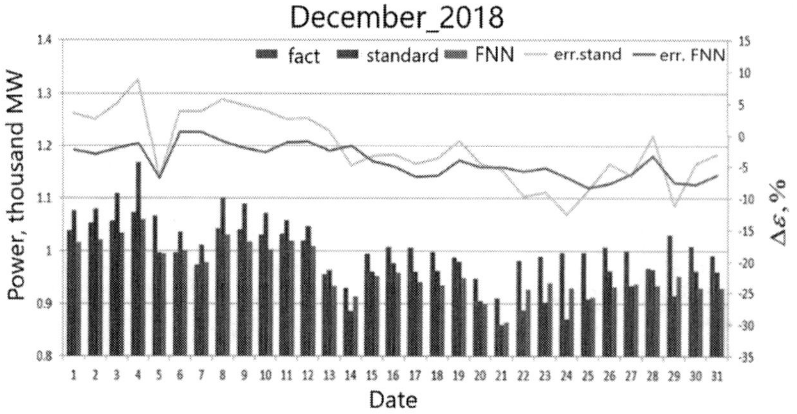

Figure 4. Forecasting the dynamics of energy consumption of the FECF (December 2018): the fact is, the standard (regression), FNN – thousand MW; and an error in the functioning of forecast systems based on the normative method and the application of the FNN - %.

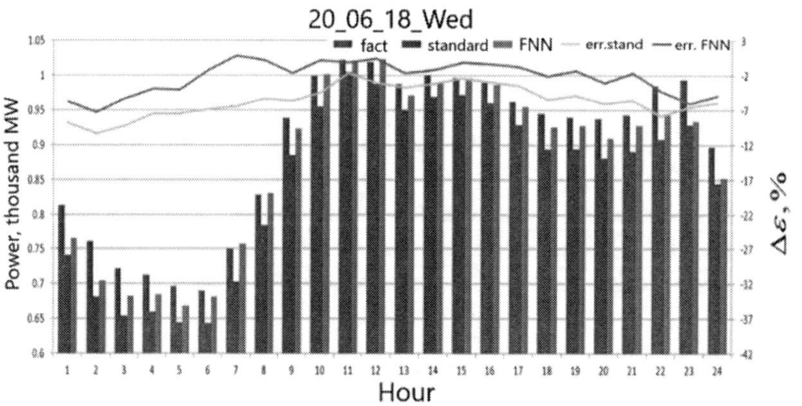

Figure 5. Forecasting the dynamics of FECF energy consumption (20.06.18 – business day/weekdays): the fact is, the standard (regression), FNN – thousand MW; and an error in the functioning of forecast systems based on the normative method and the application of the FNN - %.

The analysis of the functioning of the model [4, 5], built on the basis of FNN, allows us to speak about its high accuracy characteristics, significantly exceeding the accuracy of existing regression (classical) implementations, as follows from figures 3 - 5.

5. Discussion

Thus, the following results were obtained [4, 5], confirmed by the graphs in figures 1, 2:

- the use of classical / normative (based on regression dependencies) implementation of the process of forecasting and subsequent optimization of FECF modes is justified in the case of a high degree of constancy of factors affecting the functioning of groups of energy consumers;
- also, regression models are useful as idealized (target) structures, while the forecast error in them is fundamentally not lower than the level (3-6) %;

- the use of models for optimizing the production indicators of FECF (in particular, the process of redistributing power flows) makes it possible to achieve forecast errors of no more than (1.5-2.8) % with full consideration of the distribution trend of weakly formalized (including uncertain) factors [2, 4, 5].

6. Conclusion

The article notes the work on the implementation of the model of optimization of FECF production indicators based on the FNN. At the same time:

1) the objective optimization function for the minimum of active power losses ($F_j = \sum_{i=1}^{k} W_i \to \min$) is defined together with the restrictions typical for the FECF. The dependence of the distribution of active and reactive power on undefined and poorly formalized factors: $\xi_P(U)$, $\xi_Q(U)$, that affect the quality of the realized process of optimization of FECF is taken into account.

2) the development of a modified Newton – Raphson algorithm based on the FNN (with the possibility of implementing variations) in order to improve the efficiency of functioning is Noted. This algorithm allows us to fully take into account uncertain and weakly formalized factors: $\xi_P(U)$, $\xi_Q(U)$.

3) estimation of the accuracy of the most optimal variant of the model implementation has the following characteristics: number of layers / structure of layers (number of neurons): 4 / (3/10/25/1); type of optimization algorithm: Mamdani; distribution of terms of belonging to 1 layer: ; activation function of the first layer: psigmf.

At the same time, the following quality indicators were obtained: relative learning error-1.624 %; network learning time – 2.57 s (for the case of short – term / medium-term forecast). The simulation is based on the Intel Core i5-7400k – Core Processor (3.0 GHz). The freely configurable structure of network settings is noted, which allows using it equally effectively for tasks of short-term/medium-term and long-term forecasting and optimization of FECF production indicators.

References

[1] Zhelezko Yu S 2009 *Poteri elektroenergii. Reaktivnaya moshchnost'. Kachestvo elektroenergii: Rukovodstvo dlya prakticheskih raschetov* [Power losses. Reactive power. Power quality: a Guide for practical calculations] (Moscow: ENAS) [In Russian]

[2] Nikolenko S I, Kadurin A A and Arhangel'skaya E O 2019 *Glubokoe obuchenie* [Deep learning] (St. Petersburg: Piter)

[3] Kudinov Yu I, Kudinov I Yu and Suslova S A 2007 *Nechetkie modeli dinamicheskih processov* [Fuzzy models of dynamic processes] (Moscow: Nauchnaya Kniga)

[4] Rutskov A L, Burkovsky V L, Sidorenko E V and Krysanov V N 2020 Implementation of a SMART GRID in industrial and residential complexes based on fuzzy neural networks *J. of Mechanics of Continua and Mathematical Sciences: Special Issue* **8** 251-63

[5] Rutskov A L, Burkovsky V L and Sidorenko E V 2020 Optimization of electric power systems using fuzzy neural network algorithms *J. of Mechanics of Continua and Mathematical Sciences: Special Issue***8** 264-76.

[6] Çevik H H and Çunkaş M 2015 Short-term load forecasting using fuzzy logic and ANFIS *Neural Computing and Applications* **26(6)** 1355-67

[7] Shi B, Yu-Xia L I and Xin-Hua Y U 2009 Short-term load forecast based on modified particle swarm optimizer and back propagation neural network model *J. of Computer Applications* **29(4)** 1036-9

[8] Monteleoni C, Schmidt G A, Saroha S and Asplund E 2011 Tracking climate models Statistical *Analysis and Data Mining* **4(4)** 372-92

[9] Li P, Li Y, Xiong Q, Chai Y and Zhang Y 2014 Application of a hybrid quantized elman neural network in short-term load forecasting *Int. J. of Electrical Power and Energy Systems* **55** 749-59

[10] Nedellec R, Cugliari J and Goude Y 2014 Gefcom2012: Electric load forecasting and backcasting with semi-parametric models *Int. J. of Forecasting* **30(2)** 375-81
[11] Devaine M, Gaillard P, Goude Y and Stoltz G Forecasting electricity consumption by aggregating specialized experts *Machine Learning* **90(2)** 231-60
[12] Migliore M et al. 2006 Parallel network simulations with NEURON *J. of Computational Neuroscience* **21(2)** 119-29
[13] Berenji H R 1992 A Reinforcement Learning- Based Architecture for Fuzzy Logic Control *Int. J. of Approximate Reasoning* **6** 267-92
[14] Davison A P, Brüderle D, Eppler J, Kremkow J, Muller E, Pecevski D, Perrinet L and Yger P 2008 PyNN: A Common Interface for Neuronal Network Simulators *Frontiers in Neuroinformatics* **2** 11
[15] Alneyadi S, Sithirasenan E and Muthukkumarasamy V 2016 A survey on data leakage prevention systems *J. of Network and Computer Appl.* **6(62)** 137-52

Analysis of biogas content influence on the flame properties

E S Leonov and P A Trubaev

Department of Energy Engineering of Heat Technologie, Belgorod State Technological University named after V.G. Shukhov, Kostyukov St., 46, Belgorod, 308012, Russia

E-mail: trubaev@gmail.com

Abstract. The paper has examined the combustion of gas with variable methane content where it was supplied separately from the air. The analytical one-dimensional model of the diffusion straight-flow flame in the unrestricted and restricted space and a numerical three-dimensional model of the turbulent flame in a swirled airflow were used in this research. It has been established that in furnaces (in a restricted space) the maximum flame temperature depends only on the methane content in fuel and in average lower than the temperature of clean methane combustion by 20-70°C for conventional biogas and by 100-200°C for biogas with low methane content. The methane content decrease in gas leads to the diffusion flame length decrease in comparison with clean methane combustion by approximately two-fold for conventional biogas and fourfold for combustion of biogas with low methane content. For the turbulent flame in a swirled air flow, the flame length decrease is insignificant and is around 10% for conventional biogas and around 40% for biogas with low methane content. It can be concluded that by changing the methane content in biogas the heating plant operating mode can be stable on the condition that the biogas flow rate is controlled to ensure constant heat production during its combustion and turbulent flame with a swirled airflow. Combustion, in this case, can be done using the same burner.

1. Introduction

The last two decades showed an increased demand for energy and a large number of problems around nonrenewable energy sources which triggered the interest of the researchers in alternative energy sources study [1].

Every year amount of organic waste in cities and rural communities, as well as the production of a thousand tons of sludge from wastewater and wastewater at different agricultural and food factories, is growing. Such a state of matter leads to serious economic and ecological difficulties, but at the same time, this is a great potential for biogas production [2].

Biogas is a gas that is the result of an organic matter anaerobic fermentation. Biogas can be made from solid municipal waste, wastewater, animal manure, and agricultural waste. Biogas content heavily depends on the assimilated substrate characteristics and operating conditions of its source. In general, biogas consists of CH_4, CO_2, N_2, O_2, H_2S, and H_2. Considering methane's presence in biogas, it can be used as a fuel for heat and electricity production [3].

Biogas is a natural product and is increasingly considered as a partial alternative choice against fossil fuel, which is associated, mainly, with the fact that fossil fuel accelerates global warming and depends on geopolitical issues. However, as the author [4] says the main characteristics of biogas combustion such as the ignition delay period and its roles in the polluting substance production, like aromatic compounds and soot, are still unclear.

The most common and accessible way of direct biogas use is combustion due to its low cost and low maintenance expenses. This kind of utilization does not require the removal of H2S and moisture from the biogas. Usually, the conversion efficiency of biogas into heat in the dual-fuel burner is around 80-90% [5].

Raw biogas cannot be directly used in commercial burners instead of natural gas or liquefied petroleum gases (propane, butane). When biogas under pressure is supplied into the conventional burner, which is normally used for natural or liquefied gas, the air-to-fuel ratio is not sufficient for stable combustion due to CO_2 high content in the biogas. Therefore, biogas combustion requires the installation of a new burner with a separate instrumentation system. Considering the fact that the biogas components have corrosive nature (H_2O and H_2S,) the condensate formation needs to be prevented by keeping furnace temperature higher than the dew-point temperature. This means that the furnace should be heated by natural or liquefied gas to receive a higher operating temperature [6].

A study in the paper [6] has analyzed biogas combustion, mostly flameless in the simulated combustion chamber, and demonstrated volume dependence of conventional combustion of biogas and CH_4 (Figure 1, 2).

The burner design depends on the type of fuel that is being used. Different types of fuel have different rates of burning and different requirements for oxidizer which leads to changes in nozzle and burner head design.

The flame velocity plays an important role in burner designing when biogas is burned in a conventional way. In order to prevent flame blow off it is required to control the flow rate of fuel and air into the chamber. CH_4 concentration in the biogas components is by one order lower than in the natural gas, therefore, conventional biogas combustion provides lower flame velocity. This means that the air and fuel flow rates should be decreased to prevent a flame blow off. The maximum flame velocity is achieved at the stoichiometric air-to-fuel ratio. Because of the non-combustible components in biogas, the flame temperature during conventional biogas combustion is lower than when natural gas is used [7].

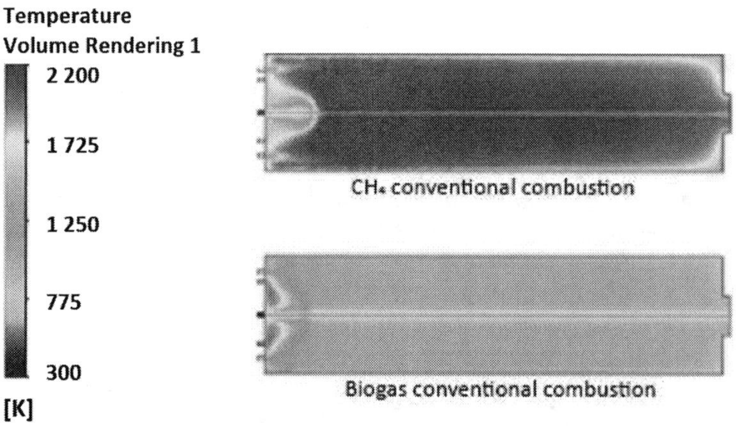

Figure 1. Temperature distribution during biogas and CH_4 combustion [6].

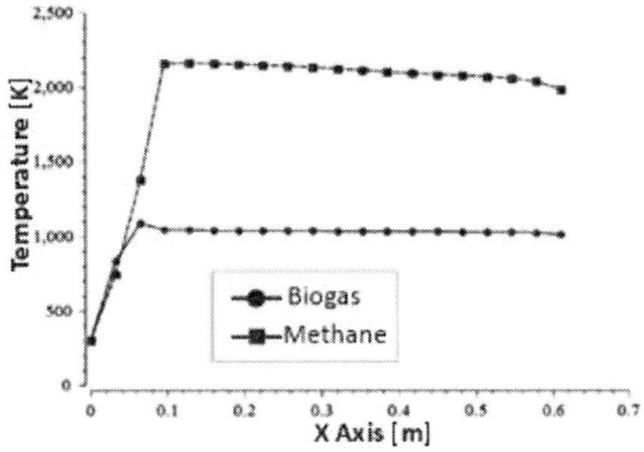

Figure 2. Temperature change at the central line of the furnace during methane and biogas combustion [6].

The paper [8] demonstrates combustion temperature dependence on CO_2 content (Figure3).

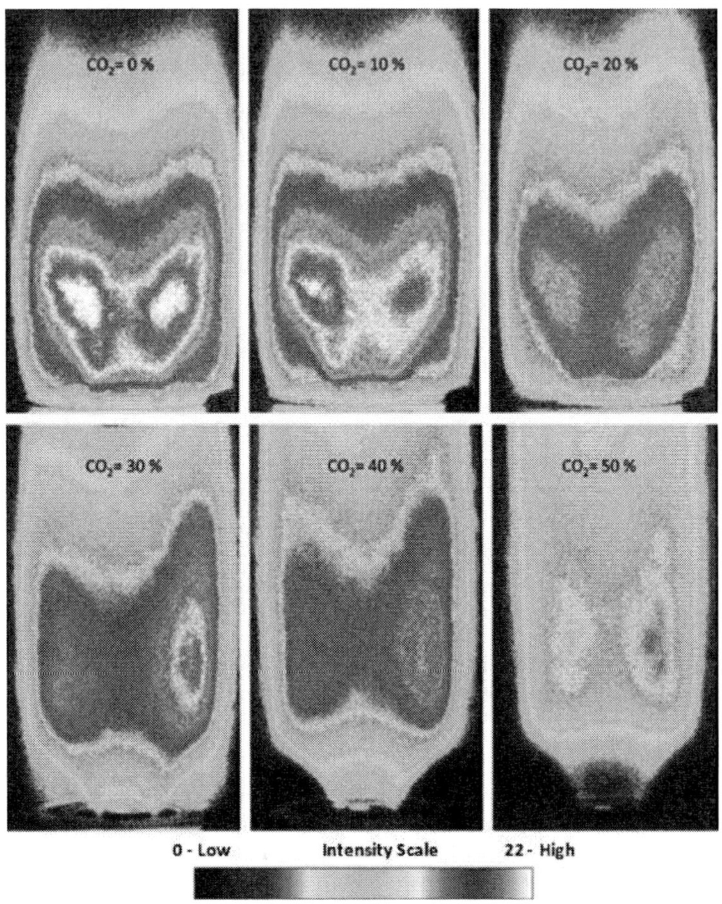

Figure 3. Gaseous fuel combustion with different CO_2 concentration [8].

There is one more way of energy use of biogas, flameless combustion, efficiency research on which is presented in the paper by the author [9]. Reactions during flameless combustion proceed at special combustion mode conditions when the volume and hot temperature of the medium are maintained higher than the temperature of self-ignition. This method does not produce any visible flame or light effect; therefore, this mode was named “flameless oxidation”. The opposite effect is present in the conventional combustion systems where reactions are concentrated in a narrow flame front. The flameless burners can burn fuel with variable composition and unsteady quality or liquid-crystal fuel. This is their major advantage [10].

Based on the author`s studies the regenerative or recuperative heat exchangers are the best equipment for heat recuperation for the flameless combustion method but it is very expensive. To keep the air temperature higher than the biogas self-ignition temperature, during the laboratory experiment [11], an additional electrical heater was used. The necessary air for combustion is supplied by the fan and the incoming air is pre-heated by the furnace exhaust gases in the heat exchanger. The air and gas flow rate regulators, gas analyzer, corresponding control valves, digital temperature meter, biogas, and CH4 capsules are the rest of the equipment that was used in this experiment.

Simple burner design is one of the prime advantages of flameless combustion systems.

The same study [6] has demonstrated the dependence of flameless combustion forming on oxygen content (Figure 4).

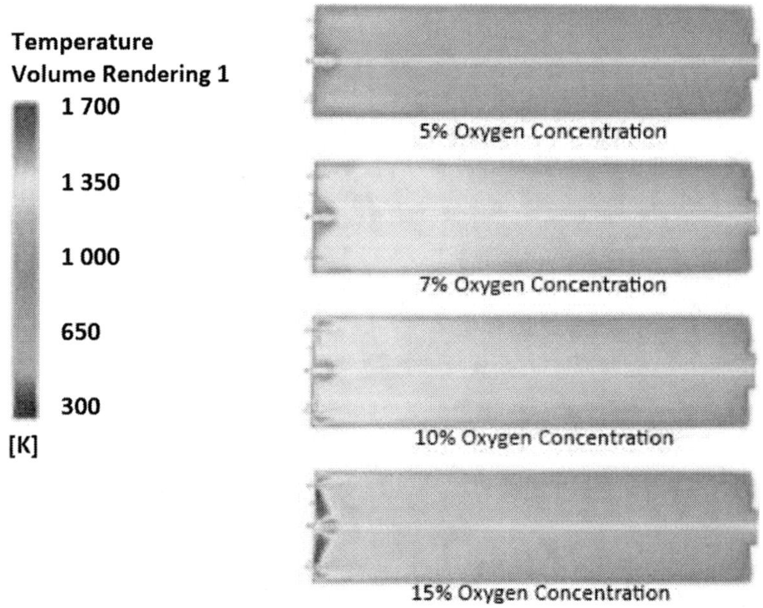

Figure 4. Biogas flameless combustion forming in relation to different oxygen concentration [6].

Conclusions of the chapter:

- there are several ways of biogas utilization, more specifically conventional and flameless combustion, modernization of biogas to methane purity of 95%, and its further use as a fuel;
- biogas modernization processes are very costly and most of the time economically unprofitable;
- biogas requires special combustion conditions.

2. Materials and Methods

Currently, commercial furnace designing, most of the time, is based on the approximative empirical dependences [12]. The objective of the paper is the study of gas combustion with variable methane content for use in conventional fuel facilities.

The research has used three mathematical models of gaseous fuel combustion during its separate feeding with air:

- straight-flow diffusion flame in an unrestricted space (one-dimensional analytical model);
- straight-flow diffusion flame in a restricted space (one-dimensional analytical model);
- turbulent flame in a swirled airflow (three-dimensional model).

2.1. Straight-flow diffusion flame in an unrestricted space

At diffusion combustion, a gaseous fuel, which is flowing out of the burner nozzle, involves surrounding air into the movement and creates a turbulent stream (Figure 5). The volume of the dragged in the surrounding air is limited only by the turbulent stream size. The diffusion combustion intensity depends on the speed of the gaseous fuel turbulent mixing with air. Chemical reactions of combustion in the diffusion flame proceed with a speed far much higher than the turbulent mixing and, therefore, practically doesn`t influence the diffusion flame size.

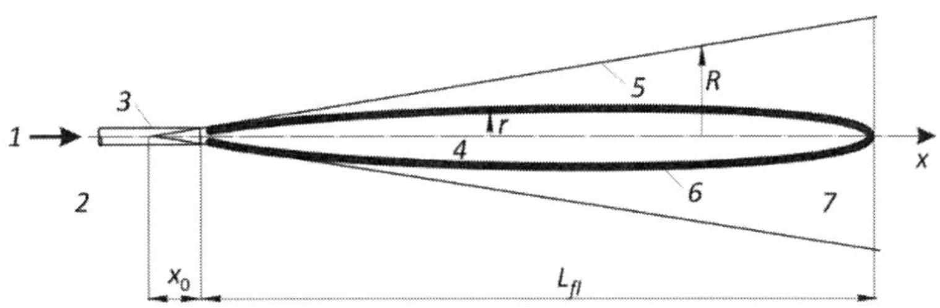

Figure 5. Diffusion flame in an unrestricted space [13]: L_{fl} – flame length; x_0 – distance from a pole of a stream to a burner outlet; x – coordinate; r – flame diameter; R – turbulent stream diameter; 1 – fuel; 2 – air; 3 – burner; 4 – an internal area of a flame; 5 – turbulent stream boundaries; 6 – flame front (area of a chemical reaction during combustion); 7 – an outer area of a flame.

The calculations were done by using the one-dimensional analytical model presented in the monograph [13].

2.2. Straight-flow diffusion flame in a restricted space

The gaseous fuel combustion in a space limited by the cylindrical walls of the furnace was analyzed. In this case, the amount of air going for combustion is limited. The modeling was carried out using the excess air coefficient 1.1.

The flame can be divided into two areas: the area of streamflow which ends when all air is dragged in into a stream and the area of an afterburning (Figure 6).

At the streamflow section, around 80% of natural gas is burned. The rest of the fuel burns out in the second section of the diffusion flame as it mixes with the air that was dragged into the turbulent streamflow movement. In the second section the lower the amount of unburnt fuel in the flame the lower the combustion intensity. As a result, the flame length is longer than during combustion in the unrestricted space.

For the flame calculation, the one-dimensional analytical model was used which is presented in the monography [13].

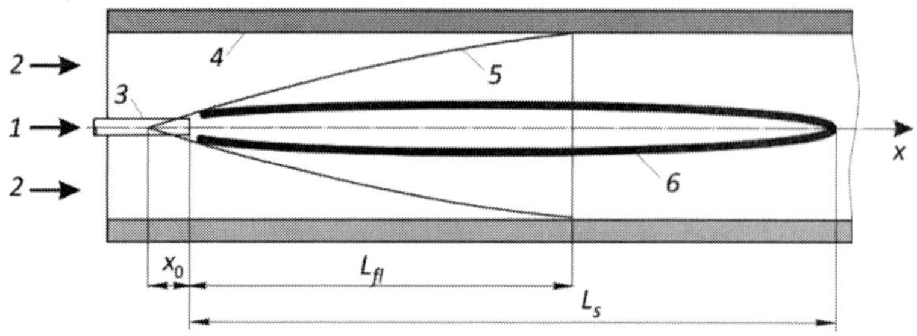

Figure 6. Diffusion flame in a restricted space [13]: L_{fl} – flame length; L_s – streamflow length; x_0 – distance from a pole of a stream to a burner outlet; x – coordinate; r – flame diameter; R – turbulent stream diameter; *1* – fuel; *2* – air; *3* – burner; *4* – furnace lining; *5* – turbulent stream boundaries; *6* – flame front (area of a chemical reaction during combustion).

2.3. Turbulent flame in a swirled airflow

The mathematical model of natural gas combustion and heat transfer in the vertical cylindrical furnace of the water heater included the system of differential equations of the turbulent motion, combustion, and radiative-convective heat exchange. The equations were solved using the numerical finite volume method. For every point of the calculation grid (Figure 7) were determined components of velocity in all three axes, fuel and air concentration, the temperature of the gaseous medium.

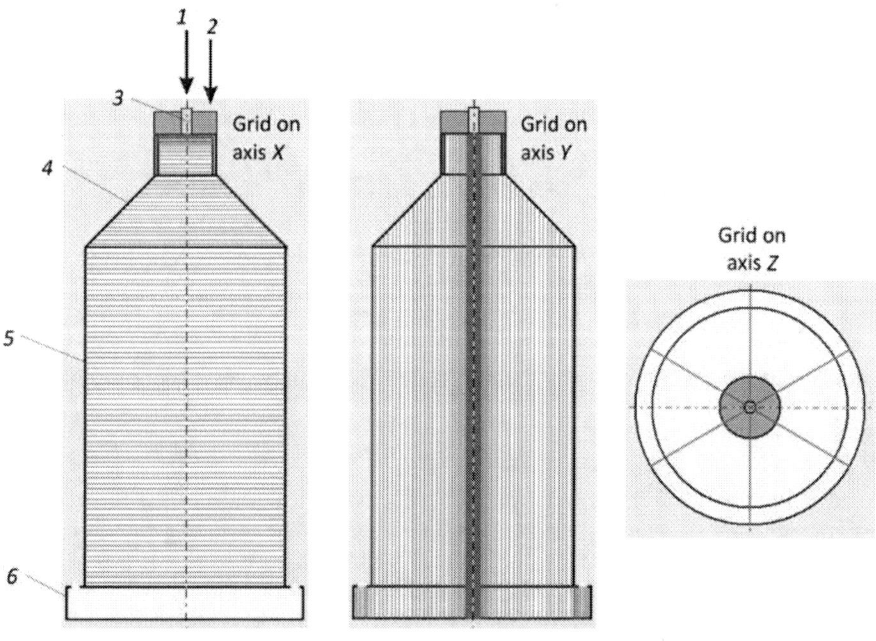

Figure 7. Calculation grid for numerical combustion modeling in the cylindrical furnace: *1* – fuel; *2* – air; *3* – burner; *4* – conical section of the furnace; *5* – cylindrical section of the furnace; 6 – an outlet for exhaust gases.

Semiempirical Prandtl's theory of turbulence [14] application has allowed correctly formulate the boundary conditions on the coarse grid [15] for the discretized equations, which are applied in the mathematical models for calculation of velocity, eddy viscosity, temperature, and convective heat transfer to the walls that limit turbulent gas flow [16]. The improved differential equations of the diffusion mathematical model [17] of the radiative heat transfer in the attenuation medium have allowed to formulate the correct boundary conditions on the enclosure wall surface at the diffuse reflection of the incident radiation [18].

3. Results and Discussion

The landfill gas is characterized by a variable composition [19]. Major components of biogas are methane, nitrogen, and carbon dioxide [20]. The modeling has used biogas of two different compositions: normal with 60% of methane content and with a low methane content of 35%. The biogas combustion was compared to methane combustion.

Three variants were considered for modeling (Table 1):

a) variable composition gas combustion with the same volumetric flow rate of gas supplied into the burner;

b) combustion with a constant heating value (heat released during its combustion) using the same burner;

c) combustion with a constant heating value and burner diameter changing to ensure constant gas velocity.

Table 1. Calculation variants.

Gas composition (by volume)			Constant gas flow rate			Constant heating value ($P = 1$ MW)				
						Flow rate Q, m^3/h (m^3/sec)	For the same burner		Burner changed to keep constant gas velocity	
CH_4	CO_2	N_2	Flow rate Q, m^3/h (m^3/sec)	Burner diameter d, mm	Exhaust velocity v, m/s		Burner diameter d, mm	Exhaust velocity v, m/s	Burner diameter d, mm	Exhaust velocity v, m/s
100%	–	–	100 (0.0278)	13.3	200	–	–	–	–	–
60%	20%	20%	-"-	-"-	-"-	167 (0.0463)	13.3	334	17.2	200
35%	43%	22%	-"-	-"-	-"-	286 (0.0794)	13.3	572	22.5	200

3.1. Modeling of gas combustion in a straight-flow diffusion flame in an unrestricted space

Figure 8 and Table 2 present the results obtained during the gas combustion modeling in the unrestricted space.

Conclusions of the chapter:

a) the maximum inside flame temperature depends only on the methane content and doesn't depend on the fuel's exhaust velocity;

b) the methane content in gas reduction leads to the flame length decrease, but for the same burner it stays constant and doesn't depend on the fuel's exhaust velocity.

3.2. Modeling of gas combustion in a straight-flow diffusion flame in a restricted space

Figure 9 and Table 2 present the results obtained during gas combustion modeling in the space limited by walls of the cylindrical furnace. Temperature curve behavior in Figure 9 reveals lengths of the streamflow section which ends up when all air is dragged into the flow and section of fuel afterburning.

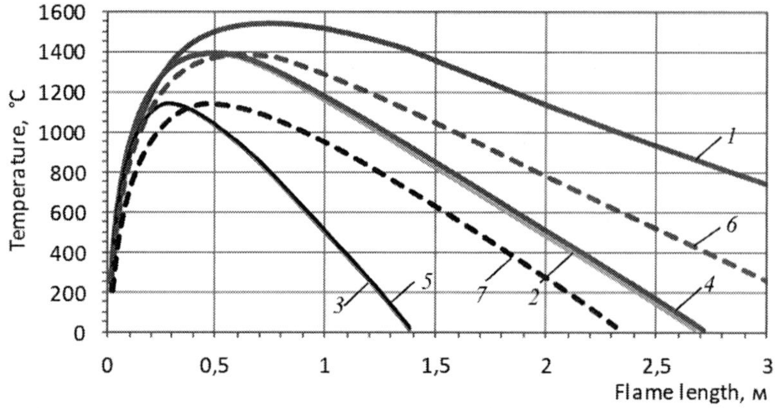

Figure 8. Flame temperature along its length:
a) at constant flow rate Q = 100 m^3/h:
1 – CH4=100% (——); 2 – CH4=60% (——); 3 – CH4=35% (——);
b) at constant heating value P = 1 MW, and fuel nozzle diameter d = 13.3 mm:
4 – CH4=60% (——); 5 – CH4=35% (——);
c) at constant heating value P = 1 MW and exhaust gas velocity v = 200 m/s: 6 – CH4=60% (- - -); 7 – CH4=35% (- - -).

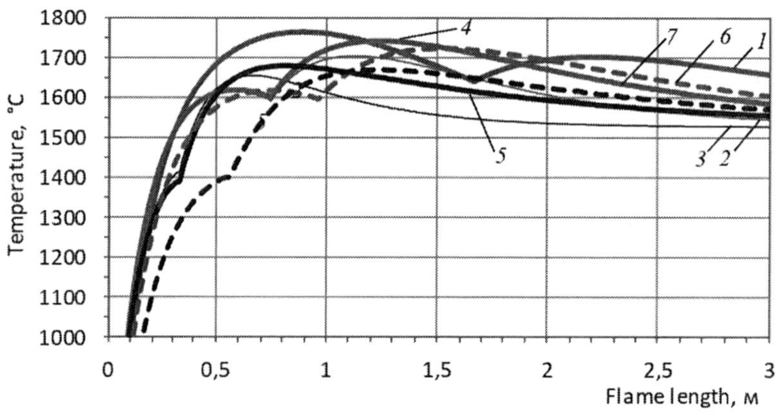

Figure 9. Flame temperature along its length:
a) at constant flow rate Q = 100 m^3/h:
1 – CH_4=100% (——); *2* – CH_4=60% (——); *3* – CH_4=35% (——);
b) at constant heating value P = 1 MW, and fuel nozzle diameter d = 13.3 mm:
4 – CH_4=60% (——); *5* – CH_4=35% (——);
c) at constant heating value P = 1 MW and exhaust gas velocity v = 200 m/s:
6 – CH_4=60% (- - -); *7* – CH_4=35% (- - -).

Conclusions of the chapter:

a) the maximum inside flame temperature notably less dependent on the methane content in gas, for conventional biogas it is lower than the clean methane combustion temperature by 20-60°C, for biogas with low methane content – by 85-110°C;

b) the flame length gets smaller when methane content in gas decreases, but at the same time for the same burner it stays constant and doesn`t depend on the output gas velocity;

c) biogas flame length decrease value in comparison with methane flame for unrestricted and restricted space are the same.

Table 2. Flame length L_{fl}, m.

Variants of modeling	Combustion in the unrestricted space			Combustion in the restricted space		
	CH_4=100%;	CH_4=60%	CH_4=35%	CH_4=100%;	CH_4=60%	CH_4=35%
a) constant flow rate Q = 100 m^3/h	2.74	1.35 (49%[a])	0.69 (25%[a])	3.08	1.44 (53%[b])	0.71 (26%[b])
b) heating value P = 1 MW, nozzle diameter d = 13.3 mm	-“-	1.36 (50%[a])	0.69 (25%[a])	-“-	1.45 (53%[b])	0.7 (26%[b])
c) heating value P = 1 MW, exhaust velocity v = 200 m/s	-“-	1.75 (64%[a])	1.16 (42%[a])	-“-	1.87 (68%[b])	1.2 (44%[b])

[a] – of 2.74 m; [b] – of 3.08 m

3.3. *Modeling of gas combustion in a turbulent flame in a swirled airflow*

Figure 10 and Table 3 present the results of numerical modeling of methane and biogas combustion in the vertical furnace of the water heater.

Fuel and air concentration field.

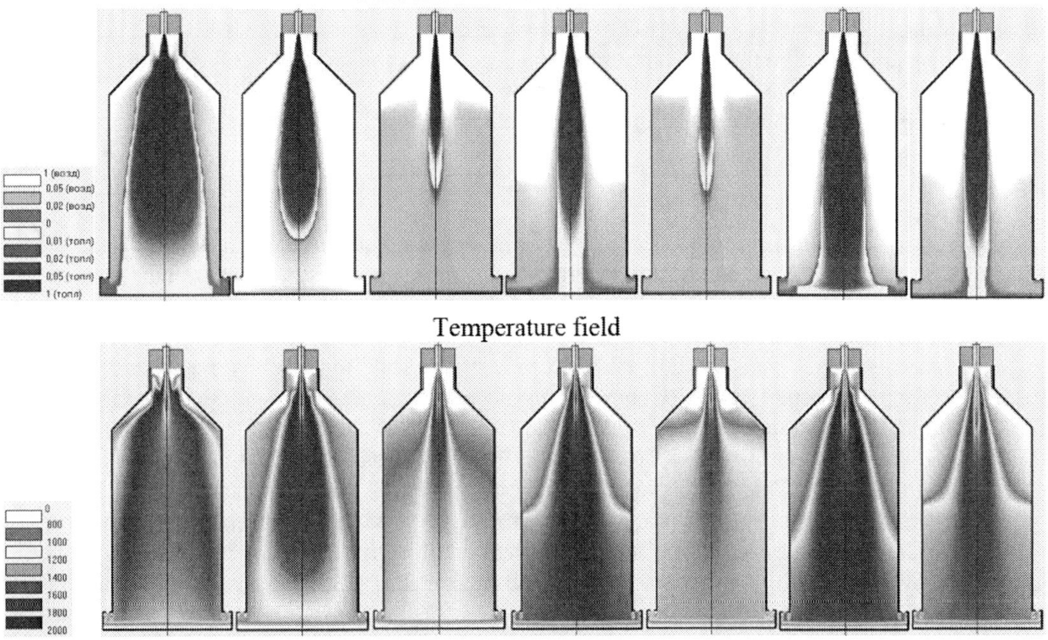

CH_4	100%	60%	35%	60%	35%	60%	35%
Q, m^3/h	100			167	286	167	286
P, MW	1.0	0.6	0.35	1.0			
d, mm	13.3					17.2	22.5
m, m/s	200			334	572	200	

Figure 10. Gas, air, and temperature concentration fields for methane and biogas combustion in the vertical furnace of the water heater.

Table 3. Flame length Lfl, m, and gas temperature during methane and biogas combustion in the vertical furnace of the water heater.

Variants of modeling	Flame length Lfl, m; Maximum temperature; The mean temperature at the nozzle area:		
	CH_4=100%;	CH_4=60%	CH_4=35%
a) constant flow rate Q = 100 m^3/h	1.86 m 1877°C 1276°C	1.67 m (89%[a]) 1807°C 960°C	1.07 m (58%[a]) 1668°C 983°C
b) heating value P = 1 MW, nozzle diameter d = 13.3 mm	-"-	1.69 m (91%[a]) 1840°C 1401°C	1.06 m (57%[a]) 1673°C 1212°C
c) heating value P = 1 MW, exhaust velocity v = 200 m/s	-"-	2.17 (117%[a]) 1817°C 1414°C	1.77 (95%[a]) 1668°C 1306°C

[a] – of 1.86 m.

Conclusions of the chapter:

a) the maximum temperature inside the flame for conventional biogas is lower than the temperature of clean methane combustion by 40-70°C, for biogas with low methane content – by 200°C;

b) the temperature at the furnace exit during biogas combustion can be even higher than during methane combustion because of a smaller amount of supplied air;

c) decrease of the methane content in gas leads to a much smaller decrease of the flame length than for the diffusion flame.

4. Conclusion

The paper has examined the combustion of gas with variable methane content where it was supplied separately from the air. The analytical one-dimensional model of the diffusion straight-flow flame in the unrestricted and restricted space and a numerical three-dimensional model of the turbulent flame in a swirled airflow were used in this research.

The research has examined combustion of clean methane and biogas with a methane content of 60 and 35% in three different variants: a) combustion of variable content gas with the same gas volumetric flow rate; b) combustion with a constant amount of heat, released during fuel combustion, using the same burner which required gas flow rate increase; c) combustion with a constant amount of heat, released during fuel combustion, using different diameter burners to ensure constant gas velocity.

It has been established that in furnaces (in a restricted space) the maximum flame temperature depends only on the methane content in fuel and in average lower than the temperature of clean methane combustion by 20-70°C for conventional biogas and by 100-200°C for biogas with low methane content. The methane content decrease in gas leads to the diffusion flame length decrease in comparison with clean methane combustion by approximately two-fold for conventional biogas and fourfold for combustion of biogas with low methane content. For the turbulent flame in a swirled air flow, the flame length decrease is insignificant and is around 10% for conventional biogas and around 40% for biogas with low methane content.

It can be concluded that by changing the methane content in biogas the heating plant operating mode can be stable on the condition that the biogas flow rate is controlled to ensure constant heat production during its combustion and turbulent flame with a swirled airflow. Combustion, in this case, can be done using the same burner.

5. Acknowledgments
The article was prepared within a development program of the Base University on the basis of BSTU named after V.G. Shoukhov.

References

[1] Isci A and Demirer G N 2007 Biogas production potential from cotton wastes *Renew. Energy Pergamon* **32** 750–7

[2] Taleghani G and Kia A S 2005 Technical-economical analysis of the Saveh biogas power plant *Renew. Energy. Pergamon* **30** 441-6

[3] Hoteit A, Chandel M K, Durecu S and Delebarre A 2009 Biogas combustion in a chemical looping fluidized bed reactor *Int. J. Greenh. Gas Control* **3** 561-7

[4] Fischer M and Jiang X 2017 Numerical studies of CO formation during biogas combustion *Energy Procedia Elsevier B. V.* **142** 426-31

[5] Kapoor R, Ghosh P, Tyagi B, Vijay V K, Vijay V, Thakur I S, Kamyab H, Nguyen D D and Kumar A 2020 Advances in biogas valorization and utilization systems: A comprehensive review *J. Clean. Prod. Elsevier Ltd* **273** 123052

[6] Hosseini S E, Bagheri G and Wahid MA 2014 Numerical investigation of biogas flameless combustion *Energy Convers. Manag. Pergamon* **81** 41-50

[7] Hosseini S E and Wahid M A 2014 Development of biogas combustion in combined heat and power generation *Renewable and Sustainable Energy Reviews Elsevier Ltd* **40** 868-75

[8] Mordaunt C J and Pierce W C 2014 Design and preliminary results of an atmospheric-pressure model gas turbine combustor utilizing varying CO_2 doping concentration in CH_4 to emulate biogas combustion *Fuel. Elsevier* **124** 258-68

[9] Hosseini S E and Wahid M A, Abuelnuor A A A 2014 Biogas flameless combustion: A review *Appl. Mech. Mater* **388** 273-9

[10] Colorado A F, Herrera B A and Amell A A 2010 Performance of a Flameless combustion furnace using biogas and natural gas *Bioresour. Technol. Elsevier Ltd* **101** 2443-9

[11] Hosseini S E and Wahid M A 2013 Biogas utilization: Experimental investigation on biogas flameless combustion in lab-scale furnace *Energy Convers. Manag. Pergamon* **74** 426-32

[12] Kuznetsov V A and Trubaev P A 2018 Resources and Problems of the Mathematical Simulating Thermo-Technological Processes *J. Phys.: Conf. Ser* **1066** 012024

[13] Kuznetsov V A and Trubaev P A 2017 Matematicheskie modeli teplomassoperenosa v vy`sokotemperaturny`x ustanovkax [Mathematical models of a heatmass transfer in high-temperature unit] (Belgorod: BSTU)

[14] Kuznetsov V A 2020 Semiempirical model of convective heat transfer of turbulent gases *J. Eng.Phys. Thermophys.* **93** 543-50

[15] Kuznetsov V A 1986 Refinement of wall-turbulence hypotheses *J. Eng. Phys. Thermophys.* **50(6)** 640-4

[16] Kuznetsov V A, Trubaev P A and Ryazancev O A 2020 An elaborated diffusion mathematic model of radiative transfer in an extinction medium *IOP Conf. Ser.: Mater. Sci. Eng.* **791** 012015

[17] Kuznetsov V A 2017 Computational simulation of convective heat transfer of turbulent gas flows *Theor. Found. Chem. Eng.* **51(6)** 1063-9

[18] Kuznetsov V A 2017 Mathematical model of the radiative heat exchange in the selective gases of a diffusion flame *J. Eng.Phys. Thermophys.* **90(2)** 357-65

[19] Trubaev P A, Verevkin O V, Grishko B M, Tarasyuk P N, Shchekin I I, Suslov D Yu and Ramazanov R S 2018 Investigation of Landfill Gas Output from Municipal Solid Waste at the Polygon *J. Phys.: Conf. Ser* **1066** 012015

[20] Zubkova O A and Sergeeva O A 2018 Control of environmental hazards of thermal-sough products determining modern finishing polymeric building materials *Chemical Bulletin* **1(2)** 31-37

Analysis of the possibility of landfill gas use as additional fuel for combustion of solid municipal waste with different moisture content

N V Kornilova[1] and P A Trubaev[2]

[1]Department of Electric Power Engineering and Automation, Belgorod State Technological University of V.G. Shukhov, Kostyukov St., 46, Belgorod, 308012, Russia
[2]Department of Energy Engineering of Heat Technologie, Belgorod State Technological University of V.G. Shukhov, Kostyukov St., 46, Belgorod, 308012, Russia

E-mail: trubaev@gmail.com

Abstract. The paper suggested the calculation method of combustion for solid and gaseous fuel cofiring. The calculation is suggested to carry out for per unit of generated during combustion total capacity instead of per unit of weight or volume of fuel. This method was tested during combustion parameters examination at cofiring of solid municipal waste, which is the main fuel, and landfill gas (biogas) which is the additional fuel for thermal mode stabilization. The calculation has been conducted for waste with the moisture content of 50%, 30% и 10%, which are solid fuel, methane, as a reference fuel, common biogas with a methane content of 60%, and biogas with a low methane content of 35%. It was established that additional input of biogas is effective for the combustion of unprepared waste with high moisture content, moreover, it is possible to use common biogas as well as biogas with low methane content. The required combustion product temperature (900-1000°C) for unprepared waste can be achieved at the excess air coefficient of 116 or more. In summary, solid fuel and biogas cofiring allows operative control of the combustion process and sustain combustion product temperature and heat output of units at a constant level. The combustion process regulation should be based on the given temperature of combustion products by changing two parameters: landfill gas flow rate and flow rate of air supplied for combustion.

1. Introduction

Currently, there is no technology for solid municipal waste utilization that combines economic efficiency and environmental safety. Because of this, most of the waste management systems in Europe employ low-cost ways of utilization [1]. Combustion, which is the most common way of utilization, is difficult to use in medium or small size communities due to inadequate output production of plants [2].

The paper [3] has analyzed four variants of combustion of leftovers which are formed after a useful fraction, that goes for recycling, is removed from the municipal waste. The research studied combustion of waste on the grate without preparation, mechanically pre-grinded, prepared from the RDF leftovers, and firing in the fluidized bed combustion chamber. It has been established that the mechanical pre-processing and increase of the heat value have a minor effect on the energy efficiency of the thermal waste recycling plant, but significantly increase the cost of the technology and energy consumption. The most energy-saving is achieved when waste is incinerated without pre-processing at the large-scale plants.

The paper also states another conclusion that the pre-processing of waste before combustion doesn`t have any energy and ecological advantages. This is because expenses for grinding and drying cannot be covered by the energy production from waste and by an improvement of the environmental indexes [4]. Besides, the preparation of waste for RDF production requires its grinding and drying which creates a large number of toxic gases [5]. The paper [6] points out that because of the much lower mean heating value the untreated solid municipal waste combustion creates a high risk of toxic matter emission due to incomplete combustion and much lower potential recuperation of energy. The paper [5] presents data that the food waste content doesn`t impact the hazardous emission composition, but it worsens when wood construction parts are incinerated (windows, doors, beams, and so on, that resulted after building demolitions).

The exhaust gas cleaning from hazardous particles removes them only from the gas but harmful particles still stay in the cleaning waste [7]. During the waste combustion, the decrease of the amount of emissions depends on the control of processes in the combustion chamber but not on the waste composition [8]. In summary, the most efficient way of harmful emission amount reduction is a combustion mode control during waste incineration to ensure the required temperature level and complete burning [6, 9]. By the temperature and excess air values adjustment, using CO content in the exhaust gas as the controlled parameter, it is possible to determine and sustain the optimal combustion conditions that result in dioxin and furan emissions minimization [10].

To sustain combustion without using a secondary fuel, the incoming waste should have a moisture content not more than 50%, operational ash content lower than 60%, and combustible material content in the as-fired mass of more than 25%. The excess air coefficient should be at 1,6-1,8 [11].

During biomass combustion for combustion temperature increase and mode stabilization, when biomass quality fluctuates, the cofiring with natural gas is used [12, 13]. The efficiency of the plant operation during several fuel types cofiring heavily relies on the fuel combustion mode and conditions, and the necessary thermal modes should be provided within the entire range of burnt fuel mixes [14]. During several fuels cofiring the hazardous emission content, heating surface fouling tendency, and bed sintering nonlinearly depend on variations of fuel mixtures. Often, the only way to obtain information on fuel mixtures cofiring specifics are studies in the experimental or commercial plant [14].

The primary target of waste and gaseous fuel cofiring is a stabilization of heat release in the furnace and boiler ignition process simplification and reduction of the increased emissions which are common for this process [7]. Also, an increase of the gas flow temperature leads to complete incineration of the inside flow solid fuel particles and organic compounds which are formed during a low temperature or incomplete combustion [15]. The load stabilization requires an input of natural gas in the amount of up to 20% from the plant`s heat output to decrease emission content by 10-15% [7]. The cofiring significantly decreases emissions that are formed during waste combustion [16].

The municipal solid waste landfills are the source of the landfill gas – biogas that contains components such as methane CH_4 (35-60%), carbon dioxide CO_2 (20-40%), and nitrogen N_2 (up to 40%). The landfill gas also requires utilization [17] because, according to the various estimates, the methane greenhouse effect exceeds the carbon dioxide greenhouse effect by 25-70 times [18]. The low biogas heating value, a large amount of CO_2, and impurities are the main obstacles that prevent its use in power generating plants [19, 20]. As far back as 1990, the Gas Research Institute (USA) has developed the combustion system for SMW furnaces that utilizes a small amount of natural gas and landfill gas cofiring [7]. This has allowed to control the dioxin and furan emissions and simultaneously stabilize the plant`s heat output. The research was conducted on two steam boilers that were burning waste with a consumption rate of 13,600 t/h and 90,7 t/h. Initially, the boilers were designed for RDF and coal cofiring but the operation has demonstrated the inefficiency of this approach due to abrupt and uncontrolled load fluctuations because of the RDF content variations up to 20-25%, ignition difficulties, and a large amount of hazardous emissions. The two-level burner was developed for RDF, landfill, and natural gas cofiring. The burners for natural gas were placed at the furnace`s lower part above the grate and provided 10% of the heat load. The burners for landfill or natural gas combustion were placed in the middle of the furnace which also provided 10% of the load (Fig. 1).

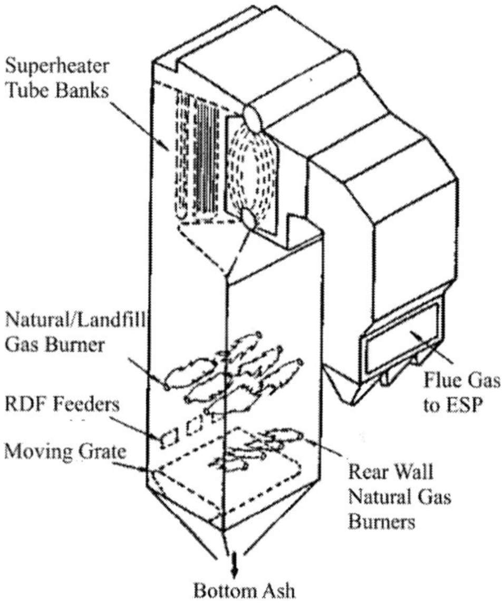

Figure 1. Gas cofiring system design [7].

Presently, the commercial furnaces designing is carried out, most of the time, using approximate empirical equations [21]. The paper [22] has conducted an analysis of the SMW combustion temperature at the low power water heating boiler with a grate in relation to SMW moisture content, excess air coefficient, the temperature of the supplied for combustion air and amount of the extracted heat and also the parameter range was determined that helps to maintain the required combustion temperature of 900…1100°C, which leads to a complete distraction of inside waste organic hazardous substances but prevents ash melting and formation of sintered mass. It was concluded that the necessary temperatures can be maintained within a wide range of the mode parameters and SMW moisture content. The paper [23] has analyzed biogas combustion and the conclusion was made that in order to maintain the combustion parameters it is necessary to use swirl burners.

The paper`s objective is the combustion calculation method development for waste (solid fuel) and biogas (gaseous fuel) cofiring, the examination of the combustion parameters for different levels of waste preparation (moisture content), methane content in biogas, and the amount of gaseous fuel supplied for burning.

Therefore, to determine the possibility of unprepared or prepared waste and biogas cofiring it is necessary to establish the possibility of obtaining the necessary temperatures.

2. Materials and Methods

2.1. Units of specific calculation and flow rate of fuels

It is suggested to carry out the calculation for per unit of generated during combustion total capacity. In this case, the mass specific flow rate of the solid fuel:

$$m_s = (1 - q_g)/Q_s, \text{ kg/MW}; \quad (1)$$

a volumetric specific flow rate of landfill gas:

$$v_g = q_g/Q_g, \text{ m}^3\text{/MW}, \quad (2)$$

where:
q_g – part of the energy released during landfill gas combustion in overall energy;
Q_s – a lower heating value of solid fuel, MW/kg;
Q_g – a lower heating value of landfill gas, MW/m^3.

If the solid fuel mass flow is given then the landfill gas volumetric flow rate can be calculated using the equation

$$v_g = (1 - m_s Q_s)/Q_g, \tag{3}$$

in the opposite case

$$m_s = (1 - v_g Q_g)/ Q_s. \tag{4}$$

2.2. *Calculation of moisture content in biogas*

Biogas has a high moisture content. This is due to the fact that it forms in an environment where the temperature is 40-60°C and moisture content is 40-50%. When biogas is cooled to the outside temperature the excessive moisture condenses and the water vapors in gas, normally, are in the saturated state when the relative humidity of biogas is $\varphi = 100\%$.

Gas moisture content d, kg/kg, is defined by the relation of a water vapor mass to a mass of dry part of gas. For 1 kg of dry gas and d kg of vapors the equations of state can be presented as follows:

$$p_w V = dR_w T, \tag{5}$$

$$p_{dg} V = R_{dg} T; \tag{6}$$

where:
V – volume;
T – temperature;
p_w and p_{dg} – water vapor and dry gas pressure, Pa, the sum of their values equal to the barometric pressure p_b, Pa;
R_w and R_{dg} – gas constants for water vapors (R_w = 461.520 kJ/(kg·K)) and dry gas, the gas constant of a gas mixture is defined on the volume fraction of components v_i and their gas constants R_i:

$$R_{dg} = \frac{1}{\Sigma \frac{v_i}{R_i}}. \tag{7}$$

When we divide the equation (5) by (6) we obtain the expression for gas moisture content determination with respect to the gas relative humidity $\varphi = p_w/p_s 100\%$:

$$d = \frac{p_{dg}}{p_b - p_{dg}} \frac{R_{dg}}{R_w} = \frac{\varphi p_s}{100 p_b - \varphi p_s} \cdot \frac{R_{dg}}{461.520}. \tag{8}$$

For the water vapor saturation pressure calculation p_s, Pa, in relation to temperature t, °C, the regression equation was obtained which is applicable within the temperature range –30…100°C, the mean square deviation is 138:

$$p_s = \exp((6.4 + 0.10128 \cdot t)/(1 + 0.00434 \cdot t)) \tag{9}$$

For the air with R = 287.520 kJ/(kg·K) the equation (8) can be written as follows

$$d_{\text{air}} = 0{,}622 \frac{\varphi p_s}{100 p_b - \varphi p_s}. \tag{10}$$

2.3. *Calculation of solid and gaseous fuel cofiring*

The waste chemical composition is presented by the following elements:

$$C^r, H^r, O^r, N^r, S^r, A^r, W^r, \text{mass \%}$$

Gaseous fuel content:

$$CH_4, C_2H_6, C_3H_8, C_4H_{10}, C_5H_{12}, CO_2, H_2\ H_2S, O_2, N_2, \text{volume \%}$$

The theoretical volume of air necessary for combustion, m^3/W:

$$V^0_{air} = (1 / 29.87) \cdot (2.67C^r + S^r + 8H^r - O^r) \cdot m_s + \\ + 0.0476 \cdot (0.5H_2 + 2CH_4 + 3.5C_2H_6 + 5C_3H_8 + 6.5C_4H_{10} + 8C_5H_{12} + \\ + 1.5H_2S - O_2) \cdot v_g \quad (11)$$

Actual volume of moist air necessary for combustion, m^3/MW:

$$V_{air} = \alpha\ V^0_{air}\ (1 + 1.293 / 0.805 \cdot d_{air}) \quad (12)$$

Combustion product composition, m^3/MW:

$$V_{cp\ CO_2} = 0.01\ (1.87C^r + 0.7S^r) \cdot m_s + \\ + 0.01(CO_2 + CH_4 + 2C_2H_6 + 3C_3H_8 + 4C_4H_{10} + 5C_5H_{12} + H_2S) \cdot v_g; \quad (13)$$

$$V_{cp\ H_2O} = (0.111H^r + 0.012W^r) \cdot m_s + \\ + 0.01(H_2 + 2CH_4 + 3C_2H_6 + 4C_3H_8 + 5C_4H_{10} + 6C_5H_{12} + H_2S) \cdot v_g + \\ + \alpha V^0_{air} \cdot 1.293 / 0.805 \cdot d_{air} + \rho_g / 0.805 \cdot d_g \cdot v_g \quad (14)$$

$$V_{cp\ N_2} = 0.79\alpha V^0_{air} + 0.008\ N^r \cdot m_s + 0.01N_2 \cdot v_g \quad (15)$$

$$V_{cp\ O_2} = 0.21(\alpha - 1)\ V^0_{air}, \quad (16)$$

where:
d_{air} – moisture content of air, kg of moisture/kg of dry air;
d_g – moisture content of gaseous fuel, kg of moisture/kg of gas;
ρ_g – gaseous fuel density, kg/m^3;
α – excess air coefficient.

The presented expressions (11) - (16) can be also applied in per unit of fuel calculation. In the calculation of the specific values for 1 kg of solid fuel the flow rates are as follows:

$$m_s\text{=1 kg;} \quad v_g = \frac{q_g}{1 - q_g} \frac{Q_s}{Q_g}, \quad \frac{\text{m}^3 \text{ of gas}}{\text{kg of solid fue'}} \quad (17)$$

and in calculation for 1 m^3 of biogas:

$$v_g=1\ m^3;\quad m_s=\frac{1-q_g}{q_g}\frac{Q_g}{Q_s},\ \frac{\text{kg of solid fuel}}{m^3\text{of gas}},\tag{18}$$

3. Results and Discussion

For the research, several average composition types of solid municipal waste with different moisture content were used: unprepared with the moisture content of 50%, partially dried waste with the moisture content of 30% and 10%, which are typical for RDF (Table 1). To recalculate composition with reference to the required moisture content the method presented in the paper [24] was used.

Table 1. Solid fuel characteristics

Item	Chemical composition, mass %							Heating value
	C^r	H^r	O^r	N^r	S^r	A^r	W^r	$Q^p_н$, kJ/kg
RDF	34.4	4.6	23.0	0.9	0.3	26.9	10.0	13 655
Partially dried MSW	26.8	3.6	17.9	0.7	0.2	20.9	30.0	10 063
Unprepared waste	19.1	2.5	12.8	0.5	0.1	15.0	50.0	6 470

In the paper [25], based on the examination of biogas release at the MSW landfills, it was established that the composition of landfill gas is very uneven. Therefore, three variants of additional gaseous fuel were considered: methane as the reference fuel, common biogas with a methane content of 60%, and biogas with a low methane content of 35% (Table 2). The biogas composition was given on the basis of mean values obtained using the instrumental measurements. Only three components were considered due to the insignificant presence of the rest in gas.

Table 2. Gaseous fuel characteristics.

Item	composition, volumetric %			Density ρ,	Heating value
	CH_4	CO_2	N_2	kg/m^3	$Q^p_н$, kJ/m^3
Methane	100	0	0	0.716	35 820
Common biogas	60	20	20	1.072	21 492
Biogas with low methane content	35.0	43	22	1.370	12 537

Biogas contains a large amount of moisture [26]. When biogas is cooled down from the temperature of the landfill body (40-60°C) to the atmosphere temperature, which takes place during pipeline transportation to the point of use, the excessive moisture turns into condensate and the relative humidity at the point of use becomes 100%. Figure 2 presents biogas moisture content influence on the theoretical combustion temperature which takes into account a dissociation of triatomic gases. The highest influence can be seen for biogas with low methane content due to its high density.

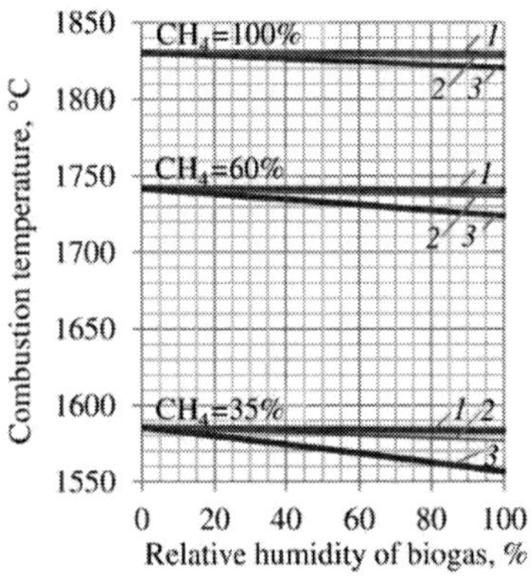

Figure 2. Theoretical combustion temperature for methane and biogas with different moisture content at α = 1.1 and gas temperature: *1* – 0°C; *2* – 20°C; 3 – 40°C.

When waste is dried in the rotary drier 6x20 m with a 1 t/h output capacity of waste with a moisture content of 50% the natural gas flow rate for drying agent heating is 50 m^3/h. Therefore, the heating value of the obtained fuel is 1.8 MW (6.5 GJ/h), and the heat energy of the gas used for drying is 0.5 MW (1.8 GJ/h). In view of this, waste recycling into RDF involves high energy consumption. However, the necessary combustion temperature (900-1000°C) can be achieved during the combustion of waste even with a moisture content of 50%.

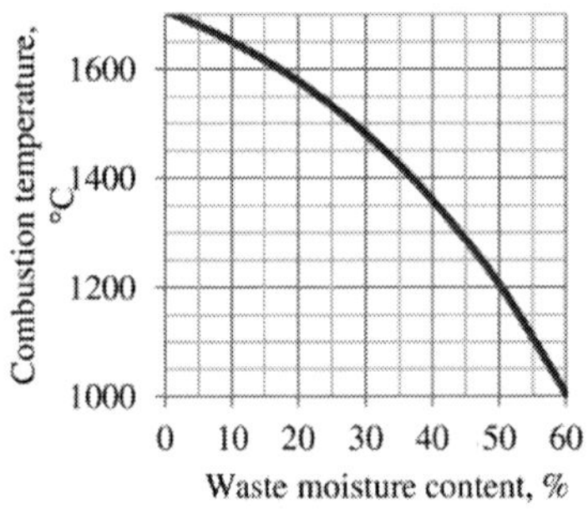

Figure 3. Theoretical combustion temperature dependence on MSW moisture content (excess air coefficient α = 1.25).

The study (Figure 4) has demonstrated that additional combustion of natural gas always increases combustion temperature. Common biogas increases combustion temperature for the unprepared waste with high moisture content, for the dried waste its combustion leads to a slight temperature decrease. Biogas with low methane content can increase the combustion temperature only for unprepared waste.

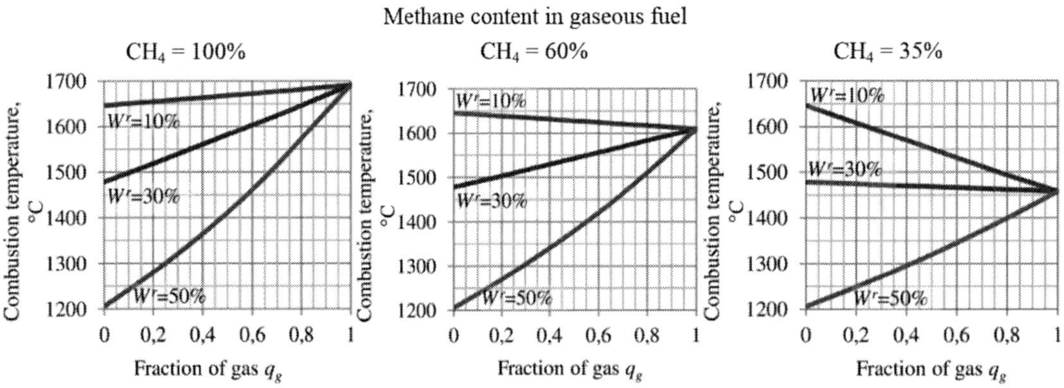

Figure 4. Theoretical combustion temperature of MSW with variable moisture content, natural gas, and biogas cofiring (excess air coefficient α = 1.25)

Since waste and biogas combustion at the excess air coefficient close to the lower limit can provoke underburning the study was carried out to determine at what values of α the combustion product temperature is 1000°C (Figure 5).

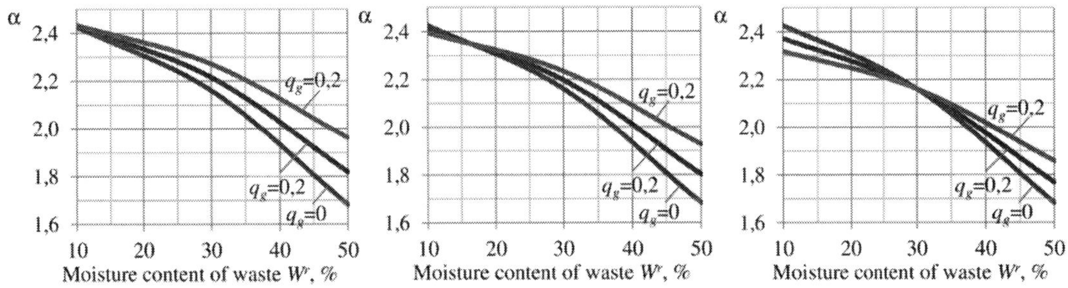

Figure 5. The excess air coefficient α that allows reaching the combustion temperature of 1000°C when waste moisture content variates.

According to the obtained dependences even for the moist waste, the required combustion temperature can be provided at an excess air coefficient of 1.6 without the additional combustion of gas.

4. Conclusion

The solid fuel and biogas cofiring allow the operative combustion process control and sustain combustion product temperature and heat output of units at a constant level. The additional input of biogas is useful for the combustion of unprepared waste with high moisture content, besides, it is possible to use common biogas as well as biogas with low methane content. The combustion process control should be based on the given combustion product temperature value by changing two parameters: landfill gas flow rate and flow rate of air supplied for combustion.

5. Acknowledgments

The work is realized in the framework of the Program of flagship university development on the base of the Belgorod State Technological University named after V.G. Shoukhov.

References

[1] Malinauskaite J et al 2017 Municipal solid waste management and waste-to-energy in the context of a circular economy and energy recycling in Europe *Energy* **141** 2013-44

[2] Fernández-González J M, Grindlay A L, Serrano-Bernardo F, Rodríguez-Rojas M I and Zamorano M 2017 Economic and environmental review of Waste-to-Energy systems for municipal solid waste management in medium and small municipalities *Waste Management* **67** 360-74

[3] Consonni S, Giugliano M and Grosso M 2005 Alternative strategies for energy recovery from municipal solid waste: Part A: Mass and energy balances *Waste Management* **25(2)** 123-35

[4] Consonni S, Giugliano M and Grosso M 2005 Alternative strategies for energy recovery from municipal solid waste: Part B: Emission and cost estimates *Waste Management* **25(2)** 137-48

[5] Edo M, Ortuño N, Persson P-E, Conesa J A and Jansson S 2018 Emissions of toxic pollutants from co-combustion of demolition and construction wood and household waste fuel blends *Chemosphere* **203** 506-13

[6] Chang Y-H, Chen W C and Chang N-B 1998 Comparative evaluation of RDF and MSW incineration *J. of Hazardous Materials* **58(1-3)** 33-45

[7] Moyeda D K, England G C, Seeker W R and Linz D G 1990 Design of advanced combustion systems for co-firing of natural gas with refuse derived fuels and landfill gas *In Proceedings of National Waste Processing Conference* (New York: ASME) pp 203–10
[8] Costa M, Dell'Isola M and Massarottica N 2012 Temperature and residence time of the combustion products in a waste-to-energy plant *Fuel* **102** 92-105
[9] Kilgroe J D 1989 Combustion control of trace organic air pollutants from municipal waste combustors *Environmental Impact Assessment Review* **9(3)** 199-222
[10] Hasselriis F 1987 Optimization of combustion conditions to minimize dioxin emissions *Waste Management & Research* **5(3)** 311-26
[11] Morcos V H 1989 Energy recovery from municipal solid waste incineration – A review *Heat Recovery Systems and CHP* **9(2)** 115-26
[12] Sondreal E A, Benson S A, Hurley J P, Mann M D, Pavlish J H, Swanson M L, Weber G F and Zygarlicke C J 2001 Review of advances in combustion technology and biomass cofiring *Fuel Processing Technology* **71(1-3)** 7-38
[13] Walter A, Souza M R and Faaij A 2005 Cofiring Biomass and Natural Gas - Boosting Power Production from Sugarcane Residues *Bioenergy - Realizing the Potential* (Oxford: Elsevier) pp 125-40
[14] Hupa M 2005 Interaction of fuels in co-firing in FBC *Fuel* **84(10)** 1312-9
[15] Moyeda D K, Seeker W R, England G C and Linz D G 1990 The formation and control of PCDD/PCDF from RDF-fired combustion systems *Chemosphere* **20(10-12)** 1817-24
[16] Werther J 207 Gaseous emissions from waste combustion *J. of Hazardous Materials* **144(3)** 604-13
[17] Striūgas N, Paulauskas R, Skvorčinskienė R and Lisauskas A 2020 Investigation of waste biogas flame stability under oxygen or hydrogen-enriched conditions *Energies* **13(18)** 4760
[18] Meyer L A et al 2014 *Climate Change 2014: Synthesis Report Contribution of Working Groups I, II and III to the Fifth Assessment Report of the Intergovernmental Panel on Climate Change* ed R K Pachauri and L A Meyer (Geneva: IPCC)
[19] Hosseini S E and Wahid M A 2014 Development of biogas combustion in combined heat and power generation *Renewable and Sustainable Energy Reviews* **40** 868-75
[20] Qin W, Egolfopoulos F N, Tsotsis T T 2001 Fundamental and environmental aspects of landfill gas utilization for power generation Chemical Engineering J. **82(1-3)** 157-72
[21] Kuznetsov V A and Trubaev P A 2018 Resources and Problems of the Mathematical Simulating Thermo-Technological Processes J. Phys.: Conf. Ser **1066** 012024
[22] Kornilova N V and Trubaev P A 2018 Analysis of MSW Combustion Temperature in a Hot Water Boiler with the Low-Capacity J. Phys.: Conf. Ser 1066 012003
[23] Suslov D Yu, Ramazanov R S and Shvydkaya M A 2018 To the Question of Biogas Combustion in Vortex Burners J. Phys.: Conf. Ser **1066** 012028
[24] Kornilova N V and Trubaev P A Method development for the thermotechnical calculations utilizing reduced characteristics of the RDF fuels *IOP Conf. Ser.: Mater. Sci. Eng.* **552** 012029
[25] Trubaev P A, Verevkin O V, Grishko B M, Tarasyuk P N, Shchekin I I, Suslov D Yu and Ramazanov R S 2018 Investigation of Landfill Gas Output from Municipal Solid Waste at the Polygon J. Phys.: Conf. Ser **1066** 012015
[26] Pogromsky A S and Anikanova T V 2018 The effect of long-term storage of electric steel smelting slags in dumps on their properties *Construction Materials and Products* **1(1)** 32-39

Development of a method for calculating a anaerobic digester with bubbling mixing

D Y Suslov

Department of heat and gas supply and ventilation, Belgorod State Technological University named after V.G. Shukhov, Kostyukov St., 46, Belgorod, 308012, Russia

E-mail: Suslov1687@mail.ru

Abstract. Currently, there is a development of biogas technologies designed to obtain an alternative source of biogas energy and utilize organic waste. When developing and designing biogas plants, an important task is to determine the design and technological parameters of the anaerobic digester in order to obtain the maximum biogas yield. For the research, an anaerobic digester equipped with a bubbling mixing system was used. The bubble tube is made in the form of a spiral and forms a cone with the base downward. A calculation method has been developed that makes it possible to determine the physical and chemical properties of the biomass, the geometric parameters of the digester, the specific and total volume of biogas produced, the design parameters of the digester heating system, the design and technological parameters of the bubbling-type mixing system. With the help of the developed methodology, the design and technological parameters of the anaerobic digester of the biogas station for the disposal of organic waste from a pig breeding enterprise and a meat processing plant were determined. As a result of calculations, it was established that the station contains 3 anaerobic digester tanks with a volume of 200 m3 each, utilizes 40 m3 of waste per day and produces 1600 m3 of biogas per day.

1. Introduction

Recently, biogas stations and biomethane plants have been introduced in the agro-industrial sector, the main task of which is the production of a renewable energy source - biomethane [1-4]. In addition, organic waste from various industries is utilized and biological fertilizers are obtained. Utilization of organic matter occurs in digesters due to the active vital activity of microorganisms that digest proteins, fats and carbohydrates. To provide a comfortable environment for reproduction and life, microorganisms need to maintain a uniform temperature level and distribution of bacteria throughout the biomass. For this, various mixing systems are used, while in biogas plants the mechanical method is most widely used [5-7]. However, during the operation of mechanical mixers, intensive mixing of the biomass occurs only in the horizontal plane in the area of movement of the mixer blades. And in the upper parts of the digester, little intensive mixing of the biomass is observed [8, 9].

The solution to these problems is possible through the use of bubble-type mixing systems [10, 11]. Recently, much attention has been paid to the bubbling method of mixing biomass, which, in contrast to the mechanical method, does not require the installation of rotating units and elements [12-15]. It is known that rotating assemblies are the most fragile element in the operation of mechanisms.

During the operation of bubbling mixing systems, an upward gas flow is formed, which carries away the biomass. This leads to the circulation of biomass from the lower part of the digester to the

IOP Conf. Series: Materials Science and Engineering 1089 (2021) 012034 doi:10.1088/1757-899X/1089/1/012034

upper part, as a result of which intensive and effective mixing of the entire volume of the digester occurs.

However, at present the process of obtaining biogas in digesters with a bubbling-type mixing system is poorly studied.

The purpose of this work is to develop a method for calculating the design and technological parameters of the digester equipped with heating and bubbling mixing systems.

2. Materials and Methods

A digester equipped with a bubbling-type mixing system has been developed. The model of the digester of the developed design is shown in Figure 1. The novelty of the digester lies in the design of the bubbling mixing system, consisting of a supply pipeline and a bubbling pipe. The bubble tube is made in the form of a vertical spiral with holes for gas outlet. The spiral is made in the form of a cone and is located with the base down (Figure 1, a). To maintain a stable temperature of the biomass in the digester, a heating system with hot water is adopted. The heating system contains a heat exchanger made in the form of a coil located on the outer wall of the digester (Figure 1, b).

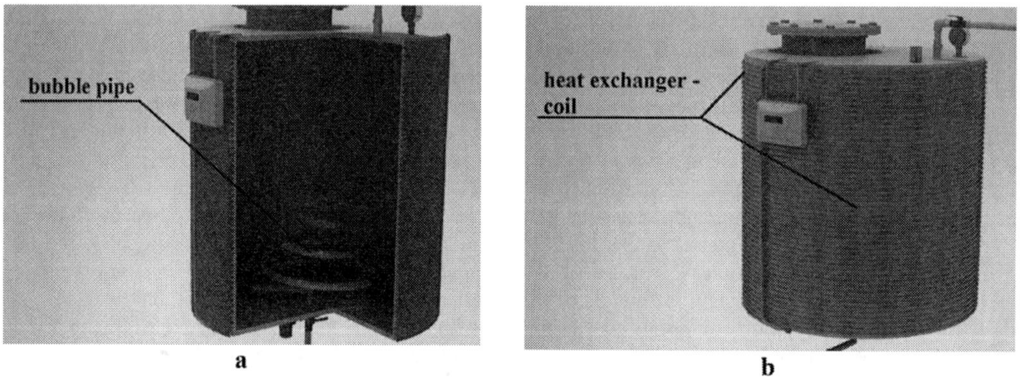

Figure 1. Digester with a bubble-type mixing system: a – the location of the bubbling pipeline in the digester; b – location of the heat exchanger - coil on the digester wall.

To determine the yield of biogas during anaerobic fermentation in a mesophilic temperature regime, the dependence of Chen and Hashimoto is used [16]. Chen and Hashimoto's model most fully describes the biogas production process depending on the most important parameters of the anaerobic fermentation process.

To determine the design and technological parameters of heating and mixing systems, dependencies describing hydrodynamic [17] and heat exchange processes are used.

3. Results and Discussion

On the basis of a set of theoretical and experimental studies of the biogas production process, a method for calculating a digester with bubbling mixing has been developed. The calculation provides for two tasks - "direct" and "reverse". The algorithm for calculating the digester is shown in Figure 2.

"Direct" task: the volume and the properties of the initial substrate for processing are set, it is necessary to determine the design dimensions of the digester, heating and bubbling mixing systems in order to obtain the maximum biogas yield.

"Inverse" task: the design dimensions of the digester and the technological parameters of the biomass in the digester are known; it is required to determine the efficiency of the digester - biogas output.

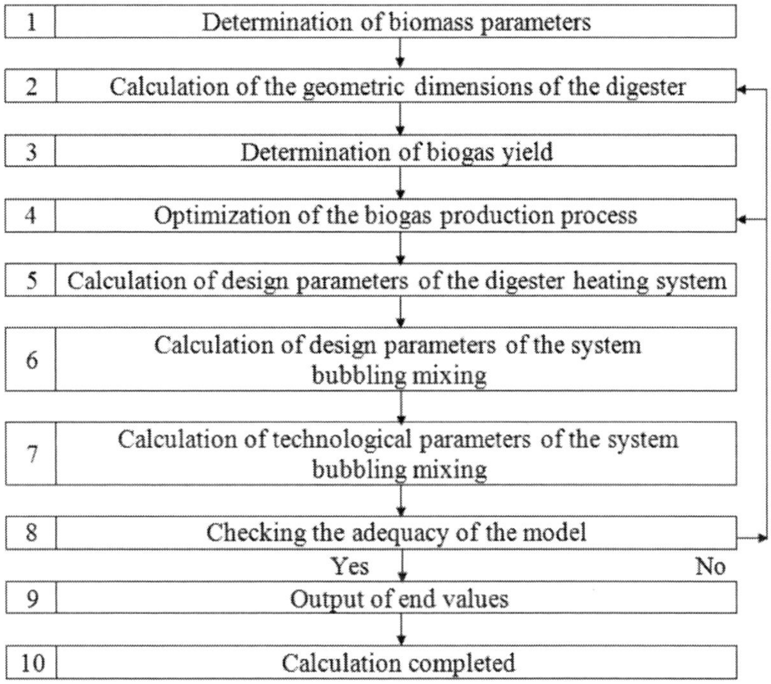

Figure 2. Algorithm of the methodology for calculating the design and technological parameters of the digester with bubbling mixing.

Consider the methodology for calculating the "direct" problem. In this case, the initial data are variable and constant factors.

Variable factors include: daily volume of the initial substrate (manure yield from the farm), $v_{s.s}$, м^3; moisture content of biomass in digester W,%; temperature of biomass in digester T, K; hydraulic retention time (duration of the fermentation process) τ, days.

Constant factors include:

- moisture content of the original substrate W_s, %;
- density of the dry part of the substrate ρ_d, kg / m^3;
- ash content of the dry part of the substrate A,%;
- gas velocity in the bubble pipe υ_{bp}, m/s;
- gas flow rate from the holes of the bubble pipe υ_{hol}, m/sec;
- gas consumption for mixing Q_g, m^3 / min;
- biogas pressure above biomass in the digester p_0, Pa;
- the density of the bubbling gas ρ_g, kg / m^3;
- kinematic viscosity of biogas γ_g, m^2 / s;
- coolant speed ω_h, m/sec.

3.1. Calculation of biomass parameters.

The density of biomass depends on the moisture content and density of the dry part of the original substrate and is determined by the expression, kg / m^3:

$$\rho = W\rho_V + (1-W)\rho_d, \quad (1)$$

where ρ_v is the density of water, kg / m^3.

The viscosity of biomass depends on the viscosity of water and moisture, Pa s:

$$\mu = \mu_V(1 + 2,5(1 - W)), \tag{2}$$

where μ_v – dynamic viscosity of water, Pa·s.

The concentration of the substrate depends on its composition and the content of volatile solids (VS) in the total biomass, kg VS / m^3:

$$S = \rho(100 - W) \times (100 - A) \times 10^{-4} \tag{3}$$

3.2. Calculation of the geometric parameters of the digester.

The volume of the digester depends on the volume of the substrate and the technological parameters of the biogas production process:

$$v = v_s + v_v, \tag{4}$$

where:
v_s is the volume of the substrate in the digester, m^3;
v_v is the volume of water required to achieve the required biomass moisture, m^3.

The volume of substrate in the digester depends on the hydraulic retention time of the biomass in the digester and the daily volume of the initial substrate:

$$v_s = \tau \cdot v_{s.s} \tag{5}$$

The volume of water added to the digester to achieve optimum moisture content can be determined by the formula:

$$v_v = m_s\left(\frac{W - W_s}{100 - W}\right), \tag{6}$$

where m_s – substrate weight, kg.

3.3. Determination of biogas yield.

The main parameters to be determined are the maximum specific growth rate of microorganisms, the biogas yield rate and the total biogas yield.

Maximum specific growth rate of microorganisms, day^{-1}:

$$\mu_m = \mu_0 + \mu_1 \cdot T + \mu_2 \cdot T^2, \tag{7}$$

where μ_0, μ_1, μ_2 are temperature coefficients.

The biogas release rate is determined by the equation of Chen and Hashimoto, m^3 / day • m^3:

$$V = \frac{B_0 \cdot S}{\tau} \cdot \left(1 - \frac{K}{\mu_m \cdot \tau - 1 + K}\right), \tag{8}$$

where:
B_0 is the limiting biogas yield from a unit of volatile solids of a given composition with an infinite duration of the process, m^3 / kg;
K is the kinetic parameter.

The total biogas yield of the digester is determined by the expression, m^3 / kg VS:

$$V_{tot} = V \cdot v \tag{9}$$

Further, *the biogas production process is optimized* with the introduction of variable parameters. At the same time, such values of the variable factors are determined at which the biogas yield will be maximum.

3.4. Calculation of the design parameters of the heating system.

The purpose of calculating the design parameters of the heating system is to determine the total length and diameter of the heat exchange pipeline - the coil.

The heat load of the digester is determined from the heat balance, W:

$$Q_h = Q_{out} + Q_{out}^{g} + Q_{out}^{bg} + Q_l - Q_{in} - Q_{in}^{bg} - Q_r , \tag{10}$$

where:
Q_{out} is the amount of heat leaving with the processed biomass, W;
Q_{out}^{g} is the amount of heat leaving with the produced biogas, W;
Q_{out}^{bg} is the amount of heat leaving with bubbling biogas, W;
Q_l is the amount of heat lost to the environment, W;
Q_{in} is the amount of heat supplied with the original substrate, W;
Q_{in}^{bg} is the amount of heat supplied with bubbling biogas, W;
Q_r is the amount of heat released in the exothermic reaction of anaerobic fermentation, W.

The area of the heat transfer surface is determined from the heat transfer equation, m^2:

$$F_h = \frac{Q_h}{k_h \cdot \Delta T}, \tag{11}$$

where k_h is the coefficient of heat transfer from the coolant through the coil wall to the biomass, W / (m^2·K); ΔT - temperature difference between hot and cold heat carrier, K.

The coefficient of heat transfer from the heat carrier through the walls of the heat exchanger and the digester to the biomass depends on the material of the digester and the mode of movement of the coolant and biomass:

$$k_h = \frac{1}{\dfrac{1}{\alpha_1} + \sum r_w + \dfrac{1}{\alpha_2}}, \tag{12}$$

where:
α_1 is the coefficient of heat transfer from the coolant to the coil wall, W / (m^2·K);
α_2 is the coefficient of heat transfer from the wall surface to the biomass, W / (m^2·K);
Σr_w is the sum of thermal resistances of all layers that make up the wall, m^2·K / W.

The coolant flow rate is determined by the expression, m^3 / s:

$$V_h = \frac{Q_h}{c_h \cdot (T_1 - T_2)}, \tag{13}$$

where:
c_h - specific heat capacity of the hot coolant, J / (kg · K);
T_1 - initial temperature of the coolant, K;
T_2 is the final temperature of the coolant, K.

The cross-sectional area of the heat exchange pipeline can be determined from the flow continuity equation, m^2:

$$S_c = \frac{V_h}{\omega_h}, \tag{14}$$

where ω_h is the speed of the coolant, m/s,

IOP Conf. Series: Materials Science and Engineering 1089 (2021) 012034 doi:10.1088/1757-899X/1089/1/012034

whence the diameter of the heat exchange pipeline will be, m:

$$D_c = \sqrt{\frac{4 \cdot S_c}{\pi}} \tag{15}$$

and the total length of the heat exchange pipe of the coil will be, m:

$$L_c = \frac{F_h}{\pi \cdot D_c} \tag{16}$$

The length of one coil of the heat exchanger is determined by the expression, m:

$$l_{c1} = \sqrt{\frac{\pi^2 \cdot D_M^2 \cdot L_c^2}{L_c^2 - H_M^2}}, \tag{17}$$

where:
D_M is the diameter of the digester, m;
H_M - digester height, m.
Distance between axes of adjacent coil turns, m:

$$h = \frac{H_M \cdot l_{c1}}{L_c} \tag{18}$$

Number of coil turns:

$$n_c = \frac{L_c}{l_{c1}} \tag{19}$$

3.5. Calculation of the design parameters of the bubbling mixing system.
The diameter of the bubble tube depends on the speed and flow rate of the bubbled gas, m:

$$d_{bp} = \sqrt{\frac{4 \cdot Q_g}{\pi \upsilon_{bp}}} \tag{20}$$

Length of the bubbling pipeline, m:

$$l_{bp} = \frac{\pi}{h_h}\left(R_1\sqrt{h^2 + R_1^2} - R_2\sqrt{h^2 + R_2^2} - h^2 \ln\frac{R_2 + \sqrt{h^2 + R_2^2}}{R_1 + \sqrt{h^2 + R_1^2}} \right), \tag{21}$$

where:
R_1 - radius of the largest turn of the bubble pipe, m;
R_2 - radius of the smallest turn of the bubble pipe, m;
h_h - horizontal step between the axes of adjacent turns, m.
Horizontal step between the axes of adjacent turns, m:

$$h_h = \frac{R_1 - R_2}{n_{t.bp}}, \tag{22}$$

where $n_{t.bp}$ – the number of turns of the bubble pipe.
Vertical pitch between the axes of adjacent turns, m:

$$h_v = \frac{H_{bp}}{n_{t.bp}}, \tag{23}$$

where H_{bp} – height of the bubble pipe made in the form of a spiral, m.

IOP Conf. Series: Materials Science and Engineering 1089 (2021) 012034 doi:10.1088/1757-899X/1089/1/012034

Number of holes in the bubble tube for gas outlet:

$$n_{hol} = \frac{l_{bp}}{h_{hol}}, \tag{24}$$

where h_{hol} – step between adjacent holes, m.

Diameter of one hole, m:

$$d_{hol} = \sqrt{\frac{4 \cdot Q_g}{\pi n_{hol} \upsilon_{hol}}}, \tag{25}$$

where υ_{hol} – gas velocity from the holes of the bubble pipe, m/s.

3.6. Calculation of technological parameters of the bubbling mixing system.

The purpose of the technological calculation of the mixing system is to determine the pressure in the holes of the bubbling pipeline.

The gas pressure in the holes of the bubbling pipeline is determined from the expression, Pa:

$$p_{hol} = \frac{\rho_g \cdot \upsilon_{hol}^2}{2} \tag{26}$$

One of the main tasks in determining the required pressure is to determine the pressure loss in the bubbling pipeline.

Taking into account the design of the bubbling pipeline made in the form of a vertical spiral in the form of a cone, the pressure loss is determined from the equation, Pa:

$$\Delta p_{bp} = \left(\frac{l_{bp}}{d_{bp}} \left[\frac{1}{3} \lambda \left(1 + \frac{\upsilon_e}{\upsilon_s} + \left(\frac{\upsilon_e}{\upsilon_s} \right)^2 \right) + 4 \frac{S_{bp} - S_{bp} \frac{\upsilon_e}{\upsilon_s}}{S_w} (1 + \frac{\upsilon_e}{\upsilon_s}) \right] + \right.$$
$$\left. f(\frac{\upsilon_{bp}}{\upsilon_{hol}}) + 0{,}0175 \cdot \delta_{sl} \cdot \frac{5}{\mathrm{Re}^{0.45}} \cdot \left(\frac{d_{bp}}{2 \cdot r_{t.bp}^{av}} \right)^{0.275} \frac{r_{t.bp}^{av}}{d_{bp}} 4 \cdot n_{t.bp} \psi \right) \frac{\rho_g \cdot \upsilon_{bp}^{\ 2}}{2}, \tag{27}$$

where:

λ is the coefficient of friction;

υ – flow rate at the beginning of the pipeline, m/s; υ_e – flow velocity at the end of the pipeline, m/s;

S_{bp} – pipeline cross-sectional area, m^2;

S_w – is the area of the pipeline walls, m^2;

δ_{sl} – angle of smooth rotation, °;

$r_{t.bp}{}^{av}$ – average radius of the turns of the bubbling pipeline, m;

ψ – is a coefficient that takes into account the mutual influence of smooth turns.

Then, the adequacy of the results obtained is checked and the final values are displayed.

The "Inverse" task is similar to the method for calculating the direct problem, but the design parameters of the digester are not calculated.

4. Conclusion

A method for calculating the structural and technological parameters of a digester of an original design equipped with a heating system and a bubbling mixing system has been developed. The technique contains direct and inverse tasks. The direct task is to determine the parameters of the digester in order to obtain the maximum yield of biogas from a certain volume of the initial substrate. The inverse task is to determine the biogas yield with known parameters of the digester and biomass.

The technique makes it possible to determine the physical and chemical properties of the biomass, the geometrical parameters of the digester, the volume of biogas produced, the design parameters of the digester heating system, the design and technological parameters of the bubbling mixing system.

The developed methodology was used to design and calculate a biogas station with a total volume of digesters of 600 m^3. The biogas station processes 40 m3 / day of organic waste from the pig breeding complex and meat processing plant, and produces 1600 m3 of biogas per day.

5. Acknowledgments

The study was carried out in the framework of the scholarship of the President of the Russian Federation SP-180.2019.1.

References

[1] Sokolov A V, Belyaeva N M 2011 *E`ffektivnost` vnedreniya biogazovy`kh ustanovok v sel`skokhozyajstvenny`kh organizacziyakh Kostromskoj oblasti Razvitie APK* [The effectiveness of the introduction of biogas plants in agricultural organizations of the Kostroma region Development of the agro-industrial complex] *Regional'naya ekonomika: teoriya i praktika* **43(226)** 38-41 [In Russian].

[2] Scarlat N, Dallemand J-F and Fahl F 2018 Biogas: Developments and perspectives in Europe *Renewable Energy* **129(A)** 457-72

[3] Chinnici G, Selvaggi R, D'Amico M and Pecorino B 2018 Assessment of the potential energy supply and biomethane from the anaerobic digestion of agro-food feedstocks in Sicily *Renewable and Sustainable Energy Reviews* **82(1)** 6-13

[4] Siegmeier T, Blumenstein B and Moller D 2015 Farm biogas production in organic agriculture: System implications *Agricultural Systems* **139** 196-209

[5] Karim K, Hoffmann R, Klasson K T and Al-Dahhan M H 2005 Anaerobic digestion of animal waste: Effect of mixing *Water Research* **39** 3597-606

[6] Jegede A O, Zeeman G and Bruning H 2019 Effect of Mixing Regimes on Cow Manure Digestion in Impeller Mixed, Unmixed and Chinese Dome Digesters *Energies* **12(13)** 2540

[7] Kalia A K, Singh S P 2001 Effect of Mixing Digested Slurry on the Rate of Biogas Production from Dairy Manure in Batch Fermenter *Energy Sources* **23** 711-5

[8] Suslov D Yu and Temnikov D O 2016 Komp'yuternoe modelirovanie peremeshivaniya substrata v biogazovyh ustanovkah [Computer simulation of substrate mixing in biogas plants] *Vestnik BGTU im V. G. Shukhova* [Bulletin of BSTU named after V G Shukhov] **4** 153-6 [In Russian].

[9] Bergamo U, Viccione G, Coppola S, Landi A, Meda A and Gualtieri C 2020 Analysis of anaerobic digester mixing: comparison of long shafted paddle mixing vs gas mixing. *Water Science and Technology* **81(7)** 1406-19

[10] Onychin E M, Medyakov A A, Kamenskih A D and Anisimov P N 2012 E`ksperimental`ny`e issledovaniya malogo bioreaktora s sistemoj barbotazhnogo peremeshivaniya i kataliticheskogo obogreva [An experimental research of the small bioreactor with the bubble mixing system and catalytic heating] *Nauchny`j zhurnal KubGAU* **78(04)** 1-14 [In Russian]

[11] Kuschev L A, Suslov D Yu and Alifanova A I 2017 Computer modeling movement of biomass in the bioreactors with bubbling mixing *IOP Conf. Series: Earth and Environ. Sci.* **52** 012025

[12] Wu B 2010 CFD simulation of gas and non-Newtonian fluid two-phase flow in anaerobic digesters *Water research* **44** 3861-74

[13] Wu B 2014 CFD simulation of gas mixing in anaerobic digesters *Computers and Electronics in Agriculture* **109** 278-86

IOP Conf. Series: Materials Science and Engineering 1089 (2021) 012034 doi:10.1088/1757-899X/1089/1/012034

[14] Dapelo D and Bridgeman J 2016 Mixing optimization of gas-mixed anaerobic digestion using CFD *European Biosolids and Organic Resources Conf.* (Edinburgh)
[15] Dapelo D 2016 *Gas mixing in anaerobic digestion: PhD thesis* (Birmingham: University of Birmingham)
[16] Hashimoto A G 1982 Methane from Cattle Waste: Effects of Temperature Hydraulic Retention Time and Influent Substrate Concentration on Kinetic Parameter *Biotechnology and Bioengineering* **24** 2039-52
[17] Dakhno A V, Smirnov Yu P, Mashkov A V and Potapenko O V 2018 To an assessment of size of diffusive losses of coolant through a rubber camera – cover *Chemical Bulletin* **1(1)** 18-26.

Forecasting of the group of power system's technical condition composite index and demand for heat energy by office buildings using an artificial neural network

S V Guzhov, T Y Andreeva and A P Sofronitsky

Department of Heat and Mass Transfer Processes and Installations, National Re-search University MPEI, Krasnokazarmennaya St., 14, Moscow, 111250, Russia

E-mail: GuzhovSV@yandex.ru

Abstract. Support of a sustained reduction of energy resource consumption by implementing advanced energy-efficient technologies, including alternative and renewable energy sources, requires a well-reasoned selection of the cost-effective measures so it can be replicated at the «green construction» sites. One of the basic approaches in forming of the present-day policy in the ecofriendly and energy-saving trends became a scaling effect of what has been achieved from the «green construction» technologies implementation at similar sites in Moscow. The actuality and scientific novelty of this paper lie in the development of the evaluation algorithm of energy and as a result of echo-economic efficiency of different measures for energy-saving, the adaptation of the forecasting methodology for building's technical condition index application, review of the alternative and renewable energy source technologies implementation at the «green construction» sites in the Moscow region, analysis of the technical and ecological efficiency actual values of the utilized energy-saving methods and technologies. Monitoring and dispatching of a large number of separate power plants require special ways of collecting, processing, and analyzing information about the condition of water, heat, fuel, and electricity supply systems to ensure reliability and stability of the equipment functioning. Failures necessitate maintenance works at facilities by way of equipment repair as well as replacement. The consumers may impose penalties on the heat supply company for violating heat energy supply obligations as well as on the building's owner for incorrect equipment maintenance. The objective of this paper is to determine an approach that helps to develop a software program, based on the neural network programming, to calculate the forecasting composite index of building's energy system, «green construction» facilities ecological and economic efficiency examination and feasibility evaluation of their further replication in Moscow region.
Keywords: composite index of technical condition, energy efficiency, renewable energy sources, green construction, software product, deterministic approach, statistical analysis, artificial neural network, office building, forecasting, demand for heat energy, heat energy consumption specific value.

1. Introduction

In order to improve the reliability of the heat substations and heat networks equipment a set of administrative and technical measures has been implemented:

- technical diagnosis and equipment examination including safety review procedure;
- record-keeping of the heat supply system damages;
- development, approval, and fulfillment of the repair schedules;

- maintaining a ledger of the finished repairs that restore performance characteristics during upgrading, modernization, project's capital construction;
- pipeline internal metal corrosion rate monitoring;
- different tests of heat supply networks;
- back-up of crucial equipment and networks;
- outdated equipment replacement.

Since equipment of heat substations and networks inside buildings are commonly available then collection, generalization, and evaluation of data on their current technical status and working conditions is a vital task for the group of similar buildings in Moscow agglomeration [1]. Monitoring should be done in relation to the reliability indexes and functioning stability of main and supporting equipment as long as the routine maintenance is done on a regular basis.

Most foreign information sources focused on the condition index and don't take into account the technical condition. The equipment could be new but without proper maintenance, it will work with parameters deviation. On the other hand, worn-out equipment in a skilled hand can be reliable and effective. The heat supply specific is in fact that preparation for the heating season in Moscow is quite long and usually takes up to 5 months. Quality of this preparation and the following work of stuff is as important as equipment's upgrade; therefore, the index of performance quality should be taken into consideration.

The «green construction» facilities include buildings that have certificates from international systems such as BREEAM, LEED, DGNB, and so on. A total of 195 buildings have been analyzed (Figure 1).

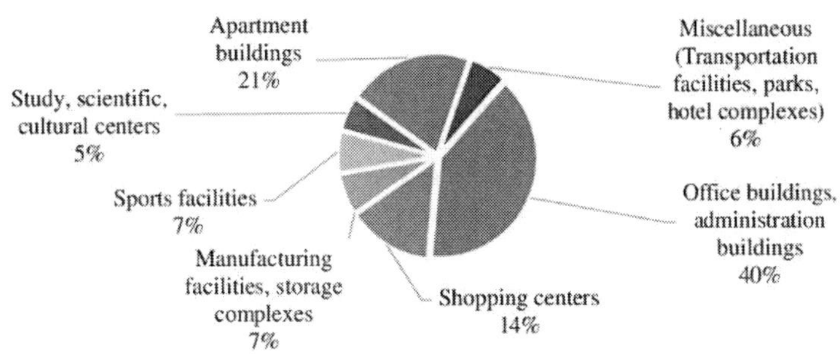

Figure 1. Structure of green buildings in Moscow.

The selected and analyzed facilities are mostly modern, recently built or redeveloped buildings and facilities that are originally designed and built with reference to [2, 3]:

- eco-management requirements;
- selection of suitable site and infrastructure;
- orientation toward sustainable use of water;
- selection of architectural and engineering designs to optimize lighting, minimize local heating and so on;
- implementation of energy conservation equipment and energy-efficient technologies that ensure environmental safety [4];
- use of eco-friendly construction and finishing materials, the possibility of separate waste collection;
- providing living environmental comfort and so on.

The preliminary data analysis allowed to select facilities that have installed alternative and renewable energy sources. Solar panels and photovoltaic cells [5] are installed at 33 facilities (Figure 2). The number of facilities that utilize new and renewable energy sources (NRES) is presented in Figure 3.

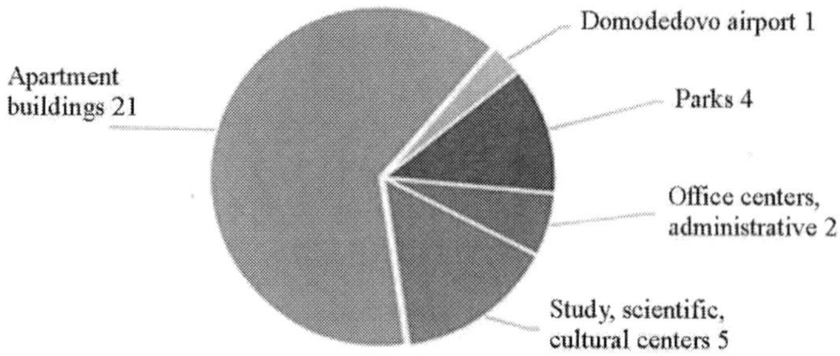

Figure 2. Facilities which utilize solar sources of heat and electricity.

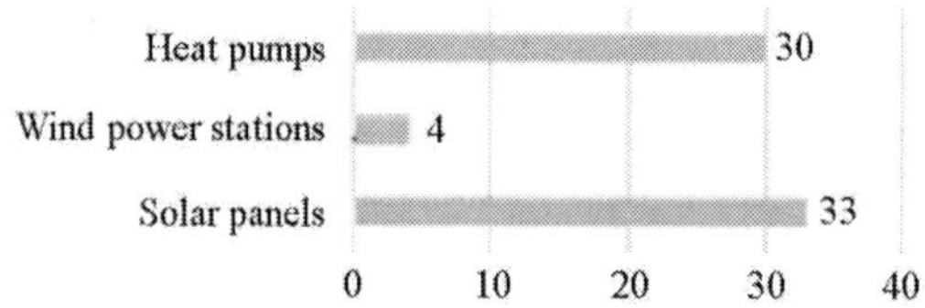

Figure 3. Number of facilities which utilize NRES.

2. Materials and Methods

Generally speaking, energy conservation measures in the system of electrical power equipment of an office building [6] can include (Figure 4):

a) rpm control for engines with unbalanced load;

b) units for reactive power compensation;

c) units for normalization of electrical power quality indexes;

d) replacement of equipment for electrothermal food heating, snow melting, supply air heating, pavement, and floor heating, and so on for equipment with alternative operating principle;

In the electrical lighting system:

a) sources of light with a luminous efficacy of no less than 100 lm/W;

b) automation of electrical lighting systems;

c) installation of solar tubes and optical fibers in the Moscow region can save 30-70% of electric power that is used for artificial lighting.

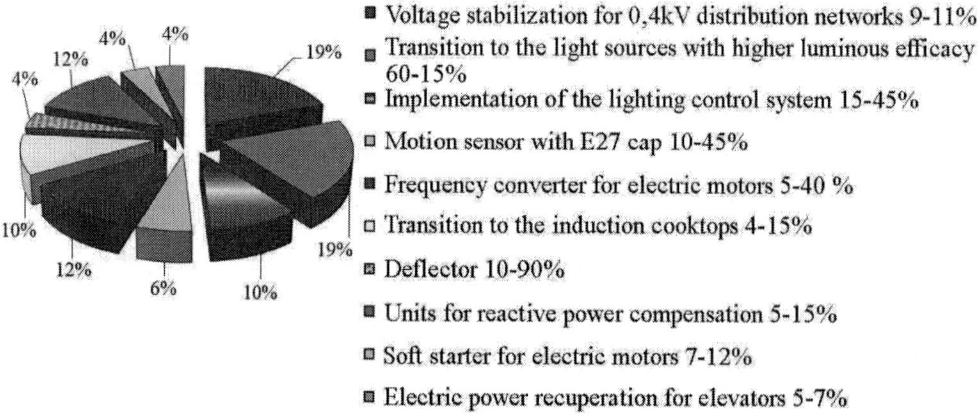

Figure 4. Composition and contribution of technologies that save an electrical power.

In the heat supply system (Figure 5):

a) automated terminals for measurement and control of the heat transfer agent temperature in the heating system or individual heat substations;

b) units for exhaust air low-grade heat utilization;

c) «air-to-air» heat pumps including with an intermediate heat transfer agent.

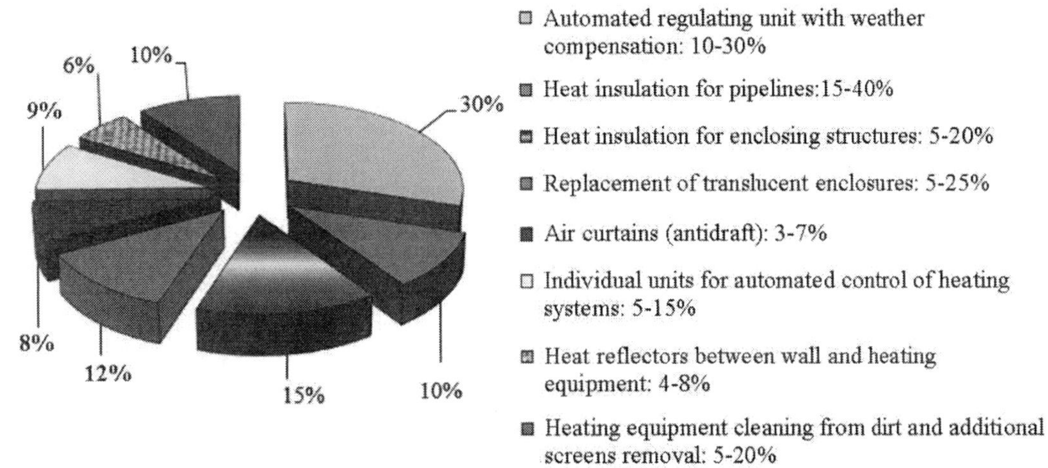

Figure 5. Composition and contribution of technologies that save heat energy.

In the water supply and sewerage systems (Figure 6):

a) effective water intake equipment which allows saving 15-20% of water resources;

b) push faucets for handwashing facilities and sensors for urinals which decrease water consumption by 4-6 times;

c) faucet aerators which allow to save water flow up to 6 l/min and decrease consumption of cold water for 30-35%, hot – for 15-20%;

d) contact-free faucets and so on.

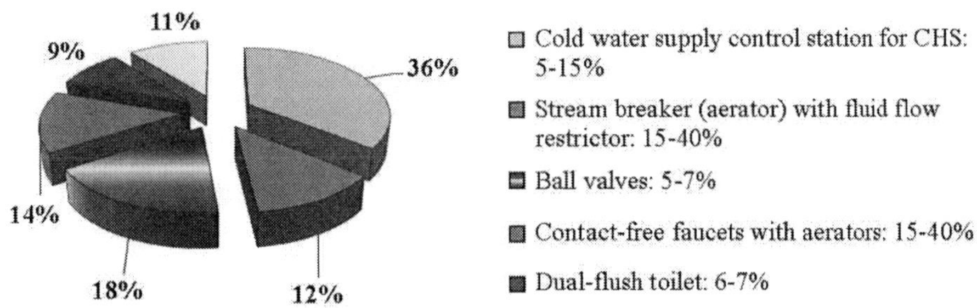

Figure 6. Composition and contribution of technologies that save water resources.

The conducted analysis clearly demonstrated a considerable variety of energy saving measures. Buildings have different functional use which means that one type of energy-saving measure gives different outcomes for different building types and sizes even of the same type. Therefore, for the building that is already in operation and had been constructed with specific energy-saving features, it is very difficult to pinpoint a part that provides savings using already implemented measures. Moreover, the operational facility doesn't allow to conduct completely controlled experiments. From this perspective, the only possible way is a deterministic approach that utilizes retrospective statistical data on a building's condition from its previous history.

This paper suggests the assessment process of the composite index of technical condition (CITC) of the technological equipment at the being examined facilities conduct according to the Order of the Ministry of Energy of the Russian Federation of 26.07.2017 No. 676. The equipment's technical condition assessment is conducted by comparison of the actual parameter values of the component's technical condition against the maximum allowable values, and also compliance with the requirements that are stated by the regulatory and technical documentation and (or) design (project) documentation of the manufacturers and then CITC of the components and equipment determination in general [7]. CITC can take values on a 100-point scale, where 0 – the lowest value, 100 – is the best value. The ITC calculation is done by the formula:

$$ITC = \sum(P_i \cdot ITC_i)/\sum(P_i), \quad (1)$$

where:
ITC_i – ITC for i-th equipment;
P_i – heat capacity of the equipment.

ITC determination for i-th equipment can be done by the empirical method. The conducted calculations demonstrate a possibility of automated calculations by using software which will allow calculating the forecasting composite index of technical condition with the accuracy of no less than 80%, the number of learning rates no more than 200, training time no more than 90 seconds.

Since equipment of heat substations and networks inside the building are commonly available then collection, generalization, and evaluation of data on the current technical status and working conditions is a vital task for a group of similar buildings in Moscow agglomeration. Monitoring should be done in relation to reliability indexes [8], efficiency and functioning stability of main and supporting equipment as long as the routine maintenance is done on a regular basis. Forecast of the CITC value change (Figure 7), during the transition from the current state (point 1) to the forecasted (point 2), should be done considering a change of all three scales: efficiency [9-11], reliability, stability.

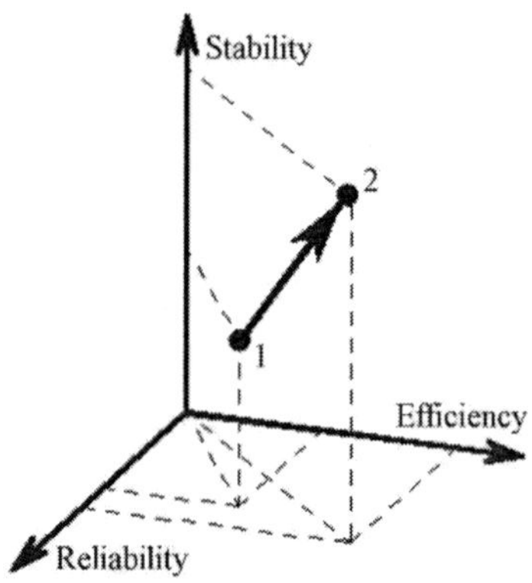

Figure 7. Model of CITC change forecasting.

For the office buildings which were constructed using an individual design or based on a significantly repurposed building, for example, from an industrial to office, calculation of the saving that was achieved from implementation at the construction or operation phase is a difficult and multifaceted problem. Let's take a look at the group of buildings that owns JCS «O1 Standard».

The analysis of an energy-saving equipment implementation that results in increased energy resource utilization efficiency, decreased environmental impact [12] and improved comfort in buildings showed the following. At the facilities certified to the international standards the most common measures are:

Energy saving in the heat supply systems:

1) heating system independent with IHS [13]. Mainly hot-air-heating with recuperation and preliminary air preparation. The northern side of the building is heated by the radiator hot water heating (for example, for the office building Ducat Palace 3), power of which is 15-20% of the whole heating. The rest of the load is covered by the underfloor water convectors (for the office building Lighthouse);

2) the ventilation system is general by using supply and exhaust units with heating from IHS and cooling by chillers with mechanical initiation;

3) the air conditioning system is central with chillers. Ducat Palace 3 office building uses the air cooler;

4) the building uses a general system of heat from exhaust air utilization. Some supply and exhaust units are using a system of recuperation with intermediate heat medium;

5) parking lot exhaust system is combined with the degassing system and made with zone intake and activated by the car exhaust gases;

6) parking lot heating is done by the supply air heating;

7) high-efficiency insulation of the air ducts and pipelines to minimize heat exchange with the environment;

8) use of the air conditioning system which controls the temperature in rooms individually;

9) use of the «Freecooling» system for air conditioning utilizing refrigeration units;

10) use of individual controllers which optimize heat consumption for room heating and ventilation;

11) for the purpose of utilizing sun radiation in building's heat balance the translucent enclosing structures are made translucent with increased heat-shielding values.

12) Energy-saving methods for light pollution reduction:

13) lighting equipment shouldn't be pointed toward the window;

14) street lamps installed along the building's front side and in the surrounding area have reflectors pointed down;

15) every leaseholder uses window blinds of a certain model so the light doesn't radiate outside during evening and night time.

16) Methods for energy consumption reduction:

17) it is necessary to use LED light sources with a color temperature of 4000 K or similar fluorescent lamps;

18) where safety rules allow the internal doors and non-load bearing enclosing structures in buildings should be made of translucent materials to use natural light as much as possible;

19) it is necessary to use at the building's reception desks and in the rooms of every leaseholder at least one lamp that is on all the time which allows to partially dim the corridor lights considering the presence of the translucent doors;

20) general lighting at the front desk during the off hours (from 8 p.m. to 7 a.m.) should be off;

21) the outside facade lighting at the office center White Stone had been removed;

22) the outdoor and indoor lighting, except lights to assist moving cars, facade lighting, emergency lights, can be automatically shut down by the lighting automatic control software;

23) the non-essential rooms and corridors have motion and presence detectors and also dusk-to-dawn sensors installed;

24) some supply and exhaust air systems have the rpm frequency control installed on the fan's electric motors (office center iCUBE);

25) installation of the electric power recuperation systems in the elevators;

26) implementation of the elevator's movement rationalization system.

27) Methods for water consumption reduction:

28) toilets are equipped with the dual-flush system: less than 6 l (large flush) and less than 3 l (small flush);

29) public bathrooms have aerators installed in the faucets which limit flow rate: for example, less the 5 l/min (kitchen faucet) and less than 4,5 l/min (bathroom faucet);

30) urinals which don't require flush after every use or with limited flow rate, less than 0,3 l/sec;

31) use of showerheads with a flow rate of less than 6 l/min;

32) installation of technical record-keeping equipment at the leaseholders and also at the consumers that have high water consumption;

33) dishwashing machines with a flow rate of less than 13 l per cycle;

34) washing machines with a flow rate of less than 40 l per cycle;

35) rainwater is not used for watering of plants on the surrounding territory;

36) automatic disablers of water flow in bathrooms are installed.

37) Waste collection:

38) availability of separate waste collection inside the building and nearby, availability of electric batteries collection, waste compactor, availability of the medical waste collection in view of the pandemic;

39) implementation of condenser water collection system;

40) the building lacks a snow melting system that works from the inhouse heat emitters such as chiller exhausts, general ventilation, sewage pipelines, and so on;

41) the building lacks a waste treatment system and equipment for industrial water reuse;

42) advanced building management systems (BMS) with a high level of automation had been implemented in the building: control of temperature, humidity, and CO2 content in the inhouse air.

From the point of view of the acting regulatory documentation on the engineering of a new or under reconstruction buildings, there are approved specific values of energy consumption on which the design documentation development is based (Table 1).

Table 1. Regulatory specific values of the consumed electrical energy, heat energy, and water resources in Moscow city.

No	Index of specific values	Value
	Electrical energy	
1.	kW/sq m (RD 34.20.185-94)	0.109
2.	kW/per person (RD 34.20.185-94)	~0.270
3.	kW/per flat (РД RD 34.20.185-94)	0.620125
4.	For a single person living in a flat with a gas stove, kWh/per person (Moscow GD of 20.12.1994 N 1161)	50
5.	For a single person living in a flat with an electrical stove, kWh/per person (Moscow GD of 20.12.1994 N 1161)	80
6.	For a family living in a flat with a gas stove, kWh/per person (Moscow GD of 20.12.1994 N 1161)	45
7.	For a family living in a flat with an electrical stove, kWh/per person (Moscow GD of 20.12.1994 N 1161)	70
	Heat energy	
8.	Heat energy consumption norm for housing unit heating, Gcal for total sq m of housing	0.016
9.	Heat energy consumption norm for water heating	0.294
10.	Building's specific annual heat consumption norm, Gcal annually/sq m	0.192
	Water consumption	
11.	Apartment hotels with water supply system, gas, and hot water system, m^3/1 person per month.	7.31
12.	Apartment building with water supply system, wastewater system, gas pipeline, and no bathtub, m^3/1 person per month.	4.57
13.	Water consumption for plant watering at the home garden during the growing season: May-September, 153 days, 18 waterings per month, m^3/sq m	0.16
14.	Floor cleaning in the administrative and public buildings, l/per sq m per cleaning	2.40
15.	Administrative buildings and rooms considering working hours at least 235 days per year, l/1 working day	28

The examined in this paper office centers: Lighthouse, Ducat Palace 3, White Stone, Silver City, iCUBE are owned by one managing company and constructed considering a unified approach to the implemented technologies [14]. Therefore, one can say that energy systems and subsystems have an adequate degree of similarity properties and can be combined into one group. In such a case, it becomes possible to form a unified artificial neural network (ANN) based on similarity properties [15] of buildings under examination. Trained ANN will allow formulating energy resources forecasting demand for energy systems of the buildings.

3. Results

The forecasting factor (Y) is a heat energy consumption by a subject which is the office building, Gcal. Climate data: average monthly outside temperature, °C; average monthly air humidity, %, - obtained from the public database. For the training purpose has been chosen a three-layer artificial neural network with 4 input neuron layers and 1 output layer, the hidden layer consists of 20 neurons. The activation function has a sigmoid shape with curvature 2.

In the capacity of the training set the following measures have been taken such as heat, electricity, and water consumption at the office centers Lighthouse, Ducat Palace 3, iCUBE from January 2016 to July 2020. A total of 684 units of primary information had been used for ANN training. In the capacity of the test set the following data have been taken such as heat, electricity, and water consumption at the office center Krugozor from January 2016 to July 2019. The total number is 220 units of information. Subjects of comparison are the forecasting and actual heat consumption from August 2019 to July 2020.

4. Discussion

Training of the artificial neuron network took 200 epochs, the maximum error is 0.17565, mean error – 0.019673 (Figure 8).

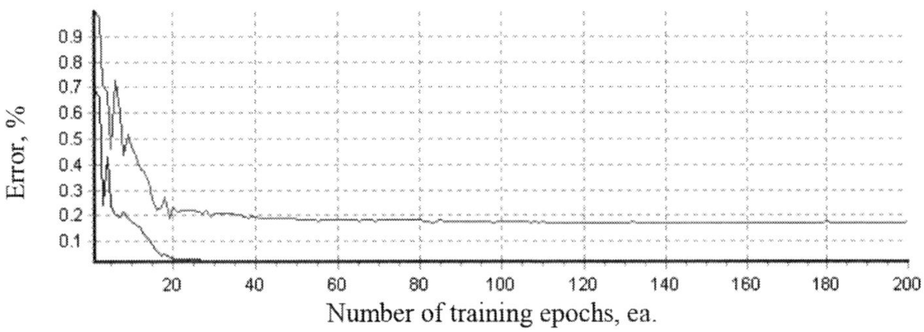

Figure 8. Process of artificial neuron network training on the training set of 684 units of primary information.

Results of forecasting which was realized on the training set showed a deviation of the forecasted value from the actual which equals 19.02%; on the test set – 28.65% (Figure 9).

Based on the conducted calculations it is possible to make a conclusion that the forecasted heat consumption values during the heating season are close to the actual heat consumption values considering the lower bound of the calculated value errors. Most of the forecasted value deviations from the actual can be observed during the summer season which tells us that the model is quite accurate in terms of forecasting heat consumption by the building's heating system and availability of an additional potential to improve calculation accuracy of heat energy consumption for hot water supply system needs. In spite of error presence, the forecasting function demonstrates entirely sufficient accuracy for further use in forecasting.

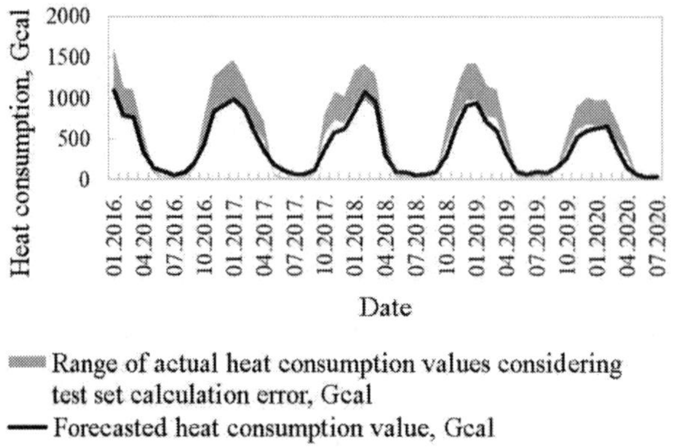

Figure 9. Results of the energy consumption forecasting at the office center Krugozor.

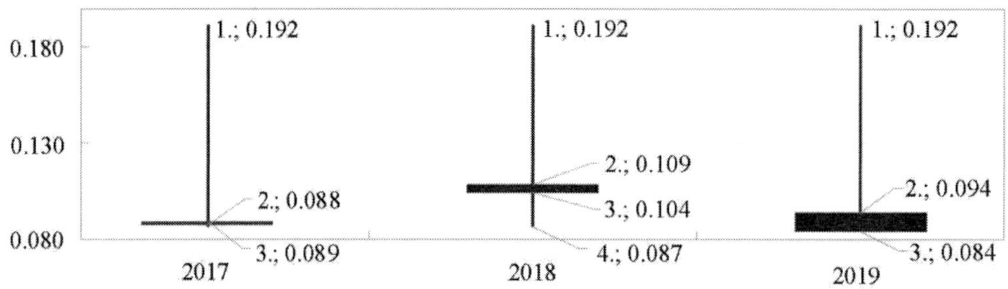

Figure 10. Comparison results of the annual specific values of heat energy consumption by the office center Krugozor:
1. Normative heat consumption = 0.192, Gcal/sq m per year;
2. Actual heat consumption, Gcal/sq m per year;
3. Forecasted heat consumption = 0.087, Gcal/sq m per year;
4. Heat consumption calculated by the Ermolaev formula = 0.087, Gcal/sq m per year.

As an example of the achieved heat energy saving determination can serve a comparison of the annual actual specific values of the obtained energy consumption values with the calculated and forecasted values (Figure 10).

5. Conclusion

This paper presents the conducted analysis and shows the structure of the operated «green construction» facilities in Moscow. It has been demonstrated that 40% of all facilities with implemented green technologies are office buildings.

The paper suggests the algorithm of energy efficiency evaluation of implemented at the facility energy-saving group of measures. The results of algorithm implementation allow to conduct a comparison of the annual specific heat energy consumption values, obtained using the actual and regulatory values, and also based on the consolidated deterministic calculations.

The paper demonstrated the applicability of methodology for calculating the current and forecasted building's index of technical condition for the considered group of problems. Automation of calculation by using the software will allow to calculate the forecasting composite index of technical condition with the accuracy of no less than 80%, the number of learning levels no more than 200, training time no more than 90 seconds.

The ANN model of energy-efficient green building was formulated. Application of this model by way of approximation on any office type building will allow to obtain forecasting results which demonstrate how the energy consumption of the building changes after its upgrade considering implementation of mentioned above energy-saving technologies. The results of forecasting which was realized on the training set showed a deviation of the forecasted value from actual which equals 19.02%, on the test set – 28.65%.

6. Acknowledgments

The investigation was carried out within the framework of the project "Integral index of energy systems of buildings based on groups of indicators of efficiency and reliability" with the support of a grant from NRU "MPEI" for implementation of scientific research programs "Energy", "Electronics, Radio Engineering and IT", and "Industry 4.0, Technologies for Industry and Robotics in 2020-2022".

References

[1] Kuznetsov V A and Trubaev P A 2018 Resources and Problems of the Mathematical Simulating Thermo-Technological Processes *J. Phys.: Conf. Ser.* **1066** 012024
[2] Sivachenko L A, Sevostyanov V S and Il'ina T N 2018 Problem Tasks in the Field of Resource and Energy Saving Technologie *J. Phys.: Conf. Ser.* **1066** 012021
[3] Romanov D O, Yavorovsky Y V and Smirnov V V 2019 Application of thermohydraulic dispatcher for existing and new district heating systems *IOP Conf. Ser.: Mater. Sci. Eng.* **552** 012025
[4] Kozhukhova N I, Lebedev M S, Vasilenko M I and Goncharova E N 2018 Fly Ash Impact from Thermal Power Stations on the Environment *J. Phys.: Conf. Ser.* **1066** 012010
[5] Zatsarinnaya J N, Amirov D I and Zemskova LV 2019 Analysis of the environmental factors influence on the efficiency of photovoltaic systems *IOP Conf. Ser.: Mater. Sci. Eng.* **552** 012033
[6] Guzhov S V, Gasho E G and Shepel V A 2020 Preparation of the forecasted fuel and energy balance of a boiler in an inadequate data condition *IOP Conf. Ser.: Mater. Sci. Eng.* **791** 012012
[7] Gasho E G and Gasho I A 2018 Economics and Ecology Symbiosis: Transition to Principles of Best Available Techniques *J. Phys.: Conf. Ser.* **1066** 01200
[8] Sultanov M M, Truhanov V M and Gorban Yu A 2019 Parametric model for predicting the reliability of heat power equipment of TPP *IOP Conf. Ser.: Mater. Sci. Eng.* **552** 012006
[9] Guzhov S V 2020 Ob ob"yedinenii determinirovannogo i stokhasticheskogo podkhodov pri prognozirovanii teplovogo balansa zdaniya dlya zanyatiy vodnymi vidami sporta [About combining determinated and stochastic approaches for prediction of the heating balance of the building for water sports] *Izvestiya VUZ Problemy Energetiki* **22(1)** 103-12 [in Russian]
[10] Guzhov S V 2020 Prognozirovaniye sprosa na elektricheskuyu energiyu izolirovannoy energeticheskoy sistemoy goroda [Forecasting Electricity Demand by an Isolated City Power System] *Obshchestvenno-Delovoy Nauchnyy Zhurnal «Energeticheskaya politika»* **6(148)** 50-7 [in Russian]
[11] 11. Guzhov S V 2020 O prognozirovanii sprosa na elektroenergiyu energosistemy regionov Rossiyskoy Federatsii s primeneniyem iskusstvennykh neyronnykh setey [On forecasting the demand for electricity by the power systems of the regions of the Russian Federation using artificial neural networks] *Izvestiya Transsiba* **1(41)** 133-40 [in Russian]
[12] Avilova I P, Naumov A E and Krutilova M O 2019 Methodology of GHG emissions assessment caused at the construction of energy facilities. Case study: Hydropower *IOP Conf. Ser.: Mater. Sci. Eng.* **552** 012018
[13] Bychikhin S A, Svirin M V and Trubaev P A 2019 Evaluation of energy-saving effect during the installing process of automated heating stations in educational institutions *IOP Conf. Ser.: Mater. Sci. Eng.* **552** 012019

[14] Vasilenko N A 2019 Parametric optimization of an architectural object's form as a method to improve its energy efficiency *IOP Conf. Ser.: Mater. Sci. Eng.* **552** 012027
[15] Gasho E G, Guzhov S V and Krolin A A 2019 On the forecasting of trends of specific mixed loads of electric consumers of microdistricts and cities change in the conditions of inadequate data *IOP Conf. Ser.: Mater. Sci. Eng.* **552** 012013

IOP Conf. Series: Materials Science and Engineering 1089 (2021) 012036 doi:10.1088/1757-899X/1089/1/012036

Mathematical description of wave propagation in order to assess the efficiency of the wave energy converter and find its optimal characteristics

E A Bekirov, M M Asanov and E R Murtazaev

Electrical Energy and Electrical Engineering Department, V.I. Vernadsky Crimean Federal University, Academician Vernadsky Ave., 4, Simferopol, 295007, Russia

E-mail: asanov.m.m@cfuv.ru

Abstract. The potential for using wave energy is very high. According to various estimates, the energy of the world's ocean waves can range from 146 TWh/year to 29,500 TWh/year. A number of mathematical models have been developed for the wave parameters analysis. The mathematical description of wave behaviour is a very complex process. Not least this is due to the complexity of the interaction of the wind that forms the waves with the water surface. As for the Black Sea, the mean annual wave power flux in this region can reach 4.8 kW/m. Such conclusions were made on the basis of modelling; the input was data on bathymetry of the region, wind speed and its direction. The obtained results on significant wave height, wave length and period were compared with the data obtained from the satellite. Many devices that convert wave energy into electrical energy have been developed. The choice of a particular device is based on a number of factors, including the installation site characteristics, weather conditions and economic opportunities. The article proposes a simple mathematical model of a wave, representing it in the form of a sinusoid. It is possible to estimate the wave energy conversion theoretical potential in a given region by analysing the obtained values of the energy received by the converter working surface (for example, turbine blade) after interacting with the wave. Calculations were carried out using the parameters of the Black Sea waves near the Crimean peninsula in winter.

1. Introduction

The energy of the world's ocean waves is estimated at about 146 TWh/year when converted into electrical energy, taking into account the conversion efficiency of 40%. This data is presented on the basis that only 2% of the entire 800,000 km coastal zone will be used, where the wave power density is higher than 30 kW/m. Other forecasts are even more optimistic. The theoretical potential is estimated at 2,000 – 4,000 TWh/year, and some authors say about 29,500 TWh/year, taking into account all areas where the wave power density is higher than 5 kW/m [1].

For comparison, the installed capacity of renewable power plants in the world as of 2018 amounted to 2,356,065 MW, and the electric energy generated by them was 6,586,124 GWh [2]. This indicates the high potential of wave energy. The advantages of using wave energy converters are:

- absence of environmental pollution during the production of electrical energy;
- the ability to ensure the energy independence of coastal regions;
- power plants are located in the sea, and as a result there is no need to allocate land for them.

2. Literature Analysis

2.1. Mathematical models of waves
The mathematical description of wave propagation is a very complex process [3]. Waves are formed under the influence of wind, while a nonlinear interaction of two substances occurs, the density of which differs by three orders of magnitude. The description of the interaction process becomes even more complicated with an increase in the difference in the speeds of these substances.

The created numerical wave models are based on the energy balance equation [4]. On the left side of the equation is the total time derivative of the two-dimensional wave spectrum S. It, in turn, depends on the frequency f, the direction of wave propagation θ, geographical coordinates x and time t. On the right side of the equation – source function S, which is mainly characterized by the sum of the following parameters: *In* is the mechanism of energy exchange between the atmosphere and waves; *Nl* is the energy-saving mechanism of nonlinear wave-wave interaction; *Dis* is scattering, the mechanism of energy loss by the wave due to wave-breaking processes and interaction of waves with turbulence of the upper water layer. The main difference between the existing models is in the description of the parameters mentioned above. Among the third-generation mathematical models of waves the WAM model [5], WAVEWATCH [6], TOMAWAC [7], the SWAN model [8], and the CREST model [9] can be mentioned.

2.2. Parameters of the Black Sea waves
The energy potential of the Black Sea has been studied for a long time. It is ranked among the inland seas or also called semi-enclosed [10] and the air currents above it mainly arise due to cyclones and anticyclones that form over Europe [11]. The presence of the Crimean and Caucasian mountains also affects the weather in the region [10].

The SWAN (Simulating Waves Nearshore) mathematical model [12] is widely used to estimate wave parameters and their seasonal variations [13]. The wind speed and its direction for the period from 1949 to 2010 from the NCEP / NCAR reanalysis were taken as the initial data. In addition, bathymetry data received from the Ministry of Defense of the Russian Federation in 1996 is used to create a digital model of the Black Sea floor. The storm that occurred on November 11, 2007 was analyzed in detail. Data on the average and maximum values of significant wave height, wave length and period for four seasons is obtained as a result of modeling. It is obvious that the parameters of the waves are maximum in winter. Moreover, the maximum values sometimes significantly exceed the average. Thus, the maximum significant wave height values in the period from December to February exceed the average by 0.95 m.

For a detailed analysis of the Black Sea wave parameters, along with the SWAN modeling program, the WRF (Weather Research and Forecasting) model is used to simulate the wind parameters in the region [11]. Data on wind speed at a height of 10 m above sea level, wave height and mean zero-crossing wave period is obtained. Both data recorded by the meteorological stations (Sulina, Chornomors'ke, and at the Gloria Platform) and those obtained from satellites provided by the altimeter missions (ERS-2, TOPEX, JASON-1 and GEOSAT Follow-On (GFO) is used to check the adequacy of the data obtained as a simulation result. Such statistical parameters as bias, the root mean square error, the scatter index, the linear coefficient correlation and the symmetric slope are used to assess the consistency of the results with real data. The simulation results showed good agreement with those obtained experimentally.

The ALADIN software package [14] can be used to simulate weather conditions. Wave parameters modeling was carried out using the SWAN model. Based on the comparison of the obtained data on significant wave height and data from a number of other sources, the parameters calculated using the ALADIN model showed less discrepancy with the real ones. Obtained parameters were compared with data from the Jason 1 and Envisat satellites for the period from 01/20/2012 to 02/10/2012. Data on mean annual wave power flux and mean annual significant wave height were obtained for the period from June 2011 to May 2015 using ALADIN. The maximum values of these parameters were 4.8

kW/m and 1 m, respectively. Modeling of mean annual wave power flux changes for 110 years from 1900 to 2010 was carried out. The SWAN model was used for this, the wind data were taken from the ERA-20C.

Wave parameters can be calculated using the WAM (WAve Model) model, which was first developed by ECMWF [10]. The ECMWF database (from 1979 to 2016) and AVISO Satellite Measurements (from 2010 to 2016) were selected as the initial data on weather conditions in the region. Wave height data from the Gloria drilling platform (from 2003 to 2009) is used to validate the results. Statistical parameters such as bias, root-mean-square error, scatter index and Pearson correlation index were used for the analysis. A table was built, in which the state of the sea for 14 points of the Black Sea and 1 point of the Sea of Azov for the period from 1979 to 2016 is assessed as a percentage according to the Beaufort scale. Sufficient agreement of the results obtained on the basis of various databases with experimental ones was demonstrated.

2.3. Wave energy converters

About 95% of the wave energy is located between the surface of the water and the upper one-fourth of the wavelength [1]. This energy can be converted in a number of different ways, which has led to a variety of devices.

The developers of such devices are faced with a number of problems [15]. The devices must have a reliable fastening system and an automatic safety system that protects them from possible excessive loads. Such devices must have the maintainability property, which consists in their adaptability to prevent and detect the causes of failures, maintain and restore an operable state through repairs and maintenance. An equally important property is the efficiency of converting wave energy into electrical energy. And finally, they should not pollute the environment during the energy conversion process.

Today there are more than 100 different devices for converting wave energy into electrical energy [1, 16]. All of them are divided into several types: an oscillating body converter, an oscillating water column, and an overtopping device [17, 18].

The first type includes devices on or below the surface of the water that use wave motion to generate electricity. Among them are point absorbers, terminators, attenuators. Point absorbers are small devices compared to wavelength that convert the vertical movement of the wave (the "heave") into electrical energy. Terminators and attenuators have large horizontal dimensions and are located parallel and perpendicular to the wave propagation, respectively. The first ones mentioned convert the horizontal front/back movement of the wave (the "surge"), the second ones – the horizontal side to side (the "sway"). More than half of the existing oscillating body converters, about 53 %, are point absorbers, 33 % terminators, and 14 % attenuators. The advantages of these devices include their size and certain versatility, independence from the location. However, their power conversion efficiency is relatively low and there are problems with mooring such devices [1].

Oscillating water column is a partially submerged chamber open to the sea below the water level, in which there is an air column above the water column. The water column moves up and down under the action of the incident waves, thereby displacing and drawing air into the chamber. The moving air rotates an air turbine that generates electrical energy. The list of the most major oscillating water column contains about 9 devices, whose power reaches 500 kW [19]. These devices are quite simple and reliable to operate. A complex control system is required to improve the efficiency of such devices.

The overtopping device reservoir is located above the mean free water surface. It is filled by oncoming waves, and the water returns back to the sea through a turbine that generates electrical energy. Some overtopping devices are described in [19], among which is Tapchan located in Norway. Its rated power is 350 kW. The operation principle of these devices is quite simple. The disadvantages include the low pressure of the water passing through the turbine, and, as a result, low productivity, as well as the large size of such devices.

It is difficult to single out any one of the listed technologies for converting wave energy at the moment. The choice of a particular device will always depend on many factors, including weather conditions, installation site, economic opportunities, technology availability, etc. An indicative capture width ratio on the absorbed power from the waves can be noted as a criterion for comparing the efficiency of transducers. For an overtopping device, it is equal to 17%, for an oscillating water column – 29%, for a point absorber – 16% [15].

3. Materials and Methods

The aim of the work is a mathematical description of the interaction of waves with a wave energy converter. The mathematical model should be as simple as possible, use only linear operators and the parameters of the waves and the energy converter itself should be used as input data.

Let the wave have a sinusoidal shape. Then it can be described by the following expression:

$$f = A_m \sin(\omega t + \varphi),$$

where:
A_m is the amplitude of a wave;
ω is the angular frequency of a wave;
φ is the initial phase of a wave.

Let us transform the expression for describing the wave in a sinusoidal form at zero initial phase:

$$f = A_m \sin(\omega t) = A_m \sin\left(\frac{2\pi}{T}\frac{x}{v}\right),$$

where:
$\omega = \frac{2\pi}{T}$;
$t = \frac{x}{v}$;
x is the distance traveled by the wave in time t;
v is the wave speed;
T is the wave period.

The transverse area of the wave can be found from the following expression (figure 1):

$$S = \int_0^{\frac{vT}{2}} A_m \sin\left(\frac{2\pi}{T}\frac{x}{v}\right) dx = \frac{A_m vT}{\pi}.$$

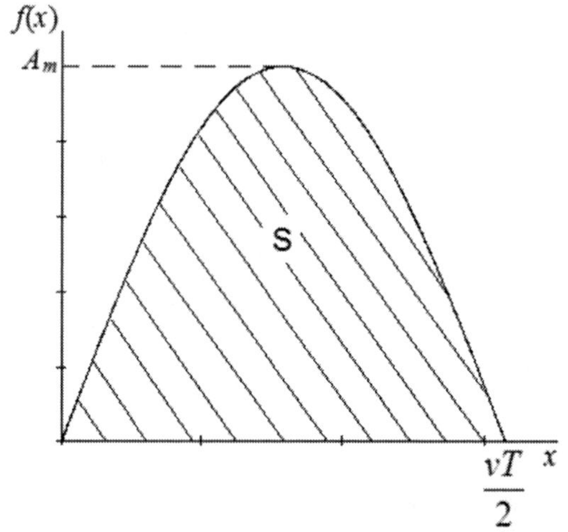

Figure 1. Graph showing transverse wave area.

The volume of the wave and its mass, in this case, can be found from the following expressions:

$$V = \int_0^b S dx = \frac{A_m vTb}{\pi};$$

$$m = \rho V = \frac{\rho A_m vTb}{\pi},$$

where:
b is the width of the wave;
ρ is the specific density of water.

The interaction of a wave with the working surface (for example, turbine blade) of a device for converting wave energy into electrical energy will be considered as an elastic collision of two bodies, in which, during a short-term contact, the bodies move at the same speed, then they fly apart and continue to move at different speeds [20]. In this case, guided by the law of conservation of energy, there is the following equation:

$$\frac{m_s v_s^2}{2} + \frac{m_w v_w^2}{2} = \frac{m_s v_s'^2}{2} + \frac{m_w v_w'^2}{2},$$

where:
m_s, m_w are the masses of the converter working surface and the wave, respectively;
v_s, v_w, v_s', v_w' are the speeds of the converter working surface and the wave before and after the collision;

$$v_s' = \frac{(m_s - m_w)v_s + 2m_w v_w}{m_s + m_w}; \quad v_w' = \frac{(m_w - m_s)v_w + 2m_s v_s}{m_w + m_s}.$$

4. Results

In figure 2 – figure 6 the graphs of the dependences of the energy received by the converter working surface after interaction with the wave on the wave parameters and the characteristics of the surface itself are presented.

The calculations were made for the Black Sea near the Crimean peninsula in winter. In the calculations, the following parameter values were taken: A_m = 0.9 m; v_w = 6.1 m/s; T = 2.7 s; ρ = 1000 kg/m^2; b = 1 m; m_s = 100 kg; v_s = 0 m/s [11, 13]. The average wind speed at an altitude of 10 meters above sea level was taken as the wave propagation speed v_w.

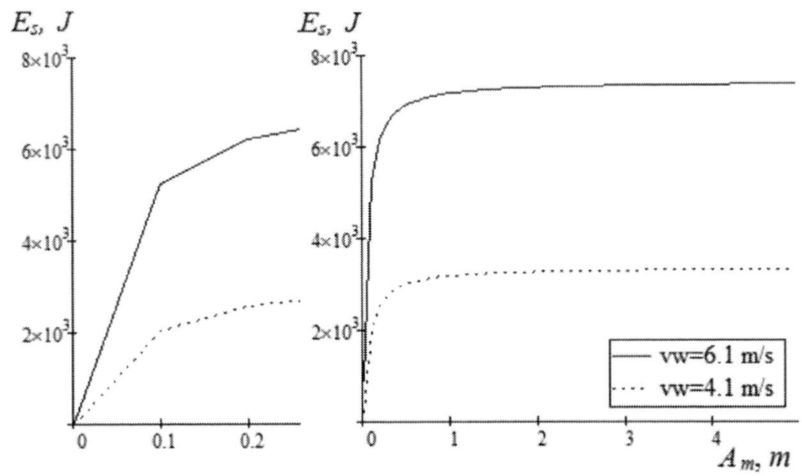

Figure 2. Dependence of the energy received by the converter working surface on the wave amplitude at different values of the wave speed.

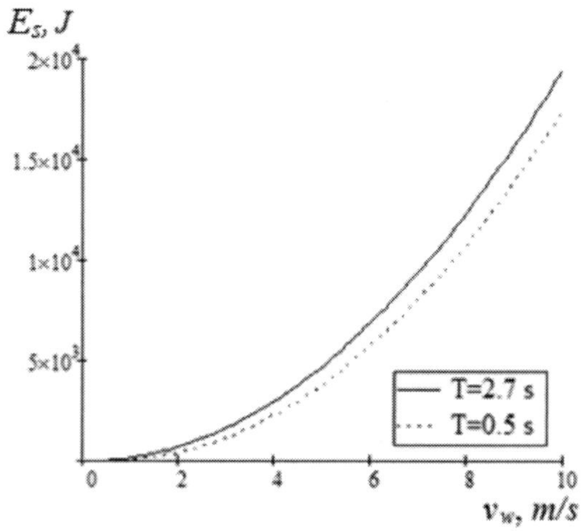

Figure 3. Dependence of the energy received by the converter working surface on the wave speed at different values of the wave period.

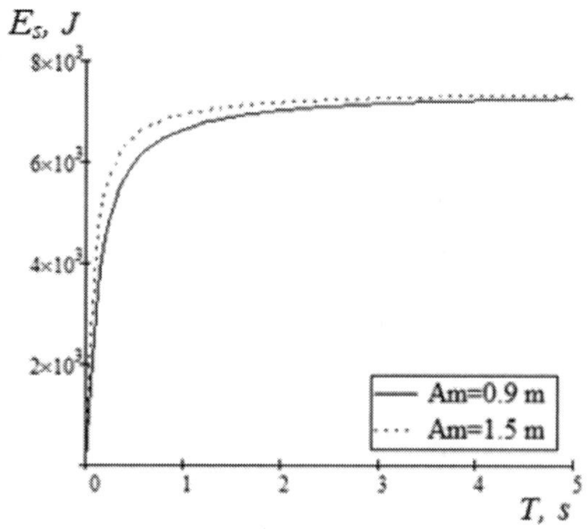

Figure 4. Dependence of the energy received by the converter working surface on the wave period at different values of the wave amplitude.

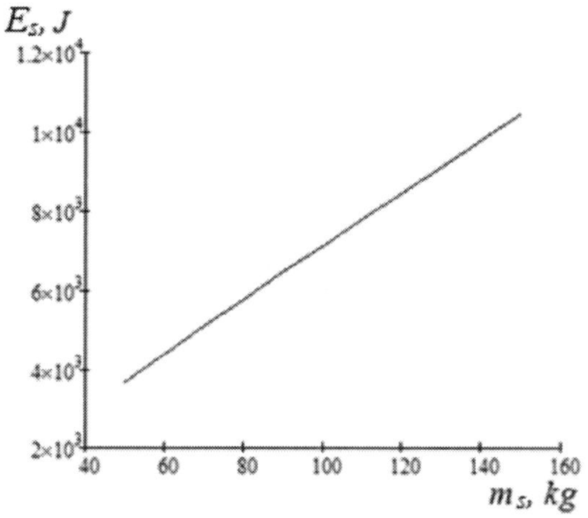

Figure 5. Dependence of the energy received by the converter working surface on its mass.

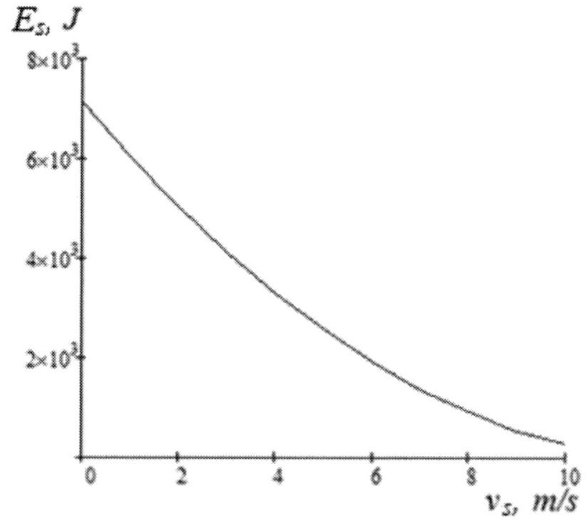

Figure 6. Dependence of the energy received by the converter working surface on its speed.

5. Discussion

The figure 2 – figure 4 show that with an increase in the values of the wave parameters, namely, the amplitude (height) of the wave A_m, the wave propagation velocity v_w and the wave period T, the energy that the converter working surface receives E_s also increases. As for the $E_s(A_m)$ and $E_s(T)$ dependences, in the region of small values of the wave parameters a sharp rise in energy values occurs. A further increase in the amplitude and period of the waves leads only to a barely noticeable rise in the energy received by converter working surface. The $E_s(v_w)$ dependence has a different character. In the region of small values of the wave propagation velocity, the increase in energy values occurs very slowly. Only in the region of large v_w values the energy rises rapidly.

From the graphs of the dependences of the converter working surface energy on the characteristics of the working surface itself (figure 5 – figure 6), it follows that a surface with a larger mass will receive a larger amount of energy when interacting with a wave. With an increase in the speed of the working surface, the energy received by it decreases.

In general, the presented results are consistent with those obtained earlier.

6. Conclusion

The existing numerical models for describing wave propagation are analyzed. Investigations of the Black Sea waves parameters are presented. The modern advances in the development of efficient devices for converting wave energy into electrical energy are described.

A simple mathematical model is proposed to describe wave propagation. The wave in the model is represented in the form of a sinusoid. It is possible to conclude about the effectiveness of installing wave energy converters in a specific location on the coast based on the results obtained using the model. The initial data of the model are such wave parameters as height, period, propagation velocity, as well as the parameters of the converter itself: the mass of the converter working surface and its speed. The energy that the working surface receives after interacting with the wave is the output of the model. For example, when the wave height changes from 0 to 5 m, the energy received by the working surface changes from 0 to $7.386 \cdot 10^3$ J. The values of the remaining parameters, in this case, are equal to $v_w = 6.1$ m/s; $T = 2.7$ s; $\rho = 1000$ kg/m^2; $b = 1$ m; $m_s = 100$ kg; $v_s = 0$ m/s.

As for the parameters of the converter itself, it can be concluded based on the simulation results that its working surface should be made as heavy as possible. And the rotation speed or the movement speed of the surface was as low as possible.

Modeling is carried out using the parameters of waves in the Black Sea near the Crimean peninsula in winter.

References

[1] Kempener R and Neumann F 2014 *Wave Energy: Technology brief* (Abu Dhabi: IRENA)

[2] Whiteman A, Rueda S, Akande D, Elhassan N, Escamilla G and Arkhipova I 2020 *Renewable Energy Statistics* (Abu Dhabi: IRENA)

[3] The WISE Group 2007 Wave modelling – The state of the art *Progress in Oceanography* **75** 603-74

[4] Thomas J T and Dwarakish G S 2015 Numerical wave modelling – A review *Aquatic Procedia* **4** 443-8

[5] WAMDI Group 1988 The WAM model – a third generation ocean wave prediction model *J. of Physical Oceanography* **18** 1775-809

[6] Tolman H L 1991 A third generation model for wind waves on slowly varying, unsteady, and inhomogeneous depths and currents *J. of Physical Oceanography* **21** 782-97

[7] Benoit M, Marcos F and Becq-Girard F 1997 Development of a third generation shallow-water wave model with unstructured spatial meshing *Proc. of the 25th Int. Conf. on Coastal Engineering (Orlando)* vol 1 (Orlando: American Society of Engineers Publications) pp 465-78

[8] Booij N, Ris R C and Holthuijsen L H 1999 A third generation wave model for coastal regions, Part I: model description and validation *J. of Geophysical Research* **104** 7649-66

[9] Ardhuin F, Herbers T H C and O'Reilly WC 2001 A hybrid Eulerian–Lagrangian model for spectral wave evolution with application to bottom friction on the continental shelf *J. of Physical Oceanography* **31** 1498-1516

[10] Onea F and Rusu L 2017 A Long-Term Assessment of the Black Sea Wave Climate *Sustainability* **9** 1875

[11] Rusu L, Bernardino M and Guedes Soares C 2014 Wind and wave modelling in the Black Sea *J. of Operational Oceanography* **7** 5-20

[12] SWAN (Simulating Waves Nearshore) URL http://swanmodel.sourceforge.net/

[13] Arkhipkin V S, Gippius F N, Koltermann K P and Surkova G V 2014 Wind waves in the Black Sea: results of a hindcast study *Nat. Hazards Earth Syst. Sci.* **14** 2883-97
[14] Galabov V 2015 The Black Sea Wave Energy: The Present State and the Twentieth century Changes URL https:/arxiv.org/pdf/1507.01187.pdf
[15] Pecher A, Kofoed J P, Folley M, Costello R, Todalshaug J H, Bergdahl L, Tetu A and Alves M 2017 *Handbook of Ocean Wave Energy* ed Pecher A and Kofoed J P (Switzerland: Springer)
[16] Joubert J R, van Niekerk J L, Reinecke J and Meyer I 2013 *Wave Energy Converters (WECs)* (Matieland: Centre for Renewable and Sustainable Energy Studies)
[17] Blaabjerg F and Ionel D M 2017 *Renewable Energy Devices and Systems with Simulations in MATLAB® and ANSYS®* (Boca Raton: Taylor & Francis Group)
[18] Hodge B K 2017 *Alternative Energy Systems and Applications* (Chichester: John Wiley & Sons, Ltd)
[19] Aderinto T and Li H 2019 Review on Power Performance and Effiency of Wave Energy Converters *Energies* **12** 4329
[20] Benenson W, Harris J W, Stocker H and Lutz H 2006 *Handbook of Physics* (New York: Springer-Verlag New York, Inc.)

Mathematical model for determining the optimal location of the biomethane plant

D Y Suslov and D O Temnikov

Department of Heat and Gas Supply and Ventilation, Belgorod State Technological University named after V.G. Shukhov, Kostyukov St., 46, Belgorod, 308012, Russia

Email: Suslov1687@mail.ru

Abstract. The work is devoted to the actual problem of using an alternative energy source - biomethane in gas supply systems. When developing and designing gas supply systems using biomethane, an important task is to determine the optimal location for biomethane plants. For the research, the methods of graphs and parametric optimization were used from the condition of minimizing the total transport costs. A mathematical model has been developed to determine the coordinates of the location of the biomethane plant depending on the production capacity and location of agricultural enterprises. Using the developed model, the location of a biomethane plant with a total volume of digesters of 5000 m3, located in the Veidelevsky district of the Belgorod region, was determined. This unit can process 315.9 t / day of organic waste and produce 504.64 m3 / h of biomethane with a methane content of more than 98%. As a result of the calculations, the coordinates of the location of the biomethane plant were determined (X = 11.82; Y = 22.35), and the location of the biomethane plant was adjusted, taking into account the relief conditions of the terrain and the presence of a road transport system.

1. Introduction

At present, the level of gasification of the Russian Federation with natural gas is 70.1%. Gasification of cities and urban-type settlements is 73%, and rural areas - 61.8%. At the same time, the priority program for the development of the gas supply system is gasification of the country's regions placed far from the main gas pipelines [1, 2].

A promising direction in the development of gas supply systems is the use of biomethane. Biomethane is biogas purified to form natural gas. The production and consumption of biomethane has become widespread in Europe and other countries of the world [3-6]. The largest amount of biomethane is produced in Germany, France, Great Britain, Sweden, the Netherlands, Denmark and Switzerland [7-9]. The number of biomethane plants (BMP) by country is shown in Figure 1. Most of the biomethane plants supply biomethane to the main or distribution gas networks [10, 11]. However, the majority of biomethane plants in Finland, Sweden, Norway and Iceland do not supply biomethane to gas networks, but are used as fuel for vehicles and other needs [11, 12]. To supply biomethane to transport gas pipelines, it is necessary to provide large volumes of gas with high excess pressure (5-7 MPa). This requires the installation of an additional compressor station and will lead to increased capital costs. Considering that when receiving biomethane, the pressure of the gas leaving the station reaches 0.3 MPa, it is more expedient to supply biomethane to medium pressure distribution gas pipelines.

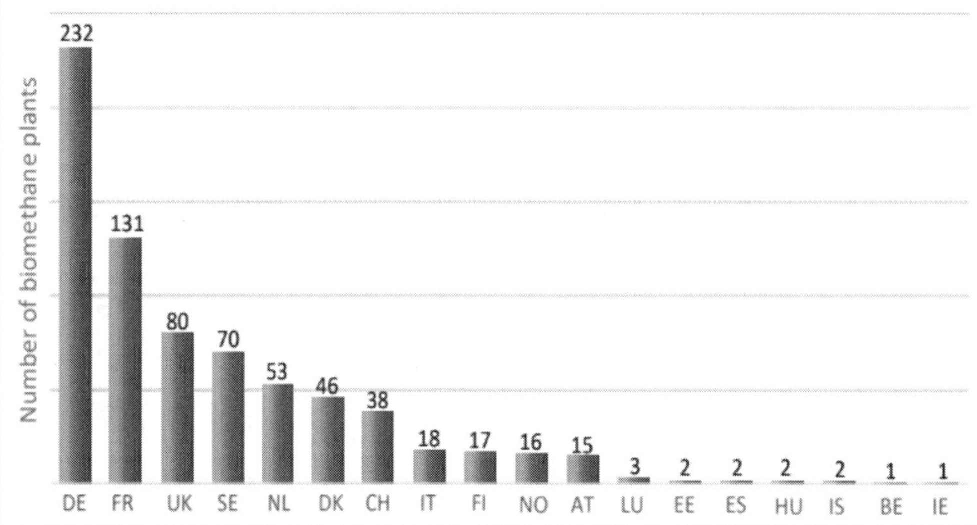

Figure 1. Number of plants per country DE – Germany, FR – France, UK - United Kingdom, SE – Sweden, NL – Netherlands, DK – Denmark, CH – Switzerland, IT – Italy, FI – Finland, NO – Norway, AT – Austria, LU – Luxembourg, EE – Estonia, ES – Spain, HU – Hungary, IS – Iceland, BE – Belgium, IE – Ireland.

Currently, 2 main schemes for obtaining and supplying biomethane are used: centralized and decentralized [13, 14].

The choice of the scheme depends on the productivity and location of agricultural enterprises and gas networks. The centralized scheme provides for one large biomethane plant that recycles waste from most agricultural businesses. This scheme is used in areas with a large number of agricultural and processing enterprises located close to each other. The centralized scheme is characterized by high costs for the delivery of the initial substrate from agricultural enterprises to the biomethane plant. At the same time, waste, as a rule, is delivered by road transport in tank tractors.

When agricultural enterprises are located at a great distance from each other, a decentralized scheme is used, in which a biomethane plant is located near each agricultural enterprise, which supplies biomethane to the nearest gas pipeline. At the same time, the costs of building biomethane plants are increasing, which significantly increases the cost of biomethane produced.

For the Belgorod region and other developed agro-industrial regions of the country, it is advisable to use a centralized scheme for obtaining and supplying biomethane.

When developing and designing gas supply systems using biomethane, an urgent problem is to determine the optimal location for biomethane plants. Domestic and foreign scientists were engaged in finding the optimal location of objects in gas supply systems [15-21].

The purpose of this work is to develop a design scheme and a mathematical model to determine the optimal location of biomethane plants.

2. Materials and Methods

When developing a mathematical model to determine the optimal location of biomethane plants on the territory of the administrative region, the theory of graphs, methods of mathematical modeling and parametric optimization from the condition of minimizing the total transport costs were used.

To solve the mathematical model, we used the engineering mathematical software Mathcad, which allows you to make calculations and visualize the results of calculations.

To determine the location of the biomethane plant on the territory of the Veidelevsky district of the Belgorod region, the coordinates of the location and the amount of organic waste of agricultural enterprises were used.

3. Results and Discussion

The rational location of the biomethane plant is to find a location in which the cost of transporting the initial substrate from the agricultural enterprise to the biomethane plant will be minimal.

Let's draw up a calculation scheme to determine the coordinates of the location of the biomethane plant (X, Y). Agricultural enterprises with the corresponding coordinates are denoted by $1(x_1,y_2)$, $2(x_2,y_2)$, $3(x_3,y_3)$, $N(x_i,y_i)$ (Figure 2).

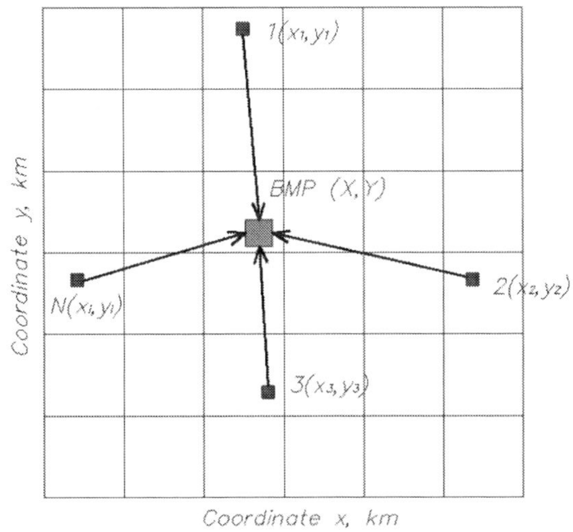

Figure 2. Calculation scheme for choosing the location of a biomethane plant.

Weber's model has been widely used to determine the optimal location of objects [22]. According to Weber's model, the coordinates of the location of the BMP biomethane plant (X; Y) are determined from the system of equations:

$$\begin{cases} A = c\sum a_i l_i; \\ l_i = \sqrt{(X - x_i)^2 + (Y - y_i)^2}, \end{cases} \tag{1}$$

where:
A - total costs of waste delivery, rubles;
c – is the cost of waste delivery, rubles;
a_i – is the volume of waste, t;
l_i – is the distance from agricultural enterprises to the biomethane plant, km;
x_i, y_i – coordinates of agricultural enterprises.

Waste delivery costs depend on the type of road transport and are determined by fuel consumption. The standard fuel consumption for a truck is determined according to the methodological recommendations "Norms of consumption of fuels and lubricants in road transport" (Order of the Ministry of Transport of Russia dated March 14, 2008 N AM-23-r):

$$Q_n = 0{,}01 \cdot (H_{san} \cdot S + H_w \cdot W) \cdot (1 + 0{,}01 \cdot D), \tag{2}$$

where:
S – vehicle mileage, km;
H_{san} – is the rate of fuel consumption per vehicle mileage, l / km;
H_w – is the rate of fuel consumption for transport work, l / 100 km;
W – is the volume of transport work, thousand km;
D – correction factor (total relative increase or decrease) to the% rate.

The fuel consumption rate is determined taking into account the mass of the trailer:

$$H_{san} = H_s + H_g \cdot G_{tr}, \quad (3)$$

where:
H_s – is the base rate of fuel consumption per vehicle mileage, l / 100 km;
H_g – fuel consumption rate for additional trailer weight, l / 100 thousand km;
G_{tr} – unladen weight of the trailer, t.

The volume of transport work is determined from the expression:

$$W = G_w \cdot S_w, \quad (4)$$

where:
G_w – is the mass of waste, t;
S_w – run with waste, km.

The required number of towing vehicles is determined based on the volume of waste transported by one vehicle:

$$k_i = \frac{G_w}{m_{nom}}, \quad (5)$$

where m_{nom} – is the nominal volume of waste transported by one trailer, t.

Using equations (1-5), we compose a system of equations to determine the optimal location of the biomethane plant:

$$\begin{cases} A = K \cdot \sum_{i=0}^{n} (k_i \cdot (H_s \cdot 2 + H_g \cdot G_{tr} \cdot 2 + H_w m_{nom})) \cdot l_i \\ l_i = \sqrt{(X - x_i)^2 + (Y - y_i)^2} \end{cases}, \quad (6)$$

where $K = 0{,}01 \cdot (1 + 0{,}01 \cdot D)$ – correction factor.

Using the developed mathematical model, we will determine the optimal location of a biomethane plant for processing organic waste from agricultural enterprises [23] located in the Veidelevsky district of the Belgorod region.

As the initial data, 6 poultry factories of the enterprise of OOO "Veidelevsky Broiler", as well as 2 enterprises for raising cattle: AO "Dolzhanskoe" and ZAO im. Kirov. The volumes of waste and the coordinates of the location of enterprises on the ground are presented in Table 1.

The total amount of organic waste generated in the Veidelevsky district is 315.9 tons / day. To process this amount of waste, we will use a biomethane plant, including 2 digesters with a volume of 2500 m3 each. The plant produces 504.64 m3 / h of biomethane with a methane content of more than 98%.

As a result of the calculations, the coordinates of the optimal location of the biomethane plant were determined (X = 11.82; Y = 22.35) and the design scheme was constructed (Figure 3). To transport the substrate to the biomethane plant, the number of trip was determined: from poultry factories 0.5 flights / day, from OOO "Dolzhanskoye" 2.5 flights / day, from ZAO im. Kirov 2.5 flights / day.

IOP Conf. Series: Materials Science and Engineering 1089 (2021) 012037 doi:10.1088/1757-899X/1089/1/012037

Table 1. Design parameters.

Company number	Name Company	Coordinates	Type waste	Amount of waste, t / day
1	Poultry factory number 1	x=17.63 y=14.26	Bird droppings	20.7
2	Poultry factory number 2	x=17.78 y=12.75	Bird droppings	19.4
3	Poultry factory Victoro-pol	x=13.97 y=6.42	Bird droppings	19.4
4	Poultry factory Belyi Kolodez	x=26.63 y=1.69	Bird droppings	20.7
5	Poultry Factory Kubraki	x=34.4 y=17.58	Bird droppings	19.4
6	Poultry actory Degt-yarnoe	x=27.07 y=26.33	Bird droppings	19.4
7	AO "Dolzhanskoye"	x=12.46 y=28.71	Cattle manure	100
8	ZAO im. Kirov	x=2.1 y=13.34	Cattle manure	96.9
	Total			315.9

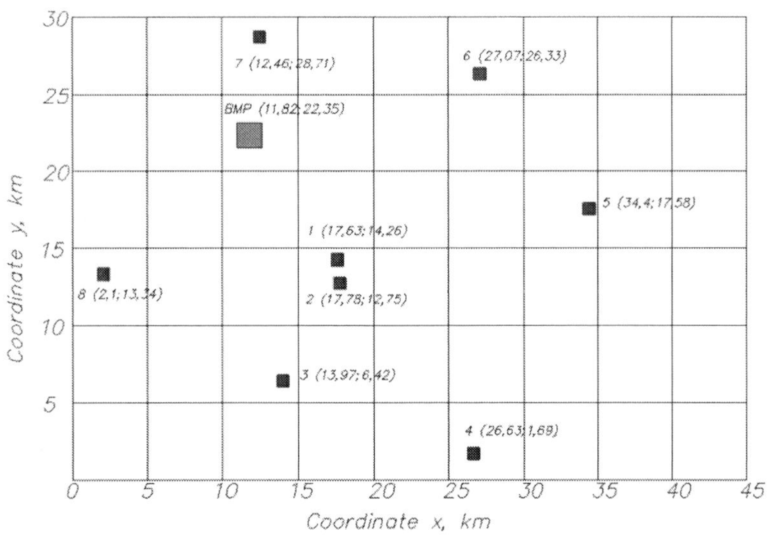

Figure 3. Location map of agricultural enterprises and biomethane plant.

However, this scheme has disadvantages, since it does not take into account the natural conditions of the area and the road transport network.

To clarify the calculations, we will draw the resulting scheme on the geographical map of the Veidelevsky region. Then we determine the actual position of the biomethane plant, taking into account the local conditions. To do this, we will move the biomethane plant 2.1 km to the south, closer to the village of Bolshiye Lipyagi. Near the village of Bolshiye Lipyagi there is a road with an improved surface (Figure 4).

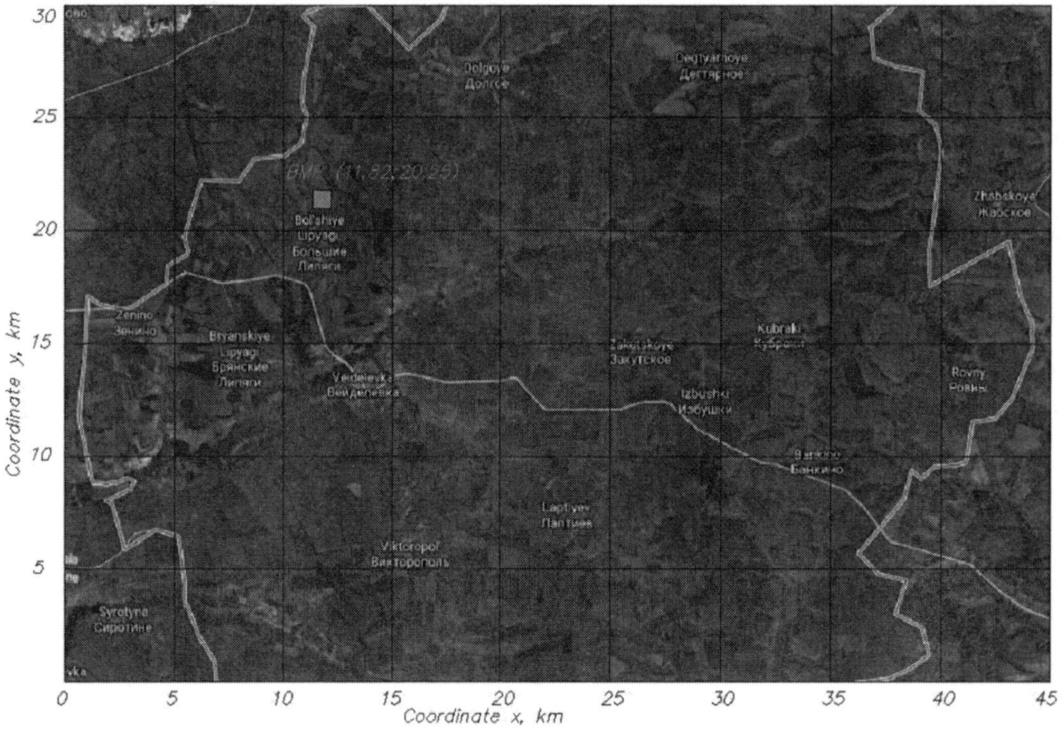

Figure 4. The actual location of the biomethane plant in the Veidelevsky district of the Belgorod region.

4. Conclusion

An alternative source of energy, biomethane, has become widespread in gas supply systems in developed countries and is one of the promising areas for the development of the gas supply system of the Russian Federation. For the Belgorod region and other developed agro-industrial regions of the country, it is advisable to use a centralized scheme for obtaining and supplying biomethane.

A mathematical model has been developed for determining the optimal location of biomethane plants on the territory of the administrative district. The model allows you to determine the optimal location of the biomethane plant, depending on the productivity and location of agricultural enterprises and the location of the gas pipeline.

The developed model was used to determine the location of a biomethane plant with a total volume of digesters of 5000 m3, located in the Veidelevsky district of the Belgorod region. The unit can process 315.9 t / day of organic waste and produce 504.64 m3 / h of biomethane with a methane content of more than 98%. As a result of solving the mathematical model, the coordinates of the location of the biomethane plant were determined (X = 11.82; Y = 22.35). Then the location of the biomethane plant was adjusted taking into account the natural conditions of the area and the road transport network (X = 11.82; Y = 20.25).

5. Acknowledgments
The study was carried out in the framework of the scholarship of the President of the Russian Federation SP-180.2019.1.

References
[1] Gazifikaciya [Gasification] URL https://www.gazprom.ru/about/production/gasification/
[2] Suslov D Y 2014 Razrabotka sistemy gazosnabzheniya sel'skohozyajstvennogo predpriyatiya s ispol'zovaniem biogaza [Development of a gas supply system for an agricultural enterprise using biogas] *Vestnik BGTU im V G Shukhova* **4** 183-6 [In Russian]
[3] Chinnici G, Selvaggi R, D'Amico M, Pecorino B 2018 Assessment of the potential energy supply and biomethane from the anaerobic digestion of agro-food feedstocks in Sicily *Renewable and Sustainable Energy Reviews* **82 (1)** 6-13
[4] Suwansri S, Moran J C, Aggarangsi P, Tippayawong N, Bunkham A, Rerkkriangkrai P 2014 A biomethane solution for domestic cooking in Thailand *Energy for Sustainable Development* **23** 68-77
[5] Hoo P Y, Patrizio P, Leduc S, Hashim H, Kraxner F, Tan S T, Ho W S 2017 Optimal Biomethane Injection into Natural Gas Grid – Biogas from Palm Oil Mill Effluent (POME) in Malaysia *Energy Procedia* **105** 562-9
[6] Repele M, Ramanis M and Bazbauers G 2016 Biomethane Supply Support Policy: System Dynamics Approach *Energy Procedia* **95** 393-400
[7] Scarlat N, Dallemand J-F and Fahl F 2018 Biogas: Developments and perspectives in Europe *Renewable Energy* **129(A)** 457-72
[8] Suslov D Y 2018 Application of Biomethane for Gas Supply Within the Settlements *J. Phys.: Conf. Ser.* **1066** 012004
[9] Geletukha G G, Kucheruk P P, Matveev Yu B 2014 Perspektivy proizvodstva i ispol'zovaniya biometana v Ukraine [Prospects of production and use of biomethane in Ukraine] *Analiticheskaya Zapiska BAU* **11** 14 [In Russian]
[10] Königsberger S et al 2018 *EBA Statistical Report 2018* (Brussels: EVA)
[11] The 'European Biomethane Map 2020' shows a 51% increase of biomethane plants in Europe in two years | European Biogas Association URL https://www.europeanbiogas.eu/the-european-biomethane-map-2020-shows-a-51-increase-of-biomethane-plants-in-europe-in-two-years/
[12] Cucchiella F and D'Adamo I 2016 Technical and economic analysis of biomethane: A focus on the role of subsidies *Energy Conversion and Management* **119** 338-51
[13] Wu B, Zhang X, Di Bao, Xu Y, Zhang S and Deng L 2016 Biomethane production system: Energetic analysis of various scenarios *Bioresour Technol* **206** 155-63
[14] Beil M and Beyrich W 2013 Biogas upgrading to biomethane *The Biogas Handbook* (Cambridge: Woodhead Publishing Limited) pp 342-77
[15] Tabunshchikov YU A, Prohorov V I, Bryuhanov O N, ZHila V A, Klochko A K 2012 Celi i zadachi optimizacii setej gazoraspredeleniya [Goals and objectives of optimization of gas distribution networks] *Vestnik MGSU* **4** 73-7 [In Russian]
[16] Medvedeva O N 2015 *Razrabotka effektivnyh gazoraspredelitel'nyh sistem* [Development of efficient gas distribution systems] (Saratov: SGTU im Gagarina YU A) [In Russian]
[17] O'Shea R, Wall D M, Kilgallon I, Browne J D and Murphy J D 2017 Assessing the total theoretical, and financially viable, resource of biomethane for injection to a natural gas network in a region *Applied Energy* **188** 237-56
[18] Weidenaar T D 2014 *Designing the biomethane supply chain through automated synthesis: PhD thesis* (Enschede: University of Twente)
[19] Hussey B 2013 *Biogas injection into the Natural Gas Grid* (Dublin: CER)
[20] Wu Y, Lai K K and Liu Y 2007 Deterministic global optimization approach to steady-state distribution gas pipeline networks *Optimization and Engineering* **8(3)** 259-75

[21] Temnikov D O and Suslov D Yu 2018 Algorithms of the Minimum Spanning Tree in Gas Supply Systems *J. Phys.: Conf. Ser.* **1066** 012001
[22] Lukinckij V S and Lukinckij V V 2007 *Modeli i metody teorii logistiki* [Models and methods of logistics theory] (St Peterburg: Piter) [In Russian]
[23] Shachneva E Yu 2018 Methods of sorption concentration of surface active substances *Chemical Bulletin* **1(2)** 24-30

Optimization of operating modes of hydropower plants with determination of the price for a hydro resource using complex criteria of ecological and economic efficiency

Yu A Sekretarev, T V Myatezh and A V Gorshin

Department of Industrial Power Supply Systems, Novosibirsk State Technical University, K. Marx Ave,30, 630073, Russia

E-mail: tatianamyateg@mail.ru

Abstract. At present, the influence of technologies on the environment became comparable with natural processes at the Earth. Therefore, the use of renewable energy sources, including water resources, is very important for the energy market operated as the Internet of things. The present paper proposes a universal method for analyzing the efficiency of technical systems using the combination of an optimization method and a method of marginal utility evaluation. Based on the comparison of water volume at a hydropower plant (HPP) and fuel amount at a combined heat and power plant (CHPP) that is used for generation of 1 kW power, it is possible to determine a water price for a HPP. Using the examples of Novosibirsk HPP and CHPP, it is expected to develop an estimation of economic effect from the implementation of the developed criteria, the proposed method of determination of a water price at HPP, and the method of separating fuel costs at CHPP. As a result of implementing the developed method for theHPP, theprice of sold electricity in the flexible energy market will be comparable with the price for sold electricity produced at theCHPP, being equal to approximately 330 RUB/MW·h. Moreover, CHPP are expected to have savings of specific fuel consumption for electricity generation in the amount of 5 g/kW·h and, in some case, 10 g/kW·h. In terms of money, this amounts about 7,350,000 – 14,700,000 RUB per month.

1. Introduction

This paper provides the review and development of the method for estimating the price of water at a HPP during optimization of operating conditions based on complex criteria of ecological-and-economic efficiency under modern conditions. At present, the influence of technologies on the environment became comparable with natural processes at the Earth. Therefore, the use of renewable energy sources, including water resources, is very important for the energy market operated as the Internet of things [1, 2]. One of the universal methods of analyzing the efficiency of technical systems is the optimization method in combination with the theory of marginal utility. This allows determining a water price for a HPP based on the comparison of water consumed at a HPP and fuel amount required at a CHPP for generation of 1 kW power. It is possible only to reduce exergy of these flows to the reasonable limit determined by damping properties and elasticity of nature for their transformation into intermediate and conditionally end products, which are human- and environment-friendly, in other words, by electricity co-generation at a HPP and a CHPP [3, 4].

2. The Model of Functioning Management for Power Plants

This investigation proposes an ecological-and-economic profit maximization criterion for optimal load dispatching (or management) of hydropower and thermal power plants under modern conditions, and a strategy for controlling a water price at a HPP using the theory of marginal utility [5, 6].

According to the profit maximization criterion, a producer will maximize its profit when producing goods at the point where marginal revenue equals marginal costs [7-9]. This leading principle of profit maximization is called as the rule of equality between marginal revenue and marginal costs. This principle is illustrated in Figure 1.

Marginal revenue (MR) is determined by a differential curve of energy demand, while marginal costs are influenced by a derivative component of energy production costs. The latter may be represented by an incremental fuel costs characteristic for thermal power plants [10, 11]. All these values have the same dimension (price/prod.unit), thus they may be comparable for calculations [12].

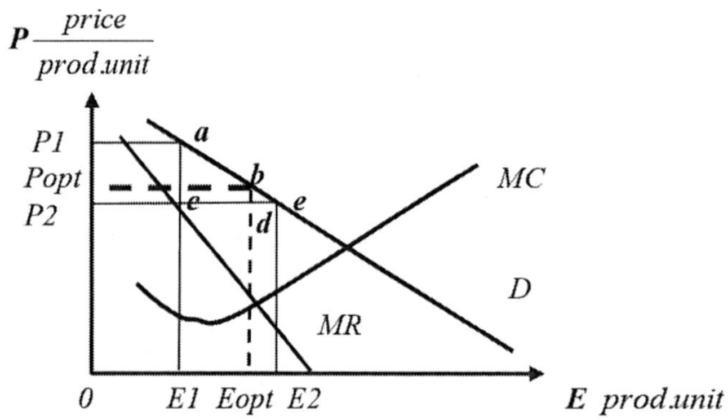

Figure 1. Determination of the optimal production output: D – energy demand for a specified time interval; E_{opt} – optimal output for a specified time interval; P_{opt} – the optimal price at the optimal production output.

The optimal energy production output (E_{opt}) allows an energy enterprise to maximize its profit. Assume that less goods E_1 are produced in comparison with an optimal output at higher price P_1. In this case, marginal revenue of the producer exceeds marginal costs. If production output raises up to E_{opt} (when profit from selling one additional product unit equals to zero), the producer will increase the total profit by the square abc.

Production output E_2 being greater than the optimal level does not also maximizes profit, because marginal costs exceeds marginal revenue. Increase of profit due to reduction of production output down to Eopt instead of E2 corresponds to the value, which is equal to the square bde.

Under modern conditions, each business unit is interested in the increase of its own profit, therefore the proposed method allows determining production outputs at a power plant with a glance to modern features of Russian power engineering [13].

Exergy allows combining the three aspects of optimization: thermodynamic, technical-and-economic and ecological – within the unified optimization system [14, 15]. Water resources are very important for saving primary energy resources and material resources at the input of the technical system. In the end, they reduce the harmful impact on humans and the environment. The general rule can be stated as follows: it is more profitable to save primary energy and material resources during implementation of technological processes as close to end sections as possible. In other words, 1 kW·h of electricity (in RUB) is more expensive at the end of the technological chain, than at the beginning.

During the scientific research, the following general problems were solved: 1. The criterion instruments for complex estimation and optimization of electricity and heat generation at a CHPP and electricity at a HPP were proved on the basis of the exergy approach with justification of the conceptual system for "environmental parameters". 2. The method for calculation of a water price for a HPP was developed that allows not only improving ecological situation in the region, but also increasing competitiveness of power plants. 3. The urgency of the concept of "marginal utility" used for calculation of a water price for HPP was substantiated. 4. The mathematical instruments were developed for exergy optimization of the complex efficiency for processing primary resources on the basis of interconnected thermodynamic and ecological-and-economic criteria in order to generate electricity and heat, and a fundamentally new method of separating fuel costs between the considered types of energy at CHPP was presented. 5. The developed exergy-and-economic and technical criteria were tested to compare various technologies for generating electricity and heat based on the optimization of operating conditions of CHPP and HPP. 6. The model was developed for launching HPP and CHPP to the competitive electricity market on the basis of solving two important and interrelated problems for managing power plant operating conditions.

The formulated problems and stages of their implementation are presented in Figure 2.

Based on the developed methodology, the power plant can solve two important and interrelated problems during the optimization of operating conditions: 1) Determine the optimal range of electricity generation at power plants at an electricity tariff, which is formed in the flexible market; 2) Based on the optimal electricity generation at power plants, it is possible to substantiate the value of a posted price in the competitive market.

3. The model for evaluating the price of a hydro resource for HPP operation in the power system

At present, a problem of optimal load distribution between HPP and CHPP in the power system is solved in the following way:

$$\begin{gathered} b = \lambda q, \\ Q_{HPP} = Q_{GIV}, \\ P_S = N_{CHPP} + N_{HPP}, \\ N_{CHPP\,min} \leq N_{CHPP} \leq N_{CHPP\,max}, \\ N_{HPP\,min} \leq N_{HPP} \leq N_{HPP\,max}, \\ \lambda = const \end{gathered} \tag{1}$$

where:

b and q – incremental rate characteristics for water and fuel at a CHPP and a HPP respectively;

P_S, N_{CHPP}, N_{HPP} – power system load, values of power served by a CHPP and a HPP respectively;

$N_{CHPP\,min}$, $N_{CHPP\,max}$, $N_{HPP\,min}$, $N_{HPP\,max}$ – minimum and maximum power for a CHPP and a HPP respectively;

Q_{GIV} – permissible water flow rate at a HPP determined by water-power calculations;

Q_{HPP} – water flow rate at a HPP;

λ – the Lagrangian multiplier.

In this case, b and q have different dimensions. Therefore, it is necessary to experimentally select the value of λ taking into account the limited hydro resources at the HPP. These circumstances lead to a serious complication of calculations associated primarily with an increase in the number of iterative procedures and the convergence of this process.

The paper proposes the other method for solving this problem.

In relation to our research, we will consider water and fuel as products used for generation of the same amount of electricity.

The above-mentioned considerations regarding marginal utility can be presented in a very simple graphical form (see Figure 3).

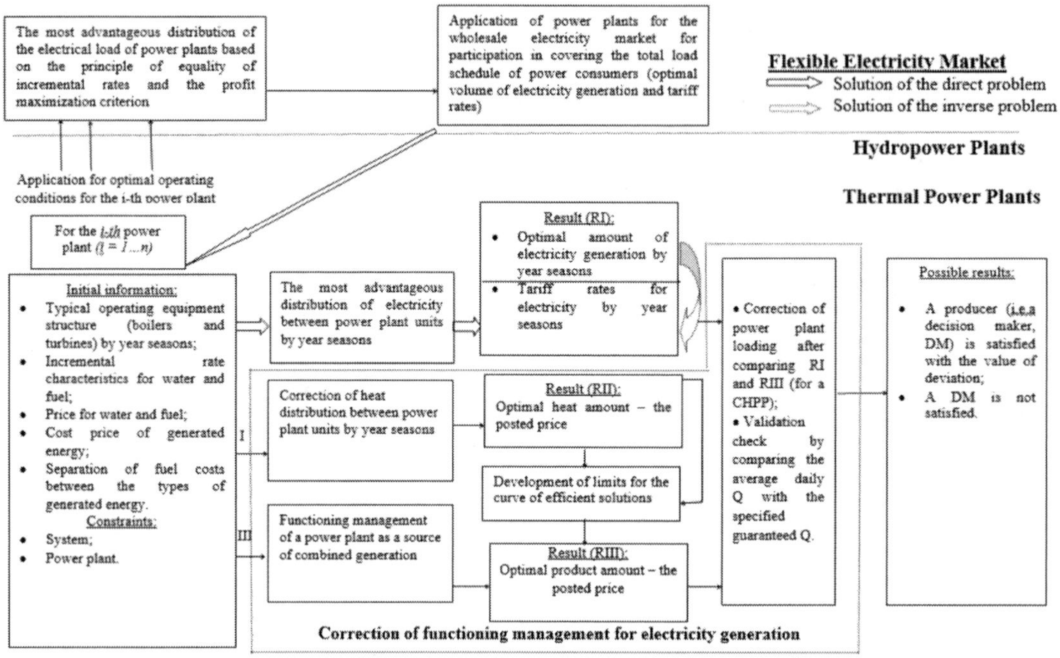

Figure 2. Block diagram for functioning management in the flexible market.

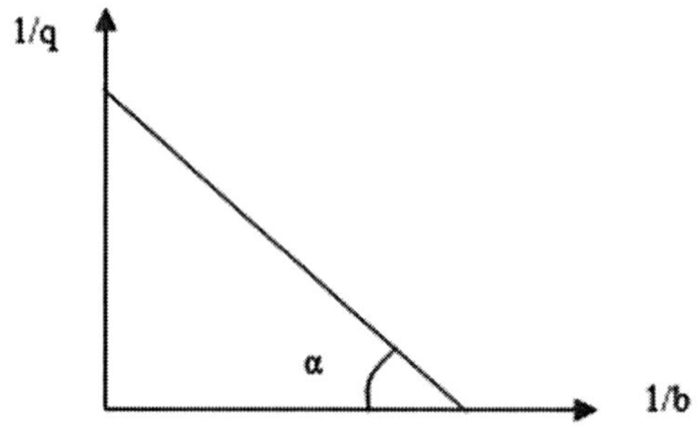

Figure 3. Indifference curve.

Moreover, it should be noted that it is necessary to convert physical quantities 1/q [s*kW/m^3] and 1/b [kW*h/ton of coal equivalent] into relative units, since this will allow comparing the indifference curve represented in relative units and the incremental rate characteristics for water at the HPP (q) and fuel at the CHPP (b) in one diagram (see Figure 4). When plotting the incremental rate characteristics for water at the HPP and fuel at the CHPP, the current values of incremental rates were divided by the average incremental rates of water and fuel, respectively.

Combining three curves in one diagram (i.e. the indifference curve and the incremental rate characteristics for water at the HPP and fuel at the CHPP), we obtain a new rule for the transition from the incremental water rate to the incremental fuel rate without using the Lagrangian multiplier λ. The value of the incremental water rate at the HPP q' locating on the indifference curve (see Figure 4) will be equal to the corresponding value of the incremental fuel rate at the CHPP locating on the same curve, since water and fuel in this case will have the same importance to the consumer.

Moreover, in order to obtain the corresponding power value for a particular hour according to the daily load schedule, it is necessary to shift the indifference curve in parallel to itself.

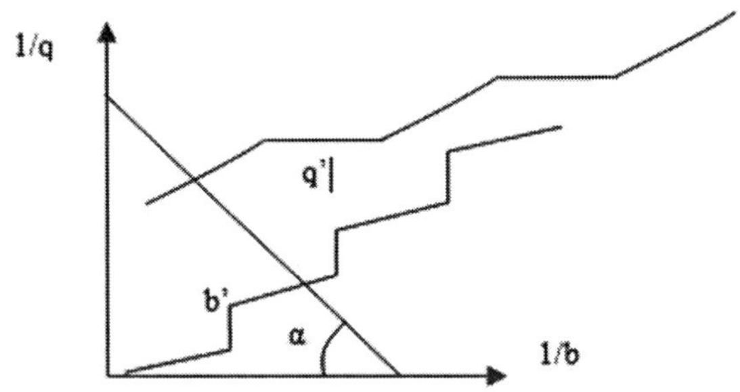

Figure 4. Incremental rate characteristics for water at the HPP and fuel at the CHPP combined with the indifference curve.

Then, using the indifference curve for marginal costs of water and fuel consumption, it should be noted that the slope of the indifference curve remains exactly the same.

Hence, the water price for a HPP can be determined using the following curve (see Figure 5) and expression (2).

$$U_q = w_p \cdot q \tag{2}$$

where:
U_q – marginal costs at a HPP;
q – incremental water rate at a HPP;
w_p – water price for a HPP.

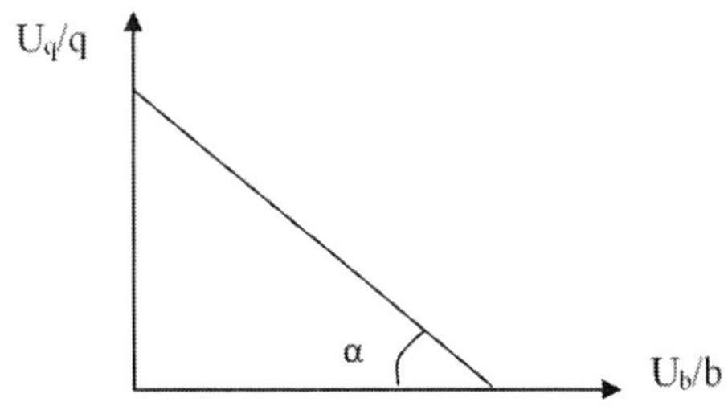

Figure 5. Indifference curve (for marginal cost): where U_q – marginal costs at a HPP, U_b – marginal costs at a CHPP, b – incremental fuel rate at a CHPP, q – incremental water rate at a HPP, wp – water price for a HPP.

Figure 6 shows the indifference curve and the incremental rate characteristics for water at the HPP and fuel at the CHPP in relative units using the example of the high-water period.

The indifference curve in relative units for the high-water period is illustrated in Figure 7.

Since for any HPP the daily consumption of water resource is uniquely determined, the problem of choosing the optimal HPP operating conditions is solved in the same way as for a CHPP with a given fuel consumption. According to the developed methodology, it is necessary to plot the incremental water rate characteristics for a given structure of operating equipment at the HPP for each optimization interval, which is a month or a decade for the period of filling the HPP reservoir. To verify the operability of the proposed model, the simplified HPP operating modes were used, in particular, seasonal periods of the year (Figure 8).

Taking into account the price of water obtained using the proposed model, it is necessary to plot the marginal costs characteristics for water flow rate at the HPP, and then obtain the optimal amount of electricity generation at the HPP (Figure 9).

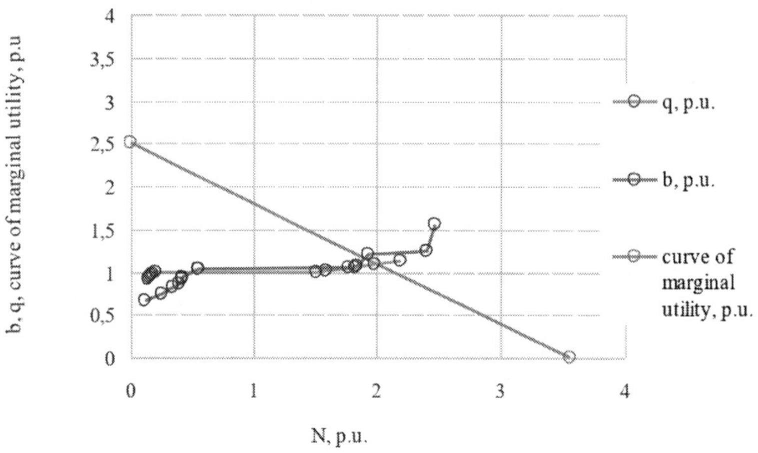

Figure 6. The indifference curve and the incremental rate characteristics for water at the HPP and fuel at the CHPP in the high-water period.

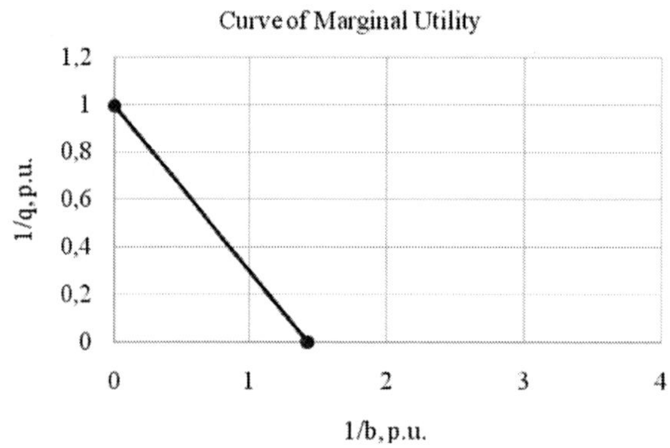

Figure 7.The indifference curve for the high-water period.

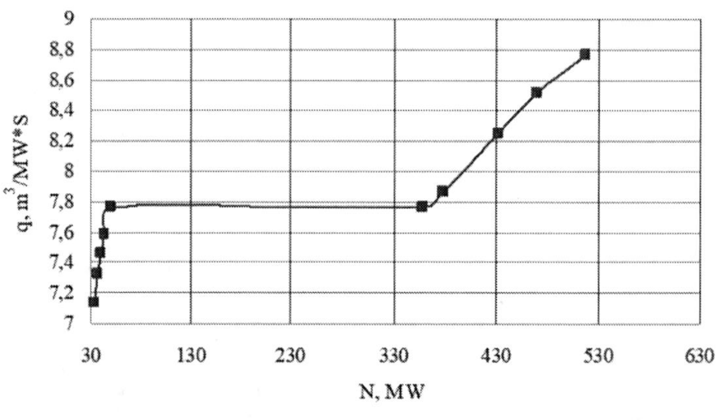

Figure 8. Incremental water rate characteristic for the Novosibirsk HPP for the high-water period.

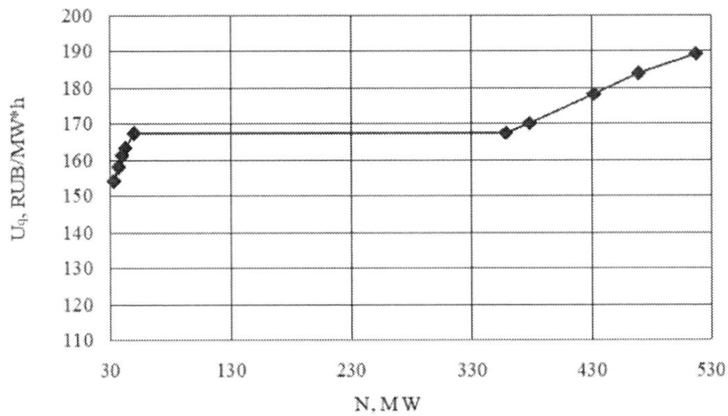

Figure 9. Marginal costs characteristics for the Novosibirsk HPP for the high-water period.

Based on the developed complex criteria of ecological-and-economic efficiency, the optimal amount of electricity generation at the HPP was obtained using the high-water period as an example (Figure 10).

Electricity generation at the Novosibirsk HPP in the high-water period varies from 100800 to 114480 MW·h. An electricity price for that period equals from 257 to 270 RUB/MW·h. As for the low-water period, electricity output is from 90000 to 100800MW·h with the posted price varied from 600 to 689 RUB/MW·h. Being operated with natural river flow, these values are equal to 93000 – 97200 MW·h and 290 – 300 RUB/MW·h.

The results of calculations are given in Table 1.

Practical testing of the developed method at the Novosibirsk HPP results in obtaining the following water prices for the HPP by year seasons (high-water period, low-water period, and operation with natural river flow): the water price for the Novosibirsk HPP in the high-water period is 21.59 RUB/m^3, in the low-water period – 30.50 RUB/m^3, for operation with natural river flow– 33.66 RUB/m^3.

Table 1. Optimal Values of Power Output for the Novosibirsk HPP by Year seasons.

Season	High-water period		Low-water period		Operation with natural river flow	
Profit rate	0%	12%	0%	12%	0%	12%
Power, MW	140	159	125	140	130	135
Electric energy, MW·h	100 800	114 480	90 000	100 800	93 600	97 200
Posted price, RUB/MW·h	257	270	600	689	290	300
Revenue, RUB	77 716 800	92 728 800	378 000 000	486 158 400	54 288 000	58 320 000
Profit, RUB		15 012 000		108 158 400		4 032 000

An analysis of the results obtained by optimizing the operating conditions of the Novosibirsk CHPP showed that it is reasonable for the Novosibirsk CHPP No.2 to produce 216 MW (348942 Gcal) in winter, 91 MW (79149 Gcal) in summer, and 146 MW (191910 Gcal) in the transition season; the optimal load for the Novosibirsk CHPP No.4 is 253 MW (356589 Gcal) in winter, 59 MW (82282 Gcal) in summer, and 210 MW (264725 Gcal) in the transition season [5, 6]. It is these operating conditions that will allow the power plants to get the maximum profit for each season of the year.

To verify the developed methodology, it is necessary to compare the found values of the average daily water flow rate, which are obtained after the distribution of the daily load of the generation company between the HPP and the equivalent CHPP according to the developed methodology, with the specified guaranteed value of the water flow rate at the HPP.

The results of the implementation of the developed methodology are given below for the example of the Siberian Generation Company (SGC), where the Novosibirsk CHPP No.2, No.3, No.4, and No.5 were taken as the equivalent CHPP and the Novosibirsk HPP was operated in the high-water period, low-water period, and with natural river flow (Figures 11-13).

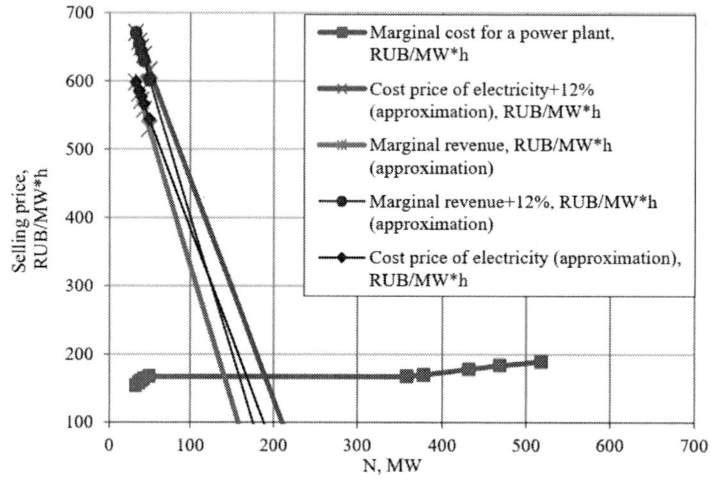

Figure 10. Curves for determination of optimal amounts of electricity generation for the high-water period at the Novosibirsk HPP.

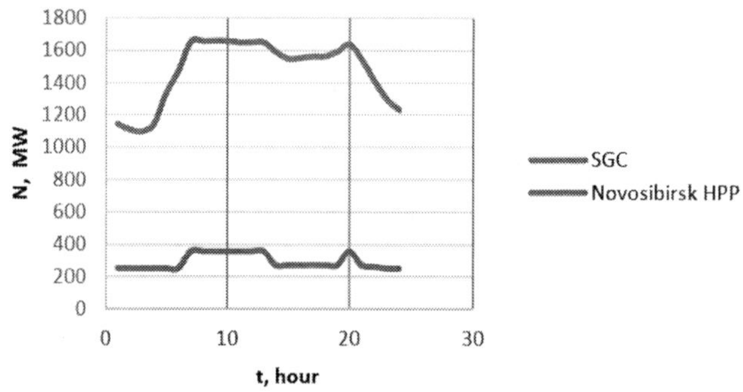

Figure 11. Schedule of daily power distribution of the SGC between the Novosibirsk HPP and the equivalent CHPP for the high-water period.

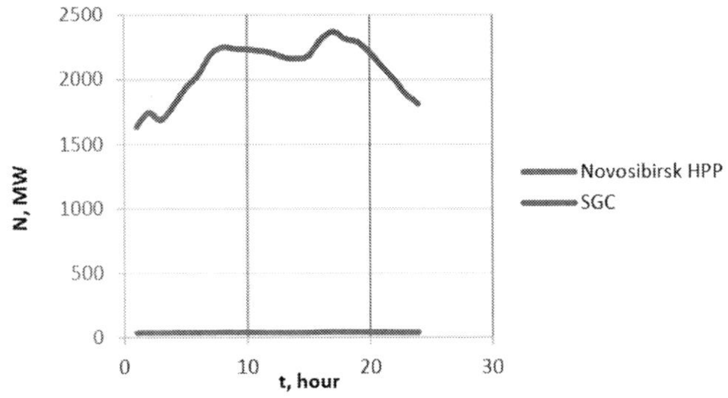

Figure 12. Schedule of daily power distribution of the SGC between the Novosibirsk HPP and the equivalent CHPP for the low-water period

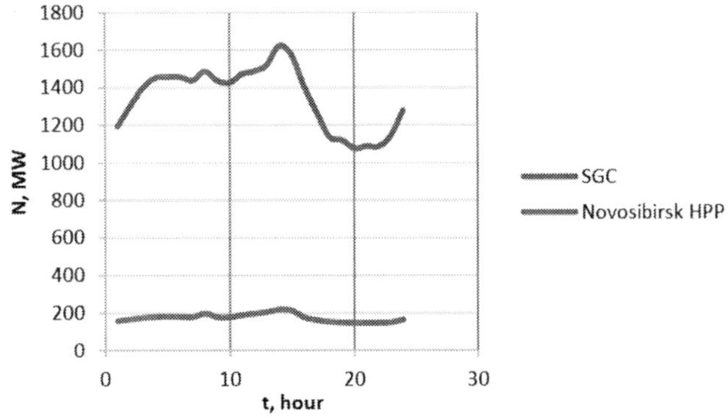

Figure 13. Schedule of daily power distribution of the SGC between the Novosibirsk HPP and the equivalent CHPP for operation with natural river flow.

In table 2 the daily power distribution of the SGC between the Novosibirsk HPP and the equivalent CHPP for the high-water period is given.

The calculation results showed that the comparison of the average daily water flow rate obtained by the developed methodology with the specified guaranteed water flow rate for each period gave an error of about 8% for the high-water period, 4% for the low-water period, and 10% for operation with natural river flow, being amounted to 2793,9m^3/s for the high-water period, 323m^3/s for the low-water period, and 1456,147 m^3/s for operation with natural river flow. This indicates the reliability of the results according to the developed methodology and allows making a conclusion about its validity and the possibility for application of the methodology to determination of the price for water as a hydro resource taking into account the operational features of HPP based on complex criteria of ecological and economic efficiency.

The results of the implementation of the Lagrangian multiplier method are given below for the example of the Siberian Generation Company (SGC), where the Novosibirsk CHPP No.2, No.3, No.4, and No.5 were taken as the equivalent CHPP and the Novosibirsk HPP was operated in the high-water period, low-water period, and with natural river flow (see Figures 14-16).

Table 2. Power Distribution in the SGC between the Novosibirsk HPP and the Equivalent CHPP for the High-Water Period.

T, hour	Q, m^3/s	N_{SGC}, MW	N_{HPP}, MW	N_{CHPP}, MW	Base spill power of HPP = 5%, MW
1	1992.77	1142.08	250.00	892.08	12.50
2	1992.77	1108.94	250.00	808.00	12.50
3	1992.77	1097.30	250.00	859.20	12.50
4	1992.77	1142.08	250.00	859.20	12.50
5	1992.77	1333.41	250.00	1093.70	12.50
6	1992.77	1475.13	250.00	1225.13	12.50
7	2853.49	1655.30	358.00	1297.30	17.90
8	2853.49	1657.00	358.00	1299.00	17.90
9	2853.49	1658.50	358.00	1378.82	17.90
10	2853.49	1658.00	358.00	1378.00	17.90
11	2853.49	1647.79	358.00	1369.79	17.90
12	2853.49	1648.91	358.00	1290.91	17.90
13	2853.49	1648.00	358.00	1290.00	17.90
14	2152.16	1589.91	270.00	1312.11	13.50
15	2152.16	1548.08	270.00	1278.08	13.50
16	2152.16	1553.21	270.00	1283.21	13.50
17	2152.16	1561.22	270.00	1291.22	13.50
18	2152.16	1562.98	270.00	1292.98	13.50
19	2152.16	1593.38	270.00	1323.38	13.50
20	2853.49	1636.90	358.00	1278.90	17.90
21	2152.16	1543.57	270.00	1304.44	13.50
22	2072.46	1414.87	260.00	1154.87	13.00
23	1992.77	1298.05	250.00	1048.05	12.50
24	1992.77	1230.14	250.00	980.14	12.50

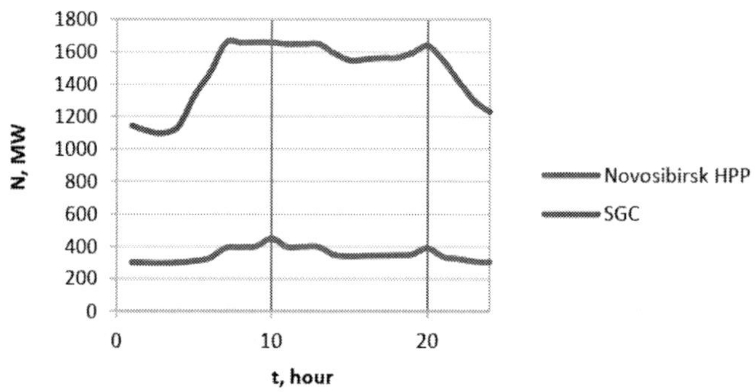

Figure 14. Schedule of daily power distribution of the SGC between the Novosibirsk HPP and the equivalent CHPP for the high-water period according to the Lagrangian multiplier method.

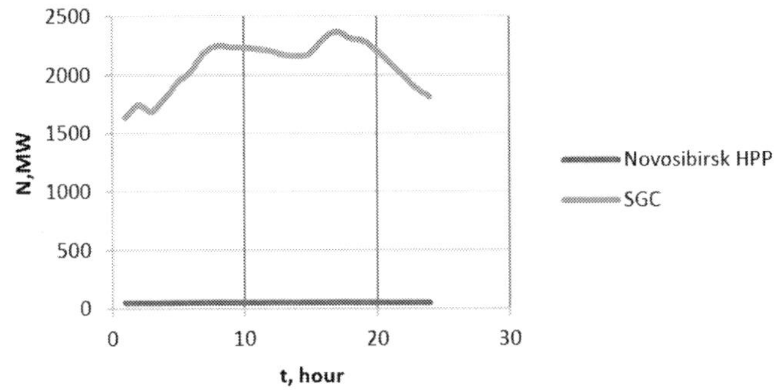

Figure 15. Schedule of daily power distribution of the SGC between the Novosibirsk HPP and the equivalent CHPP for the low-water period according to the Lagrangian multiplier method.

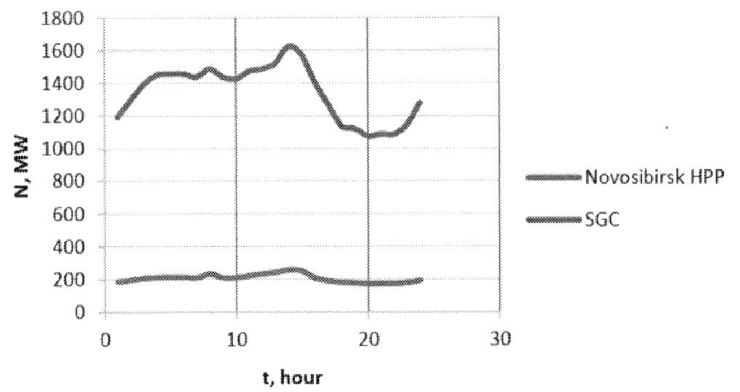

Figure 16. Schedule of daily power distribution of the SGC between the Novosibirsk HPP and the equivalent CHPP for operation with natural river flow according to the Lagrangian multiplier method.

In table 3 the daily power distribution of the SGC between the Novosibirsk HPP and the equivalent CHPP for the high-water periodaccording to the Lagrangian multiplier method is given.

Similar calculations using the conventional methodology for distributing the load of the power system between its power plants by the Lagrangian multiplier method showed comparable results. For this methodology, there is an increase in the share of CHPP in electricity production for the needs of an electricity consumer and an increase of the error when comparing the average daily water flow rate obtained by the Lagrangian multiplier method with the specified guaranteed value of water flow rate for each period. The comparison gave an error of about 10% for the high-water period, 7% for the low-water period, and 10% for operation with natural river flow, being amounted to 2329,48 m^3/s for the high-water period, 333,195 m^3/s for the low-water period, and 1182,58 m^3/s for operation with natural river flow.

It is necessary to note that the quantity of Novosibirsk HPP in covering the daily power schedule according to the developed methodology increases in 20% (20% for the high-water period, 22% for the low-water period, and 18% for operation with natural river flow) comparison with the traditional Lagrangian methodology.

Table 3. Power Distribution in the SGC between the Novosibirsk HPP and the Equivalent CHPP for the High-Water Period according to the Lagrangian Multiplier Method.

T, hour	Q, m^3/s	N_{SGC}, MW	N_{HPP}, MW	N_{CHPP}, MW	Base spill power of HPP = 5%, MW
1	1281.89	1194.46	186.67	1007.79	9.33
2	1362.28	1298.44	196.00	1102.44	9.80
3	1450.39	1394.86	206.00	1188.86	10.30
4	1513.27	1450.36	213.00	1237.36	10.65
5	1522.33	1457.01	214.00	1243.01	10.70
6	1517.80	1456.50	213.50	1243.00	10.68
7	1495.20	1438.08	211.00	1227.08	10.55
8	1698.38	1485.82	233.00	1252.82	11.65
9	1486.20	1437.60	210.00	1227.60	10.50
10	1490.70	1425.74	210.50	1215.24	10.53
11	1604.81	1472.21	223.00	1249.21	11.15
12	1698.38	1487.08	233.00	1254.08	11.65
13	1793.97	1519.81	243.00	1276.81	12.15
14	1929.29	1621.81	256.81	1365.00	12.84
15	1862.08	1573.58	250.00	1323.58	12.50
16	1459.31	1404.59	207.00	1197.59	10.35
17	1310.38	1266.40	190.00	1076.40	9.50
18	1233.90	1137.73	181.00	956.73	9.05
19	1192.11	1121.32	176.00	945.32	8.80
20	1159.05	1076.79	172.00	904.79	8.60
21	1167.28	1088.93	173.00	915.93	8.65
22	1165.64	1087.23	172.80	914.43	8.64
23	1207.93	1146.04	177.90	968.14	8.90
24	1344.90	1278.25	194.00	1084.25	9.70

Therefore, it can be concluded that the developed methodology allows not only determining a price of a hydro resource using the operational features of the HPP, but also solving energy-saving and ecological problems.

4. Results

The present paper proposes a universal method for analyzing the efficiency of technical systems using the combination of an optimization method and a method of marginal utility evaluation. Based on the comparison of water volume at a hydropower plant (HPP) and fuel amount at a combined heat and power plant (CHPP) that is used for generation of 1 kW power, it is possible to determine a water price for a HPP.

Practical testing of the developed method at the Novosibirsk HPP results in obtaining the following water prices for the HPP by year seasons (high-water period, low-water period, and operation with natural river flow): the water price for the Novosibirsk HPP in the high-water period is 21.59 RUB/m3, in the low-water period – 30.50 RUB/m3, for operation with natural river flow – 33.66 RUB/m3.

The calculation results showed that the comparison of the average daily water flow rate obtained by the developed methodology with the specified guaranteed water flow rate for each period gave an error of about 8% for the high-water period, 4% for the low-water period, and 10% for operation with natural river flow, being amounted to 2793,9m^3/s for the high-water period, 323m^3/s for the low-water period, and 1456,147 m^3/s for operation with natural river flow. This indicates the reliability of the results according to the developed methodology and allows making a conclusion about its validity and the possibility for application of the methodology to determination of the price for water as a hydro resource taking into account the operational features of HPP based on complex criteria of ecological and economic efficiency.

Similar calculations using the conventional methodology for distributing the load of the power system between its power plants by the Lagrangian multiplier method showed comparable results. For this methodology, there is an increase in the share of CHPP in electricity production for the needs of an electricity consumer and an increase of the error when comparing the average daily water flow rate obtained by the Lagrangian multiplier method with the specified guaranteed value of water flow rate for each period. The comparison gave an error of about 10% for the high-water period, 7% for the low-water period, and 10% for operation with natural river flow, being amounted to 2329,48 m^3/s for the high-water period, 333,195 m^3/s for the low-water period, and 1182,58 m^3/s for operation with natural river flow.

It is necessary to note that the quantity of Novosibirsk HPP in covering the daily power schedule according to the developed methodology increases in 20% (20% for the high-water period, 22% for the low-water period, and 18% for operation with natural river flow) comparison with the traditional Lagrangian methodology.

Therefore, it can be concluded that the developed methodology allows not only determining a price of a hydro resource using the operational features of the HPP, but also solving energy-saving and ecological problems.

5. Conclusion

The critical analysis of existing criteria for power plant management was carried out. A new approach for managing operating conditions of a power plant was proposed that considers exergy-and-ecological needs of society and takes into account the technological features of the energy industry. The mathematical model of functioning management for hydropower and thermal power plants for short-term optimization of load schedule coverage under the market conditions was developed. The use of the theory of marginal utility was substantiated for calculation of a hydro resource price using the operational features of the HPP based on the developed model.

Using the proposed model allows evaluating a price of a hydro resources at the HPP and greatly simplifying the use of algorithms for solving short-term optimization problems. Proposed approaches and methods were verified and validated mathematically and experimentally. The general statements of the methodology were realized for the specific power facilities.

6. References

[1] Sekretarev U A and Mitrofanov S V 2016 Decision support system for hydro unit commitment *2nd Int. Conf. On Ind. Eng., Applications and Manufacturing* (Chelyabinsk: IEEE) 7911424

[2] Sekretarev Y A and Panova Y V 2016 Investigations of possible using a generalized fuzzy interval for analyzing operating conditions of power equipment at hydropower plants *2nd Int. Conf. On Ind. Eng., Applications and Manufacturing* (Chelyabinsk: IEEE) 7911025

[3] Sekretarev Y A and Panova Y V 2018 Development of the intelligent decision support system for situation management of hydro units *14th Int. Sci.-Tech. Conf. on Actual Problems of Electronic Instrument Eng. APEIE* (Novosibirsk: IEEE) pp 384-8

[4] Sekretarev U A and Mitrofanov S V 2013 Preventive control taking into account of an operational condition power equipment and flowing path of hydropower plant *J. of Siberian Federal University. Engineering And Technologies* **6(1)** 3-14

[5] Moshkin B N, Myatezh T V, Sekretarev Y A and Averbukh M A 2019 Mathematical model of managing of the generating company on the criterion of the profit maximization *IOP Conf. Ser.: Mater. Sci. Eng.* **552** 012016

[6] Sekretarev Y, Sultonov S and Nazorov M 2016 Optimization of long-term modes of hydropower plants of the energy system of Tajikistan *2nd Int. Conf. on Industrial Eng.,*

Applications and Manufacturing (Chelyabinsk: IEEE) 7911428

[7] Ghamgeen I R 2018 Inflows and energy demands of dokan hydropower reservoir operation under stochastic conditions *13th Ieee Conf. on Ind. Electr. and Appl.* (Wuhan: IEEE) 8397804

[8] Malinina T V, Shulginov R N and Yushkov E S 2013 Operating efficiency evaluation of npp integrated with a pumped storage hydropower plant *Atomic Energy* **115** 64-7

[9] Berlin V V and Murav'ev O A 2015 Governing the turbine-generator unit of a small-scale hydropower plant with a long penstock *Power Technology and Engineering* **49** 240-4

[10] Lukutin B V, Shandarova E B, Matukhin D L, Makarova A F and Fuks I L 2016 Operation modes of a hydro-generator as a part of the inverter micro hydropower plant *IOP Conf. Ser.: Mater. Sci. Eng.* **124(1)** 012070

[11] Zhang L, Wu Q, Ma Z and Wang X 2019 Transient vibration analysis of unit-plant structure for hydropower station in sudden load increasing process *Mech. Syst. and Signal Processing* **120** 486-504

[12] Chang J, Wang X, Li Y, Wang Y and Zhang H 2018 Hydropower plant operation rules optimization response to climate change *Energy* **160** 886-97

[13] Balzannikov M I 2017 Research on flow in water intake of a run-of-river hydropower plant *MATEC Web Conf.* **117** 00013

[14] Abakanov T D, Begalinov A B and Abakanov A T 2016 Seismic stability of tunnels at the kapchagai hydropower plant *Soil Mechanics and Foundation Engineering* **53** 60-5

[15] Sekretarev Y, Sultonov S and Shalnev V 2015 Optimal control mode of the Vakhsh hydropower reservoirs to reduce electricity shortage in Tajikistan *Appl. Mech. and Materials* **792** 446-50

IOP Conf. Series: Materials Science and Engineering 1089 (2021) 012039 doi:10.1088/1757-899X/1089/1/012039

Prospect of development wind power plant taking into account the visual range of wind turbines location and determining the efficiency of the wind flow transition at rotors of veritical axis wind turbines.

V Y Sokolov[1] and I V Kolesnikova[2]

[1]Department of Electrical and Heat Power Engineering, Orenburg State University, Shevchenko St., 28, Orenburg, 460005, Russia
[2]Department of Philosophy, Cultural Studies and Sociology, Orenburg State University, Victory Ave., 13, Orenburg, 460018, Russia

E-mail: [1]teploosu@mail.ru, [2]ivk777@bk.ru

Abstract. Proposed method of calculating structural elements of blades of wind machines is not connected with intuitive selection of approximating functions of rigidity of structural elements. Application of the proposed design method showed necessity to introduce not only transverse reinforcing elements into the blade of the windwheel, but also longitudinal ones, which will allow to increase the generated electric energy from the small area of wind power plants location. The advantage of the proposed method for calculating the structural elements of wind machine blades is that it is not associated with an intuitive selection of approximating functions. The use of the calculation method showed the need to introduce not only transverse reinforcing elements, but also longitudinal ones into the wind wheel blade. We recommend that these elements be made in the form of a longitudinal corrugation laid between the transverse elements of the blade. The rigidity of these ribs will allow you to obtain a constant flow passing through the paddle wheel. The research conducted is not sufficient to determine the limits of the ratio of potential to electricity generation that can be achieved in practice at real wind farms. This value will depend on the stability of local air flows in the surface layer, spatial density, altitude, and the actual resistance of wind turbines. Further study of the interaction of these parameters is necessary, which is the focus of our current and future research.

1. Introduction

One of the problems of further development of wind farms is to determine the vector in which efforts should be directed. The obvious problem is simply scaling the technology.

For most horizontal wind turbines currently in operation, a reasonable distance between the supports should be 10 times rotor diameter. Industrial stations keep a distance of 7 to 10 rotor diameters if the installation is downwind relative to each other with the prevailing wind direction, 3 to 5 times normal to the prevailing direction. These are the minimum recommended distances. But even at this distance, there can still be a significant change in the continuity of the wind flow, as has been noted recently in a number of studies on the performance of wind farms [1, 2, 15, 16].

From the classical theory of the ideal rotor are the following basic provisions: maximum utilization of wind energy of an ideal propeller is equal to 0,593 (the works of N. E. Zhukovsky) and 0,687 (the works of G. K. Sabinina), loss of speed in the plane of the propeller equal to one-third the speed of wind, complete loss of wind speed for the wind wheel is two times more loss in wind speed in the plane of the wind wheel, the wind speed behind the wind wheel three times less than wind speed before the wind.

The efficiency of the wind wheel mainly depends on two time components: the time it takes for the blade to move to the distance separating it from the neighboring blade, and the time it takes for the area of strong disturbance created by the blade to move to a distance equal to the characteristic length.

The second component depends on the size and shape of the blades and changes inversely with the wind speed.

Another important point is that industrial power plants are usually installed on high ground or in open areas with a pronounced prevailing wind direction. This is not the same when the wind blows from all directions throughout the year, even with the prevailing wind direction. For this case, a minimum distance of 10 times the diameter of the rotor is the most reasonable for several separate small generators. This is quite justified aerodynamically.

But when studying the question of the visual signature, the question of the effective use of modern wind farms appears. Most of the wind energy that goes to wind farms is not used. Modern wind farms try to compensate for this disadvantage by gaining access to stable and powerful wind flows moving above the ground turbulent layer. Structures of wind farms are installed higher and with a large working diameter of the wheel. But this increase in size and height leads to increased costs, problems in engineering, and difficulties in maintaining larger equipment.

In this regard, the work of John O. Dabiri from the Laboratory of Aeronautics and bioengineering, California Institute of technology, Professor of applied science and engineering, draws attention. The scientist described the new design in the «Journal of Renewable & Sustainable Energy» published by the American Institute of physics.

Given that the efficiency of modern horizontal-axis wind turbines is approaching its maximum possible level of 59.2%, the large difference in the specific power of vertical-axis wind turbines and horizontal-axis wind turbines is an unconvincing argument in favor of horizontal machines. The results of the research of John O. Dabiri point out that in the case of vertical-axis wind turbines, we are talking about the efficiency of the entire station, and not of each individual installation, as in the case of horizontal-axis wind turbines, the efficiency of which is maximum at a sufficient distance from each other. In our case, the efficiency limit depends directly on the wind force, primarily on the speed of the horizontal wind flow and the upward turbulent flow necessary to direct the kinetic energy of the wind to the turbines on the windward side of the station. This limit is based on the characteristics of the surface atmospheric layer and surface irregularities-obstacles created for the movement of air flow by wind power plants themselves [1, 7, 8, 16]. And this indicator exceeds the theoretical limit of the efficiency of an isolated horizontal-axis wind turbine, which is the main factor determining the higher efficiency of vertical-axis wind turbines. In other words, despite the fact that an isolated vertical-axis wind turbine often shows a lower power factor than an isolated horizontal-axis wind turbine [2], this disadvantage is compensated for (and even significantly overlaps) by the fact that vertical-axis wind turbines can be located much closer to each other. Thus, the amount of wind energy not captured by the blades of one vertical-axis turbine will be intercepted by the neighboring turbine. To calculate the upper limit of the amount of energy received, it is necessary to take into account the horizontal and vertical flows of kinetic energy of wind.

The specific power of vertical-axis wind turbines of wind farms is determined by the ratio of the total power to the unit of area, which is about 2-3 Watts per sq. m. [9]. The proximity of the turbines to each other increased the efficiency of the installation compared to the efficiency of an isolated turbine to 5%, while the similar arrangement of horizontal-axial wind turbines reduces their efficiency by 20-50% [3-5, 8].

2. Materials and Methods

At the Department of electrical and thermal power engineering of Orenburg state University, taking as a basis the implemented model of building a power system consisting of solar panels and a typical design of a thermal power plant based on Orsk HEC-1 for 40 MW, we obtained results that fully confirm the preliminary results of calculations on simplified numerical models, according to which closely located vertical-axial wind turbines can improve each other's wind field. As an ideal wind farm layout, we see the use of traditional horizontal-axis wind turbines together with relatively small arrays of vertical-axis wind turbines that fill the empty space between large machines.

One of the main issues when considering the structural strength under the influence of wind flow in a sharply continental climate is the determination of the deformation of the median surface and, as a result, the change in the curvature of the streamlined profile of the wind wheel wing. This issue can be solved by using the deformation compatibility equations described in the work of A. S. Vol'mir [17]. To do this, we selected the coordinate lines of the x and y axes so that they coincide with the lines of curve of the median surface. We will count the z coordinate along the normal to the surface, counting z as positive towards the center of curve. in the case we are considering, we mean shells of positive or zero Gaussian curve. Let's denote the displacement of the points of the median surface along the lines x, y, z, respectively, through u, v, ω. For the initial curvature of the lines x and y, we introduce the notation kx и ky.

We will find expressions for the deformations of the median surface at the shell deflections comparable to its thickness. To do this, use the expressions obtained for a flat plate. It is quite natural to use additional members when working with a thin plate in the form of a shell. The offset scheme can be seen in figure 1, which shows the element of the x line in the initial and offset positions.

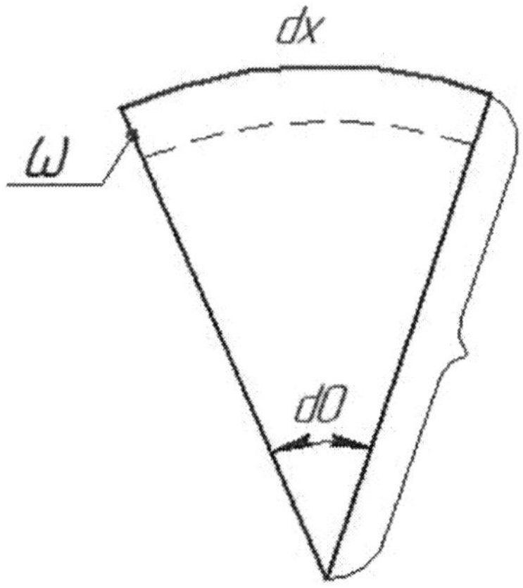

Figure 1. Determination of deformations in the middle surface during deflections.

The radius of curve of the element, initially equal $\rho_x = \frac{1}{k_x}$ to after the offset will be $\rho_x - \omega$. The relative deformation caused by moving the element to the center of curve is equal to

$$\varepsilon_x = \frac{(\rho_x - \omega)d\theta - \rho_x d\theta}{\rho_x d\theta} = -\frac{\omega}{\rho_x} = -\omega k_x \tag{1}$$

For the y axis by analogy

$$\varepsilon_y = -\frac{\omega}{\rho_y} = -\omega k_y \tag{2}$$

The final expressions for deformations of the median surface are given as:

$$\begin{cases} \varepsilon_x = \dfrac{du}{dx} - k_x\omega + \dfrac{1}{2}\left(\dfrac{d\omega}{dx}\right)^2 \\ \varepsilon_y = \dfrac{du}{dy} - k_y\omega + \dfrac{1}{2}\left(\dfrac{d\omega}{dy}\right)^2 \\ \gamma = \dfrac{du}{dy} + \dfrac{dv}{dx} + \dfrac{d\omega}{dx}\dfrac{d\omega}{dy} \end{cases} \tag{3}$$

In this case, the deformations of the blade body elements are not independent. Therefore, the following strain compatibility equation can be used for further calculations:

$$\frac{d^2\varepsilon_x}{dy^2}+\frac{d^2\varepsilon_y}{dx^2}-\frac{d^2\gamma}{dxdy}=\left(\frac{d^2\omega}{dxdy}\right)^2-\frac{d^2\omega}{dx^2}\frac{d^2\omega}{dy^2}-k_x\frac{d^2\omega}{dy^2}-k_y\frac{d^2\omega}{dx^2} \quad (4)$$

To determine the bending deformations, we use the hypothesis of straight normals. Let us denote by xx и xy the changes in the curve that will receive the coordinate lines x and y at some point on the median surface; by χ we will still denote the «curve» of the torsion of the surface.

For a closed shell the curve of the line segment dx receives a change due to the deflection ω

$$\Delta x=\frac{1}{\rho_x}-\frac{1}{\rho_{x-\omega}}=-\frac{\omega}{\rho_x(\rho_x-\omega)} \quad (5)$$

Let's approximate $\Delta x=\frac{\omega}{\rho_x^2}$ and enter this value in the expression for x_x and get

$$x_x=-\frac{d^2\omega}{dx^2}-\frac{\omega}{\rho_x^2} \ \ and \ \ x_y=-\frac{d^2\omega}{dy^2}-\frac{\omega}{\rho_y^2} \quad (6)$$

We determine the change in the curve of a circular cylindrical shell of radius R in the section made along the arc shown in figure 2, provided that the relationship between the deflection ω and the arc coordinate y has the form

$$\omega=f\sin\frac{ny}{R} \quad (7)$$

In our case, the parameter n corresponds to the number of waves formed on the surface of the closed shell of the rotating rotor elements.

Calculating the change in the surface curve we get:

$$x_y=f\frac{n^2}{R^2}\sin\frac{ny}{R}-f\frac{1}{R^2}\sin\frac{ny}{R} \quad (8)$$

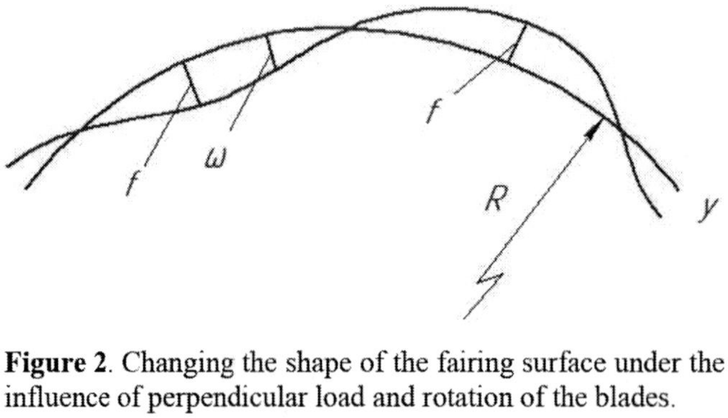

Figure 2. Changing the shape of the fairing surface under the influence of perpendicular load and rotation of the blades.

As we can see, the specific weight of the second term of the calculation equation 8 in comparison with the first one is measured by the ratio 1/n2. You can omit the second term if a sufficiently large number of waves (for example, more than four) are formed along the length of the arc of the circle. In a more General form, this requirement can be expressed in such a way that the dimensions of surface irregularities formed when the shell is deformed must be small in comparison with the radii of curvature of the structure. In problems relating to stability of shells, this is usually true, but in our case we have a long, tending to a cylindrical shell, the length of which surpasses the radius of the curve of the considered surfaces, so we consider it most correct to consider it as the shell buckling under external pressure and twisting.

Now let's assume that the shell has deviations from the ideal shape obtained during the initial manufacture or installation, characterized by initial deflections of the ω_{If}. Then, by analogy with expressions relating to a flat surface with initial deformation, we can represent the equation of compatibility of deformations as follows:

$$\frac{d^2\varepsilon_x}{dy^2}+\frac{d^2\varepsilon_y}{dx^2}-\frac{d^2\gamma}{dxdy}=\left(\frac{d^2(\omega_{\text{If}}+\omega)}{dxdy}\right)^2-\left(\frac{d^2\omega_{\text{If}}}{dxdy}\right)^2-$$
$$-\frac{d^2(\omega_{\text{If}}+\omega)}{dx^2}\frac{d^2(\omega_{\text{If}}+\omega)}{dy^2}+\frac{d^2\omega_{\text{If}}}{dx^2}\frac{d^2\omega_{\text{If}}}{dy^2}-k_x\frac{d^2\omega}{dy^2}-k_y\frac{d^2\omega}{dx^2} \quad (9)$$

The calculated dependences represent the following values: kx и ky – the main curvature of the middle surface of the shell, ω – the deflection of an arbitrary point of the middle surface, ω_{If} – the initial deflection, f – the deflection arrow, R – the radius of the middle surface of the shell under consideration, γ – the shear strain of the middle surface, ε_x and ε_y – elongation deformations along the x and y axes.

Let's write out the relations between deformations and stresses in the middle surface:

$$\left.\begin{aligned}\sigma_x&=\frac{E}{1-\mu^2}\left(\varepsilon_x+\mu\varepsilon_y\right)\\ \sigma_y&=\frac{E}{1-\mu^2}\left(\varepsilon_y+\mu\varepsilon_x\right)\\ \tau&=\frac{E}{2(1+\mu)}\gamma\end{aligned}\right\} \quad (10)$$

The constraints between the moments of curvature will remain the same as for a flat plate:

$$\left.\begin{aligned}M_x&=-D\left(\frac{d^2\omega}{dx^2}+\mu\frac{d^2\omega}{dy^2}\right)\\ M_y&=-D\left(\frac{d^2\omega}{dy^2}+\mu\frac{d^2\omega}{dx^2}\right)\\ H&=-D(1-\mu)\frac{d^2\omega}{dxdy}\end{aligned}\right\} \quad (11)$$

For transverse forces, we get expressions:

$$\left.\begin{aligned}Q_x &= -D\frac{d}{dx}\nabla^2\omega\\ Q_y &= -D\frac{d}{dy}\nabla^2\omega\end{aligned}\right\} \tag{12}$$

The equilibrium equation is automatically executed when the stress function is introduced. We substitute the expressions (11) and (12) into the equilibrium equation of the system:

$$\frac{dQ_x}{dx}+\frac{dQ_y}{dy}+\sigma_x h\left(k_x+\frac{d^2\omega}{dx^2}\right)+\sigma_y h\left(k_y+\frac{d^2\omega}{dy^2}\right)+2\tau h\frac{d^2\omega}{dxdy}+q=0 \tag{13}$$

Let's receive the following dependence:

$$D\nabla^2\nabla^2\omega=\sigma_x h\left(k_x+\frac{d^2\omega}{dx^2}\right)+\sigma_y h\left(k_y+\frac{d^2\omega}{dy^2}\right)-2\tau h\frac{d^2\omega}{dxdy}+q \tag{14}$$

Now we transform the equation of the compatibility of strains, expressing strains through stresses:

$$\begin{gathered}\frac{d^2\sigma_x}{dy^2}-2\frac{d^2\tau}{dxdy}+\frac{d^2\sigma_y}{dx^2}-\mu\left(\frac{d^2\sigma_x}{dy^2}+2\frac{d^2\tau}{dxdy}+\frac{d^2\sigma_y}{dx^2}\right)=\\ E\left[\left(\frac{d^2\omega}{dxdy}\right)^2-\frac{d^2\omega}{dx^2}\frac{d^2\omega}{dxy^2}-k_x\frac{d^2\omega}{dy^2}-k_y\frac{d^2\omega}{dx^2}\right]\end{gathered} \tag{15}$$

We introduce the stress function S into the equation and then the basic equations of flexible gentle shells take the form:

$$\frac{D}{h}\nabla^2\nabla^2\omega=L(\omega,S)+k_x\frac{d^2S}{dy^2}+k_y\frac{d^2S}{dx^2}+\frac{q}{h} \tag{16}$$

$$\frac{1}{E}\nabla^2\nabla^2 S=-\frac{1}{2}L(\omega,S)-k_x\frac{d^2\omega}{dy^2}+k_y\frac{d^2\omega}{dx^2} \tag{17}$$

To determine the nature of the curved surface obtained during loading in a system with one degree of freedom, a more accurate solution to the proposed problem is necessary, which can be solved if you approximate the curved surface with several parameters and present the initial equations for calculating in finite differences. The solution of the problem is carried out using electronic computers using the finite differences method. In this case, the windwheel blade is considered as a flat plate with sides in the plan, a, b having some initial bend.

We use dimensionless parameters

$$\bar{x}=\frac{x}{a},\quad \bar{y}=\frac{y}{b},\quad \bar{t}=\frac{th}{ab}V,\quad \bar{\omega}=\frac{\omega}{h},\quad \bar{S}=\frac{S}{Eh^2} \tag{18}$$

The calculated constraints take the form

$$\frac{d^2\bar{\omega}}{d\bar{t}^2} = \frac{d^2\bar{\omega}}{d\bar{x}^2}\left(\frac{d^2\bar{S}}{d\bar{y}^2} - \bar{p}_x\right) + \frac{d^2\bar{\omega}}{d\bar{y}^2}\left(\frac{d^2\bar{S}}{d\bar{x}^2} - \bar{p}_y\right) - 2\frac{d^2\bar{\omega}}{d\bar{x}d\bar{y}}\frac{d^2\bar{S}}{d\bar{x}d\bar{y}} - \frac{1}{12(1-\mu^2)}\nabla^4(\bar{\omega} - \overline{\omega_0}) + \frac{\bar{k}_y}{\lambda^2}\left(\frac{d^2\bar{S}}{d\bar{x}^2} - \bar{p}_y\right) + \overline{k_x}\cdot\lambda^2\left(\frac{d^2\bar{S}}{d\bar{y}^2} - \bar{p}_x\right) + \bar{q} \tag{19}$$

$$\nabla^4\bar{S} = \left(\frac{d^2\bar{\omega}}{d\bar{x}d\bar{y}}\right)^2 - \frac{d^2\bar{\omega}}{d\bar{x}^2}\frac{d^2\bar{\omega}}{d\bar{y}^2} - \left(\frac{d^2\overline{\omega_0}}{d\bar{x}d\bar{y}}\right)^2 + \frac{d^2\bar{\omega}_0}{d\bar{x}^2}\frac{d^2\bar{\omega}_0}{d\bar{y}^2} - \frac{\bar{k}_y}{\lambda^2}\frac{d^2(\bar{\omega} - \overline{\omega_0})}{dx^2} - \overline{k_x}\cdot\lambda^2\frac{d^2(\bar{\omega} - \overline{\omega_0})}{dy^2} \tag{20}$$

The calculation process begins with the creation of a calculation grid, for the nodes of which the loss is determined and the stress function is found by solving the system of difference equations by successive approximations. To do this, sequentially bypass the nodes of the grid in question. As the initial approximation of the stress function, its values can be taken at the previous step in time, if they are not equal to zero, or the value of the displacement function. The iteration ends after the required degree of accuracy has been achieved. Step on time is completed by recalculation of out-of-contour values of ω and S in accordance with accepted boundary conditions are values of $\omega^{(3)}(x, y, 3\Delta t)$ and $S^{(3)}(x, y, 3\Delta t)$, functions in out-of-contour nodes are determined. After each time step, or at random intervals, stress values must be calculated.

3. Results

Calculations showed that the static load increased in proportion to time. In the case where the loading rate is relatively small (p/t = 2.15), the intense increase in deflections occurs at the load $p \approx 1.8\ p_{max}$, then non-linear fluctuations occur with respect to the equilibrium positions. With a high loading rate (p = 5t), deflection increases sharply when the $p \approx 1.8\ p_{max}$. After that, during each period of oscillation, the shell of the rotor blade jumps from one equilibrium shape to another, so that the deflection changes the sign. With a sharp increase in deflections and subsequent oscillations, the shape of the surface changes greatly, which in turn leads to an increase in resistance in the boundary layer of the surface when the incoming air flow passes, and the amount of loading of the support elements of the structure increases.

The elements of the system that perceive the incoming wind flow can be divided into three main groups according to their structure

Elements of small curvature of small curvature ($0 < k < 50$). These elements of the wind turbine blades were deformed during loading from the very beginning by the appearance of deflections and bending stresses. The nature of the longitudinal deflection during the first loading period depended on the shape of the initial loss; when approaching the critical force, in some cases the deflections increased sharply, in others they changed their sign.

Loss of stability at the first critical load occurs in the elements of the blades of small curvature without a clear jump. However, with further loading, jumps occur: they coincide with the moments of a sharp change in the number of half-waves along the length of the panel. The dimensions of the sides of the reference element of the blade structure change: the side located along the length is elongated in the middle part of the element under consideration and reduced in the extreme parts compared to the other side having a transverse direction. This means that the deformation elements are not square: at maximum load, they are elongated along the arc.

The elements of the mean curvature (50 < k < 175). These panels differed from the previous ones in the nature of wave formation even at the very moment of loss of stability. Here, the predominant direction of deformations of the reference elements to the center of curvature was clearly expressed, while according to linear theory, the elements of the rotor blade facing the center of curvature or from the center have equal amplitudes.

In most cases, when the panels lost stability, several deflections of the shells were formed, as shown in Fig.3. More often, the option shown in Fig.3 left with one deep bulge to the center of curvature and two smaller ones from the center.

Figure 3. Wave formation on wind turbine blade elements

Further loading of the blade elements was accompanied by the expansion of the main bulge and the formation of additional dents arranged in a staggered order.

When measuring the bends of the blade shells, jumps in the effective load were noted at the moment of loss of stability of the blade element, while the devices recorded a significant decrease in the main load.

In the process of reducing the effective pressure, small deformations disappeared imperceptibly, but then a very strong jump was observed, during which the main bulge was pushed through. The value of the compressive force that this reverse jump corresponded to was much lower than the critical load. After complete unloading, residual deformations were observed for some elements of the blades. Therefore, during repeated loading, the critical force was significantly lower than the initial value.

Elements of large curvature (k >175) reacted very weakly to external disturbances.

4. Conclusion

The advantage of the proposed method for calculating the structural elements of wind machine blades is that it is not associated with an intuitive selection of approximating functions. The application of the calculation method showed the need to introduce not only transverse reinforcing elements, but also longitudinal ones into the wind wheel blade. We recommend performing these elements in the form of a longitudinal corrugation laid between the transverse elements of the blade. The rigidity of these edges will allow you to get a constant flow passing through the blade wheel.

The research conducted is not sufficient to determine the limits of the ratio of potential to electricity generation that can be achieved in practice at real wind farms. This value will depend on the stability of local air flows in the ground layer, spatial density, height, and the actual resistance of wind turbines. It is necessary to further study the interaction of these parameters, which is the focus of our current and future research.

Referens

[1] Physick W L and Garratt J R 1995 Incorporation of a high-roughness lower boundary into a mesoscale model for studies of dry deposition over complex terrain *Boundary-Layer Meteorol* **74** 55-71

[2] Hau E 2013 *Wind Turbines: Fundamentals, Technologies, Application, Economics* (Berlin: Springer)

[3] Okulov V and Sorensen J 2007 Stability of helical tip vortices in a rotor far wake *J. of Fluid Mechanics* **576** 1-25

[4] Dabiria J O 2011 Potential order-of-magnitude enhancement of wind farm power density via ounter-rotating vertical-axis wind turbine arrays *J. of Renewable and Sustainable Energy* **3** 043104
[5] Kiranoudis C T and Maroulis Z B 1997 Effective Short-Cut Modelling of Wind Park Efficiency *Renewable Energy* **11** 439-57
[6] Laursen J, Enevoldsen P and Hjort S 2007 3D CFD quantification of the performance of a multi-megawatt wind turbine *J. of Physics Conf. Series* **75(01)** 012007
[7] Bossanyi E A, Maclean C, Whittle G E, Dunn P D, Lipman N H and Musgrove P J 1980 The Efficiency of Wind Turbine *Clusters. Proc. Third Intern. Symp. Wind Energy Systems* (DK: Lyngby) 401-16
[8] Liu M, Yocke M and Myers T 1983 Mathematical model for the analysis of wind-turbine wakes *J. of Energy* **7** 73-8
[9] MacKay D 2009 *Sustainable Energy – Without the Hot Air* (Cambridge: UIT Cambridge Ltd)
[10] Magnusson M, Rados K and Voutsinas S 1996 A study of the flow downstream of a wind turbine using measurements and simulations *Wind Engineering* **20(6)** 389-403
[11] Magnusson M and Smedman A 1999 Air flow behind wind turbines *J. of Wind Eng. and Industrial Aerodynamics* **80** 169-89
[12] Mann J 1994 The spatial structure of neutral atmospheric surface-layer turbulence *J. of Fluid Mechanics* 273 141-68
[13] Mann J 1998 Wind field simulation *Probabilistic Engineering Mechanics* **13(4)** 269-82
[14] Rajagopalan R, Klimas P, Rickerl T and Propul J 1990 Aerodynamic interference of vertical axis wind turbines *J.l of Propulsion and Power* **6** 645-53
[15] Bazilevs Y, Calo V M, Hughes T J R and Zhang Y 2008 Isogeometric fluid–structure interaction: theory, algorithms, and computations *Computational Mechanics* **43** 3-37
[16] Kuraishi T, Takizawa K and Tezduyar T E (2019) Tire aerodynamics with actual tire geometry, road contact and tire deformation *Computational Mechanics* **63** 1165-85
[17] Vol'mir A 1956 Gibkie plastinki i obolochki [Flexible plates and shells] (Moscow: Gostekhteorizdat) [In Russian]
[18] Whittlesey R, Liska S and Dabiri J 2010 Fish schooling as a basis for vertical axis wind turbine farm design *Bioinspiration & Biomimetics* **5(3)** 035005

Qualitative scenario analysis of development of energy storage systems in Russia until 2035

A G Logacheva[1], Y N Zatsarinnaya[2], E G Stepanova[2], E V Rep'ev[2] and E A Fedotov[2]

[1] Department of Power supply systems of Industrial Enterprises, Kazan State Power Engineering University, St. Krasnoselskaya, 51, 420066, Kazan, Russia
[2] Department of Power stations named after V K Shibanov, Kazan State Power Engineering University, St. Krasnoselskaya, 51, 420066, Kazan, Russia

E-mail: waysubbota@gmail.com

Abstract. A stationary energy storage boom is forecast for the next two decades, according to a report by the US consulting firm Bloomberg New Energy Finance (BNEF). BNEF analysts believe that energy storage around the world will grow exponentially, from a modest 9 GW / 17 GWh commissioned by 2018 to 1,095 GW / 2,850 GWh by 2040. Experts call the ongoing global changes a "contributory revolution". The development of energy storage systems is related to trends in the energy sector, energy costs, political and environmental conditions in the world. Moreover, energy storage technologies can face both general and specific risks. The authors of the article took into account possible risks and carried out a qualitative scenario analysis of the development of energy storage systems in Russia in the future until 2035. The authors used a probability scale to avoid cognitive biases in the scenario assessment. The most likely scenario for the development of ESS today we consider the basic.

1. Introduction

Energy storage systems (ESS) are the reason for the global transformation of the electric power industry, which is being launched all over the world and is becoming relevant for Russia now. There are several reasons for the beginning of the energy transition and the rapid development of storage systems. One, the most significant of them, has been the change in the attitude of many countries towards the used energy resources in the past few years. A prerequisite for this was the adoption in 2015 of the Paris Agreement, which united the efforts of world powers to deterrence climate change on the planet. Many countries have made commitments and developed new strategies to reduce greenhouse gas emissions by using renewable energy (RES) power plants and replacing carbon-fueled cars with electric vehicles. The task of developing ESS has become relevant for several areas of application at once.

The highlighting of such a trend as RES has pushed the development of distributed generation, digitalization and intellectualization of energy systems. The emerging need for ESS led to a sharp decrease in the cost of almost all known energy storage systems to 85% in the period 2010-2018. According to BNEF, the cost of ESS will continue to decline as demand for them grows. The decline in the cost of ESS, in turn, influenced the trends in the global electric power systems. Global think tanks are looking at storage as part of a new energy fabric and smart energy technologies. "Smart storage systems" claim to change the traditional understanding of the global energy, accelerating and facilitating its digital transition: the development of energy and capacity markets; expanding the range of services, the use of aggregators, demand management (Demand Response); increase in the share and support of renewable energy sources in the energy balance, smoothing of daily and annual load schedules, replacement of grid and generating capacities.

2. Materials and Methods

A forecast of scenarios and main parameters of the technical and economic development of energy storage systems for the period up to 2035 was developed in the composition of three main options – negative, basic and innovative.

The negative scenario does not envisage radical changes in ESS technologies, it is expected that the current priorities of the energy policy of large producers will be maintained, and the pace of development of energy storage units will be rather slow. There will be no noticeable improvements in the area of storage systems.

According to the basic scenario, there is no technological revolution, but the introduction of technologies that are already being tested at the present time is expected. A gradual increase in the economic efficiency of the technologies already in use is expected. The transfer of technologies to developing countries is complicated (the average term of technology transfer remains at the level of 10-12 years). Environmental and energy efficiency trends that have emerged in recent years are helping to strengthen the position of ESS in the energy markets.

The innovative scenario implies the accelerated development of new technologies and halving the time of their transfer from developed countries to developing ones. This scenario implies the full use of the ESS potentials and the fastest increase in their market share, the accelerated development of new technologies, their cost reduction, and active support from the state.

To compile a scenario analysis in the field of energy storage systems in Russia, we applied correlation analysis. It allows you to numerically assess the influence of various factors on the development of ESS and to assess the likelihood of the occurrence of each of the three proposed development scenarios. The influencing factors are grouped according to five criteria: politics, economics, ecology, risks, technologies. Since the factors we have identified can affect both all types of ESS as a whole, as well as a specific technology, the concept of "weight" of a factor has been introduced. "Weight" is calculated as follows: "weight" = "correlation coefficient" x "share in ESS". The "share in the ESS" refers to the part of the ESS technologies that is affected by the factor under consideration. The correlation coefficient is calculated as follows:

$$r = \frac{n\sum xy - \sum x \sum y}{\sqrt{(n\sum x^2 - (\sum x)^2)(n\sum y^2 - (\sum y)^2)}}$$

where r is the correlation coefficient, and x and y are the factors themselves between which a correlation is established.

For example, there is a link between GDP (x) and ESS investment (y). In this case, the correlation coefficient can be in the range from -1 to 1, where the "+" and "-" signs show a direct or inverse dependence of factors, and the absolute value of the correlation reflects how strong the link is. If the coefficient is close to 1, then the factors are completely dependent on each other, a value close to 0 indicates the absence of a link.

Excel and Matlab software products were used for calculations. The following is a detailed description of the scenario analysis performed.

2.1. Forecasting the development of ESS based on the analysis of economic, environmental, technological, political factors and risks.

2.1.1. Economic factors.

Factor 1. Reducing the cost of ESS technologies. According to the latest BNEF forecasts, by the year 30 the global ESS market will grow 6 times in comparison with the level of 2017 [1]. The energy storage market will develop along a trajectory that is similar to the development dynamics of the photovoltaic solar energy market in the period 2000-2015. By 2035, it is predicted that the cost of electrochemical ESSs will decrease on average to $ 300–350 per kWh from the current 500–600 per kWh. Figure 1 demonstrates the value of energy storage cost reduction over the lifecycle of an LCOS.

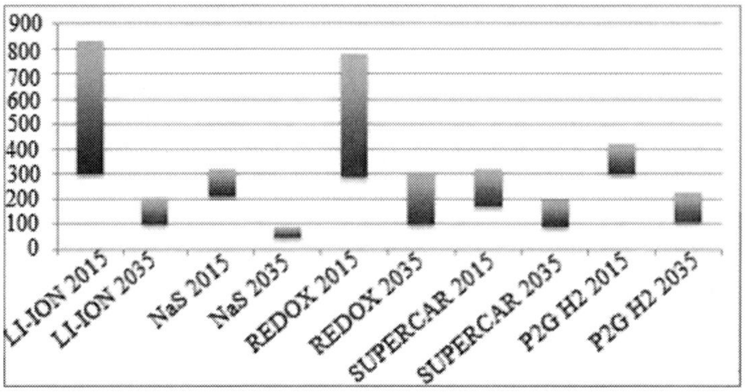

Figure 1. Forecast of the dynamics of the present value of electricity accumulation using the example of electrochemical storage.

Thus, in the long term, the most probable decrease in the cost of ESS is considered with an increase in battery life, and an accelerated growth of ESS can be predicted. Decrease economic conditions could lead to moderate growth in ESS. Stagnation in the development of ESS is the least likely, since in the period from 2010 to 2016 the cost of batteries fell by almost 4 times - from 1,000 to 273 USD per kWh - and continues to decline on average to 20% per year [2].

Factor 2. The existence of a price difference for price arbitrage on the wholesale electricity market. The cost of electricity on the market depends on the time of consumption. During off-peak load hours (night, weekends), electricity is cheaper. In the Youth Forecast of Global Energy Development in 2019, the UPS of the Middle Volga was highlighted among the regions most suitable for the installation of ESS. The difference in prices for the UPS of the Middle Volga as of 05/01/2020 [3] at night and at peak hours is 193.4 rubles.

Since until 2035 the existence of peak consumption hours in the energy system of Russia and the difference in prices at night and daytime are most likely. Energy storage devices can profit from buying electricity at a low price and selling it at a high price. The most likely scenario is the basic scenario. The observed price difference is not large and is not able to determine the accelerated growth of ESS. The situation when the schedule of energy system consumption will be absolutely flat is unlikely by 2035, therefore the price difference will remain, and the negative scenario is the least likely.

Factor 3. GDP. GDP and investment are interconnected: a small investment recovery can provide a significant increase in gross domestic product. At the same time, a change in the country's GDP also affects investment activity in ESS projects [4]. Figure 2 shows the correlation of RF GDP and investment [5] in the period 2015-2018.

In this case, the actual correlation coefficient was r = - 0.39. The sign "-" here shows that during the analyzed period, with an increase in GDP, the number of investments decreased in the proportion of 1 to 0.39.

In connection with the epidemiological situation in the Russian Federation, the International Monetary Fund predicts a decrease in Russia's GDP in 2020 by 5.5%. In 2021, the fund expects Russia's GDP to grow by 3.5%. In the long term, stabilization of GDP growth is considered most likely, and moderate growth of ESS can be predicted. On the other hand, given the epidemiological situation, the decline in GDP growth may be delayed and exceed current forecasts, which will lead to stagnation in the development of ESS. Accelerated ESS growth driven by rapid GDP growth is least likely.

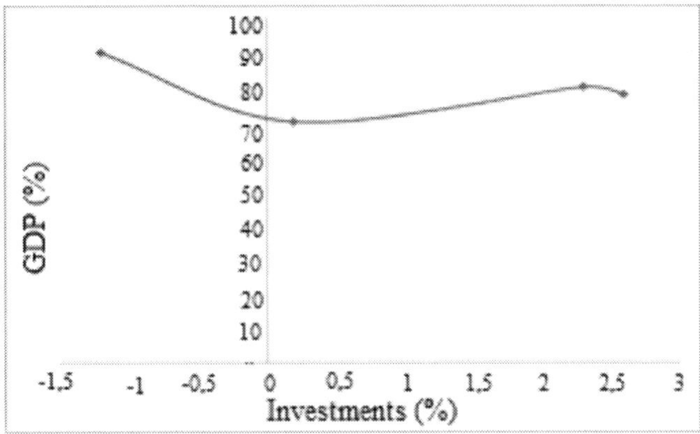

Figure 2. Correlation between GDP and investment in the Russian Federation in the period 2015-2018.

Factor 4. Supply security. According to the report of Berenberg Thematics, a team of the oldest bank in Germany, focused on researching breakthrough technologies that may be of interest to investors, there is a risk of the security of supply of metals required for the production of lithium-ion batteries [6]. Metals with a high supply risk include cobalt and natural graphite. The risk arises from the fact that 56% of world cobalt production comes from Congo and 69% of natural graphite comes from China.

The emergence of international and internal political conflicts in exporting countries can cause trade wars in the producing countries, which, in turn, can cause military clashes. Any of the wars will damage the economies of the countries as a whole and may affect the cost of production of ESS. Despite the existence of such a risk, the 2018-19 trade war between the United States and China did not lead to an increase in the cost of ESS, which indicates a low correlation between risk and the development of ESS. Therefore, in the long term, it is believed that the decrease in the cost of ESS will continue, and the basic scenario of the growth of lithium-ion storage systems can be predicted with high probability. At the same time, if a larger number of countries are involved in a trade war, as in one of the well-known stories of the 19th century trade war of the USA, England and France against China [7], then the probability of the predicted decrease in the cost of Li-ion ESS decreases, and, consequently, decreases the probability of an innovative scenario and accelerated growth in the use of lithium-ion storage. In the event of global conflicts between countries, the probability of which until 2035 is low, it can be predicted that the probability of a pessimistic scenario for lithium-ion storage devices is small.

Factor 5. Emergency situations of a global nature. Today, the whole world is experiencing a crisis associated with the COVID-19 pandemic. Against the background of the general economic downturn, the pace of implementation of ESS projects may also slow down. The pandemic has led to a decrease in the consumption of resources, and in particular of oil. For the Russian economy, which is dependent on energy prices, this scenario means a depreciation of the ruble. Since ESS technologies often require the import of components, for Russian companies this will mean an increase in the costs of projects being implemented. Analyzing the link of investments [8] in the Russian energy sector with oil prices of Brent, we obtained a correlation coefficient equal to 0.82. The graph of changes in oil prices and the volume of investments by years is shown in Figure 3. According to experts, the recovery of demand for oil products to the level preceding the pandemic will not occur until 2021.

Thus, under the influence of the current crisis until 2035, accelerated growth in the number of ESS projects can be expected with the least probability. At the same time, due to forecasts of a recovery in oil demand in 1-2 years, the probability of moderate growth in the long term in the field of new projects is higher than the probability of stagnation.

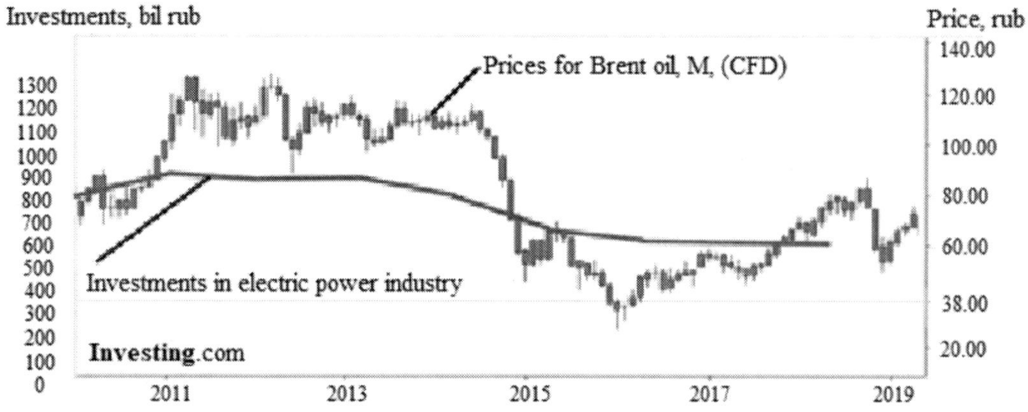

Figure 3. Dynamics of prices for Brent oil and volumes of investments in the Russian electric power industry in 2010-2019.

2.1.2. Ecological factors.

Factor 1. The volume of CO2 emissions. If the graph of CO2 changes for the last 10 years is scaled, superimposed on the graph of changes in the power accumulated by ESS [9], one can notice the correlation between them (Fig. 4 a). The linear correlation coefficient between the two data sets is 0.7, which indicates a high degree of dependence - the development of energy storage systems in renewable energy sources is dictated by environmental problems.

Factor 2. Increase in air temperature. Considered above in relation to CO2 emissions is also relevant for an increase in the average temperature over the earth (Fig. 4 b).

Based on the forecast of BP [10] and ENERS Future [11], we have compiled three scenarios. In the negative scenario, over the next 20 years, the growth rate of CO2 emissions will be approximately 3.5 times lower than the growth rate over the past two decades. The temperature will rise by 0.2-0.3^0 by 2035 and by 1^0 by the end of the century. This means that the growth of RES and ESS will remain and will be moderate, although fossil fuels will remain a competitive and attractive source of energy. Most likely scenario.

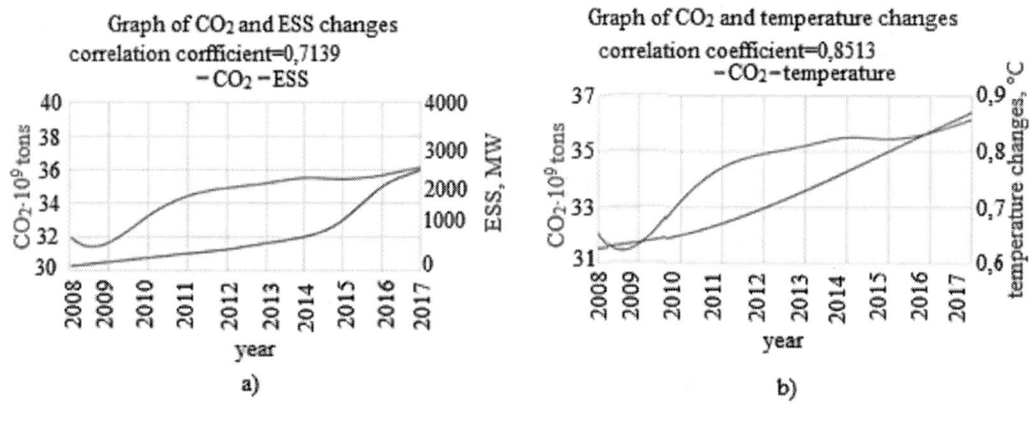

Figure 4. Graph of CO_2 and ESS changes (a); graph of CO_2 and temperature changes (b) for 2008-2017.

In the basic scenario, a 13% reduction in carbon dioxide emissions compared to 2020 would result in a cooling of about 0.2 ° C. The events that led to such a huge decrease in emissions can either be associated with a significant increase in RES and ESS. A significant increase in RES and ESS can only be achieved by investing enormous funds in these industries. But given the recent sharp decline in economies, this scenario has an average probability.

According to the innovative scenario, CO2 emissions will be reduced by up to 30% of the current level. This will cause a cooling of about 0.5 ° C. This scenario is even more optimistic when viewed from an environmental point of view. From the point of view of the ESS, the difference from the basic scenario is small. Such a global reduction in human carbon dioxide emissions into the atmosphere is unlikely due to an increase in RES and ESS. From our point of view, the probability of such a scenario is low.

2.1.3. Technological factors.

Our team is aware of a wide variety of technological characteristics, the improvement of which can significantly affect the development of ESS, we did not set ourselves the goal of considering all existing technologies. We considered the technologies of promising ESS and the most popular in 2020.

Factor 1. Increasing the efficiency of electrolysis systems for hydrogen production. According to the Concept for the Development of the Electricity Storage Systems Market in the Russian Federation, one of the priority scientific and technical tasks is the development of highly efficient water electrolyzers and systems for compact storage and transportation of hydrogen. Many researches are underway to improve the efficiency of electrolytic systems. For example, a research team from Ulsan National Institute of Science and Technology, Korea Institute for Energy Research and Sukmun University improved an already existing design called a Solid Oxide Electrolysis Cell (SOEC) by making both electrodes solid, like the electrolyte that serves as an ion conductor. They retained all the positive qualities of the existing cell and developed a new hybrid system, called Hybrid-SOEC, which is four times more efficient than the existing one [12].

The authors analyzed the correlation between hydrogen production and the growth in commissioning of ESS projects (it was 0.9), the graph is shown in Figure 5a, which indicates that the improvement of hydrogen production technology should affect the accelerated growth of ESS projects.

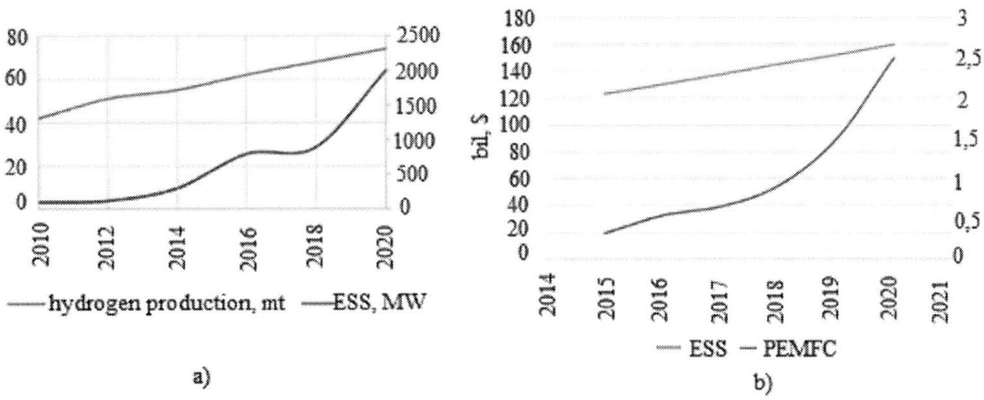

Figure 5. Change in the growth of ESS projects and hydrogen production (a) and the volume of the PEMFC cell market (b) for 2010-2020.

At the same time, the methods of producing hydrogen are diverse and only 5% of hydrogen from water by electrolysis due to the high cost of the technology. Therefore, despite significant scientific developments in the field of efficiency of electrolysis systems, the most likely scenario is the stagnation of this technology, since at the moment, historically, more profitable methods of producing hydrogen (catalytic steam reforming) have developed. In the long term, moderate development can take place, provided that the study and implementation of more technologically advanced methods of producing hydrogen (in this case, precisely by electrolysis), accompanied by an increase in efficiency, continues.

Factor 2. Hydrogen-air proton exchange membrane fuel cell (PEMFC). The second technology, the improvement of which we have considered, is also related to hydrogen ESS. We considered a hydrogen-powered electromagnetic cell with two electrodes and a polymer membrane as the electrolyte - PEMFC. The growth of the PEMFC market is mainly driven by the downward trend in greenhouse gas emissions coupled with favorable government initiatives. Similarly, the increasing demand for clean fuel vehicles to limit carbon emissions is one of the main factors driving the industry forward. PEMFCs use oxygen as an oxidizing agent and are charged with pure hydrogen, which results in the release of harmless chemicals other than heat and water vapor [13].

PEMFC systems are primarily the preferred choice for the production of hydrogen-powered vehicles due to their fast start-up combined with high energy output density for efficient engine operation, while at the same time being predicted to be used as zero-emission hydrogen ESS. Key companies are: Ballard Power Systems; Plug Power; Hydrogenics. Our team analyzed the volume of the market in billion $ PEMFC and ESS for 2015-20 (Fig. 5 b), the correlation was determined = 0.93, which indicates a high degree of influence of the development of technology on the ESS. At the same time, as in the previous factor, the share of the use of technology is currently very small, therefore it is not able to have a significant impact on the development of ESS globally, despite its significant improvement. Therefore, in the long term, a high probability of technology stagnation is predicted. The accelerated growth of this technology, as in the previous factor, is limited by the current situation in the hydrogen fuel market, the lack of infrastructure and unresolved safety issues.

Factor 3. Modernization of electrodes of lithium-ion storage devices. Lithium-ion energy storage devices are today the leaders in the world market for electrochemical storage devices. Credit Suisse estimates that by 2035 more than 75% of Li-ion battery (LIB) demand will come from the electric vehicle sector [14].

Experts from the Energy Center of the Skolkovo Business School say that with the expansion of the electric vehicle market, the LIB market will increase. In this regard, work is underway to improve LIB technologies to reduce the cost of electric cars, which should lead to greater demand for them. One of the areas of technical improvement is the modernization of electrodes.

The most promising technology in the short term is solid-state LIB technology. Storage devices made with the use of solid electrolytes have a number of advantages over storage devices with liquid electrolyte. These include an accelerated charging process, an extended life cycle, the ability to perform in different forms, the absence of the probability of electrolyte leakage.

However, today the main problem with solid-state LIBs is their high cost and small size. Solid-state batteries of small dimensions have been created, which are used as backup power in computer electrical circuits.

Due to the limitations and lack of mass commercial samples, it is impossible to achieve accelerated growth of projects with solid-state LIBs in the near future. However, TAIYO YUDEN and Samsung are already announcing designs that could take the technology to a new level in the electrochemical storage market. Due to the high interest of world leaders in the technology sector in solid state LIBs, a breakthrough in technology can be expected in the coming years. This scenario can lead to the approach of solid-state LIBs to their theoretical energy capacity of 1000 kWh / kg and their moderate growth in ESS projects.

2.1.4. Political factors.

Factor 1. Incentive instruments (R&D, grants, etc.). Currently, incentive instruments aimed at the development of ESS are actively developing, in particular, their number and funding are increasing. World experience shows that most countries (USA, Germany, etc.) have successfully used this method in the last decade. In particular, a large number of grants are allocated for the implementation of projects related to ESS [15]. The implementation of these practices over the past years shows their effectiveness and practical significance - the characteristics of technologies are increasing, thereby reducing costs for manufacturers and a gradual reduction in the cost of ESS technologies. In Russia, there are state RFBR, RSF, FFI, and other funds, which distribute grants for R&D. In recent years, large domestic companies (Rosneft, Gazprom, Mail.ru Group, etc.) have joined this practice. Thus, the most likely scenario for the development of incentive instruments is an innovative one, since the state and business are interested in the development of energy efficient technologies that can increase the reliability of energy systems and ensure the economic growth of industry.

Factor 2. Benefits (subsidies, loans, special programs, government procurement, accelerated depreciation, etc.) A variety of benefits in the area of ESS management system helps to implement pilot energy storage projects in different countries. Such incentive measures can increase the economic efficiency of the introduction of ESS in the electric power industry, especially given the high share of capital costs in the total cost of their operation in the life cycle. This will allow for wider application and faster adoption of the technology in other countries. This factor directly affects ESS. Today, in the Russian Federation, pilot projects for the application of ESS are implemented mainly by state-owned companies, which find it easier to use benefits than others. However, there is a trend in the development of ESS technologies, which will contribute to the expansion of benefits for the entire list of industrial companies in the Russian Federation. Thus, the most likely scenario for the development of benefits for ESS is a conservative one, since the actualization of ESS increases every year, which will contribute to the development of preferential conditions for ESS, at least in the coming years.

Factor 3. Integration of ESS into the system services market and into the wholesale electricity and power market. The integration of ESS into the system services market and the wholesale electricity and power market is widely discussed in Russian companies. NcP "Market Council" and JSC "ATS" are developing amendments to the regulations governing the activities of the wholesale electricity and power market entities, aimed at integrating the ESS. In 2018-2019, at a meeting of the commission for managing innovative development of the Rosseti company, analytical work was done to determine the most promising schemes and directions for introducing ESS at power facilities. As a result, the list of pilot projects planned for implementation in IDGC of Center, IDGC of Center and Volga Region, Lenenergo, IDGC of Urals was approved. Thus, the most likely scenario for the development of the integration of electricity supply systems into the market of system services and in the wholesale electricity and power market is an innovative one, since there is an interest of large energy companies of the Russian Federation and non-profit structures.

Factor 4. Customs policy instruments. To reduce the impact of customs policy on ESS technologies, work is currently underway to eliminate customs barriers within the EnergyNet roadmap (Order No. 830-r dated April 28, 2018). As an action plan to improve legislation and remove administrative barriers, the roadmap aims to introduce new technologies in the energy sector and improve the quality of energy supply services. The most likely scenario for the development of customs policy instruments is a conservative one, since the domestic political sector will facilitate the regulation of various instruments by means of legislative acts, roadmaps, etc., but more likely there will be resistance from the outside (sanctions and other complicating instruments).

3. Results

According to the analysis and the calculations performed, the most likely scenario for the development of ESS is the basic. All the data used for the calculation are presented in Table 1. Multiplying the factor weight by the probability for each scenario and then summing the values obtained within each scenario for all factors, we obtained the final scenario rating.

The negative scenario scored 15 points, the basic - 24, the accelerated - 19. Thus, the most likely scenario for the development of ESS today we consider the basic.

Table 1. Numerical assessment of the probabilities of scenarios.

Factor name	Factor significance			Scenario probability*			Factor contribution to scenario		
	Correlation	Share in ESS	factor weight	negative	basic	innovative	negative	basic	innovative
Reducing the cost of ESS technologies	1	0.5	0.5	1	2	3	0.5	1	1.5
The difference in prices for the wholesale electricity and power market	1	1	1	1	3	2	1	3	2
GDP	0.39	1	0.39	2	3	1	0.78	1.17	0.39
Supply security	0.4	0.37	0.148	1	3	2	0.148	0.44	0.3
Emergencies of a global nature	0.82	1	0.82	2	3	1	1.64	2.46	0.82
CO2 emissions	0.7	1	0.7	3	2	1	2.1	1.4	0.7
Temperature	0.7	1	0.7	3	2	1	2.1	1.4	0.7
Increasing the efficiency of electrolysis systems for hydrogen production	0.9	0.05	0.045	3	2	1	0.135	0.09	0.045
PEMFC fuel cell	0.93	0.016	0.015	3	1	2	0.044	0.015	0.03
Solid State LIB	1	0.37	0.37	1	3	2	0.37	1.11	0.74
R&D, grants, etc.	1	1	1	1	2	3	1	2	3
Privileges	1	1	1	1	3	2	1	3	2
Integration of ESS into the wholesale electricity and power market	1	1	1	1	2	3	1	2	3
Integration of ESS into the system services market	1	1	1	1	2	3	1	2	3
Customs policy	1	1	1	2	3	1	2	3	1
Scenario rating							**15**	**24**	**19**

* A qualitative scale was used: 1 - high probability of occurrence, 2 - medium probability of occurrence, 3 - low probability of occurrence.

4. Conclusion

Energy storage systems provide an energy transition characterized by the emergence of new types of electric power entities, an increase in the share of renewable energy sources in the energy balance and the spread of new types of technological solutions. The direction of accumulation systems in 2020 is extremely relevant. The highlighted risks that were included in the scenario analysis criteria, namely ecology, economics, politics and technology, were used to make forecasts for the development of ESS technology. It should be noted that two types of scenarios are most often used in world practice: the best and the worst. The authors decided to use more than two scenarios. This was a dangerous situation, because people, by their very nature, tend to give the highest degree of probability to the average of the scenarios. Therefore, it was extremely important to introduce a scale of probabilities that avoided cognitive distortion. It was thanks to this approach that it was possible to determine the confidence intervals for each of the scenarios as accurately as possible, and the introduced weight scale allowed us to determine that the most significant of the factors influencing the scenario were global emergencies, climate change on the planet, prices on energy markets and all political factors (benefits, grants, customs policy, regulations).

In the context of the transformation of the world's energy systems with deep decentralization of electricity production, the introduction of energy storage systems will determine the shape of the electricity industry of the future. To reduce dependence on oil products, Russia needs to diversify its economy by developing production, including mastering the production of energy storage devices. Such a decision on the wave of general development and the transition to a new technological order will allow Russia to maintain a competitive position in the global market.

References

[1] Bloomberg. Global Storage Market to Double Six Times by 2030 URL https://about.bnef.com/blog/global-storage-market-double-six-times-2030/
[2] Lithium-ion Battery Costs and Market URL https://data.bloomberglp.com/bnef/sites/14/2017/07/BNEF-Lithium-ion-battery-costs-and-market.pdf
[3] Otchyot o ravnovesny`kh czenakh v naibolee krupny`kh uzlakh raschyotnoj modeli [Report on equilibrium prices in the largest nodes of the calculation model] URL https://www.atsenergo.ru/nreport?rname=big_nodes_prices_pub
[4] Prognoz soczial`no-e`konomicheskogo razvitiya Rossijskoj Federaczii na period do 2036 goda [Forecast of social and economic development of the Russian Federation for the period up to 2036.] URL https://economy.gov.ru/material/file/a5f3add5deab665b344b47a8786dc902/prognoz2036.pdf
[5] Bazhutov A V and Ilyashenko V V 2010 Vzaimosvyaz` tempov inflyaczii i investiczionny`kh vlozhenij v e`konomike Rossijskoj Federaczii [The correlation of inflation and investment rates in the economy of the Russian Federation] *Izvestiya UrGEU* **2(28)** 38-44 [In Russian]
[6] Berenberg Thematics. Executive Summary. Battery adoption at the tipping point URL https://www.bacanoralithium.com/pdfs/Berenburg_lithium_report_Feb2016.pdf
[7] Ososko Ya S and Ivanova A E 2019 Torgovy`e voĭny`: istoriko-e`konomicheskiĭ aspekt [Trade wars: historical and economic aspect] *Nauchny`e Zapiski Molody`kh Issledovatelej* **4** 5-10 [In Russian]
[8] Chebotareva G, Strielkowski W and Streimikiene D Risk assessment in renewable energy projects: A case of Russia *J. of Cleaner Production* **269** 122110
[9] Abramov AYu, Bogachenko P V, Kulikov A V, Ryapin I Yu *Primenenie sistem nakopleniya e`nergii v Rossii: vozmozhnosti i bar`ery`* [Energy storage systems application in Russia: opportunities and barriers] (Moscow: Biblioteka Idea) [In Russian]
[10] Wagner L, Ross I, Foster J and Hankamer B 2016 Trading Off Global Fuel Supply, CO_2 Emissions and Sustainable Development *PLoS ONE* **11(3)** e0149406
[11] Edition of the EnerFuture World Energy Scenarios Enerdata URL https://www.enerdata.net/publications/reports-presentations/world-energy-forecasting-

scenarios-2017-edition.html
[12] Use of the Modular Helium Reactor for Hydrogen Production URL https://www.osti.gov/biblio/821808-K6FjPq/native/
[13] Guerrero Moreno N, Cisneros Molina M, Gervasio D and Pérez Robles J F 2015 Approaches to polymer electrolyte membrane fuel cells (PEMFCs) and their cost *Renewable and Sustainable Energy Reviews* **52** 897-906
[14] Zatsarinnaya Y N, Logacheva A G, Gainullin R N, Alekseeva S F, Amirov and D I 2019 Solution for renewable future *E3S Web of Conf.* **124** 04010
[15] Zatsarinnaya Yu, Logacheva A and Grigoreva M 2020 Electricity in retail markets as a commodity in smart energy systems *IOP Conf. Ser.: Mater. Sci. Eng.* **791** 012061

IOP Conf. Series: Materials Science and Engineering 1089 (2021) 012041 doi:10.1088/1757-899X/1089/1/012041

The waste-to-energy technology of the energy sector

L A Nikolaeva and R Ya Iskhakova

Department of Water and Fuel Technology, Kazan State Power Engineering University, Krasnoselskaya St., 51, Kazan, 420066, Russia

E-mail: imreginaiskh@gmail.com

Abstract. Nowadays a problem of reducing anthropogenic impact on the environment is important. In the present article a perspective method of waste-to-energy technology of the energy sector is reviewed. This article focuses on the carbonate sludge (large-tonnage industrial waste) produced by power stations that is used as a sorption material, and adsorption properties of carbonate sludge, its influence on activated sludge is considered. Sludge is a microporous sorbent which is confirmed by the sorption curves for oil products and ammonia nitrogen. The carbonate sludge increases the activated sludge volume index by improving its sedimentation properties and the efficiency of centrifugation during activated sludge dewatering due to the rupture of hydrate shells of activated sludge. Granules that can be used as a fuel are obtained at the following mass ratios: excess activated sludge - 70%, carbonate sludge - 8%, binder - 22% by weight (starch and technical lignosulfonate). The production technology of fuel granules for incineration has been developed based on the research results.

1. Introduction

The energy sector is the main one for the development of all branches of industry, transport, utilities, and agriculture since the electric and thermal energies are necessary for the production of any type of goods and services. Therefore, this industry is characterized by the highest scale of production, which inevitably leads to a significant anthropogenic impact on the environment [1, 2].

Currently, the issue of efficient waste disposal of the energy sector is a priority. The following types of waste are generated at power stations: solid (slag, ash, sludge, etc.), liquid (wastewater, chemically contaminated liquids), and gaseous (fuel combustion by-products) [3]. Therefore, the issue of developing the waste-to-energy technologies at the enterprises of the energy sector is actual, which will minimize the anthropogenic impact of the energy enterprises on the environment and increase the energy efficiency of production [4, 5].

The preparation of additional water, as well as the treatment of wastewater, generated at energy enterprises, are mandatory stages before their use and drainage into the reservoir. Traditionally, there are four stages of industrial wastewater treatment: mechanical, biological, physical/chemical, and disinfection [6, 7].

The aim of the work is to develop waste-to-energy technology to improve energy efficiency and the environmental friendliness of an enterprise in the energy sector. The technology includes wastewater treatment with dosing of carbonate sludge, sedimentation of activated sludge in a secondary clarifier, centrifugation, and molding the fuel granules for incineration with additional heat released.

Karmanovskaya GRES (state district power station) is selected as the enterprise, which has a stage of water treatment of additional water for the condensate-feed path of boilers, as well as treatment of industrial wastewater.

2. Materials and Methods

Water treatment at the Karmanovskaya GRES is carried out in preliminary treatment of additional water, including liming and coagulation, and ion exchange-chemical desalination [8]. The Karmanovskoe reservoir is a source of water supply for the station; annually, about 1 million m^3 of the reservoir's source water is used to compensate steam and condensate losses at the station.

During the preparation of additional water at the stage of liming and coagulation, carbonate sludge is formed. Carbonate sludge from water treatment is a large-tonnage industrial waste that has been stored for decades at the plant's sludge disposal sites. This leads to salinization of groundwater, alkalization of the soil, lands degradation etc. Therefore, despite the fact that carbonate sludge is safe (hazard class V), the issue of effective disposal of carbonate sludge is relevant for Karmanovskaya GRES. The qualitative phase analysis of the company diffractometer D 8 ADVANCE Bruker showed the chemical composition presented in table 1 [8].

Table 1. The chemical composition of the carbonate sludge.

Chemical composition	%
$CaCO_3$	72
$Mg(OH)_2$	9
$Ca(OH)_2$	< 1
SiO_2	0.5
Other substances	17.5

The method of gas chromatography-mass spectrometry showed that contents of humic acid in the carbonate sludge are up to 12%. The bulk density is 0.62 g/cm^3, humidity - 3%, ash content - 89%.

The treatment facilities of the station receive household wastewater from the state district power station, wash water, drainage water from sludge and sand sites, as well as wastewater from the nearby Energetik village.

The large-tonnage production waste – excess activated sludge, is generated at biological wastewater treatment plants [9]. Activated sludge is a complex community of microorganisms, the life of which is carried out due to the oxidation of organic substances, phosphorus and nitrogen-containing compounds of wastewater from industrial enterprises and the municipal sector [10].

In experimental studies, an excess of activated sludge from the biological treatment plant of the Karmanovskaya GRES is used. The characteristics of the sludge are studied: the moisture content of the original activated sludge is 99%, after the stage of compaction (sedimentation under the influence of gravitational forces) - 96%, density – 1.15 g / cm^3, ash content – 12%.

Bulk density is determined according to this method: the sorbent dried at 110 ° C is poured in portions of 20 cm^3 into a previously weighed cylinder with a volume of 100 cm^3 and a height of 240 mm. The cylinder is shaken by lightly tapping the bottom on a wooden disc for 30 seconds in an inclined (70-80°) position after which the cylinder is weighed again.

The determination of moisture content method consists in drying a sample of the product to constant weight and determining the reduction in the weight of the product.

Ash content is determined according to this method: the sample is burned in a muffle furnace heated at a certain rate to a temperature of 815 ° C, and kept at this temperature until constant weight. The platen is placed in a muffle furnace at room temperature. Within 60 minutes the oven temperature is raised to 500 ° C and then this temperature is maintained for 60 minutes. Heating is continued to (815 ± 10) ° C in the same oven and kept at this temperature for at least 60 minutes.

The concentration of ammonium nitrogen was determined in accordance with the titrimetric method using the preliminary distillation of ammonia to boric acid solution.

The residual content of petroleum products was determined using the IR spectrometric method. The method is intended for measuring the mass concentration of oil products in wastewater samples by IR spectrometry using the AN-1 and KN-1. The range of measured concentrations of oil products is from 0.05 to 50 mg / dm3. The method consists of the extraction of emulsified and dissolved petroleum products from water with carbon tetrachloride; separation of petroleum products from associated organic compounds of other classes in a column filled with aluminum oxide and measurement of mass concentration of petroleum products by IR spectrometry.

The sludge volume index is equal to the volume occupied by 1 g of dry active sludge after 30 minutes of settling in the cylinder. A sample of the activated sludge mixture is shaken, 100 cm3 is poured into the cylinder and is precipitated for 30 minutes, then the volume occupied by the activated sludge is measured. The sludge is filtered through a pre-weighed filter and dried in an oven at a temperature of 105°C. After drying, the filter is cooled and weighed. The difference between the mass of the filter before and after determination is the dry mass of the activated sludge.

To assess the moisture yield of excess activated sludge, laboratory studies were carried out on a CM-6M centrifuge, in which the separation of the dispersed and liquid phases occurs by precipitation of suspended particles under the action of centrifugal forces. Previously, the excess activated sludge was intensively mixed with the sludge in various ratios for 5 minutes. Then the mixed sediment was placed in 10 ml centrifuge cylinders and centrifuged under various conditions.

When molding granules, starch and technical lignosulfonate (LST) were used as a binder.

The experimental determination of the heat of fuel incineration is carried out by the calorimetric method. The calorific value is the total energy released as heat when a substance undergoes complete incineration with oxygen under standard conditions. The chemical reaction is typically a hydrocarbon or other organic molecule reacting with oxygen to form carbon dioxide and water and release heat. Elemental analysis of fuel pellet substances with carbon, nitrogen and sulfur content (CHNS - analysis) was carried out on an EA 3000 EuroVector analyzer..

3. Results

The sludge has adsorption properties to various substances [11]. The ability of the sludge to adsorb oil products, ammonium nitrogen and other impurities has been established. Ammonium nitrogen and phosphates are typical for domestic wastewater entering the wastewater treatment plant from the Karmanovskaya GRES and the Energetik village; oil products enter the wastewater from the station's fuel oil facilities.

The sorption capacity of petroleum products and ammonium nitrogen was determined under static conditions. The sorption isotherms of the investigated contaminants are constructed (Figure 1).

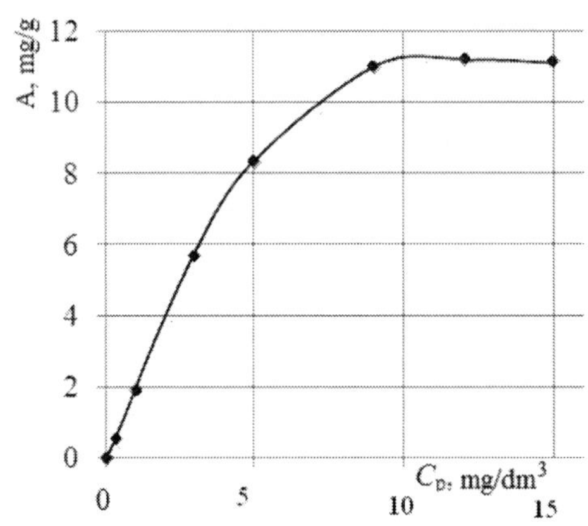

Figure 1. Oil products sorption isotherm.

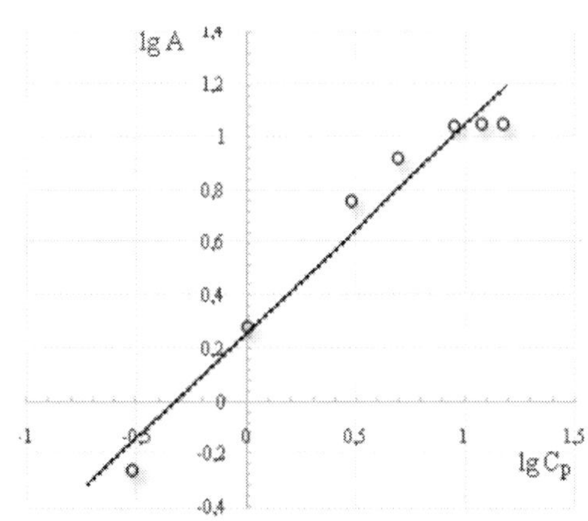

Figure 2. Logarithmic form of the oil products sorption isotherm.

The adsorption isotherm is of type I according to the BET classification. Therefore, the carbonate sludge contains micropores [12]. The initial portion of the curve is described by linear dependence of Henry's law [13]. Then the curve goes to the equilibrium, which conforms to Langmuir's theory of the monolayer formation on the sorbent surface. The isotherm (Figure 2) is described by Freundlich's equation (1):

$$A = 0.28 C_p^{0.75} \tag{1}$$

The ammonium nitrogen adsorption isotherm is shown in Figure 3.

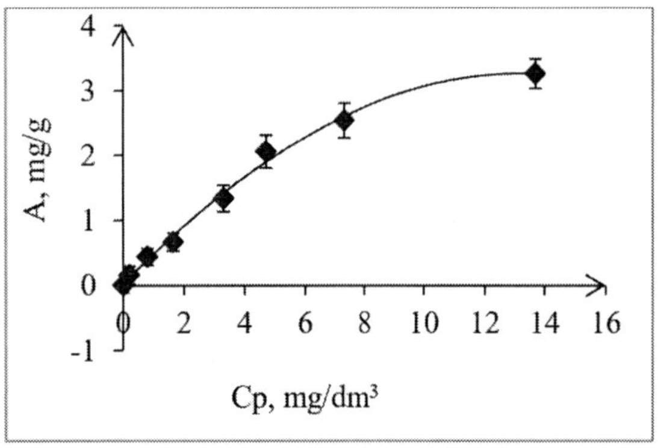

Figure 3. Ammonia nitrogen sorption isotherm.

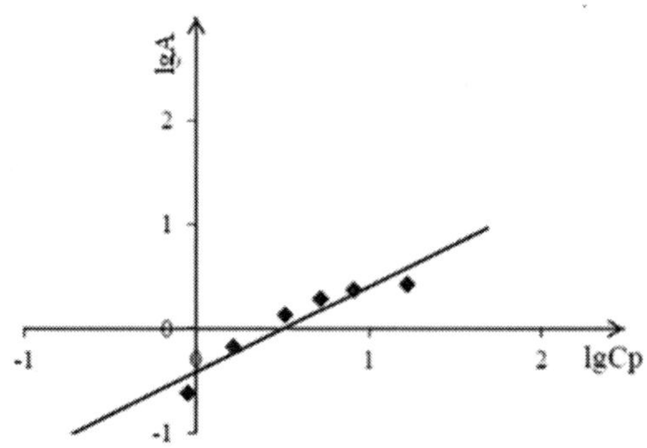

Figure 4. Logarithmic form of the ammonia nitrogen sorption isotherm.

As the ammonia nitrogen concentration increases, the sludge saturates and reaches the sorption equilibrium. The adsorption capacity of the material for ammonia nitrogen is 3.2 mg/dm^3.The sorption isotherm of ammonia nitrogen in logarithmic form is shown in Figure 4. The isotherm is described by Freundlich's equation (2):

$$A = 2.51 C_p^{0.52} \tag{2}$$

Therefore, the curves display the possibility of using sludge as a sorption material to improve the quality of biological wastewater treatment.

Dried carbonate sludge with constant humidity of 3% was dosed into the aeration tank. It was experimentally established that the optimum dose of injected sludge is 600 mg/dm^3 [14]. The carbonate sludge can be used as a sorption material for biological sorption treatment of industrial wastewater together with activated sludge microorganisms. Such a method will combine the biochemical oxidation of impurities with activated sludge and adsorption treatment with carbonate sludge. The combination of biochemical and adsorption treatment will improve the quality of wastewater treatment.

The introduction of this dose of sludge reduces the value of the sludge volume index and the concentration of suspended solids in the clarified water to the normative values (Table 2).

The obtained values are reached after settling of wastewater in the secondary clarifier for 2 hours. Dosing of carbonate sludge makes it possible to increase the efficiency of wastewater treatment in terms of suspended solids. The formation of a biofilm on its surface contributes to the weighting of activated sludge, thereby the sedimentation properties of activated sludge are increased. The research results indicate that the maximum effect of reducing suspended solids is achieved when dosing sludge in an amount of 600-900 mg / dm^3 and after the settling stage in the secondary clarifier the concentration of suspended solids does not exceed the normative values of suspended solids in the reservoir (5.65 mg / dm^3).

Table 2. Change the sludge volume index and suspended solids from the injected dose of sludge.

Dose of carbonate sludge, g/dm^3	Concentration of suspended solids, mg/dm^3	The sludge volume index , sm^3/g
0	28.14	141
0.3	10.11	112
0.6	5.62	80
0.9	5.29	74

After the secondary clarifiers, the carbonate sludge together with the excess activated sludge was centrifuged.

Excess activated sludge and a mixture consisting of excess activated and carbonate sludge in various proportions were centrifuged. The centrifugation index was used, which shows the degree of dehydration of sludge in the field of centrifugal forces. The centrifugation index U≤6-8 is considered a satisfactory value of sludge dewatering. The centrifugation times were 1, 2 and 3 minutes. The centrifuge rotation speed was between 1000 and 1500 rpm. The research results are shown in Figure 5 and Figure 6.

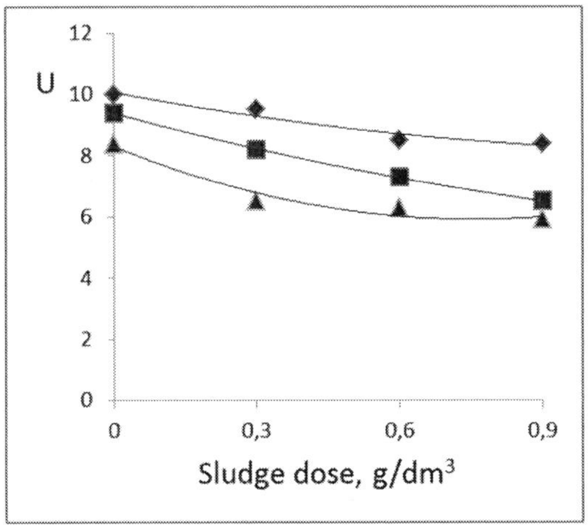

Figure 5. Dependence of the centrifugation index value on the injected sludge dose (1000 rpm) (Centrifugation time: - 1 min, ■ - 2 min, ▲ - 3 min).

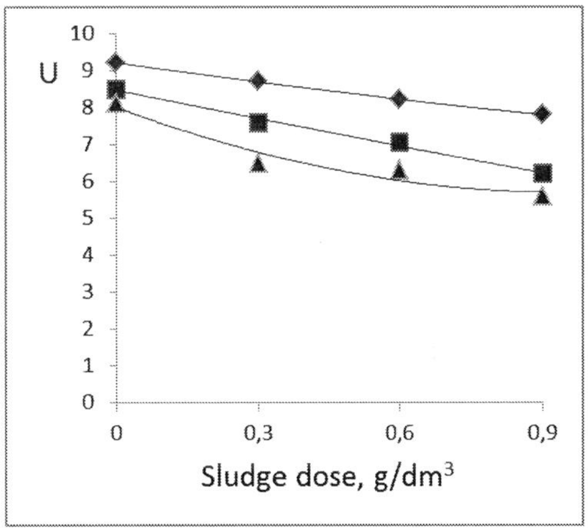

Figure 6. Dependence of the centrifugation index value on the injected sludge dose (1500 rpm) (Centrifugation time: - 1 min, ■ - 2 min, ▲ - 3 min).

The satisfactory value of the centrifugation index (Fig. 5) (U = 7.95) is achieved with the introduction of carbonate sludge with a minimum dose of 0.6 g / dm^3 (centrifuging the sediment at a speed of 1000 rpm for 1 minute) , while centrifugation of only excess activated sludge under the same conditions has a much higher index (U = 10.05) and is ineffective.

The satisfactory value of the centrifugation index (Fig.6) (U = 8.01) in the treatment of excess activated sludge without adding carbonate sludge is achieved at 1500 rpm for 3 minutes. It is the most energy-consuming process.

After centrifugation, the dewatered mixture of excess activated and carbonate sludge was molded by hand into granules with a diameter of 3-5 mm using a binder by rolling for further incineration [15]. Granules were obtained at the following mass ratios: excess activated sludge — 70%, carbonate sludge — 8%, binder — 22% by weight (starch and LST). Then, after rolling, drying was carried out at 105 ° C for 30 minutes.

Table 3 displays the general technical characteristics of the obtained granules using starch and LST as a binder: moisture, bulk density, ash content, and abrasion resistance. Calorimetric determination of the calorific value of the granules was carried out (C 2000 basic V. 1).

Table 3. General technical characteristics of the obtained granules.

Sample	with starch (22 % by weight)	with LST (22 % by weight)
Moisture,%	4.9±0.1,	3.3±0.1.
Bulk density, kg/m^3	831	792
Ash content, %	30.2	27.7
Colour of ash	light grey	light brown
Abrasion resistance, %	0.6	0.1
Heat of combustion (calorimetric bomb), MJ / kg	9672.6	10345.5

Therefore, comparing the general technical characteristics of granules using starch and LST, as well as their heats of incineration shows that the use of granules using LST as a binder is more efficient.

An elemental analysis of this sample for the content of carbon, hydrogen, nitrogen and sulfur (CHNS - analysis) was carried out on an EA 3000 EuroVector analyzer. According to the results the fuel granules have the following composition: C = 30.1%; H = 2.9%; S = 1.1%; N = 1.24%.

Based on the elemental composition according to Mendeleev's formula, the lower heat of combustion of the fuel can be calculated (3):

$$Q = 339.3C + 1256H - 109(O - S) - 25.2(9H + W) \tag{3}$$

The lowest heat of combustion of fuel using LST as a binder will be 9582.8 kJ / kg.

4. Discussion

The obtained curves (Figure 1, Figure 3) indicate the possibility of using carbonate sludge as a sorption material to improve the quality of biological wastewater treatment.

A decrease in the value of the sludge volume index indicates the effective biological treatment systems: the activated sludge has a dense structure, there are no loose small flakes. The decrease of the sludge volume index is also explained by the alkaline nature of the sludge. The increase in the pH of the water to a weakly alkaline medium helps to reduce the growth of filamentous bacteria and allows suppressing the swelling of activated sludge. The charged Ca^{2+} ions on the sludge surface contributes to the electrostatic interaction of the material with activated sludge, which, under the condition of $4 < pH < 9$, is considered negatively charged. Therefore, the carbonate sludge causes the conglomeration of activated sludge particles, which has a positive effect on the operation of secondary clarifiers.

Under the same experimental conditions, it was found that due to the addition of carbonate sludge, the moisture yield of excess activated sludge increases and the power input for the process is reduced. The moisture of the excess activated sludge after centrifugation is 84%, while the moisture content of the mixed wastes (excess activated and carbonate sludge) ranges from 69-71% (depending on the dose of injected sludge and the conditions of the experiment).

The chemical treatment of the sediment by adding carbonate sludge, which acts as a mineral coagulant is proposed for improving the moisture disengagement of excess activated sludge. The introduction of carbonate sludge makes it possible to disrupt the agglomeration stability of the excess activated sludge suspension by breaking the hydration shell enveloping the sludge particles. When centrifuging activated sludge together with carbonate sludge, strong structural bounds are broken, which makes it possible to separate colloid-bound moisture. Thus, a redistribution of the bound moisture of excess activated sludge occurs and the content of free water in the suspension increases. A significant content of Ca^{2+} forms a rigid mechanical structure of the sediment, promotes electrostatic interaction with activated sludge, which in the range 4 <pH <9 is considered as negatively charged.

The production technology of fuel granules for incineration has been developed based on the research results.

5. Conclusion

Based on the results of the research, an integrated waste-to-energy technology for the wastes disposal of Karmanovskaya GRES (carbonate sludge and activated sludge) was obtained, which includes the following stages:

- biosorption wastewater treatment using carbonate sludge;
- increasing the sludge volume index by improving the sedimentation properties of activated sludge;
- increasing the efficiency of centrifugation during sludge dewatering due to the rupture of hydrate shells of activated sludge with sludge;
- creation of granules on the basis of waste for incineration, which enables to get additional energy).

6. Acknowledgments

The study was carried out within the framework of the Russian Science Foundation grant No. 18-79-10136 «Theoretical methods of modeling and development of energy-efficient import-substituting devices for cleaning and raw hydrocarbons deep conversion at the enterprises of the fuel and energy complex.

References

[1] Nikolaeva L A and Iskhakova R Ya 2019 Adsorption of industrial wastewater from oil products with application of mathematical modeling *IOP Conf. Series: Earth and Environ. Sci.* **288** 012017
[2] Medvedev V T 2002 *Inzhenernaya Ekologiya* [Engineering Ecology] (Moscow: Gardariki) [in Russian]
[3] Timonin A S 2003 *Inzhenerno-Ekologicheskii Spravochnik* [Engineering-Ecology Reference Book] (Kaluga: Izd. N. Bochkarevoi) [in Russian]
[4] Brunner P H and Rechberger H 2015 Waste to energy – key element for sustainable waste management *Waste Management* **37** 3-12
[5] Lindberg D, Molin C and Hupa M 2015 Thermal treatment of solid residues from WtE units: a review *Waste Management* **37** 82-94
[6] Henze M, Mark van Loosdrecht M, Ekama G and Brdjanovic D 2008 *Biological Wastewater Treatment: Principles, Modeling and Design* (London: IWA Publishing)
[7] Grady C P L, Daigger G T, Love N G, Filipe C D M 2011 *Biological Wastewater Treatment* (Boca Raton: CRC Press)
[8] Nikolaeva L A and Iskhakova R Ya 2019 Integrated Wastewater Treatment for a GRES *Thermal Engineering* **66** 587-92
[9] Chang I-S, Clech P L and Jefferson B 2002 Membrane fouling in membrane bioreactors for wastewater treatment *J. of Environmental Eng.* **128(11)** 336-43
[10] Muga H E and Mihelcic J R 2008 Sustainability of wastewater treatment technologies *J. of Environmental Management* **88** 437-44
[11] Lupeiko T G, Gorbunova M O and Bayan E M 2004 Deep purification of aqueous solutions to remove iron (III) with carbonate-containing industrial waste *J. Appl. Chem.* **77** 79-82
[12] Romankov P G and Frolov V F 1990 *Massoobmennyye protsessy khimicheskoy tekhnologii. Sistemy s dispersnoy tverdoy fazoy* [Mass Exchange Processes of Chemical Technology] (Leningrad: Chemistry) [In Rissian]

[13] Wanga S and Peng Y 2010 Natural zeolites as effective adsorbents in water and wastewater treatment *Chemical Engineering J.* **156** 11-24
[14] Nikolaeva L A, Laptev A G and Iskhakova R Ya 2018 Improving the efficiency of wastewater biological treatment at chemical plants *Water Resources* **45** 231-7
[15] Parka J B K, Craggsa R J and Shilton A N 2010 Wastewater treatment high rate algal ponds for biofuel production *Bioresource Technology* **102** 35-42

Theoretical description of preheating a gas-air mixture in a gas burner with a heat divider

R S Ramazanov, D Y Suslov, L A Kushchev and A S Seminenko

Department of Heat and Gas Supply and Ventilation, Belgorod State Technological University named after V.G. Shukhov, Kostyukov St., 46, Belgorod, 308012, Russia

E-mail: suslov1687@mail.ru

Abstract. The article is devoted to the study of thermal processes occurring during the operation of a household gas burner. The preparation of gas fuel for combustion is one of the important functions that a gas burner performs. At the same time, when designing gas burner devices, it is important to determine the final temperature of the gas-air mixture inside the burner body, since an increase in the temperature of the resulting mixture makes it possible to intensify the combustion process. We have proposed a solution that allows to intensify the transfer of heat in the body of a gas burner from the lid heated due to contact with the flame to the gas flow using the thermal shape of the conical divider installed in the center on the inner side of the lid. It has been established that the location of the heat divider in the center on the inner side of the cover allows minimizing or completely eliminating the formation of a stagnant zone. The conical shape of the heat divider provides less resistance to the oncoming flow movement, and also because of its own side surface, it allows you to increase the useful area of the heat exchange surface. An expression is obtained for determining the final temperature of the gas-air mixture at the outlet from the firing holes of the burner body.

1. Introduction

Improving the life and well-being of the population requires the development of infrastructure, the development of regional energy and industry, as a result, leads to an increase in energy consumption, including natural gas.

In 2019, production of natural and associated petroleum gas in Russia amounted to 738.4 bcm, which is 1.7% more than in 2018 and is a record figure for the entire period of Russian gas production [1].

In the structure of natural gas use in Russia at the end of 2019, with a share of 60.2% and a consumption volume of 289 bcm, the communal and industrial sector is in the lead, the level of gasification reached 70.1% [2].

One of the main devices of industrial and domestic gas-using devices is a gas burner device. It is designed to convert the energy of the combustible gas process into thermal energy, while the gas must be completely burned [3-7].

In this regard, gas supply specialists are faced with new complex tasks related to the development of new systems and development and reconstruction of use, increasing their reliability, more economical use of gas and minimal environmental impact.

Investigations of combustion processes and the development of effective gas-burner devices are the subject of many works by domestic and foreign scientists. An example is the work [8], in which he investigates the efficiency of heat recirculation in the operation of a gas burner using the porous medium technology. In [9] the author was aimed to study the influence of four significant parameters, namely degrees of outer and inner ports and the number of outer and inner ports of high-pressure gas burners, to the thermal efficiency by the experiments. In [10] the author of this experimental investigation was to see the effects on using different design burner heads on the performance of LPG cooking stove. Also, burners of different material were used to study the effects of burner material on LPG stove performance. [11] Author this paper discusses the performance investigations of a conventional burner with different porous media used in LPG cooking stove.

The burner device for burning gaseous fuel has the following main functions: gives the necessary components and the flow rate of gas and air; ensures mixing of gas fuel with air, sometimes heats the resulting mixture to the ignition temperature - thereby preparing the resulting mixture for subsequent combustion; provides stabilization of combustion, creates conditions for the combustion of the mixture without separation and flame breakthrough; provides a minimum content of toxic gases in the composition of combustion products [12].

Consequently, the development of methods for increasing the efficiency of gas-burner devices, obtaining expressions for calculating the average temperature of the gas-air mixture inside the gas burner with a mathematical description of the thermal processes taking place is urgent.

2. Materials and Methods

To theoretically describe the process of heating a gas-air mixture in a burner body equipped with a heat divider, we used the equations of jet aerodynamics, as well as heat transfer equations based on the laws of thermal processes of Newton-Richman and Stefan-Boltzmann.

3. Results and Discussion

Let us consider the processes of movement of the flow of the gas-air mixture inside the burner, which have a significant effect on the distribution of heat.

In the general case, the flow pattern of the gas flow inside the burner body is similar to the pattern of motion of the shock gas jet [13-15] (Figure 1.).

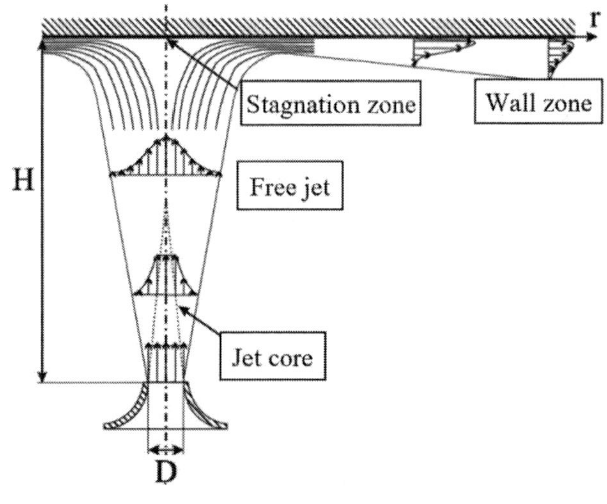

Figure 1. Shock gas jet scheme.

The gas stream flowing out of the gas burner nozzle strikes at right angles against the flat inner surface of the cover, which is distant at a distance H, as a result of which it forms a stagnant zone - a certain area of vortices that prevents further free movement of the jet (Figure 2.). This area is characterized by a lower intensity of heat transfer, and when calculating heat transfer from the plate to the incident transverse gas flow, the plate surface occupied by the stagnant zone is excluded - thereby reducing the useful heat transfer surface and, accordingly, the heat transfer efficiency. For example, in [16], in calculating heat transfer, the stagnant zone area was taken to be equal to the area limited by the diameter of the outlet part of the Venturi tube, and thus the useful area corresponded to the annular part, where the outer diameter is equal to the diameter of the burner cover, and the inner diameter is the diameter of the Venturi tube.

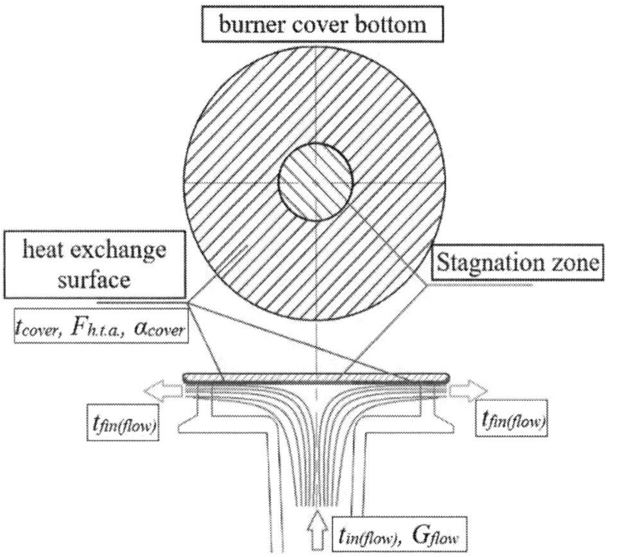

Figure 2. The movement of the gas-air mixture under the lid of a typical gas burner, not equipped with a heat divider.

We have proposed a solution that makes it possible to intensify the transfer of heat in the body of a gas burner from the lid heated due to contact with the flame to the gas flow using a conical heat divider installed in the center on the inside of the lid (Figure 3.). Placing a heat divider in the center on the inner side of the lid minimizes or completely eliminates the formation of a stagnation zone. The conical shape of the splitter provides less resistance to the oncoming flow of the flow, and also due to its own lateral surface allows to increase the useful area of the heat exchange surface.

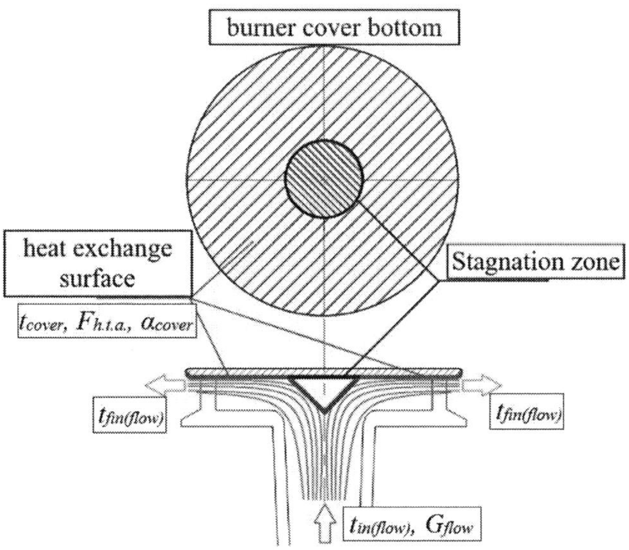

Figure 3. The movement of the gas-air mixture under the cover of the developed gas burner equipped with a heat divider.

The design of the gas stove burner with increased efficiency has been developed. The main structural elements of the proposed gas burner are: body, gas nozzle, mixer, firing holes for the outlet of the gas-air mixture, a cover with a preheating device and a primary air regulator [17-18] (Figure 4.).

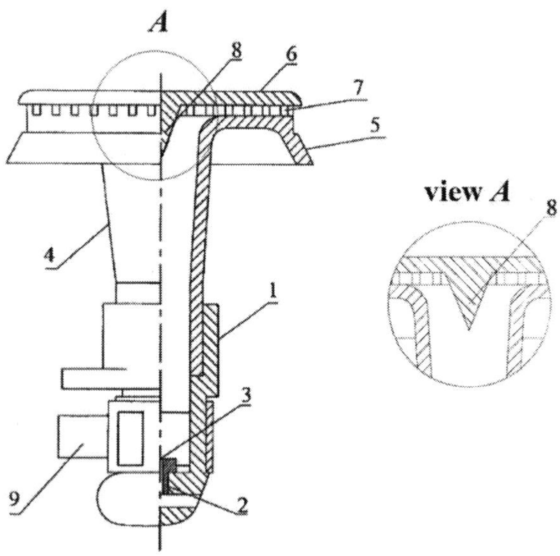

Figure 4. Original gas burner with preheating of the gas-air mixture: 1 - housing, 2 - gas nozzle, 3 - axial outlet, 4 - mixer, 5 - outlet end of the mixer, 6 - cover, 7 - fire holes, 8 - heat divider, 9 - primary air regulator.

To describe the operation of a gas-burning device, we draw up the heat balance equations, the distribution scheme of the main heat losses of a gas-burning device is represented by Figure 5.:

$$Q_\Sigma = Q_1 + Q_2 + Q_3 + Q_4 \tag{1}$$

where Q_Σ is the heat released during the combustion of gaseous fuel; Q_1 is heat losses caused by radiative heat exchange between the volume of burnt gas and the environment; Q_2 the total amount of heat absorbed by the test vessel; Q_3 the heat loss with flue gases, after heat exchange with the side surface of the test vessel; Q_4 the total amount of heat absorbed by the body of the gas burner.

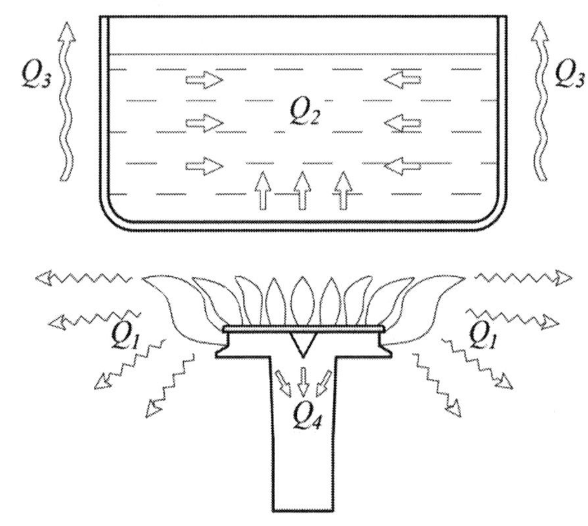

Figure 5. Scheme of distribution of heat losses of the heat balance of a gas burner device.

The total amount of heat generated by the combustion of gaseous fuel is calculated:

$$Q_{\Sigma} = G_{gas} \cdot Q_f \tag{2}$$

where G_{gas} is the mass fuel consumption, kg/s; Q_f is the lowest heat of combustion of the fuel, kJ/kg.

Heat losses caused by radiative heat exchange between the volume of burnt gas and the environment are expressed according to the Stefan-Boltzmann law by the formula [19-22]:

$$Q_1 = \varepsilon \cdot \sigma \cdot \left(t_{flame}^4 - t_{amb}^4\right) \cdot F_{flame} \tag{3}$$

where ε is the emissivity (emissivity); σ is the Stefan-Boltzmann constant (σ = 5,67·10^{-8} W/(m^2·K^4), t_{flame}, t_{amb} is the flame and ambient temperature, °C, F_{flame} is the flame surface area, m^2.

We represent the processes of heat transfer between the surfaces of solids and a gaseous medium by expressions based on the Newton-Richmann law [23-24].

The total heat absorbed by the test vessel is determined:

$$Q_2 = \left(c_{p(water)} \cdot m_{water} + c_{p(vessel)} \cdot m_{vessel}\right) \cdot \left(t_{in(water)} - t_{fin(water)}\right) \tag{4}$$

where $c_{p(water)}$, $c_{p(vessel)}$ mass heat capacity at constant pressure of water and the material of the test vessel (aluminum), kJ/(kg·°C), m_{water}, m_{vessel} is the mass of water and the test vessel itself, kg; $t_{in(water)}$, $t_{fin(water)}$ is the initial and final water temperature in the test vessel, °C.

Heat loss with flue gases after heat exchange with the side surface of the test vessel:

$$Q_3 = G_{sm.gas} \cdot c_{p(sm.gas)} \cdot \left(t_{sm.gas} - t_{amb}\right) \tag{5}$$

where $G_{sm.gas}$ mass flow rate of flue gases, kg/s; $c_{p(sm.gas)}$ is the mass heat capacity at constant pressure of flue gases, kJ/(kg·°C); $t_{sm.gas}$, t_{amb} is the temperature of flue gases after heat exchange with the side surface and ambient temperature, °C.

Heat absorbed by the body of the gas burner device:

$$Q_4 = c_{p(burner)} \cdot m_{burner} \cdot \left(t_{fin(burner)} - t_{in(burner)}\right) \tag{6}$$

where $c_{p(burner)}$ is the mass heat capacity at constant pressure of the material of the gas burner device, kJ/(kg·°C); m_{burner} is the mass of the gas burner device, kg, $t_{in(burner)}$, $t_{fin(burner)}$ is the initial and final temperature of the gas burner device, °C.

$$\eta_{(th)} = Q_2 / Q_{\Sigma} \tag{7}$$

Thus, the thermal efficiency of a gas burner device is the ratio of the useful heat absorbed by the test vessel to the total amount of heat released during the combustion of gaseous fuel.

Preparation of gas fuel for combustion is one of the important functions that a gas burner performs. At the same time, it is useful for developers of gas burner devices to determine the final temperature of the gas-air mixture inside the burner body, since an increase in the temperature of the resulting mixture makes it possible to intensify the combustion process. Since combustion is a chemical reaction of oxidation of gaseous fuel with oxygen contained in the air, this feature of the process is explained, for example, by the Van't Hoff rule, which states that when heated, the rate of most chemical reactions increases [25-27].

The process of operation of the gas burner device is accompanied by the absorption by the body of the gas burner device of a certain part of the heat released during the combustion of gas. This type of heat loss is unavoidable for atmospheric burners. At the same time, the heated surfaces of the gas burner transfer heat to the gas-air mixture flowing in the housing [28], thereby ensuring its preheating.

Let us consider the processes of heat transfer during steady state operation, occurring in the housing of a gas burner.

In steady-state operation, when the temperature of the burner cover has reached a constant value of t_{cover} and does not change over time, the amount of heat absorbed by the flow of the gas-air mixture when the flow temperature changes from $t_{in(flow)}$ to the temperature $t_{fin(flow)}$ will be expressed:

$$Q_{flow} = G_{flow} \cdot c_{p(flow)} \cdot \left(t_{fin(flow)} - t_{in(flow)}\right) \tag{8}$$

where G_{flow} the mass flow rate of the gas-air mixture, kg/s, $c_{p(flow)}$ is the mass heat capacity at constant pressure of the gas-air mixture, kJ/(kg·°C); $t_{in(flow)}$, $t_{fin(flow)}$ respectively, the initial temperature and the temperature of the gas-air mixture at the outlet from the firing holes, °C.

In this case, as a result of convective heat transfer from the burner cover with a temperature t_{cover} to the gas flow with an initial temperature $t_{in(flow)}$ the amount of heat is transferred, determined by the formula:

$$Q_{cover} = F_{h.t.a.} \cdot \alpha_{cover} \cdot \Delta t_{flow} \tag{9}$$

where $F_{h.t.a.}$ is the area of the inner surface of the cover participating in the heat exchange with the flow of the gas-air mixture, m^2, α_{cover} is the heat transfer coefficient of the inner surface of the cover participating in the heat exchange with the flow of the gas-air mixture, W/(m^2·K), Δt_{flow} is the logarithmic mean temperature difference, °C.

Logarithmic mean temperature difference

$$\Delta t_{flow} = \frac{\left(t_{cover} - t_{in(flow)}\right) - \left(t_{cover} - t_{fin(flow)}\right)}{\ln\dfrac{\left(t_{cover} - t_{in(flow)}\right)}{\left(t_{cover} - t_{fin(flow)}\right)}} \tag{10}$$

Thus, setting the equality condition, the amount of heat given off by the heated lid to the flow of the gas-air mixture and the amount of heat absorbed by the flow of the gas-air mixture ($Q_{cover} = Q_{flow}$) we obtain the following expression:

$$\ln\frac{\left(t_{cover} - t_{in(flow)}\right)}{\left(t_{cover} - t_{fin(flow)}\right)} = \frac{F_{h.t.a.} \cdot \alpha_{cover}}{G_{flow} \cdot c_{P(flow)}} \tag{11}$$

By taking the logarithm of the right-hand side of the expression and performing the appropriate abbreviations, we obtain the equation for determining the final temperature of the gas flow (temperature at the outlet from the firing holes of the gas burner):

$$t_{fin(flow)} = t_{cover} - \left(t_{cover} - t_{in(flow)}\right) \cdot \exp\left(-\frac{F_{h.t.a.} \cdot \alpha_{cover}}{G_{flow} \cdot c_{P(flow)}}\right) \tag{12}$$

This equation makes it possible to determine the final temperature of the gas flow ($t_{fin(flow)}$) after contact with the inner surface of the lid participating in heat exchange ($F_{h.t.a.}$), with the temperature (t_{cover}). Taking into account the design of the burner with a heat divider, the useful heat exchange area consists of the area of the side surface of the cone and the annular surface of the cover.

4. Conclusion

The design of a gas burner with increased thermal efficiency has been developed, equipped with a conical-shaped heat divider installed in the center on the inner side of the cover. Installation of a heat divider allows preheating of the gas-air mixture in the burner housing.

The heat balance of the burner equipped with a heat divider is compiled, taking into account the heat absorbed by the body of the gas burner device with a heat divider. The heat balance allows you to determine the thermal efficiency of the burner of the developed design.

An expression is obtained for determining the final temperature of the gas-air flow after heat exchange with the cover of the gas burner equipped with a heat divider.

It was found that the location of the heat divider in the center on the inner side of the cover allows minimizing or completely eliminating the formation of a stagnation zone. The conical shape of the heat divider provides less resistance to the oncoming flow movement, and also because of its own side surface, it allows you to increase the useful area of the heat exchange surface.

5. Acknowledgments

The work is realized in the framework of the Program of flagship university development on the base of the Belgorod State Technological University named after V.G. Shukhov, using equipment of High Technology Center at BSTU named after V.G. Shukhov.

References

[1] Statisticheskij sbornik. TEK Rossii – 2019 [Statistical compilation. FEK of Russia – 2019. Analytical center under the government of the Russian Federation] (Moscow: Analiticheskij centr pri Pravitel'stve Rossijskoj Federacii) URL https://ac.gov.ru/publications/topics/topic/13700

[2] Godovoj otchet PAO «Gazprom» za 2019 g. [Gazprom public joint stock company annual report for 2019] URL https://www.gazprom.ru/investors/disclosure/reports/2019/

[3] Bantu A A, Nuwagaba G, Kizza S and Turinayo Y K 2018 Design of an Improved Cooking Stove Using High Density Heated Rocks and Heat Retaining Techniques *J. of Renewable Energy* **2018** 9620103

[4] Decker T J 2017 *Modeling tool for household biogas burner flame port design: masters theses* (Fort Collins: Colorado State University)

[5] Turns S R 2012 *An Introduction to Combustion: Concepts and Applications* (New York: WBC McGraw-Hill)

[6] Le Corre O and Loubar K 2010 *Natural Gas: Physical Properties and Combustion Features* (London: Engineering & Technology, Physical Sciences)

[7] Hull S et al 2011 *Guidebook to Gas Interchangeability and Gas Quality* (Brussels: Published by BP in association with the IGU)

[8] Jugjai S and Rungsimuntuchart N 2002 High efficiency heat-recirculating domestic gas burners *Experimental Thermal and Fluid Science* **26(5)** 581–92

[9] Aroonjarattham P 2016 The Parametric Studied of High Pressure Gas Burner Affect Thermal Efficiency *Eng. J.* **20(3)** 33-48

[10] Khan M and Saxena A 2013 Performance Of LPG Cooking Stove Using Different Design Of Burner Heads *Int. J. of Eng. Research & Technology* **2(7)** 656-9

[11] Dahiya D, Singh L R and Bhatia P 2016 Improvement of the Domestic LPG Cooking Stoves *A Review Indian J. of Science and Technology* **9(1)** 1-8

[12] Huda Z 2020 *Metallurgy for Physicists and Engineers. Fundamentals, Applications, and Calculations* (London: Engineering & Technology, Physical Sciences)

[13] Yang Y T and Tsai S Y 2007 Numerical study of transient conjugate heat transfer of a turbulent impinging jet *Int. J. of Heat and Mass Transfer* **50** 799-807

[14] Xin Liu, Gabour L A and Lienhard V J H 1993 Stagnation-Point Heat Transfer During Impingement of Laminar Liquid Jets: Analysis Including Surface Tension *J. Heat Transfer* **115(1)** 99-105

[15] Nakoryakov V E, Pokusaev B G and Troyan E N 1978 Impingement of an axisymmetric liquid jet on a barrier *Int. J. Heat Mass Transfer* **21(9)** 1175-84

[16] Tenenev V A, Gubert A V, Mihajlov YU O and Korepanov M A 2010 Issledovanie processov v

gazovyh gorelkah dlya bytovyh plit [Research of processes in gas burners for household stoves] *Himicheskaya Fizika i Mezoskopiya* **12(1)** 45-53 [In Russian]

[17] Suslov D Y, Ramazanov R S, Temnikov D O and Lobanov I V 2019 Development and research of low pressure injection burner for biogas combustion *IOP Conf. Ser.: Mater. Sci. Eng.* **552** 012031

[18] Suslov D YU and Ramazanov R S 2020 Modelirovanie szhiganiya biogaza v inzhekcionnoj gorelke s teplovym rassekatelem [Simulation of biogas combustion in an injection burner with a heat divider] *Vestnik BGTU im V. G. Shukhova [Bulletin of BSTU named after V G Shukhov]* **4** 40-7 [In Russian]

[19] Yunus A 1998 *Heat transfer. A Practical Approach. Second Edition* (New York: WBC McGraw-Hill)

[20] Lienhard J H IV and Lienhard J H V 2019 *A heat transfer textbook* (Cambridge: Phlogiston Press)

[21] Kelleher M D, Shah R K, Sreenivasan K R and Joshi Y 2012 *Experimental Heat Transfer, Fluid Mechanics and Thermodynamics* (Amsterdam: Elsevier)

[22] Viskanta R 1993 Heat Transfer to Impinging Isothermal Gas and Flame Jets *Experimental Thermal and Fluid Science* **6(2)** 111-34

[23] Kuznetsov V A and Trubaev P A 2019 Convective heat transfer in the near-wall turbulent gas stratum *IOP Conf. Ser.: Mater. Sci. Eng.* **552** 012005

[24] Geers L F G, Tummers M J, Bueninck T J and Hanjalic K 2008 Heat transfer correlation for hexagonal and in-line arrays of impinging jets *Int. J. of Heat and Mass Transfer* **51(21-22)** 5389-99

[25] Gardiner W C Jr (Ed.) 1999 *Gas-Phase Combustion Chemistry 2nd Edition* (Dordrecht: Springer)

[26] Raju K S N 2011 *Fluid mechanics, heat transfer, and mass transfer. Chemical engineering practice* (New York: John Wiley & Sons Limited)

[27] Penumadu P S and Rao A G 2017 Numerical investigations of heat transfer and pressure drop characteristics in multiple jet impingement system *Applied Thermal Engineering* **110**1511-24

[28] Dakhno A V, Smirnov Yu P, Mashkov A V and Potapenko O V 2018 To an assessment of size of diffusive losses of coolant through a rubber camera – cover *Chemical Bulletin* **1(1)** 18-26

Cluster characteristic of water used by energy facilities by total antiocidant activity

M L Kalayda and M E Gordeeva

Department of Water bioresources and aquaculture, Kazan state power engineering university, Krasnoselskaya St., 51, Kazan, 420066, Russia

E-mail: Maria.Galeeva@gmail.com

Abstract. Energy facilities are located on most large reservoirs, therefore, the issue of water quality is especially acute in the water area of such water bodies. The article describes 9 clusters of water with values of total antioxidant activity, isolated now on the basis of numerous analyses. Water with a simplified structure is distinguished, containing 1 cluster and characterized water bodies with an abundance of life. Ordinary water with 2 clusters in the structure is water widely used by humans for household consumption. Water with 3-4 clusters observed in artesian wells. The most complex in cluster composition is spring water (5 clusters). The complex cluster structure of water made it possible to propose an index of water structure, which reflects the percentage of clusters with certain values of total antioxidant activity and the degree of water trophy. As a result, the work provides the recommendations on the values of total antioxidant activity of water at the water intake and offtake of combined Heat and Power plant.

1. Introduction

Water systems are a source of water for both people and industry. The identification of priorities for water protection and their sustainable management is now particularly relevant in the context of increased anthropogenic impacts on water resources [1]. Especially important in this context are water bodies that are objects of multi-purpose use, including for energy purposes. The Republic of Tatarstan have 7 operating electric centers (Kazan combined Heat and Power plant 1 (Kazan CHPP-1), combined Heat and Power plant 2 (Kazan CHPP-2), combined Heat and Power plant 3 (Kazan CHPP-3), Nizhnekamsk CHPP-1, CHPP-2, Naberezhnye Chelny CHPP, Urussu Hydroelectric Power Station (HPP)), 2 thermal power plants (Zainsk HPP and PSU-TPP PJSC "Nizhnekamskneftekhim"), 1 hydroelectric power plant (Nizhnekamsk GES). For functioning of the listed energy facilities the water source is need – it can be different types of water bodies. An example of using a reservoir for the functioning of a CHPP is shown in Figure 1. In most cases, water is used in significant amounts as a refrigerant. In most cases water used as a coolant in significant quantities. Thermal power plants (TPP) are the largest consumers of water with the largest consumption for cooling process equipment. The main consumption of service water at the TPP is caused by the need to remove heat from the exhaust steam in the turbine condensers – 85-95%. An average of 60-100 m^3 of water is required for condensation of 1 ton of steam [1]. The requirements for water quality used by energy facilities are sufficiently high in accordance with the Rules for technical operation of power plants and networks of the Russian Federation, primarily for water rigidity and the amount of contamination (not more than 10 mkg/l). The quality of water plays a great role for energy facilities, since substances dissolved in water cause some kind of malfunction in the operation of energy equipment, this is mainly due to the formation of corrosion, salt and biological fouling in the thermal units of the heat transfer surface of the cooled equipment. Compliance with the water quality requirements at the step of water treatment make possible to prevent emergency shutdowns of heat and power equipment. There is always a desire to find an integrated indicator of water quality that would reflect the possibilities and directions of water using: with the least amount of contaminations for technical facilities, "live" water for using in agriculture and aquaculture, without "live" objects for human drinking. One of these current integrated indicators of water quality is the total antioxidant activity (TAOA).

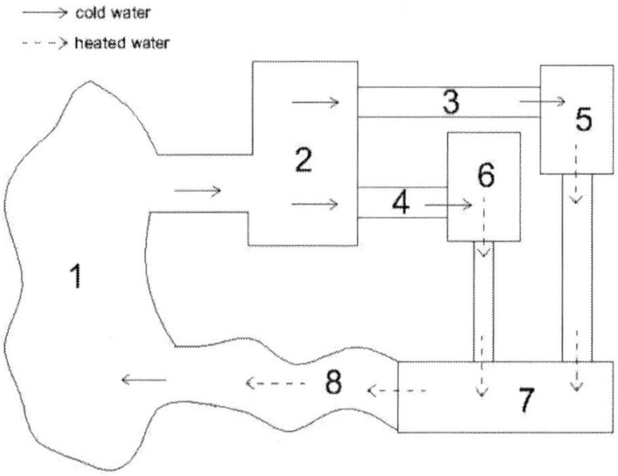

Figure 1. Plan of water using of the lake in technical water supply system of CHPP [1]: 1 – lake; 2 – outdoor gravity feed; 3 – forebay of CHPP; 3, 4 – underground water; 5-6 – turbine condensers; 7 – underground circulation channel; 8 – open offtake.

The purpose of the work is to classify the waters of basin-cooler of energy facilities according to the total antioxidant activity using a cluster approach.

2. Materials and Methods

Total antioxidant activity (TAOA) was determined by culonometric method using electrogenerated bromine. Samples were analyzed on an Expert-006 coulometer (Ekonix-Expert LLC, Russia) using a certified method [2]. Electrogeneration of bromine was carried out from a 0.2 M solution of potassium bromide in a 0.1 M aqueous solution of sulfuric acid at a constant current strength of 100.0 mA. Into the electrolytic cell was been introduced 30 ml of the background solution and when the indicating current reached a certain value, an aliquot of the aqueous extract of the test sample in amount of 100 mkl. The determination was carried out at room temperature. The device was calibrated with an alcoholic solution of the Russian standard sample (RSS) of routine, prepared according to the current State Pharmacopoeia of the XI edition. TAOA was expressed in mg of a standard rutin (Ru) sample per 1 dm^3 (l) extraction or in g of Ru per 100 g test sample.

Redox potential (Eh) values were obtained using an Ionomer I-160 Mi instrument, which also measured the pH values of the aquatic ecosystem. According to the certificate of this instrument, the Eh is measured using measuring combined electrode and a comparison electrode. The combined electrode has a glass body with a diameter of 12 mm. In its lower part there is a working membrane sensitive to hydrogen ions, which is a ball made of special glass. Porous ceramic is soldered above the ball, providing electrolytic contact between the electrolyte poured into the electrode and the analyzed solution. The upper part of the electrode ends with a bushing, from which a cable comes out with a connector for connection to the converter.

3. Results and discussions

In the specialized hydrochemical literature [3, 4] there is no water classifications according to the indicator of TAOA and no standards for TAOA for the purposes of water treatment [1] at energy facilities.

Multiple correlation-regression analysis of TAOA and other physicochemical indicators revealed weak association with dissolved oxygen and redox potential (Eh) and strong association with pH (Table 1) [1]. Analysis of literary sources [5–9] and owns researches [10] prove the presence of a strong correlation between pH and Eh: 1 unit of pH corresponds to 58.873 mV of Eh. Interest in the study of Eh [11–13] has not faded since the beginning of the 20th century [14, 15] and, the current studies prove the relevance and complexity of this indicator under the correct classification.

Table 1. Correlations between TAOA, dissolved oxygen, Eh and pH.

		pH	Dissolved oxygen	Eh	TAOA
pH	Correlation		0.3347	-0.8902	-0.8109
	Sample Size		82	82	82
	p-value		0.0021	0.0000	0.000
Dissolved oxygen	Correlation	0.3347		-0.3935	-0.3337
	Sample Size	82		82	82
	p-value	0.0021		0.0003	0.0022
Eh	Correlation	-0.8902	-0.3935		0.6966
	Sample Size	82	82		82
	p-value	0.0000	0.0003		0.0000
TAOA	Correlation	-0.8109	-0.3337	0.6966	
	Sample Size	82	82	82	
	p-value	0.000	0.0022	0.0000	

Numerous studies at the Department of water bioresources and aquaculture of the Kazan State Power Engineering University made it possible to distinguish a number of water clusters with different chemical characteristics (Figure 2) [5].

Cluster №1 – 7.5469 mg Ru per 1 dm^3;
Cluster №2 – 6.8212 mg Ru per 1 dm^3;
Cluster №3 – 6.0956 mg Ru per 1 dm^3;
Cluster №4 – 5.3699 mg Ru per 1 dm^3;
Cluster №5 – 4.6443 mg Ru per 1 dm^3;
Cluster №6 – 3.9912 mg Ru per 1 dm^3;
Cluster №7 – 3.2655 mg Ru per 1 dm^3;
Cluster №8 – 2.4673 mg Ru per 1 dm^3;
Cluster №9 – 1.8142 mg Ru per 1 dm^3.

As can be seen from the data given in Figure 2, water of the open offtake channel of the energy facility is characterized by the presence of cluster № 7 (dominance, 90%) and cluster № 8 (10%). At the same time, the water of the similar lake that does not experience the anthropogenic impact, represented in the form of discharge of heated waters of the energy facility, consists of one cluster – cluster № 7 (3.2655 mg Ru per 1 dm^3).

It should be noted that the water of the intake part of the basin-cooler is characterized by the presence of two water clusters – № 7 and № 8. Unlike the water of the open offtake channel, in the water of intake zone pf basin-cooler the dominant type of cluster changes from cluster № 7 to cluster № 8 (TAOA – 2.4673 mg Ru per 1 dm^3; occurrence 60%).

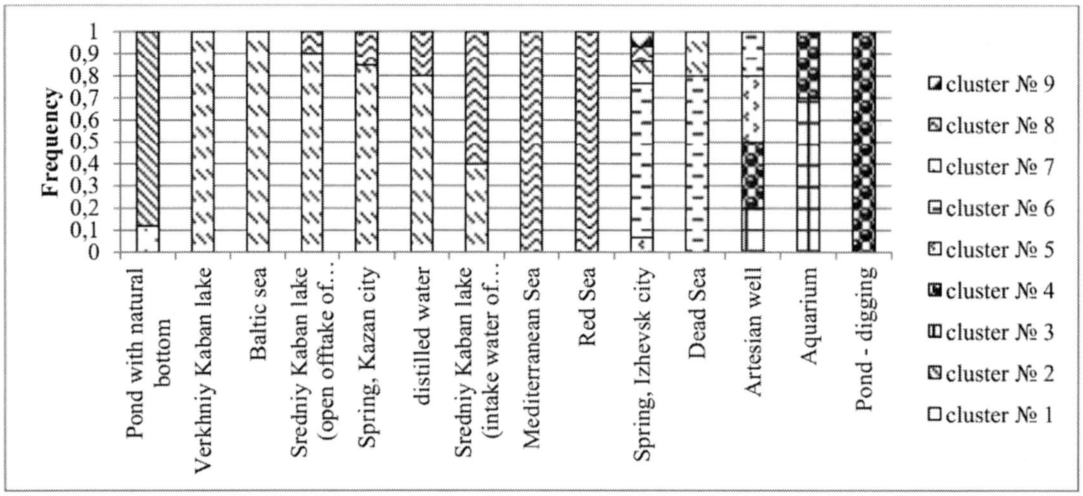

Figure 2. Water clusters of various water bodies according to the total antioxidant activity.

For comparison, turn attention to the composition of clusters in the seas. The Mediterranean and Red seas which contain enough amounts of salts (36-40 and 40-41 ‰ respectively) are characterized by existence of one cluster in structure – cluster № 8 (TAOA – 2.4673 mg Ru per 1 dm^3). The waters of the Baltic Sea are characterized by lower salinity – 3-4 ‰, in connection with which this sea is called a brackish water body. In terms of water structure, the Baltic Sea, similar to the Mediterranean and Red, consist of one type of cluster, but there is a change in TAOA values from 2.4673 mg Ru per 1 dm^3 to 3.2655 mg Ru per 1 dm^3.

Artesian and spring waters attract special attention. It is these types of water that are most complex in their cluster structure (Figure 3). In artesian water there are 4 types of water clusters (clusters: 3, 4, 5, 6, which corresponds to the values of TAOA 6.0956; 5.3699; 4.6443; 3.9912 mg Ru per 1 dm^3), in spring water – 5 types (clusters: 5, 6, 7, 8, 9, which corresponds to the values of TAOA 4.6443; 3.9912; 3.2655; 2.4673; 1.8142 mg Ru per 1 dm^3).

Compared to spring and artesian waters, the waters of lakes with anthropogenic effects have a simpler cluster structure (the presence of 2 clusters of water instead of 4-5). At the same time, the waters of seas (except for Dead Sea) and fishery reservoirs are characterized by the presence of only one cluster of water (Figure 3).

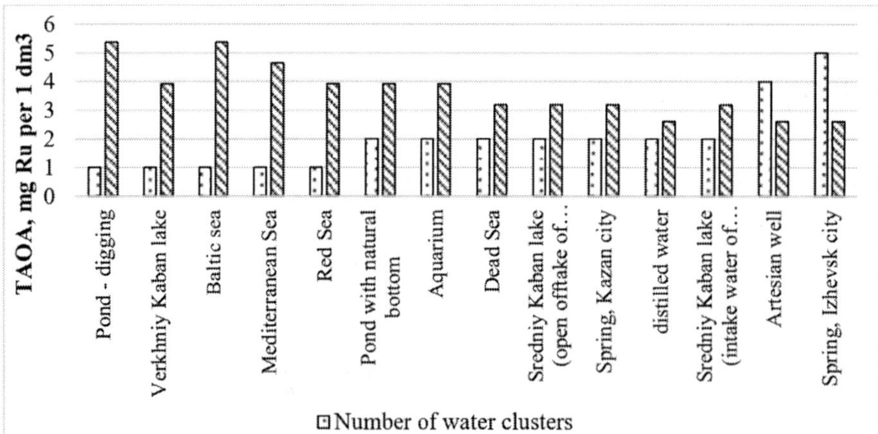

Figure 3. Number of clusters in water from different water bodies.

From the data presented on Figure 3, it can be seen that the simplified (one cluster of water) water structure is possessed by the waters of fish ponds with spring feeding, lakes with weak anthropogenic effects and seas (Baltic Sea, Mediterranean Sea and Red Sea), while the mode of TAOA is the maximum of the researched. The ordinary water structure (the presence of two clusters) is characterized by the waters of ponds with spring feeding located in the city, lakes with anthropogenic impact, including basin-coolers of energy facilities, seas (Dead Sea), spring and distilled waters. Complicated cluster structure (3-4 water clusters) is characterized by artesian waters, and complex structure – spring waters (5 clusters), while the mode of TAOA values is minimal.

Since the structure of natural waters differed in the different percentage of clusters, which were characterized by different values of the TAOA of water, a water structure index is proposed. This index was determined by the equation 1:

$$Iws = (F \times TAOA) \times It/N, \quad (1)$$

where:
Iws – index of water structure;
F – frequency of water cluster occurrence (in shares);
TAOA – total antioxidant activity (mg Ru per 1 dm^3);
It – indicator of trophy (0,5 – dystrophic reservoirs; 1 – oligotrophic reservoirs; 2 – mesotrophic reservoirs; 3 – eutrophic reservoirs);
N – number of occurrence clusters in researched sample of water.

Figure 4 shows the diversity of water characteristics from different places according to the water structure index (Iws).

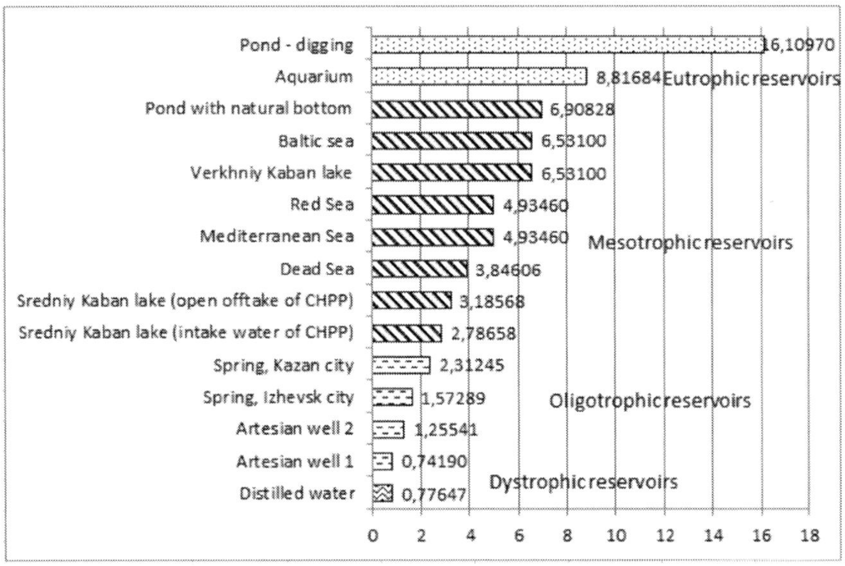

Figure 4. Water classification according to the Index of water structure.

As can be seen from the data on Figure 4, the minimum index values are indicated for distilled water (0.7765).

Artesian water and spring water belong to the second group with index values from 0.8 to 2.5.

The following group of waters is characterized by index values from 2.5 to 7. This group includes: various types of lakes, rivers and seas.

The maximum values of the index, from 7 and above, belong to industrial reservoirs with a high trophy level.

Of particular note are the waters of the Sredniy Kaban Lake with the index values 2.7865-3.1856. In the 1920s, the waters of this lake were recognized as the best according to the salt composition and softness for use in thermal power engineering. As can be seen from the data on the water structure index, this water is close to distilled and spring water.

Like any physicochemical indicators, the values of TAOA change throughout the year (Figure 5).

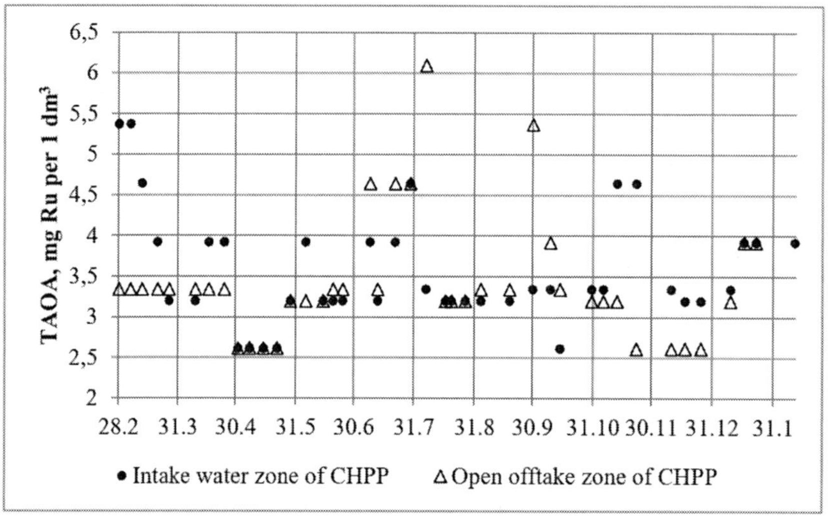

Figure 5. Annual dynamics of TAOA (mg Ru per 1 dm^3) in water of basin-cooler of energy facility.

From the data given in Figure 5, it can be seen that the range of changes of TAOA values in the water of intake zone is from 2.6124 to 5.3699 mg Ru per 1 dm3, in the water of open offtake channel from 2.6124 to 6.0956 mg Ru per 1 dm^3 [16]. Thus, the range of the variation series in the zone of the open offtake channel of the CHPP is 1.3 times higher.

An increase in water the TAOA values are observed in July at maximum levels of biota development. In the water intake zone, an analogous increase of TAOA values is also observed in February, in the area without glaciation, in the zone of maximum mixing of water with a high concentration of oxygen. The annual number of water clusters encountered both in the water intake zone and in the open offtake channel zone are 5 types.

The range of changes in the water values of Eh and pH in the water of intake zone is from -16 to -129 mV and from 5.61 to 12.5 respectively, in the water of open offtake channel from -17 to -135 mV and from 6.24 to 9.34 respectively [1].

Similar to the values of TAOA, the maximum increase of Eh values in the water is observed in summer in the open offtake channel (up to -16 mV) and in spring in the water intake zone (up to -17 mV). Maximum reduction of Eh values is observed in April in two zones (up to -129 mV in the water intake zone and up to -135 mV in the open offtake channel) [1].

4. Conclusion

Water generally has a complex structure with different values of TAOA, which allow isolating clusters. Water with simplified structure containing 1 cluster is isolated. These waters are characteristic of reservoirs with an abundance of life – fish ponds, seas. Water with ordinary structure is characterized by the presence of 2 clusters. These are waters used by humans widely for household consumption. Waters with 3-4 clusters are observed in artesian wells. The most complex in cluster composition is spring water (5 clusters), maybe this makes it so attractive to people.

The analysis revealed that energy facilities, both hydro- and thermal power engineers, use water containing, as a rule, 2 clusters with the value of TAOA mg Ru per 1 dm3. If for water intake for thermal power facilities preference should be given with TAOA values – 2.4673 mg Ru per 1 dm^3, then for water offtake into natural reservoirs the values should be – 3.2655 mg Ru per 1 dm^3.

The complex cluster structure of water made it possible to propose the index of water structure, which reflects the percentage of clusters with certain values of TAOA and the degree of water trophy:

$$Iws = (F \times TAOA) \times It/N,$$

where Iws – index of water structure; F – frequency of water cluster occurrence (in shares); TAOA – total antioxidant activity (mg Ru per 1 dm3); It – indicator of trophy (0,5 – dystrophic reservoirs; 1 – oligotrophic reservoirs; 2 – mesotrophic reservoirs; 3 – eutrophic reservoirs); N – number of occurrence clusters in researched sample of water.

The waters used for drinking (artesian, spring) are characterized by index values in the range from 0.8 to 2.5. With an increase in trophy, the index values increase, so in mesotrophic reservoirs the index value lies in the range from 2.5 to 7, and in eutrophic ones above 7. The maximum value of the index encountered belongs to the pond-digging used for fisheries purposes (Iws = 16.1097). In case of technical use, water with the lowest index values can be selected. There is a tendency to reduce the number of water clusters with an increase in the water structure index.

References

[1] Gordeeva M E and Kalayda M L 2013 *Kompleksnaya ocenka sostoyaniya e`kosistemy` ozer. Urbanizirovanny`e territorii* [Integrated lake ecosystem assessment. Urban areas] (Astraxan`: AGTU) [In Russian]

[2] Galeeva M E, Lapin A A, Chugunov Yu V and Kalayda M L 2012 Antioksidantnaya aktivnost` – perspektivny`j integral`ny`j pokazatel` dlya opredeleniya indeksa kachestva vod [Antioxidant activity – a promising complex indicator for determining the water quality index] *Butlerovskie Soobshheniya* **29** 110-9 [In Russian]

[3] Fairbridge R 1972 *The Encyclopedia of Geochemistry and Environmental Sciences* (New York: VNR)

[4] Lehr J, Keeley J and Lehr J 2005 *Water Encyclopedia: Domestic, Municipal, and Industrial Water Supply and Waste Disposal* (New Jersey: John Wiley & Sons, Hoboken)

[5] Zhang Z and Furman A 2020 Soil redox dynamics under dynamic hydrologic regimes – A review *Science of The Total Environment* **763** 143026

[6] Frohne T, Rinklebe J, Diaz-Bone R A and Du Laing G 2011 Controlled variation of redox conditions in a floodplain soil: impact on metal mobilization and biomethylation of arsenic and antimony *Geoderma* **160(3-4)** 414-24

[7] Buresh R and Patrick Jr W 1981 Nitrate reduction to ammonium and organic nitrogen in an estuarine sediment *Soil Biol. Biochem* **13** 279-83

[8] Xu Ya, Feng J and Li H 2021 Water management increased rhizosphere redox potential and decreased Cd uptake in a low-Cd rice cultivar but decreased redox potential and increased Cd uptake in a high-Cd rice cultivar under intercropping *Science of The Total Environment* **751** 141701

[9] Frohne T, Rinklebe J, Diaz-Bone R A and Laing Du 2011 Controlled variation of redox conditions in a floodplain soil: impact on metal mobilization and biomethylation of arsenic and antimony *Geoderma* **160(304)** 414-24

[10] Lapin A A, Gordeeva M E and Kalayda M L 2019 Klasternaya xarakteristika vod po velichine ix summarnoj antioksidantnoj aktivnosti [Cluster characteristics of waters in terms of their total antioxidant activity] *Butlerovskie Soobshheniya* **60(10)** 67-73 [In Russian]

[11] Tadesse I, Green F B and Puhakka J A 2004 Seasonal and diurnal variations of temperature, pH and dissolved oxygen in advanced integrated wastewaterpond system treating tannery effluent *Water Research* **38(3)** 645-54

[12] Domínguez-Villar D, Arteaga C, García-Giménez R, Smith E and Pedraza J 2008 Diurnal and seasonal water variations of temperature, pH, redox potentialand conductivity in gnammas (weathering pits): implications for chemical weathering *Catena* **72(1)** 37-48

[13] Minamikawa K and Sakai N 2006 The practical use of water management based on soil redox potentialfor decreasing methane emission from a paddy field in Japan *Agriculture, Ecosystems and Environment* **116(3-4)** 181-8

[14] Wang Z P, DeLaune R D, Masscheleyn P H and Patrick W H 1993 Soil redox and pH effects on methane production in a flooded rice soil *Soil Sci. Society America J.* **57(2)** 382-5

[15] Masscheleyn P H, DeLaune R D and Patrick Jr W H 1993 Methane and nitrous oxide emissions from laboratory measurements of rice soil suspension: effect of soil oxidation-reduction status *Chemosphere* **26(1-4)** 251-60

[16] Gordeeva M E 2019 Integrated use of lake-type basin-coolers for recreational purposes *IOP Conf. Ser: Earth and Environmental Sci.* **288** 012041

Development of technical vision hardware and software complex to improve concentration tables' operational efficiency and range of applications

M A Velichko[1], E V Esina[2], Yu P Gladkikh[1], V A Esin[3] and I B Kostina[1]

[1] Department of Computer Science, Natural Sciences and Teaching Methods, Belgorod State University, 85, Pobeda St., Belgorod, 308015, Russia
[2] Department of Foreign Languages and Professional Communication, Belgorod State University, 85, Pobeda St., Belgorod, 308015, Russia
[3] Department of Mathematics, Belgorod State University, 85, Pobeda St., Belgorod, 308015, Russia

E-mail: velichko@bsu.edu.ru

Abstract. The work is devoted to development of technical vision system for the concentration tables. It is shown that new technologies applied to the design of concentration tables will significantly help companies, specializing in extraction and handling of common minerals in deep processing of construction sands in order to obtain more expensive products. Technical vision hardware and software complex is considered as one of automation systems, that can greatly improve productivity of concentration tables, provide energy, human and essential mineral resource saving (e.g. of such minerals as titanium-zirconium raw material). Described is the mechanism of different fractions dividing on the concentration table. Shown is the process of dividing line recognition by technical vision system. Depicted is the system structure and the used elements. Pictured is the software general operation including camera image analysis before making decisions and hardware control.
Key words: technical vision, computer vision, image processing, concentration table, tailings processing, environmental management program.

1. Introduction

Technical vision is the application of computer vision to industry and manufacturing. Computer vision is a general set of methods that allows computers to get information from the images and in such a way to "see". Technical vision is as an engineering field that deals with digital input/output devices, computer networks and other tools designed to control production equipment. Technical vision is directly related to computing, optics, mechanical engineering and industrial automation.

One of the most common applications of technical vision is inspection of agricultural and industrial products such as microcontrollers, automobiles, food and medicines. While people working on the assembly lines examine parts of the product, making conclusions about the performance quality, technical vision systems use digital smart cameras, as well as image processing software to perform similar checks [1].

Vision systems are programmed to perform highly specialized tasks, such as counting or separating objects on the conveyor, reading codes or serial numbers, searching for surface defects or foreign inclusions. The benefits of such inspection systems based on technical vision include significant reduce of operational costs, high speed, possibility of 24-hour operation, accuracy of repeated measurements. Once installed technical vision systems usually don't need permanent maintaining. They operate according to the established algorithms in the absence of fatigue, illness, or inattention. Modern vision systems can be monitored remotely using WiFi, LoRa, Internet technologies, etc. They can be installed in the rigid areas where human labor can't be used.

Having a variety of advantages, computer image processing systems are usually designed to perform single, repetitive tasks, and despite significant improvements in this area, no technical or computer vision system can yet match some of the capabilities of human vision in terms of image understanding, tolerance for lighting change and image degradation.

Thus, technical vision has various directions for future development – both in terms of improving the technologies and algorithms used, and also in terms of expanding the range of applications.

This paper describes the development of hardware and software complex of technical vision for the processing (concentrating) tables.

The concentration table is intended to separate minerals in the water environment by their density when processing ores of non-ferrous, ferrous, rare and precious metals [2]. It is used in mining and metallurgical industries.

Concentration tables are characterized by a fairly high degree of enrichment (ratio of the initial sample mass to the concentrate mass), which can be 20...50, depending on the size of the material and the content of heavy minerals in it (when extracting free gold, the degree of concentration reaches 1000 or more).

However, the processing tables have a very small production capacity, despite the rather large concentration area. To increase their performance, new technologies should be introduced in the design of processing tables. Automation of some or most enrichment processes would help to enlarge their effectiveness. They will be able to function 24 hours a day without the need for constant monitoring. Automation will not only decrease the use of financial and human resources, but also minimize the risk of errors, industrial injuries, etc. Reducing the cost of enrichment will also expand the range of possible applications of concentration tables.

The introduction of a technical vision system is one of such automation processes. Machine vision will allow the processing table to "see" the lines between different fractions of material and, in accordance with the information received, to make decisions about moving the receiving trays, changing the engine frequency, water jet pressure, etc [3].

Actual application of concentration tables is the sand tailing procession [4]. In the European part of Russia, an increasing number of companies, specializing in the extraction and handling of common minerals, are deep processing construction sands in order to obtain more expensive products (primarily glass and molding sands). This trend is positive and corresponds to the State interests under the environmental management program.

Russia has a huge resource potential of titanium-zirconium raw materials both in the form of recorded (proven) reserves and in the form of a product that is extracted along the way. Despite the presence of a large number of explored objects, many of them are still not involved in operation for various reasons. At the same time, the need for import substitution, primarily of titanium raw materials (95% is imported from abroad), to a lesser extent of zirconium, increases every year.

Improving characteristics, introducing new technologies in production of concentration tables, using technical vision and other automation systems will ultimately help processing companies to reduce the titanium-zirconium raw material shortage.

2. Materials and Methods

Separating particles by concentration tables is carried out in a thin layer of water flowing along a slightly inclined flat surface of the deck, which performs reciprocating movements in a horizontal plane perpendicular to the water flow direction [5]. The deck of the concentration table usually has a trapezoidal or rhombic form and is made of wood, aluminum or fiberglass.

IOP Conf. Series: Materials Science and Engineering 1089 (2021) 012044 doi:10.1088/1757-899X/1089/1/012044

The deck's surface is covered with special rubber or plastic and contains narrow strips called riffles. The length of the riffles is reduced in the direction of the loading tray, which is fed with raw materials in the form of pulp. The asymmetric course of the table deck in the direction of its longitudinal axis is provided by a spring. Deck's slight tilt (1-10°) perpendicular to the direction of movement is regulated by the handwheel. Water enters the water tray and is evenly distributed across the deck[6].

Each grain on the table deck is simultaneously affected by two forces, that are the hydraulic pressure of the flush water flowing across the deck, and the inertia that occurs when the deck moves back and forth and is directed along the table deck. When the deck moves forward (from the drive) with a gradual increase in speed, all the material on the deck moves with it to the end of the forward stroke[7].

The deck moves faster in the backward direction (under the action of the spring) that leads to the manifestation of significant inertial forces exceeding the friction of the grains on the surface of the table deck, and to their movement along the riffles. In this case, velocity of grains with different specific weights will not be the same. Higher-density grains with high inertia will move along the deck faster than less inertial grains of lower density [8].

Flushing water, on the contrary, acts with greater force on the grains of light minerals, since at the same weight of light and heavy mineral particles, the cross-sectional area that determines the hydraulic pressure of the flushing water is larger for a light mineral particle, so velocity of light mineral grains across the deck is greater than that of heavy mineral grains. As a result of these phenomena, a fan of grains of different densities diverging from the loading point is formed on the table deck. Grains of the heaviest minerals (heavy fraction) are concentrated in the zone farthest from the loading tray, grains of the lightest minerals (light fraction) are concentrated closer to the tray, and grains of minerals with intermediate density or aggregates of heavy and light minerals (industrial product) are located between them.

Slime particles are removed by flushing water at the beginning (at the outer side) of the table deck. Products of different densities can be sent to the appropriate receivers using dividing partitions [9].

Recognizing the dividing line is a complex problem. Its solution makes it possible to upgrade various aspects of the concentration table operation. This paper describes the solution of one part of this problem, namely the automatic movement of the drain hole, which gets one of the fractions of the enriched rock.

3. Results

In this work visual selection of useful rock from the rest of the sand mass by the difference in their colors was implemented. When the concentering table is installed properly, all its mechanisms are engaged and turned on, software is tested, alignment is completed, the deck of the concentrating table starts moving with the given motor frequency, the water starts flowing from the water tray. Shortly after the movement of the processing table deck starts, the line between useful and empty rock appears on the deck's glossy reflective coating. Our main task was to automatically determine the location of this line for the subsequent movement of the drain hole to it. We also had to take into consideration permanent oscillations of the table's deck.

It was decided to use a personal computer (PC) and a webcam as the base of our technical vision hardware complex [10]. The webcam operated in the visible range and had resolution of 640 by 480 pixels, PC ran a 64-bit operating system and was equipped with the LPT interface for transmitting control signals to the stepper motor driver. The corresponding LPT driver was installed onto the PC and tested before starting software development. LPT was chosen for the hardware complex prototype because of its simplicity and reliability (for the same reasons, it is still widely used in some coordination machine applications) [11].

The camera was installed so that its lens was focused on the edge of the processing table's deck, located along the water drain line.

To solve the image processing problem, the new software application was developed.

The program reads the image from the camera and puts it in the clipboard. The camera focus point can be observed in the application main window. Before the deck starts moving the image in the application exactly repeats what the camera 'sees'. At that stage, when the deck is stable, it is recommended to make all adjustments of the vision system. Shortly after the start of the deck's moving, the color difference between useful and empty rocks, that is, line of rocks separation can also be observed in the application window. The program allows to use the mouse to select points on both sides of this line to visually determine the colors in its vicinity. Now when dividing line moves the camera sight follows it and so does the drain hole.

The image pixels, that have been put into clipboard are analyzed by the software app to identify the colors selected above. According to the specified algorithm, only sections of the image are selected where the pixels of the reference colors are next to each other. The width of the image being searched can be also adjusted in the app. Color comparison is made with a tolerance to optimize the operation of the complex. The result of processing a series of images from the camera is the application's decision to transmit the control signal to the stepper motor.

4. Discussion

Software application main window is shown in figure 1. Human face is captured by the camera to clarify colors definition.

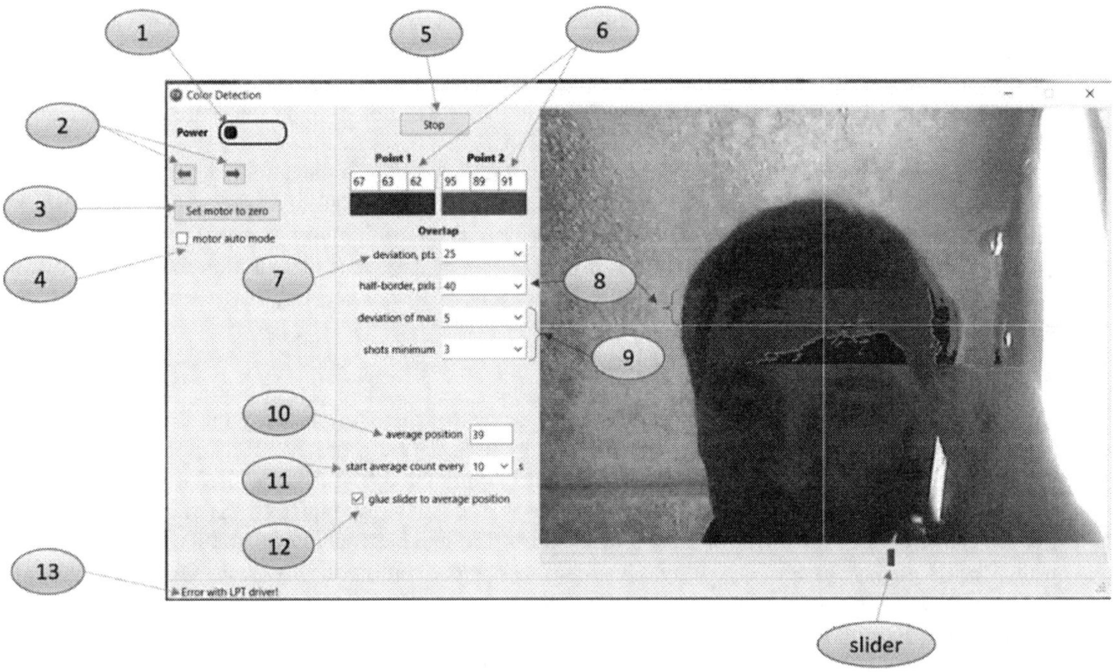

Figure 1. Technical vision software application main window.

Annotation to the numbers on the image are given below. They explain the software working principle.

1. Motor start / stop button.
2. Drain hole single shift in the left or right direction. The movement of the drain hole is provided by a stepper motor using a drive belt. Single shift corresponds to the given number of steps made by the motor. This number is adjusted in the application and depends on the table size, camera position, system alignment, etc.
3. Setting up the zero position of the drain hole to start shifts counting.

IOP Conf. Series: Materials Science and Engineering 1089 (2021) 012044 doi:10.1088/1757-899X/1089/1/012044

4. This box should be checked to synchronize the movements of the drain hole and the slider under the camera image. Otherwise, the drain hole is shifted only when the control arrows are pressed (see p. 2).

5. Start/stop capturing images from the camera.

6. Left and right mouse buttons are used to choose 2 reference colors on the camera image. All pixels that meet the requirements for matching color # 1 or color #2 and lie within the selected band will change their color to red or blue, respectively. Pixels that are located on the border of the blue and red areas are colored green. The next stage of image analysis applies only to the green area.

7. Acceptable deviation of the pixel color from the reference color (point 1 or point 2). Its calculation is based on the deviation of RGB color components.

8. The half-width of the band, within which colors are analyzed.

9. In the horizontal direction, the band within which the image pixels are analyzed is divided into n segments. If the total number of green pixels (located at the intersection of red and blue pixels) in a particular segment (when calculating from the left to the right edge of the band) meets the conditions set in the program, the slider moves to this segment. To avoid reacting to every change in the number of green pixels, certain conditions to the slider movement can be applied, such as the minimum required number of matches and the minimum required change in the number of matches. If these conditions are not met, the slider remains in the previous position (on the previous segment).

10. Slider position under the image from the camera. This parameter is read-only and corresponds to one of n segments of the band for pixels analysis.

11. Averaging period. The table's deck is 'vibrating' back and forth during processing, so the image from the camera is also 'trembling'. The green pixels on the image also follow deck's vibrations. The goal of our vision system is to determine the shift of the partition line between fractions in relation to the table that is robustly fixed but not to the vibrating deck. As image jitter hinders our task it should be eliminated or minimized. In this work we solved the problem by summing and averaging the amount of matching pixels. Averaging period can be changed in this editable field.

12. If this box is checked the slider follows the average value of matched pixels, otherwise it follows the current value.

13. If the LPT driver is installed correctly, ' LPT driver OK' string appears.

The concentration table in our work consists of one horizontal level and has axial symmetry. The axis of symmetry is located along the vibration vector of the deck. Along the axis, in turn, is a row of hoses with taps that make up the water tray. The traps act as regulators of water pressure and, thus, are one of the parameters of the enrichment system. Subsequently, this parameter is also planned to be automated when improving the technical vision system, in particular, for more efficient use of water.

The presence of the axial symmetry of the deck allowed us to develop a prototype vision system for only one half of the concentration table. The prototype included a single camera located on a pole perpendicular to the surface of the deck. Since the deck carried out vertical oscillating movements, the pole with the camera was attached to a fixed part of the table (on one side). It was assumed that the raw material comes from one side of the deck, symmetrically on both sides of the axis of symmetry of the table (this process also subsequently involves the introduction of automation elements). Even if the raw material is unevenly poured on both sides of the axis, after some time after the start of the deck movement, the dividing lines (to the right and left of the axis) are shifted approximately the same distance along the specified axis. Therefore, for the first prototype, it was decided to use one intelligent video processing system and one camera. There were only 2 motor and the corresponding drive belts that moved the drain holes on both sides of the deck, but the belts moved simultaneously.

Subsequently, to improve the accuracy of the drain hole positioning, it was decided to parallelize the system so that each side of the table was equipped with its own camera. The availability of two hardware units is questionable and depends on the performance of the personal computer and the optimization of the software. At this stage, we are considering the option of using a single PC and running 2 identical applications on it, each of which will process the video sequence of its own camera.

But, perhaps, for greater reliability, the option with 2 PCs of lower performance will be used. In any case, additional research is needed to address this issue.

When writing the software, the visual programming environment C++ Builder was used. The choice was due to the convenience of developing applications with a graphical interface in C++, which, in turn, was well suited for working with LPT libraries and video cameras, as well as for fast processing of video images. The application managed several threads, the main one being the motor driver control thread. The graphical interface is mainly needed only at the stage of preparing and adjusting the table. After you complete the configuration, you can disable the image output to the screen(s) to significantly improve system performance. In the future, it is supposed to install the system remotely using the WiFi network and / or the Internet. The system can also be monitored remotely [12, 13].

Experiments with the technical vision system were carried out indoors and outdoors under different weather conditions (with different light and wind). The use of averaging algorithms allowed for the most part to neutralize the impact of wind. But with this configuration, the system was very sensitive to the level of lighting. The soundboard is covered with a white glossy coating with a high degree of reflection. In bright daylight, the camera's illumination was critical to the system's operation. Morning and evening lighting changed quickly, and as a result, the reference colors that were originally selected changed and stopped serving as a reference for comparing image pixels with them. Various light filters were used, but in the end, the need to complete work on the prototype forced us to temporarily complete the experiments and focus on testing in a room with constant lighting. The correct selection of the zero position of the chamber and the number of steps of the motor to move the drain hole along the axis of symmetry of the table showed that the prototype is workable and can be used separately or as part of the automation system for processing tables.

5. Conclusion

We have developed the new hardware and software complex of technical vision for Russian-made concentration tables. Its prototype consisting of the web camera, stepper-motor with corresponding driver and run by the special software installed on the personal computer [14-16] was examined in laboratory conditions with various changing parameters, including environmental conditions, lighting, working frequency, camera focus, etc.

Given the change in color perception when changing the lighting, as well as taking into account the high reflective ability of the table surface, it is recommended to use the hardware and software complex in a room with constant lighting.

This prototype was successfully tested in real tasks of sand tailing procession in Voronezh State University. After some upgrade of this system it will be installed on the concentration tables used for titanium-, zirconium- and gold-containing raw materials enrichment.

References

[1] Gritskevich E V, Evdokimova L A, Nemoguchev A A and Adolf M V 2015 Razrabotka programmno-texnicheskogo kompleksa dlya issledovanij sistem texnicheskogo zreniya [Development of the software-tehnical complexes for sistems ofreseach technical vision] Intere`kspo Geo-Sibir` **5(2)** 102-6 [in Russian]

[2] Kuskov V B, Kuskova Ya V and Ananenko K E 2012 Ispy`taniya koncentracionny`x stolov razlichny`x tipov [Comparative tests of different types of concentrating tables] *Czvetny`e Metally`* **10** 28-33 [in Russian]

[3] Kuskova Y V, Erokhina O O and Simakov A S 2019 Problematics and perspectives of the development of automatic control systems for concentration tables using computer simulation *J. Phys.: Conf. Ser.* **1384** 012023

[4] Zolotareva G S, Bondarenko S V, Nenakhov S V and Spitsyn V I 2018 Texnogenny`e rossy`pi kak istochnik titan-cirkonievogo sy`r`ya [Technogenic drains as an source of titanium-

zirconium raw materials] *Vestnik VGU. Seriya: Geologiya* **4** 57-66 [in Russian]
[5] Kuskova Ya V 2012 Razrabotka novy'x konstrukcij koncentracionny'x stolov dlya povy'sheniya e'ffektivnosti obogashheniya polezny'x iskopaemy'x [Development of new designs of concentration tables to improve the efficiency of mineral processing] *Gorny'j Informacionno-Analiticheskij Byulleten'* **9** 393-6 [In Russian]
[6] Vasilyev A M and Kuskov V B 2017 Zakonomernosti razdeleniya tonkozernisty'x materialov na koncentracionnom stole [Regularities of fine-grained materials separation process on concentrating table] *Obogashhenie Rud* **3(369)** 63-8 [in Russian]
[7] Soshkin G S, Vaniev E' Sh, Kotlyarov V V and Borodin E V 2018 Upravlenie gravitacionny'm processom obogashheniya zolotosoderzhashhix sul'fidny'x rud [Control of the gravitation process of enrichment of gold-containing sulfide ore] *Czvetny'e Metally'* **3** 8-13 [in Russian]
[8] Ryabinin V F, Vusikhis A S and Kudinov D Z 2016 Issledovanie vozmozhnosti gravitacionnogo razdeleniya xvostov flotacii tal'ka [Investigation of the possibility of talcum flotation tailings] *Obogashhenie Rud* **3(363)** 52-6 [in Russian]
[9] Kokoulin A N, Yuzhakov A A, Kiryanov D A, Stanisavljevic N and Batinic B 2019 Hierarchical object recognition system in machine vision *Int. Conf. High-Tech and Innovations in Research and Manufacturing* **1353** 012054
[10] Manukyan R L and Petrashin I V 2019 Mashinnoe zrenie. Osnovny'e komponenty' mashinnogo zreniya [Machine vision. Basic components of machine vision] *Fundamental'ny'e i Prikladny'e Nauchny'e Issledovaniya: Aktual'ny'e Voprosy', Dostizheniya i Innovacii* (Penza: Nayka i Prosveshhenie) pp 295-7 [in Russian]
[11] Makarkin D A, Fedorinina N A and Gorshkova K V 2020 Mashinnoe zrenie v robototexnicheskix sistemax [Machine vision in robotic systems] *Fundamental'ny'e i Prikladny'e Nauchny'e Issledovaniya: Aktual'ny'e Voprosy', Dostizheniya i Innovacii* (Penza: Nayka i Prosveshhenie) pp 141-3 [in Russian]
[12] Velichko M A, Krasovskaya L V, Starovoytov I S, Korenkova N A and Galtseva I N 2019 The development of radio and 3g based telemetry system for the remote gas accounting and control nodes *Int. J. of Innovative Technology and Exploring Eng.* **8(9S3)** 1417-21
[13] Velichko M A, Gladkikh Y P and Kostina I B 2020 The development of solar-powered gas metering units equipped with radio or 3G based telemetry systems *IOP Conf. Ser.: Mater. Sci. Eng.* **791** 012060
[14] Velichko M A, Kostina I B and Gladkikh Y P 2019 The use of modern highly precise ultrasonic gas flowmeters and etalon stands to enlarge industrial resource efficiency *IOP Conf. Ser.: Mater. Sci. Eng.* **552** 012028
[15] Velichko M, Satler O, Krasovskaya L, Bronnikova M, Belyaeva I, Kostina I and Gladkikh Yu 2018 Calibration of Ultrasonic Flow Meter on Wi-Fi Network Using a Web Browser *J. Advanced Research in Dynamical and Control Systems* **8** 1593-6
[16] Velichko M, Satler O, Krasovskaya L, Erina T, Belyaeva I and Gladkikh Yu 2017 Using Drone as WiFi Access Point During Infrared Thermography for Subsidiary Data Acquisition *J. Fundam. Appl. Sci.* **9(1S)** 1279-88

Feasibility of construction and installation site of differential pressure sensors for gas energy efficient plant for coarse and fine purification

A P Usachev[1], A L Shuraits[2], A V Rulev[1], A V Biryukov[2] and A A Pikalov[3]

[1]Department of Heat and Gas Supply, Ventilation, Water Supply and Applied Fluid Dynamics, Yuri Gagarin State Technical University of Saratov, Polytechnicheskaya 77, Saratov, 410054, Russian Federation
[2]Gipronigaz JSC, prospect. Kirova S.M., 54, Saratov, 410012, Russian Federation
[3]Revision Control Consulting LLC, Engels, Russian Federation

E-mail: ritamitrofanova@yandex.ru

Abstract. The paper substantiates the construction and location of differential pressure sensors for a coarse and fine gas energy-saving plant. Energy saving is achieved by installing filtering cylindrical elements of fine and coarse purification one inside the other, optimizing their shape and substantiating the location of the differential pressure sensors. The results of the studies show that the total specific heat consumption for heating and ventilation of purification plants from all energy-saving measures for the proposed option is 1.85 times lower than the existing analogue of the non-optimized shape and measuring the pressure drop at the inlet and outlet branch pipes. The article also proved that the static gas pressure in the lower part of filter elements of the coarse and fine purification will always be higher than in their upper part. This results in a proportionally higher amount of inlet gas and therefore more solids settling in the bottom of the filter element. Measurement of the pressure drop in the lower part of the filter element, the most clogged with solid particles, makes it possible to receive a signal at the dispatching console earlier and provides a time reserve for technical personnel to timely remove accumulated blockages from coarse purification elements and prompt replacement of fine purification elements.
Key words: gas purification, installation site, sensor, pressure drop, filter element.

1. Introduction. Introductory part and relevance of the development of methodological provisions for substantiating the location of differential pressure sensors located in one body of a gas purification plant

Equipping gas networks with modern high- throughput capacity gas reduction points (GRP) is primarily due to the high demand for natural gas due to the harsh climatic conditions of the Russian Federation. At present, more than fifty gas reduction points (GRP) equipped with high-precision gas equipment of high throughput capacity are successfully operated.

Reliable operation of high- throughput capacity gas control equipment currently used in the Russian Federation, according to modern requirements [1], is ensured through the use of cylindrical two-stage units (CTU) of coarse and fine purification, which are installed in separate buildings (Figure 1). An example of a reduction line for head GRPs with a total throughput capacity of 500 thousand m^3/h with a high-throughput capacity CTU for coarse and fine purification installed on it is shown in Figure 1.

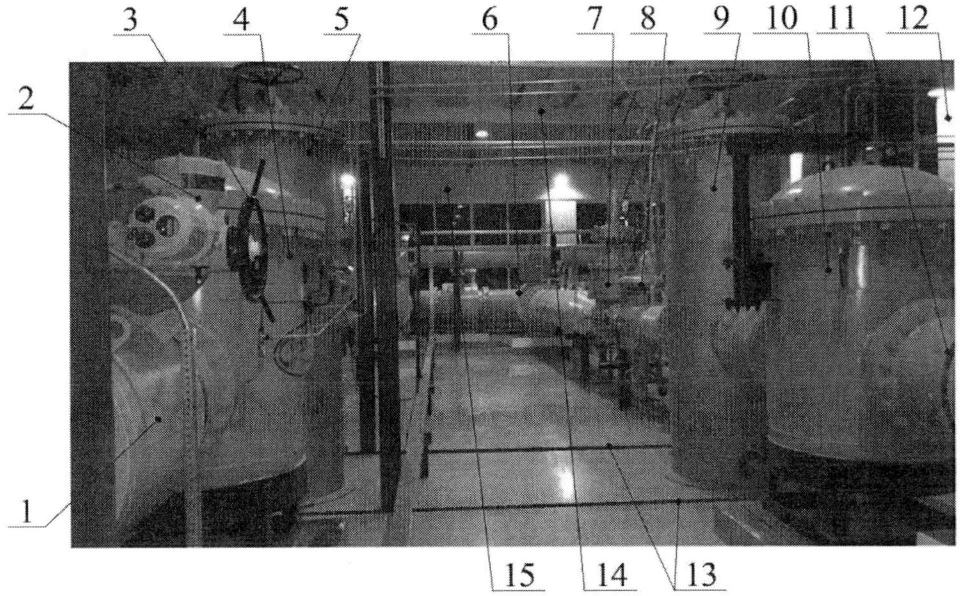

Figure 1. Gas equipment of a modern domestic GRP, including cylindrical units of coarse and fine purification for the main and reserve reduction lines, located in separate buildings. 1 – ball valve with electric 2 and manual 3 drives at the entrance to the main reduction line; 4, 5 – installations for coarse and fine purification, located on the main reduction line; 6, 11 – ball valves with electric and manual drives at the inlet and outlet of the reserve reduction line; 7, 8 – working and reserve gas pressure regulators located on the reserve reduction line; 9, 10 – installations of fine and coarse purification, which are part of the reserve reduction line; 12 – one of the windows; 13 – red lines that limit the area of the floor structure required to accommodate one purification unit; 14 – ceiling structures; 15 – one of the walls.

At present, priority attention in the design of reduction points, including gas purification plants, is given to the issues of energy and resource conservation associated with the reduction of heated area and the use of renewable energy sources [2-7]. In order to reduce the capital and energy consumption of purification plants and building fences allocated for their placement, as well as the cost of heating and ventilation of additional premises for their placement (red lines in Figure 1 limit the floor area required to accommodate one body of a one-stage purification plant), JSC "Giproniigaz" is proposed to place in the internal volume of one body 1 (Figure 1), filtering cylindrical elements (FCE) of coarse purification 7 of the mesh and FCE of fine purification 8 of the filter cloth, installed one inside the other [8] and enclosed in corrugated protective nets to prevent deformation when the pressure drop on them rises above the design value. In order to further reduce the capital and energy consumption of the purification plants the optimal ratio of the height H of the body 3 to the diameter D, equal to (H/D) opt = 3.3 [1] was substantiated (Figure 1), at which its dimensions, area and thermal losses decrease.

Clogging of the filter meshes corresponds to a strictly defined pressure drop across the FCE, which is determined by measuring the latter using differential pressure gauges that receive a pulse from differential pressure sensors. In the known literature [9-14] there are no clear recommendations on the place of installation and design of differential pressure sensors, as a result of which they are installed at arbitrary points along the height of filter elements of the coarse and fine purification and along the horizontal distance to the FCE. So for the existing technical solutions [1], differential pressure sensors are installed at the inlet and outlet branch pipes of the CTU, which significantly increases the distance between the two housings of the purification plants, limited by red lines 13 in Figure 1, and, consequently, the material and energy consumption of the purification plants and for their placement of building fences.

In this regard, an urgent task in the development of energy-saving two-stage purification plants of high throughput capacity is the substantiation of the installation site and the structural arrangement of differential pressure sensors for coarse and fine purification FCE.

2. Materials and Methods. Feasibility of the location of differential pressure sensors for filtering cylindrical elements of coarse and fine purification, located in a single body of a two-stage purification unit

In order to select the location of the differential pressure sensors, let us analyze the flow pattern from the gas inlet (point S) to its outlet (point K), shown on the left side of the CTU (Figure 2). We split the entire length of the FCE into a number of sections $\delta = a, b, ..., j-1, j$ of equal length $L_a, L_b, ..., L_{j-1}, L_j$. The section boundaries in Figure 2 are shown with solid red lines.

The gas flow in the CTU is carried out due to the pressure difference created by taking it into the outlet branch pipe 6, when the pressure at the outlet (point K) will always be less than at the inlet (point S). The total pressure drop (Ps - Pк) determines the distribution of gas pressures along its entire path from point S to K as follows.

For example, for the upper part of the CTU: 1) Ps> Pm - for crude gas in space 10; 2) Pm> Pc - for crude gas during its flow through the upper part of FCE 7 in space 10; 3) Pc> Pn – for a coarsely purified gas during its flow through the upper part of FCE 8; 4) Pn> Pк - for a finely purified gas during its flow in space 12. Hence, the value Ps - Pk is defined as:

$$Ps - Pk = (Ps - Pm) + (Pm - Pc) + (Pc - Pn) + (Pn - Pk). \quad (1)$$

Thus, when the gas moves through each cross-section along the height of the CTU, approximately the same flow pattern takes place. For example, when gas moves through the upper cross section, (points S $\rightarrow$ M $\rightarrow$ C $\rightarrow$ N $\rightarrow$ K), the gas flow rises vertically up from point S to point M, then turns 90 °, passes along a horizontal line through a filter element 7 of coarse purification from point M to point C, continues to move along a horizontal line through the filtering cylindrical element 8 of fine purification from point C to point N, after which it turns 90o and moves vertically downward from point N to point K. In Figure 2, the gas flow pattern in individual sections is shown with brown solid line. Let us consider the flow pattern in its individual sections for crude, coarsely purified and finely purified gas.

For the crude gas during its flow in space 10 along the vertically installed filtering element 7 (Figure 1), the location of the differential pressure sensor was determined based on the analysis of the pressure losses of the gas flow in this area.

Figure 1 shows that the gas flow in the annular space 10 is due to the created pressure drop (Ps-Pk) along the filtering element from bottom to top from point S to point M, which leads to friction losses. In this case, part of the gas in space 10 is continuously separated from the total flow and passes through the coarse filter element 7.

The value of the pressure of the crude gas in the lower point S is determined according to [16]:

$$P_s = P_m + \Delta P_{tr}, \quad (2)$$

where: Pm - quantity of pressure of the crude gas in point M of the annular space 10, daPa;
ΔPtr - gas-dynamic friction losses during the flow of a crude gas in the section between point S and point M, daPa.

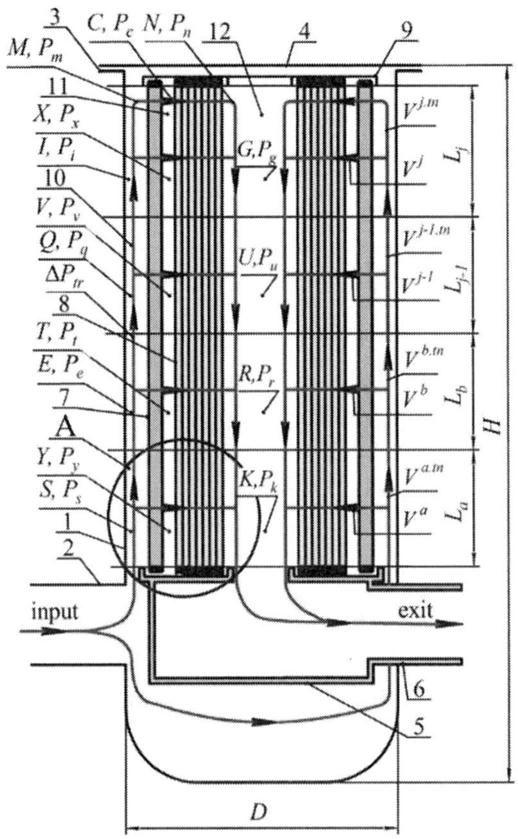

Figure 2. Scheme of natural gas purification from mechanical impurities. 1 - vertical cylindrical body; 2, 6 - inlet and outlet branch pipes; 3, 4 - flange and cover of CTU; 5 - glass; 7, 8 - FCE for coarse and fine purification; 9- cover for FCE 7 and 8; 10 - annular space between the inner and outer surfaces of body 1 and FCE 7; 11 - the gap between FCE of coarse 7 and fine 8 purification; 12 - internal space of FCE fine purification.

Gas-dynamic frictional losses and local resistances in formula (2) occurring during the flow of a crude gas flow in the section between point S and point M are determined according to [16]:

$$\Delta P_{tr} = \left(\sum_{\delta=1}^{\delta=j} \frac{\lambda \cdot L_{\delta=j}}{d_{ek}} + \zeta_{\delta=j} \right) \frac{\left(V^{\delta.tn}/F \right)^2 \cdot \rho}{2 \cdot g}, \tag{3}$$

where: λ is the coefficient of roughness of the walls of the space 10, located between the inner surface of the CTU body and the outer surface of FCE 7;

L_δ is the length of the section δ located along the FCE 7, m;

d_{ek} is the equivalent constant diameter of the inner annular space 10 between the inner and outer surfaces of the body of the CTU 1 and FCE 7, m;

$\sum_{\delta=1}^{\delta=j} \zeta_{\delta=j}$ is the total quantity of the coefficient of local resistance in the sections $a, b, ..., j-1, j$ for the case of a vertical flow along the body with uniform extraction of a part of the gas in the FCE for coarse purification with a rotation of 90 °;

$V\delta.tn$ and F are transit flow rate and flow section of the annular space between the inner and outer surfaces of the body of CTU 1 and FCE 7;

ρ is gas density in the internal space 10 at its actual pressure, kg/m^3.

Figure 2 shows that the gas from the inlet branch pipe 2 spreads over the entire lower part of the flow section of the gap between the housing 1 and FCE 7, from here rising upwards, it evenly enters the filter meshes along the FCE 7 height.

We will replace the uniformly distributed flows passing through the filtering surfaces of the FCE with a number of concentrated $V^{\delta} = V^{a}, V^{b},..., V^{j-1}, V^{j}$ within each section $a, b, ..., j-1, j$... Moreover, all concentrated flows are equal to each other $V^{a} = V^{b} = V^{j-1} = V^{j}$.

The total flows will be written as:

$$V = V^{a} + V^{b} + ... + V^{j-1} + V^{j}. \quad (4)$$

Transit flows for each section are calculated as:

$$V^{a.tn} = V - 0.5V^{a}; \quad (5)$$

$$V^{b.tn} = V - (0.5V^{b} + V^{a}); \quad (6)$$

$$V^{j-1.tn} = V - (0.5V^{j-1} + V^{a} + V^{b}); \quad (7)$$

$$V^{j.tn} = V - (V^{a} + V^{b} + \ldots + V^{j-1} + 0.5V^{j}). \quad (8)$$

Formula for determining the concentrated gas flow rate within each section $a, b, ..., j-1, j$ is

$$V^{\delta} = \frac{V}{\delta}. \quad (9)$$

An analysis of formulas (5) - (8) shows that the values of the transit flow rate as the gas flow rises upward uniformly decreases. So, for example, in the presence of four sections - a, b, j-1, j, j and under the assumption of uniform gas distribution along the entire path from a to j, the quantity of the transit flow rate in section j, according to (8), will be $V^{j.tn} = 0.5\ V^{j} = 0.5(V/4) = 0.125V$, that is, approximately 12.5% of the total flow rate gas V.

An analysis of formula (3) shows that the larger the total length L and the smaller the diameter d_{ek},, are the higher the friction loss ΔPtr is. According to the results of research for new designs of the CTU, for example, manufactured by AO (JSC) "Giproniigaz", the minimum metal consumption and costs for a cylindrical filter are achieved when the ratio of the height of its body to the diameter is equal to 3.3 [1]. Then the length L, which is equal to the total height of FCE 7, will be 65% of the height of the CTU body. In this case, the quantity of friction losses ΔPtr, according to formula (3), significantly increases in comparison with the existing designs of the CTU, which have a small ratio of the body height to the diameter.

An analysis of formula (3) also shows that the pressure of the crude gas Ps in the lower part of the FCE 6 (point S) will always be higher than in the upper part (point M), that is, $P_s > P_m$ by the amount of pressure losses due to friction ΔPtr in the section the annular space 10 between the inner and outer surfaces of the filter housing 1 and FCE 7 of coarse purification.

For the crude gas, according to the general picture described above, its flow for each cross section passes in the horizontal direction through FCE 7. For example, in the upper cross section, the gas flow moves along a horizontal line from point M to point C through FCE 7.

Since in the annular space 10, due to the difference (Ps - Pm), the gas pressure in each higher cross section, counting from point S, will decrease, then the pressure drops through the FCE of coarse purification in these sections will also decrease. For example, in the upper cross-section with a gas flow along a horizontal line from point M to point C, the pressure drop (Pm - Pc) will be less than in lower cross-sections, including the pressure drop (Ps - Py) in the lowest cross-section from point S to point Y.

This leads to a larger amount of coarsely purified gas flowing, and, consequently, a higher amount of solid impurities settling in the lower part of the FCE 7 of coarse purification and, as a consequence, a greater degree of clogging of its filter mesh.

It follows that the measurement of static pressures in the lower part of the FCE 7 at point Y will show a higher value than in the upper part (point C).

In this case, the maximum allowable pressure drop in the lower part of the filtering device 7 will be reached at an earlier period than at other higher measuring points.

For coarsely purified gas, according to the general picture described above, its flow through the FCE of fine purification 8 for each cross section, as well as for the FCE of coarse purification, runs in the horizontal direction. For example, in the upper cross-section, the gas flow moves along a horizontal line from point C to point N through FCE of fine purification 8.

Since in the annular space 11 the pressure difference (Ps - Py) between the lowest points S and point Y is the highest in space 11, the gas pressure in each higher cross section, counting from point Y, will decrease. The pressure drops across the filter elements of fine purification in these sections will also decrease. For example, in the upper cross-section with a gas flow along a horizontal line from point C to point N, the pressure drop (Pc - Pn) will be less than in lower cross-sections, including the pressure drop (Py - Pk) in the lowest cross- section from point Y to point K.

This leads to a greater amount of finely purified gas flowing, and, consequently, a higher amount of solid impurities settling in the lower part of FCE 7 and, as a consequence, a greater degree of clogging of its filter cloth. Hence it follows that the measurement of static pressures in the lower part of the FCE 7 at point Y will show a higher value than in the upper part (point C). In this case, the maximum allowable pressure drop, ΔPmd = 35.0 kPa [12] in the lower part of the filtering device 7 will be reached in an earlier period than at other higher measurement points.

For a finely purified gas, the flow also passes under the action of a total pressure drop (Ps - Pk) along a vertical line from top to bottom from point N to point K along the filter element of fine purification 8 inside space 12. Analysis shows that the pressure of the finely purified gas Pk in the lower part (point K) of the filter element 7 will always be more than in the upper (point N), that is, Pk > Pn, as well as in section 1, by the amount of pressure loss due to friction in the inner space 12 of the FCE fine purification 8.

It follows that the measurement of static pressures in the lower part of the filter element 7 at point K will show a higher value than in the upper part (point N).

3. Results. Development of the design of sensors and determination of the pressure drop filter elements of coarse and fine purification located in a single building of the CTU

Based on the results of the analysis of formulas (3) - (8) and the analysis of the gas flow patterns in all the sections considered above, devices have been developed that measure with a minimum error the static pressure drop directly on the filter elements of coarse and fine purification (Figure 3), which, according to the law Pascal impacts all walls of the CTU and the pipelines and parts located in it with the same force. An important feature of the method is the elimination of the influence of the flow rate at the point of static pressure measurement. In order to eliminate this effect, the gas flow is directed strictly along the lateral surface or strictly tangentially to the lateral surface of the sensor tube, on which the holes for measuring static pressure are located.

Let us consider separately the development of devices for measuring the pressure drop across the FCE for coarse and fine purification.

The pressure drop across the FCE of coarse purification, according to the proposed method (Figure 3), is measured using two sensors: a sensor tube 13 and a sensor tube 14. The sensor tube 13 designed to measure the static pressure of the crude gas Ps is located in the lower part of the FCE of coarse purification 7 and is connected by its hole with the space 10. This allows the crude gas to flow through the impulse tube to the differential pressure gauge. The tube-sensor 14 for measuring the static pressure of the coarsely purified gas Py is located in the lower part of element 7 of coarse purification and is connected by its hole 16 on the lateral surface of the tube 14 with the space 11. This allows the coarsely purified gas to flow through one of the impulse tubes to differential pressure gauge. The hole 16, designed to communicate with the space 11 at the location of the filter element 7, is located on the lateral surface of the sensor tube 14 at the location mark of the sensor 13 inlet for measuring the static pressure of the crude gas. In this case, the flow of coarsely purified gas is directed strictly tangentially to the lateral surface of the sensor tube 14 at the hole 16 location for measuring the static pressure. The upper end of the sensor tube 14 is sealed with a plug 17.

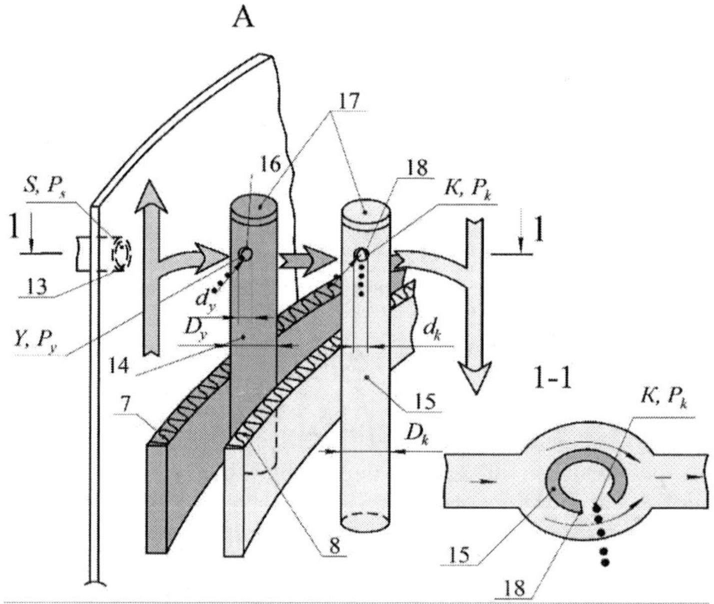

Figure 3. Axonometric diagram of the installation of differential pressure sensors 13, 14 and 15, as well as their design.

By the values of the pressures Ps and Py by means of a differential pressure gauge, their difference is determined, that is, the difference (Ps-Py) on element 7 of the coarsely purified gas. The diameter Dy of the sensor tube 14, as well as the diameter of the hole 16 in its lateral surface, are taken on the basis of design considerations.

The pressure drop across the filtering element of fine purification, according to the proposed method (Figure 3), is measured using two sensors: a sensor tube 14 and a sensor tube 15. The sensor tube 14 designed to measure the static pressure of the coarsely purified gas Py is located in the lower part element of fine purification 8 and is connected by its hole with the space 11. This allows the coarsely purified gas to flow through one of the impulse pipes to the differential pressure gauge.

The tube-sensor 15 for measuring the static pressure of the finely purified gas Pk is located in the lower part of the element 8 of fine purification and is connected by its hole with the space 12. This allows the finely purified gas to flow through one of the impulse pipes to a differential pressure gauge The hole 18, designed to communicate with the space 12 of the filter element 8, is located on the lateral surface of the sensor tube 15 at the location mark of the sensor 13 inlet for measuring the static pressure of the crude gas and at the mark of the location of the hole 16 of the sensor tube 14. In this case, the flow of finely purified gas is directed strictly tangential to the lateral surface of the sensor tube 15 at the location of the hole 18 for measuring static pressure. The upper end of the sensor tube 15 is sealed with a plug 17.

According to the values of the pressures Py and PK, by means of a differential pressure gauge, their difference is determined, that is, the difference (Py-Pk) on the element 8 of a finely purified gas. The diameters Dk of the sensor tube 15, as well as the diameter dk of the holes 18 in its lateral surface, are taken on the basis of design considerations.

Thus, one of the differential pressure gauges measures the difference between the static pressures of the crude and coarsely purified gas coming from sensors 13 and 14, before and after FCE 7, that is, the pressure drop across this FCE; the second differential pressure gauge measures the difference between the static pressures of the coarsely purified and finely purified gas coming from sensors 14 and 15, before and after the FCE 8, that is, the pressure drop across this FCE.

The proposed principles for measuring the pressure drop across FCE of the coarse and fine purification make it possible to significantly increase the possibilities for the timely removal of mechanical impurities from the FCE of coarse purification and replacing the FCE of finely purification with a new one, as well as to prevent deformation and destruction of the FCE due to an increase in the pressure drop on the CPE in excess of the calculated one.

For the purpose of experimental verification of the theoretical provisions (3) - (8) to substantiate the location of the pressure loss sensors, an experimental sample of the CTU was manufactured and tested. The results of tests on the change in pressure losses along the length of the gas flow in the annular space 10 are shown in Figure 4 as separate experimental points. Here, for comparison, the theoretical values obtained from formulas (3) - (8) are shown in the form of a solid line 1. The experimental points shown in the graph are taken as the average value over four measurements (n = 4). The Student's coefficient for four measurements made (n = 4) with a given reliability of the results obtained equal to $\alpha = 0.9$ is $t_{\alpha=0,9} = 0.9 = 2.35$. The average deviation of the calculated and experimental values ΔPtr is 18.3%.

Figure 4 also shows that the growth dynamics of pressure losses along the length of the filter element 7 is significantly reduced. This situation is explained by a decrease in the transit gas flow rate $V^{\delta.tn}$ in the formula (3) along the length of the filter element 7.

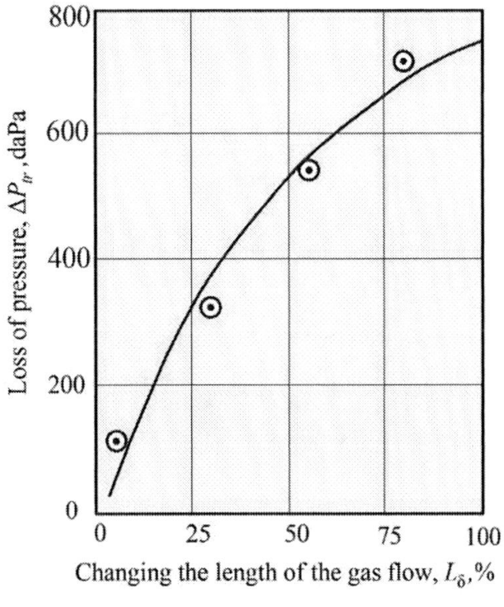

Figure 4. Change in pressure losses along the length of the gas flow in the annular space 10.

IOP Conf. Series: Materials Science and Engineering 1089 (2021) 012045 doi:10.1088/1757-899X/1089/1/012045

4. Conclusion

1. Formulas (5) - (8) are proposed for determining the transit gas flow rates in the annular space between the CTU housing and the filtering element of coarse purification within each section $a, b, \ldots, j-1, j$ in the presence of a uniform flow distribution along the FCE length. The analysis of formulas (5) - (8) showed that the values of the transit flow rate decrease uniformly as the gas flow rises upward. So, for example, in the presence of sections $a, b, j\text{-}1, j$ and under the assumption of uniform distribution of gas along the entire path from a to j, the value of the transit flow in section j, according to (8), will be $V^{j.tn} = 0.5V^{j} = 0.5(V/4) = 0.125V$, that is, 12.5% of the total gas consumption V.

2. Formula (9) has been obtained for determining the concentrated gas flow rate within each section a, b, …, j–1, j. Analysis of formula (9) shows that the values of the concentrated flow rate as the gas flow rises upward in each section a, b, …, j–1, j under the assumption of uniform distribution of the flow along the entire path of movement from a up to j remain practically constant.

3. The calculation results, according to formulas (2) and (9), show that the maximum pressure drop across the FCE takes place in its lower part and it is always greater than in its higher located parts. This leads to a proportionally higher amount of flowing gas, and hence a larger amount of solid particles settling in the lower part of the FCE and, as a consequence, a greater degree of clogging of its filtering material.

4. Measurement of the pressure drop in the lower part of the filter element, which is the most clogged with solid particles, makes it possible to receive a signal at the dispatching console earlier and provides a time reserve for technical personnel to timely remove accumulated clogs and timely replace the filter FCE of fine purification.

5. The gas pressure sensor is proposed to be made of a bent steel tube, installed in the lower part of the inner volume of cylindrical filter elements of coarse and fine purification so that the hole for connection with the inner space is located on their lateral surface, at the same mark as the sensor hole crude gas pressure. In this case, the upper end of the pressure sensor tube is sealed with a plug.

6. The results of the experiments (Figure 4) show higher experimental values of pressure drops for the lower sections of the FCE in comparison with the theoretical quantities for the same sections. This is explained by the better conditions for the flow into the lower sections of the FCE, which have a shorter length, and, therefore, lower gas-dynamic resistance. The figure also shows that the dynamics of the growth of pressure losses along the length of the filtering element 7 is significantly reduced. This situation is explained by a decrease in the transit gas flow rate $V^{\delta.tn}$ in the formula (3) along the length of the filter element 7.

7. The coincidence of theoretical and experimental data on ΔPtr with an average deviation of 18.3% allows us to recommend formulas (3) - (9) to determine the pressure loss of the gas flow along the length of the annular space 10.

8. The results of the calculations show that the total specific heat consumption for heating and ventilation of purification plants from all energy-saving measures for the proposed option with the location of the coarse and fine purification FCE inside one building is 1.85 times lower compared to the existing CTU of the non-optimized form and measuring the pressure drop across the inlet and outlet branch pipes.

References

[1] Usachev A P, Gustov S V and Shuraits A L 2013 *Teoreticheskie i prikladny`e osnovy` povy`sheniya e`ffektivnosti i bezopasnosti e`kspluatacii ustanovok gruboj ochistki prirodnogo gaza ot tverdy`x chasticz v sistemax gazoraspredeleniya* [Theoretical and applied foundations of increasing the efficiency and safety of operation of coarse purification plants for natural gas from solid particles in gas distribution systems] (Saratov: SGTU)

[2] Oral A.Y, Banu B Z and Ozer M 2014 *Int. Congress on Energy Efficiency and Energy Related Materials ENEFM2013* (Antalya: Springer)

[3] Khalil H A E E and Khalil E E 2015 *Energy Efficiency in the Urban Environment* (Boca Raton: CRC Press)
[4] Doty S and Turner W C 2012 *Energy Management Handbook* (Lilburn: Fairmont Pres)
[5] Doty S 2016 *Commercial Energy Auditing Reference Handbook* (Lilburn: Fairmont Press)
[6] Kreith F and Goswami D Y 2015 *Handbook of Energy Efficiency and Renewable Energy* (Boca Raton: CRC Press)
[7] Usacheva E Yu, Shurayts A L and Rulev A V 2016 Assessing Energy Efficiency of Compression Heat Pumps in Drying Processes when Zeotropic Hdrocarbon Mixtures are Used as Working Agents *MATEC Web Conf.* **73** 02015
[8] Mokhatab S and Poe W A 2012 *Handbook of Natural Gas Transmission and Processing.* (Oxford: Elsevier)
[9] Liu Z, Shang J, Chen H, Liu Yu and Wang R 2018 Improved design of two-stage filter cartridges for high sulfur natural gas purification *Separation and Purification Technology* **198(8)** 155-62
[10] Hutten I M 2007 *Handbook of Nonwoven Filter Media* (Butterworth: Elsevier)
[11] Purchas D B and Sutherland K 2002 *Handbook of Filter Media* (Oxford: Elsevier)
[12] Sparks T and Chase G 2016 *Filters and Filtration Handbook* (Oxford: Butterworth-Heinemann)
[13] Guo B and Ghalambor A 2012 *Natural Gas Engineering Handbook* (Houston: Gulf Publising Company)
[14] Thomas D, Charvet A, Bardin-Monnier N and Appert-Collin J-C 2016 *Aerosol Filtration* (London: ISTE Press Ltd)
[15] Usachev A P, Rulev A V, Shurayts A L and Khomutov A O 2018 Method description for arranging coarse and fine gas purification filtering cartridges within one device *AIP Conf. Proc.* **2007(1)** 200310
[16] Chugaev R R 1982 *Gidravlika. Uchebnik dlya vuzov. Izdanie 4-e* [Hydraulics. Textbook for universities Ed 4th] (Leningrad: Energoizdat) [in Russian]

Identification of the energy consumption macromodel for the energy complex in oil and gas producing enterprise into the generalized golden section metric

V I Kolesov[1], G A Khmara[2] and M I Khakimyanov[3]

[1]Department of Road transport, construction and road machines, Tyumen Industrial University, Volodarskogo St., 38, Tyumen, 625000, Russia
[2]Department of Electric power industry, Tyumen Industrial University, Volodarskogo St., 38, Tyumen, 625000, Russia
[3]Department of Electrical Engineering and Electrical Equipment, Ufa State Petroleum Technological University, Kosmonavtov St., 1, Ufa, 450062, Russia

E-mail: hakimyanovmi@gmail.com

Abstract. The problems of increasing the energy efficiency of oil and gas production are becoming very urgent. The cost of purchasing electricity greatly affects the cost of hydrocarbons. The cost of electricity is an important contributor to the price of oil and gas. This is due to low oil and gas prices and expensive electricity. Reliable power supply to oil and gas fields is a big problem. Oil companies are starting to use gas turbine power plants, solar panels and wind turbines. Under these conditions, oil companies are trying to optimize their energy costs. Most of the electricity is consumed by downhole production and injection wells. Development of mathematical models allows minimizing power consumption. All technological processes of oil and gas production are interconnected. Water injection into the reservoir affects the production pumps. The operation of borehole pumps affects the oil treatment and pipeline transportation systems. The tools' synthesis for modelling development strategies in the electric power field into oil and gas producing enterprises (OGPE) is considered. The structural and parametric identification of the macromodel for energy consumption in the OGPE energy complex into the generalized Golden section metric is performed. The calculations' results are given.

1. Introduction

Oil production is an important industry for many CIS countries. The amount of oil produced per year is shown in the table 1. The diagram of oil production by the CIS countries is shown in Figure 1 [1].

Substantiation of the development strategy for large-weights technological systems is a fundamental state's aim. At the same time, the main difficulty is to understand the development mechanism itself. It ultimately determines the working algorithm of the management process.

Such kind tasks arise, in particular, during the electrotechnical complexes' efficiency managing in oil and gas producing enterprises (OGPE) [2-4]. Unfortunately, there are no unified approaches and complete clarity in the issues of the efficient approach substantiation for their solution [5, 6]. It makes difficult to achieve an effective result. This article attempts to fill the gap with modern technologies base.

Table 1. The amount of oil produced, million tons/year.

Country	Oil production
Russia	504.08
Kazakhstan	81.21
Azerbaijan	50.25
Turkmenistan	9.42
Uzbekistan	4.6
Ukraine	3.8
Belarus	1.56
Kyrgyzstan	0.07
Tajikistan	0.02
Moldova	0.02

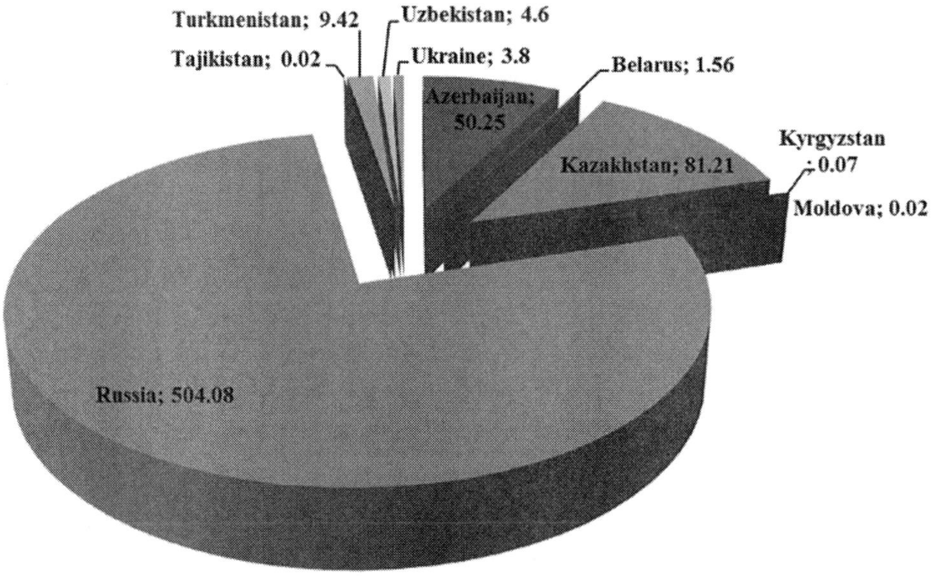

Figure 1. Oil production by the CIS countries, million tons in year.

The structure of electricity consumption by various technological processes at the field is shown in figure 2.

IOP Conf. Series: Materials Science and Engineering 1089 (2021) 012046 doi:10.1088/1757-899X/1089/1/012046

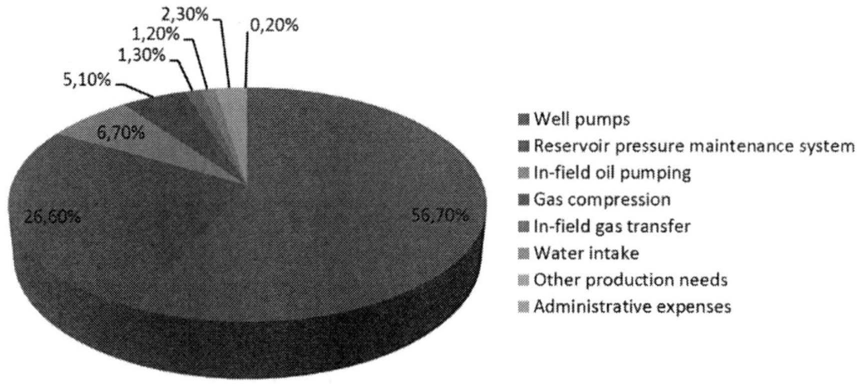

Figure 2. The structure of electricity consumption by various technological processes at the field.

The following technological processes are the most energy intensive [7]:

- well pumps (56.70%);
- reservoir pressure maintenance system (26.60%);
- in-field oil pumping (6.70%);
- gas compression (5.10%);
- in-field gas transfer (1.30%);
- water intake (1.20%);
- other production needs (2.30%);
- administrative expenses (0.20%).

2. Problem solution

Electrotechnical complexes on the weights of oil and gas production enterprises belong to the class of complex systems. It is necessary to apply scientific methods based on a systematic approach [8, 9] to improve the efficiency of their management. A well-known authority in the field of complex systems management is the Institute of Management Problems by V. A. Trapeznikov into the Russian Academy of Sciences. This Institute develops a number of scientific methods for improving the efficiency of management that are wildly used in practice [10]. One of them is the "Golden section" method (or "Golden proportion"). It is the basis of the modern scientific direction "F-technology" (F – Fibonacci) [11, 12].

Practice shows that in the vast majority of situations, the method of the "generalized Golden section" (GGS) works when a proportion is performed between the parts of single segment [13]:

$$(1/x)^q = x/(1-x), \tag{1}$$

from which it follows

$$x^g + x - 1 = 0, \tag{2}$$

where:
x – is the dominant;
$1 - x = x_c$ – is the subdominant;
q and g – are indicators ($g = q + 1$).

It is important to note that the classical "Golden proportion" is only a special case of the ratio (1) when $q = 1$.

The root of equation (2) x_{opt} characterizes the share of the dominant in the total unit result (figure 3).

Taking into account that $+x_c = 1$, the ratio (1) can be written as $1 - x = x^g$ or

$$x = 1 - (1 - x_c)^g. \tag{3}$$

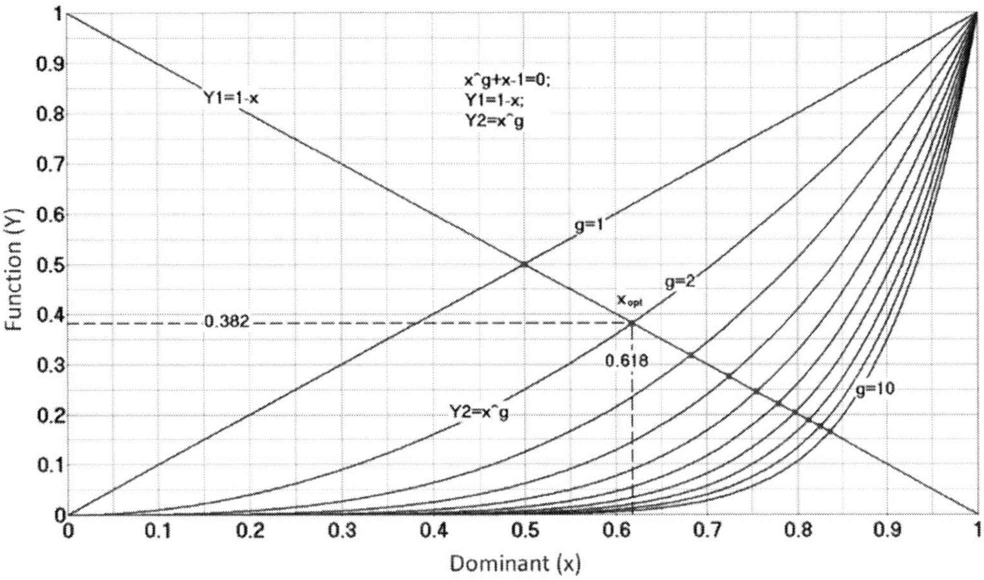

Figure 3. Solution of equation $x_{opt}^{g} + x_{opt} - 1 = 0$.

In practice, this relationship between the dominant x and the subdominant x_c is widely used in the "Pareto" ABC analysis [14]. This analysis is focused on identifying priority components in a complex multi-component system. It turned out that the Pareto diagram is almost perfectly approximated by the function:

$$y = 1 - (1 - x)^{g(x)}, \tag{4}$$

where $x = r/r_{max}$; r and r_{max} – are the current and maximum rank, respectively.

As you can see, the ratio (4) is similar to the ratio (3). The indicator g does not depend on x in a complex system which is ideally suited to the GGS.

As you know, the Pareto diagram is based on the rank distribution of the "weights" w_k for the complex system components. It is essentially their cumulate, the increasing sum of the "weights" $y = \sum_{k=1}^{n} w_k$ (here k - is the rank number; n – is the number of the system components).

In a complex system that perfectly corresponds to the GGS, when $y = 1 - (1 - x)^g$ the "weights" structure is clearly determined and subdues the regularity

$$w_k = \left(1 - \frac{k-1}{n}\right)^g - \left(1 - \frac{k}{n}\right)^g. \tag{5}$$

In this case, the senior " weights" w_1 is associated with the indicator g by the ratio

$$w_1 = 1 - \left(1 - \frac{1}{n}\right)^g, \tag{6}$$

it allows to easily determine the indicator

$$g = \frac{\ln(1 - w_1)}{\ln(1 - 1/n)} \tag{7}$$

by the known value w_1 and then build a rank distribution of "weights" $w_k(r)$.

According to the studies [15, 16], the senior "weights" is estimated at 0.5 level with the energy efficiency analysis for the seven-component electrical complex of OGPE (figure 4).

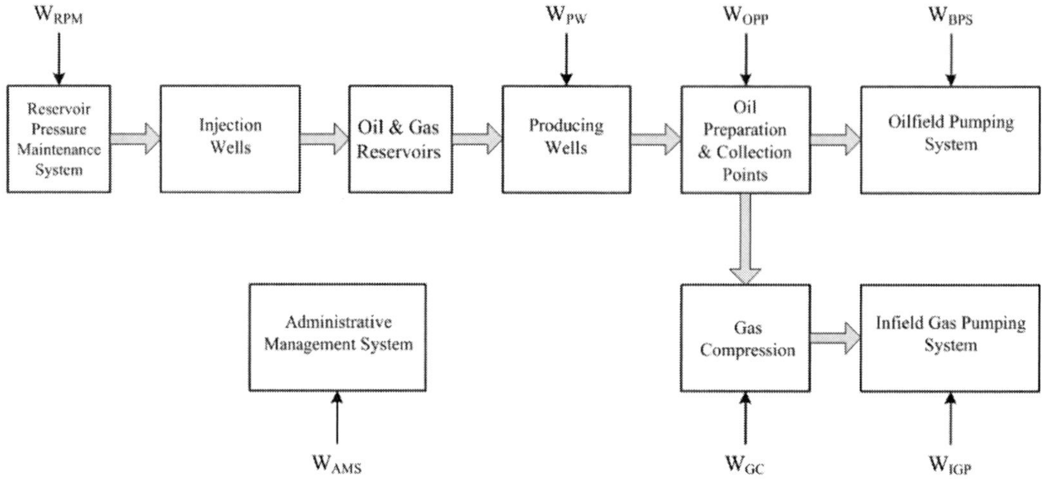

Figure 4. OGPE structure as an electricity consumer: W_{RPM} – electricity used to maintain reservoir pressure; W_{PW} – electricity consumed by producing wells; W_{OPP} – electricity consumed by oil preparation points; W_{BPS} – electricity consumed by booster pumping stations; W_{AMS} – electricity consumed by the administrative management system; W_{GC} – electricity used for gas compression; W_{IGP} – electricity used for infield gas pumping.

Then in the GGS metric the indicator g is:

$$g = \frac{\ln(1-w_1)}{\ln(1-1/n)} = \frac{\ln(1-0.5)}{\ln(1-1/7)} = 4.4966. \tag{8}$$

The total energy is

$$W_\Sigma = \sum_{k=1}^{7} W_k \tag{9}$$

and the "weights" of the components is

$$w_k = W_k / W_\Sigma . \tag{10}$$

At the same time, the rank distribution $g = 4.4966$ of "weights" w_k and the Pareto diagram

$$y(r) = 1 - \left(1 - \frac{r}{7}\right)^{4.4966} \tag{11}$$

based on it should have the form shown in figure 5. Comparison of calculated and actual "weights" gives reason to believe that the real results do not contradict the accepted hypothesis about the mechanism operation of "generalized Gold division" in practice.

3. The "weights" structure of the system components

The "weights" structure of the system components carries information about its organizational level. The structural entropy

$$H = -\sum_{k=1}^{n} w_k \ln w_k \tag{12}$$

or its normalized value

$$H_n = H/\ln(n) \tag{13}$$

acts as a numerical characteristic. The relative entropy of a binary system (in which $n = 2$) is usually used as a reference standard.

The relative entropy of a real system is compared with it. It corresponds to a situation where

$$-\sum_{k=1}^{7} w_k \cdot \ln(w_k)/\ln(7) = -\left[w_p \cdot \ln(w_p) + w_h \cdot \ln(w_h)\right]/\ln(2), \tag{14}$$

where w_p and w_h – are shares of order and chaos that meet the normalization condition $w_p + w_h = 1$.

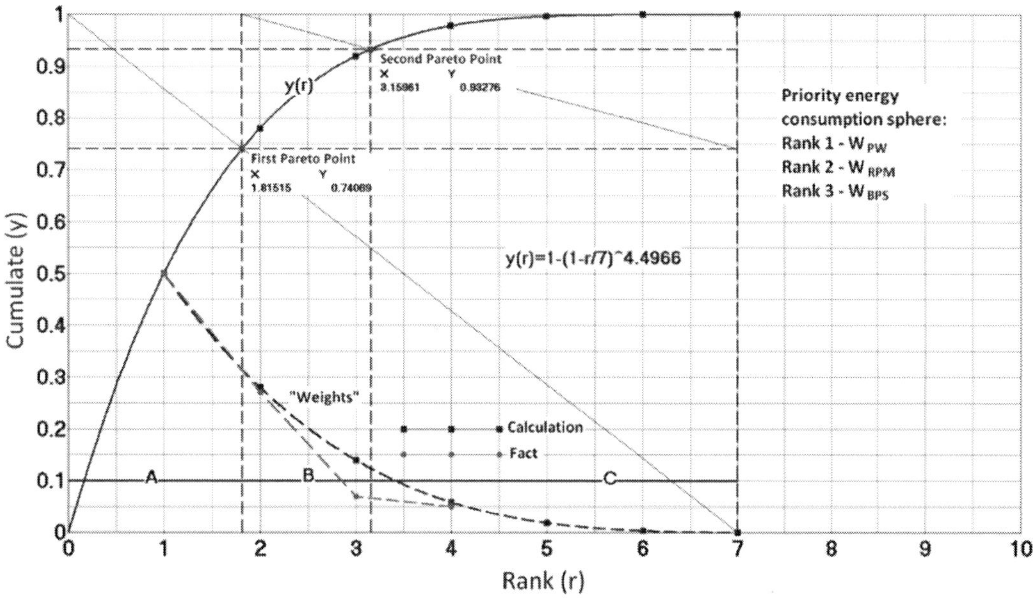

Figure 5. Rank distribution of "weights" and Pareto diagram.

The "weights" data is required to calculate the relative entropy of a seven-component electrotechnical complex. Their values are shown in the table 2. They are calculated in the GGS metric.

Entropy is $H = 1.238$. Relative entropy is $H_n = 0.6362$. The $H_n = 0.6362$ level in the binary system (reference standard) corresponds to the share of order – $w_p = 0.84$ and the share of chaos $w_h = 0.16$. It means that the OGPE electrotechnical complex belongs to the class of open systems with a high organizational level.

Table 2. The "weights" data of a seven-component electrotechnical complex.

Rank (r)	w_k	$w_k \cdot \ln(w_k)$
1	0.5	0.347
2	0.28	0.356
3	0.14	0.275
4	0.059	0.166
5	0.019	0.074
6	0.003	0.019
7	$1.585 \cdot 10^{-4}$	0.001

In addition, the results of the Pareto ABC analysis are presented in figure 5. Its specifics are described in [17, 18] point in detail. The key role of ABC analysis is the priority objects' identification [19, 20]. They make a major contribution to the output product. The main objects are those whose ranks are in the A+B area (see figure 5). There are three of them: W_{PW} (electricity consumed by producing wells), W_{RPM} (electricity used to maintain reservoir pressure) and W_{BPS} (electricity consumed by booster pump stations).

4. Conclusion

Thus, the following conclusions can be made on the results of the conducted researches:

1. The actual structure of the energy consumption weighting coefficients for the electrotechnical complex OGPE does not contradict the law "generalized Golden section".
2. The rank distribution of " weights " indicates that the analysed complex system belongs to the class of open systems with a high organizational level.
3. The number of priority ranks does not exceed three ranks in the analysed seven-component complex system. It determines the prospective direction for further work to improve management efficiency.
4. The producing wells, the system for maintaining reservoir pressure and the system for infield oil pumping are the objects with priority ranks to improve the energy efficiency of OGPE.

References

[1] Rzayev A H, Rezvan M H and Khakimyanov M I 2013 Automation artificial lift systems in the CIS *Oil & Gas Business* **5** 19-29

[2] Khakimyanov M I and Shafikov I N 2013 Analiz potrebleniya elektroenergii pri mekhanizirovannoy dobyche nefti elektrotsentrobezhnymi nasosami [Analysis of energy consumption for mechanized oil production electric submersible pumps] *Elektrotekhnicheskiye i Informatsionnyye Kompleksy i Sistemy* **9(3)** 37-41 [In Russian]

[3] Kolesov V I, Savinykh Yu A and Khmara G A 2014 Perspektivnyye tekhnologii v neftegazovoy otrasli v usloviyakh ustoychivogo innovatsionnogo razvitiya regiona [Advanced technologies in the petroleum complex in the conditions of stable innovation development of the region] *izvestiya Vysshikh Uchebnykh Zavedeniy. Neft' i Gaz* **6** 23-8 [In Russian]

[4] Latypov I S, Khmara G A and Sushkov V V 2017 Podkhod k obosnovaniyu vybora energoeffektivnoy formy vitogo neizolirovannogo provoda vozdushnoy linii elektroperedachi klassa napryazheniya 6 - 35 kV [Approach to substantiating the choice for the twisted non-insulated wire' energy-efficient form of an overhead power transmission line with voltage class 6-35 kV] *Promyshlennaya Energetika* **4** 8-12 [In Russian]

[5] Kolesov V I, Khmara G A and Portnyagin A L 2015 Modelirovaniye energoeffektivnosti bureniya neftyanykh i gazovykh skvazhin [Modeling of power-efficiency of oil and gas wells drilling] *Izvestiya Vysshikh Uchebnykh Zavedeniy. Neft' i Gaz* **4** 56-64 [In Russian]

[6] Menezes G S, Silva-Filho A G, Souza V L, Medeiros V W, Lima M E, Gandra R and Bragança R 2012 Energy estimation tool fpga-based approach for petroleum industry *In 2012 41st Int. Conf. on Parallel Processing Workshops* (Pittsburgh: IEEE) pp 600-1

[7] Ivanovsky V N, Sabirov A A, Degovtsov A V, Tretyakov O V, Mazein I I, Merkushev S V, Krasnoborov D N 2019 Improving the efficiency of oil well operation with small diameter lateral holes *Delovoy zhurnal NEFTEGAZ.RU* **6S(90)** 44-51

[8] Kolesov V I, Khmara G A and Portnyagin A L 2018 System aspects of price responsive energy consumption *IOP Conf. Ser.: Earth Environ. Sci.* **181** 012015

[9] Chen L Y and Wang T C 2009 Optimizing partners' choice in IS/IT outsourcing projects: The strategic decision of fuzzy VIKOR *Int. J. of Production Economics* **120(1)** 233-42

[10] Aleksanov A A, Bochkov A F, Prangishvili I V and Yadykin I B 2004 System aspects in creating a corporative information control system for a regional power company *Thermal Engineering* **51(9)** 754-62

[11] Prangishvili I V, Burkov V N, Gorgidze I A, Javakhadze G S and Khurodze R A 2009 *Systemic*

Regularities and Systemic Optimization (Nova Science Publishers, Inc.) pp 1-153
[12] Chen T L, Cheng C H and Teoh H J 2007 Fuzzy time-series based on Fibonacci sequence for stock price forecasting *Physica A: Statistical Mechanics and its Applications* **380** 377-90
[13] Pronzato L, Wynn H P and Zhigljavsky A A 1998 A generalized golden-section algorithm for line search *IMA J. of Mathematical Control and Information* **15(2)** 185-214
[14] Tykocki J and Jordan A 2014 Pareto-ABC Analysis of High Voltage Single Core Cable Temperature *Przegląd Elektrotechniczny* **90(10)** 172-8
[15] Khakimyanov M I, Shafikov I N and Khusainov F F 2015 Control of sucker rod pumps energy consumption *In 2015 Int. Siberian Conf. on Control and Communications SIBCON* (Omsk: IEEE) pp 1-4
[16] Khakimyanov M I, Shafikov I N and Khusainov F F 2018 Electric Submersible Pumps in Oil Production and Their Efficiency Analysis *In Titel: Proc. of the 4th Int. Conf. on Applied Innovations in IT* (Hochschule Anhalt) pp 35-8
[17] Chen Y, Li K W and Liu S F 2008 A comparative study on multicriteria ABC analysis in inventory management *In 2008 IEEE Int. Conf. on Systems, Man and Cybernetics* (Singapore: IEEE) pp 3280-5
[18] Chu C W, Liang G S and Liao C T 2008 Controlling inventory by combining ABC analysis and fuzzy classification *Computers & Industrial Engineering* **55(4)** 841-51
[19] Zasadzień M 2014 Using the Pareto diagram and FMEA (Failure Mode and Effects Analysis) to identify key defects in a product *Management Systems in Production Engineering* **4(165)** 153-6
[20] Hematyar S 2011 Fuzzy classification of gas power plant spare parts by combination statistical classification technique, SAW, ABC analysis *In 2011 IEEE Int. Conf. on Industrial Engineering and Engineering Management* pp 1800-4

Logistic and mathematical model of application efficiency liquefied natural gas

O N Medvedeva and S D Perevalov

Department of Heat and Gas Supply, Ventilation, Water Supply and Applied Hydro and Gas dynamics, Yuri Gagarin State Technical University of Saratov, 77 Politekhnicheskaya st., Saratov, 410054, Russia

E-mail: medvedeva-on@mail.ru

Abstract. Today, natural gas, with certain advantages over other fossil fuels, is one of the main energy resources for various industries and housing and communal services. As a rule, when choosing a gas transportation method for different categories of consumers, several options are considered depending on the mutual removal of the source and the consumer – construction of a gas pipeline or the creation of a liquefied gas infrastructure. Analysis of energy supply problems in some constituent entities of the Russian Federation showed that a reasonable choice of technological parameters of a gas distribution system based on liquefied natural gas (LNG) is an urgent task that requires preliminary modeling. The paper proposes a solution to a particular problem of supplying consumers with liquefied natural gas, a logistic model for the optimal functioning of the system for the complex: LNG plant – a consumer. Based on the results of calculations for the conditions of the Republic of Karelia, recommendations are proposed for the optimal location of the source (LNG plant) and optimization of the transport component.

1. Introduction

The Russian Federation, as the largest energy country, has 19.1% (38 trillion m^3) of the world's confirmed natural gas reserves and ranks second in terms of production – 679 billion m^3 (based on the results of 2019). In Russia as a whole, according to Federal Agency for Mineral Resources (Rosnedra), at the end of 2019, the increase in gas reserves due to geological exploration is tentatively estimated at 1.354 trillion m^3. In 2019, the total gas production (natural and associated petroleum) in the Russian Federation reached a record level over the past 19 years – 737.8 billion m^3. For many years, Russian pipeline gas has held a leading position in European countries (in 2019, Russian gas imports exceeded 200 billion m^3, or 45 % of demand), entered the markets of the Asia – Pacific region, in addition, the production and delivery of liquefied natural gas, the production level of which is planned to reach 80 – 120 million tons per year by 2035.

Natural gas is a leading component of the country's fuel and energy balance. It forms the basis of energy consumption of tens of thousands of industrial enterprises, it generates over 35 % of electric and thermal energy. The total supply of gas to consumers under the Unified Gas Supply System is more than 75 billion m^3, in addition, 17 million tons of liquefied hydrocarbon gases (LHG) and 29.5 million tons of LNG are produced and supplied.

The main consumers of gas in the industry are enterprises of the electric power industry, metallurgy, chemical industry, which account for more than half of the gas consumed in Russia.

The most important consumer of natural gas is also the housing and utilities sector of the country, which accounts for 20 % of consumed gas (11 % – population, 8 % – utilities sector).

As of January 1, 2020, the average level of gasification of the population of the Russian Federation is 70.1 %, including: in cities – 71.9 %, in rural areas – 59.4 %, with an annual increase in gasified apartments in the amount of 2÷4 times. In remote and hard–to–reach areas, where the network gas supply is not feasible for economic and technical reasons, autonomous gas supply systems are used based on alternative energy sources: LNG, compressed natural gas (CNG) and liquefied petroleum gases.

If the level of gasification of the population of the European part of the country is high enough and reaches 75–80 and more percent, then gasification of Siberia and the Far East is at the level of 30–35 %. Since these data are of strategic importance for the country's development, the energy strategy of the Russian Federation for the period up to 2035 sets the following tasks:

- improvement of the domestic gas market and efficient satisfaction of domestic gas demand;
- development of the production of low-tonnage liquefied natural gas and the formation on its basis of the internal market of liquefied natural gas as a tool for ensuring energy security of territories remote from the Unified Gas Supply System;
- formation of oil and gas mineral resource centers in new regions (Eastern Siberia and the Far East) and on the continental shelf of the Russian Federation;
- development of the main gas transportation infrastructure (including the creation of gas transportation infrastructure in Eastern Siberia and the Far East);
- socially and economically feasible increase in the level of gasification of the constituent entities of the Russian Federation, taking into account the specifics of regional fuel and energy balances, including the creation of conditions for the primary supply of gas to land plots involved in circulation for housing construction, as part of the implementation of national projects and national programs (by 2024 74.7 %; by 2035 - 82.9 %)".

Despite the wide scale of gasification of the country, the level of gas supply to the rural population does not meet modern requirements and potential capabilities of the Unified Gas Supply System. More than half of the country's rural population is not covered by networked natural gas. Gas supply to apartments is reduced to the use of bottled liquefied petroleum gas. The latter is mainly implemented for the needs of food preparation, while the rest of the household and household needs (heating and hot water supply), as well as industrial needs are covered by solid and liquid fuels. In this regard, the strategic task set by the President of the Russian Federation and enshrined in Federal Law No 69 of March 31, 1999 "Gas Supply in the Russian Federation", which provides for extensive gasification based on network natural gas, as well as using alternative sources (liquefied natural gas, liquefied hydrocarbon gases, compressed natural gas) in the area of the Unified Gas Supply System (regional gas supply system) and in areas without gas supply systems, is a priority area for the development of the gas industry in general and its gas distribution industry in particular.

At present, the gas distribution systems of cities and settlements are elaborate technologically, organizationally and economically interconnected and centrally controlled complex of facilities intended for the production, transportation, storage and supply of gas, including gas networks of various pressure categories, underground gas storage facilities, gas distribution stations, reduction points, gas–using equipment. This circumstance predetermines the variety of system–forming ties and the complex nature of the interaction of factors that determine the mechanism of functioning of gas distribution systems [1, 2]. Reliability of gas supply to Russian consumers is one of the main tasks in the development of the domestic gas market [3].

Development of branch of manufacture liquefied natural gas for maintenance of internal consumption of the country and as for export abroad becomes actual in connection with a rise in prices on energy carriers and growth power consumption [4, 5].

Supply liquefied natural gas possesses a number of advantages which in some cases are defining at a choice of system of a gas–transport network [6]. To such advantages concern [7, 8]:

- an opportunity of storage under small superfluous pressure at temperature about 112 K;
- nontoxic;
- high caloric content (in comparison with other kinds of fuel);

- low temperature of boiling – a guarantee of full evaporation LNG at the lowest temperatures of air;
- efficiency and convenience of storage, transportation and consumption (at liquefied natural gas its density increases in 600 times);
- LNG can be delivered to the consumer by any kind of transport (including intercontinental transportations), etc.
- Supply of consumers based on LNG is economically justified in following conditions [9, 10]:
- small volumes of consumption of gas the object removed from the main gas main;
- a stable seismic conditions on a site of prospective construction of a gas main;
- a complex lay of land on a prospective line of construction of a gas main;
- the gasified object or a prospective line of construction of a gas main are in area with a unique natural landscape;
- presence of a formidable barrier on a prospective line of construction of a gas main (the sea, the river, mountains).

For optimum operation of all LNG–complex to build a strong relationship between its main elements [11, 12]. The liquefaction complex is an important and capital–intensive component in the processing chain of transport of natural gas, as a strong point of gas supply systems. with a large number of settlements that require natural gas supply and their significant dispersal of the definition of optimal location liquefaction plant requires a pre-feasibility studies.

2. Materials and Methods

Transportation of liquefied natural gas is significantly more economical than pipeline transportation. This applies both to the transportation of large volumes of gas over long distances by various types of transport, and to the delivery of small volumes of LNG over short distances [1, 13].

At the same time, the advantage of large–capacity transportation of gas in liquefied form begins with distances to consumers over 2500-3000 km [1, 13]. The bulk of the cost comes from handling operations, and LNG also requires more capital investments in infrastructure initially. The creation of LNG infrastructure for low–tonnage transportation will be beneficial in cases where the construction of a gas pipeline is economically unprofitable (due to the difficult terrain, significant distance between the consumer and the source, the consumer needs gas in small volumes, and other justified reasons). The choice of gas delivery option in each specific case is carried out by means of a comparative technical and economic analysis [14, 15].

The LNG infrastructure includes technically and technologically elaborate facilities for production, storage, transportation and regasification [16].

In the process of designing an LNG production, an important issue is solving the problem of locating a gas supply source [13, 17, 18] - in our case, a plant for the production of liquefied gas. The aim of the study is to build a logistic model for the functioning of the LNG complex using the example of the Republic of Karelia.

3. Search for the optimal location of the LNG plant in the Republic of Karelia

To date, the level of gasification of the republic with network natural gas does not exceed 8.5 %. The specificity of the gasification of this region lies in the fact that the gas distribution networks are far enough from the main pipelines of the network natural gas, and the process of gasification with natural gas began relatively recently, in 1996.

Determination of the optimal location for the liquefied natural gas plant is based on the document "On Approval of the Regional Program "Gasification of the Republic of Karelia for 2017–2021", as well as "Adjustment of the General Schemes of Gas Supply and Gasification of Regions of the Russian Federation. Republic of Karelia" it was found that at the time of the implementation of the regional gasification program of the Republic of Karelia, 771 settlements will remain ungasified. When calculating, it was assumed that LNG will be delivered to settlements with a population of more than 200 people, while the percentage of people who will use gas for various needs will be 80 %. The model assumes that the volume of LNG in the storage tanks of the consumer is sufficient to avoid LNG shortages throughout the winter season.

To determine the distance between cities, the coordinates were taken from GOOGLE maps and translated into the local coordinate system MSK–10.

To minimize costs, we will use the method for determining the center of mass:

$$X_{cm} = \frac{\sum_{i=1}^{n}(X_i \cdot Q_i)}{\sum_{i=1}^{n} Q_i}; \tag{1}$$

$$Y_{cm} = \frac{\sum_{i=1}^{n}(Y_i \cdot Q_i)}{\sum_{i=1}^{n} Q_i}. \tag{2}$$

where:
Y_i and X_i are the coordinates of the i-th consumer;
Q_i is the gas consumption by the i-th consumer;
X_{cm} and Y_{cm} are the coordinates of the center of gravity of freight traffic.

Based on formulas (1), (2), the optimal coordinates of the LNG plant for gas supply to 137 settlements were determined.

We assume that the cost of LNG transportation will be proportional to the linear distance between cities, then the distance between the plant and consumers will be determined by the formula [13]:

$$L = \sqrt{(x_2 - x_1)^2 + (y_2 - y_1)^2}, \tag{3}$$

where x is the coordinate of the x consumer; y is the coordinate of the y consumer.

The volume of gas consumed per year is calculated by the formula:

$$V_{\text{ann}} = \frac{N_p \cdot V_{ann}^c}{100 \cdot a_{gc}}, \tag{4}$$

where:
N_p is the number of gas-supplied population;
V_{ann}^c – annual gas consumption by one consumer;
a_{gc} is the percentage of gasification (gas coverage).

To determine the required number of flights to cover the gas demand per year, we use formula:

$$n_{ann} = \frac{V_{ann}}{V_{ct}}, \tag{5}$$

where V_{ct} is the gas capacity of one cryogenic tank.

To determine the mileage of each of the cryogenic tanks, we will use info of the distance from the proposed location of the LNG plant:

$$\ell_{ct} = \frac{L}{n_{ann}}. \tag{6}$$

Knowing the distance from the LNG plant and the average speed of the tanker truck, we calculate the time spent on one trip (both ways):

$$\tau = \frac{120 \cdot L}{\vartheta_{ct}}, \tag{7}$$

where ϑ_{ct} – average speed of the cryogenic tank.

The number of flights per month is calculated by the formula:

$$m = \frac{t \cdot \varphi}{22 \cdot \tau}, \tag{8}$$

where:
t is the time of movement of the tanker;
φ is the utilization rate of the shift.

Calculate the number of tank trucks to transport the required volume of gas:

$$n_{ct} = \frac{n_{ann}}{12 \cdot m}, \tag{9}$$

where 12 is the number of months for LNG delivery.

In addition, we will take into account the costs of refueling cryogenic tanks with diesel fuel:

$$C_{df} = P_{df} \cdot \ell_{ct}, \tag{10}$$

where P_{df} is the fuel price of one liter.

As a result, the cost of production will consist of the following values:

$$C_{LNG} = V_{ct} \cdot n_{ann}. \tag{11}$$

Based on the results of the numerical analysis, we take the production cost of one ton of LNG equal to 25000 rubles.

In addition to the costs of tankers, we will calculate the costs of wages for drivers:

$$C_d = 12 \cdot S_d \cdot n_{ct}. \tag{12}$$

where S_d is the salary of one driver.

Finally, we will calculate the future potential profit:

$$Pp_{LNG} = V_{ann} \cdot P_{LNG} - C_d - C_{df} - C_{LNG} \tag{13}$$

where P_{LNG} is the cost of one m^3 of gas.

4. Results and Discussion

The calculation results are presented in tables 1, 2 and 3. Figure 1 shows a graphical interpretation of the calculation results.

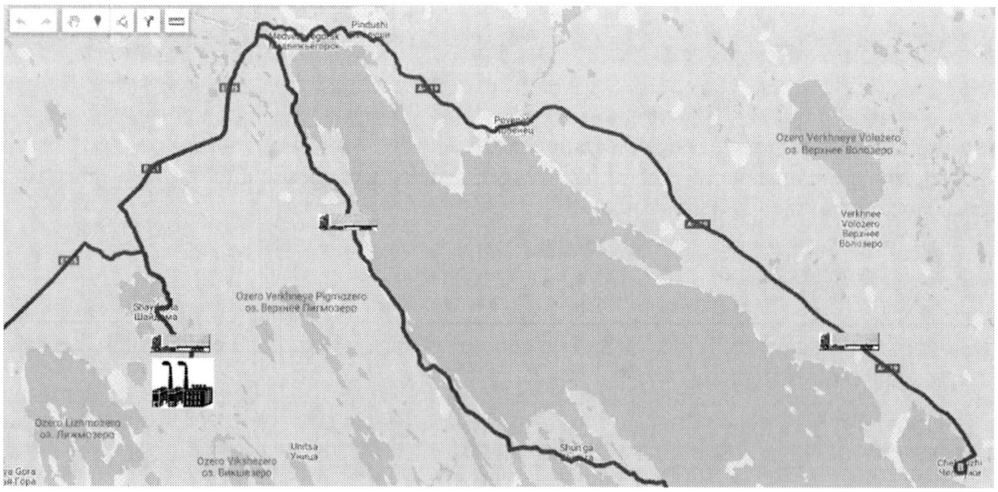

Figure 1. A graphical interpretation of the calculation results.

Table 1. Initial data for calculation.

Locality	Latitude	Longitude	Latitude	Longitude	Center of mass	Number of gas-supplied population
Besovec	61°51'40.94" N	34° 9'40.97" E	55.75	37.61	12.28	120
Vilga	61°49'5.00" N	34° 4'39.00" E	61.82	34.08	14.53	1053
Derevyanka	61°33'43.79" N	34°27'9.23" E	61.55	34.34	26.64	1622
Derevyannoe	61°37'4.44" N	34°37'51.19" E	61.62	34.62	24.83	1177
Zaozer'e	61°51'49.49" N	34°22'51.22" E	61.85	34.37	7.72	1337
Ladva	61°21'21.45" N	34°38'1.89" E	61.35	34.63	49.38	2145
Ladva–Vetka	61°21'32.89" N	34°27'19.96" E	61.35	34.45	49.03	983
Meliorativnyj	61°51'53.96" N	34°14'27.22" E	61.85	34.23	9.56	2475
Novaya Vilga	61°48'38.03" N	34° 9'3.99" E	61.80	34.15	10.57	1579
Paj	61°12'26.61" N	34°26'16.48" E	61.20	34.43	65.78	498
Rybreka	61°17'15.19" N	35°31'4.74" E	61.28	35.52	84.13	467
Uzhesel'ga	61°43'12.81" N	34°29'26.26" E	61.72	34.48	11.29	254
Chalna–1	61°54'26.31" C	34° 9'32.93" E	61.90	34.15	15.93	1866
Shyoltozero	61°22'10.20" N	35°21'24.62" E	61.36	35.35	71.62	878
Shoksha	61°27'3.11"C N	35° 2'54.07" E	61.45	35.03	53.45	167
Shujskaya	61°56'37.65" N	34°14'46.25" E	61.93	34.23	17.33	781
Shuya	61°53'55.21" N	34°14'3.02" E	61.88	34.23	12.94	3139
Petrozavodsk	61°47′46″ N	34°20′57″ E	61.79	34.35		281023

Table 2. Calculation results.

Locality	Volume of gas consumed per year [mln m^3]	Required number of flights	Distance from the LNG plant [km]	Time spent on one trip	Number of flights
Besovec	124.01	12918	158.65	25	25
Vilga	1088.17	113351	1647.33	30	21
Derevyanka	1676.18	174602	4651.75	54	12
Derevyannoe	1216.32	126699	3146.22	50	12
Zaozer'e	1381.66	143923	1111.28	16	40
Ladva	2216.65	230901	11401.58	99	6
Ladva–Vetka	1015.83	105816	5187.83	99	6
Meliorativnyj	2557.67	266424	2548.18	20	32
Novaya Vilga	1631.742	169973	1796.43	22	29
Paj	514.64	53608	3526.36	132	4
Rybreka	482.60	50271	4229.20	169	3
Uzhesel'ga	262.48	27342	308.56	23	28
Chalna–1	1928.33	200868	3199.42	32	20
Shyoltozero	907.33	94513	6768.76	144	4
Shoksha	172.58	17977	960.77	107	6
Shujskaya	807.09	84072	1456.84	35	18
Shuya	3243.85	337901	4370.91	26	24
Petrozavodsk			56470.08		

Table 3. Calculation results of the future potential profit.

Locality	Number of tanks per working day	Costs of refueling cryogenic tanks with diesel fuel [mln rub]	Costs of wages for drivers [mln rub]	Future potential profit [mln rub]
Besovec	3	7.46	2.16	679.87
Vilga	22	77.43	15.84	5956.97
Derevyanka	59	218.63	42.48	9058.44
Derevyannoe	43	147.87	30.96	6583.88
Zaozer'e	15	52.23	10.80	7618.99
Ladva	156	535.88	112.32	11676.37
Ladva–Vetka	72	243.83	51.84	5352.37
Meliorativnyj	34	119.77	24.48	14076.40
Novaya Vilga	24	84.43	17.28	8970.77
Paj	55	165.74	39.60	2656.03
Rybreka	68	198.77	48.96	2435.52
Uzhesel'ga	4	14.51	2.88	1442.03
Chalna–1	41	150.37	29.52	10541.61
Shyoltozero	96	318.13	69.12	4657.49
Shoksha	13	45.16	9.36	905.02
Shujskaya	19	68.47	13.68	4405.25
Shuya	58	205.43	41.76	17788.61
Petrozavodsk		2654.10	563.04	114805.60

According to the results of the calculation, the possible profit for the year will be 112.298 million rubles including the purchase of 15065 units of tanks for the transportation of LNG.

5. Conclusion

The lack of theoretical elaboration of the issue under consideration, the need for practical recommendations to improve an integrated approach to the creation of LNG infrastructure determined the goal and objectives of this work.

As the results show, the use of natural gas in liquefied form contributes to the diversification of energy supply, allowing to improve the structure of energy supply sources, increasing the reliability and energy security of the country, and medium– and low–tonnage LNG production can play a significant role in the development of regions of the Russian Federation, solving the problem of the availability of an efficient energy resource.

References

[1] Fedorova E B and Tyrsin A V 2020 Sovershenstvovanie normativno-texnicheskogo obespecheniya v sfere proizvodstva szhizhennogo prirodnogo gaza [Improvement of regulatory technical support in the field of liquefied natural gas production] Oborudovanie i Texnologii dlya Neftegazovogo Kompleksa 2(116) 67-70 [In Russian]

[2] Rachevsky B S 2018 Proizvodstvo i potreblenie komprimirovannogo i szhizhennogo prirodnogo gaza CNG/LNG [Production and Consumption of Compressed and Liquefied Natural Gas CNG/LNG] (Moscow: Gubkin University) [In Russian]

[3] Makarova A A and Mitrova T A 2019 Prognoz Razvitiya E`nergetiki Mira i Rossii 2019 [Forecast of energy development in the world and Russia 2019] (Moscow: Skolkovo) [In Russian]

[4] *World LNG Report* URL https://igu.org/event-list/lng2022/

[5] Baranov A Yu, Valentinova K A and Ivanov L V 2020 Modelirovanie ispareniya szhizhennogo prirodnogo gaza v mobil`ny`x rezervuarax [Modeling of liquefied natural gas evaporation in mobile reservoirs] *Nauchno-Texnicheskij Vestnik Informacionny`x Texnologij, Mexaniki i Optiki* **20(4)** 595-602 [In Russian]
[6] Arkharov A M and Semenov V Y 2016 Analysis of thermodynamic efficiency of small-scale natural gas liquefying plant operating on medium-pressure cycle *Chemical and Petroleum Engineering* **51** 656-64
[7] Mokhatab S, Mak J Y, Valappil J V and Wood D A 2014 *Handbook of Liquefied Natural Gas* (Oxford: Gulf Professional Publishing)
[8] Yamamoto K 1958 Transportation of Liquefied Natural Gas *J. of the Fuel Society of Japan* **37** 673-80
[9] Breeze P 2016 *Gas-Turbine Power Generation* (London:Academic Press) pp 9-19
[10] Pershin N V 2020 Analiz raboty` sistemy` transportirovki szhizhennogo prirodnogo gaza [Analysis of the operation of the liquid natural gas transport system] *Morskie Intellektual`ny`e Texnologii* **1-1(47)** 125-30 [In Russian]
[11] Osorio-Tejada J L, Llera–Sastresa E and Scarpellini S 2017 Liquefied natural gas: Could it be a reliable option for road freight transport in the EU? *Renewable and Sustainable Energy Reviews* **71** 785-95
[12] Mokhatab S, Poe W A and Mak J Y 2019 *Handbook of Natural Gas Transmission and Processing* (Cambridge: Elsevier)
[13] Medvedeva O N, Frolov V O and Kochetkov A V 2015 A Tank for Transporting Liquefied Natural Gas *Chemical and Petroleum Engineering* **51(3)** 257-9
[14] Lisowski E and Lisowski F 2019 Study on thermal insulation of liquefied natural gas cryogenic road tanker *Thermal Science* **23(4)** 1381-91
[15] Sanavandi H and Ziabasharhagh M 2016 Design and comprehensive optimization of C3MR liquefaction natural gas cycle by considering operational constraints *J. of Natural Gas Science and Eng.* **29** 176-87
[16] Dutta A, Karimi I A and Farooq Sh 2018 Adjusting the Heating Value of LNG using a Superstructure for Hydrocarbon Recovery *Computer Aided Chemical Eng.* **44** 1105-10
[17] Peterson T G and Weisend J G II 2019 Liquefied Natural Gas (LNG) Safety *Cryogenic Safety* (Cham: Springer) pp 181-9
[18] Kuwahara N, Bajay S V and Castro L N 2000 Liquefied natural gas supply optimisation *Energy Conversion and Management* **41(2)** 153-61

IOP Conf. Series: Materials Science and Engineering 1089 (2021) 012048 doi:10.1088/1757-899X/1089/1/012048

On the Possibility of Application Infrared Crystalline Fibers for Transfer of Temperature Signals from Bearings inside Nuclear Power Plants' Containment

M P Kharaim[1], V S Kostarev[1], N N Akif'eva[1], A A Yuzhakova[2], L V Zhukova[2] and A S Korsakov[2]

[1] Department of Nuclear Power Plants and Renewable Energy Sources, Ural Federal University named after the first President of Russia B. N. Yeltsin, Mira St., 19, Ekaterinburg, 620078, Russia
[2]Department of Physical Chemistry and Chemistry of Colloids, Ural Federal University named after the first President of Russia B. N. Yeltsin, Mira St., 19, Ekaterinburg, 620078, Russia

E-mail: maxkharaim@gmail.com

Abstract. The possibility of application flexible infrared Nanopolycrystalline (PIR) fibers made from radiation-resistant crystal systems AgBr – TlI and AgBr – $TlBr_{0.46}I_{0.54}$, for contactless measure temperature bearings of own needs mechanisms and signal temperature transmission from nuclear power plant containment is considered. Radiation resistant allows the use that crystals and PIR fibers in high radiation conditions up to 500 kGy, what opens great opportunities of involving that optical materials in atomics. The class of the presented materials is distinguished by a wide transmission range in the range from 0.4 to 60 μm for crystals and from 2.0 to 25 μm for PIR fibers without absorption window, low optical loss (up to 0,1 dB/m), high flexible. The PIR fibers proposed in this paper are made by extrusion from single crystals of silver and thallium halides and are transparent in the mid-infrared range from 2 to 25 μm, which, according to the laws of Planck and Wien, corresponds to temperatures from +1100 °C to –200 °C.

1. Introduction

Nowadays, in thermal and nuclear power engineering, resistance thermometers are used to control temperature of the bearings of own needs mechanisms, the action of which is based on the dependence of the electrical resistance of metals on temperature. Platinum and copper are used for the manufacture of sensing elements of standard resistance thermometers. Of these, only platinum possesses acceptable resistance to oxidation, which determines enough service life, and therefore platinum resistance thermometers have found wide application in thermal power engineering, including for measuring the temperature of the bearings of own needs mechanisms.

However, in nuclear power, the use of platinum as a material for sensitive elements of resistance thermometers installed in the containment of a nuclear power plant has a number of difficulties, since its activation in gamma and neutron fields leads to a change in the temperature characteristics of this material.

Natural platinum contains 6 different isotopes (190, 192, 194, 195, 196 and 198), the percentage of which in a wide range. Heat cross section of capture of natural platinum according to the estimate is 8 b [1]. Comparing this value with the value of the radiation capture cross-section of various structural materials (Al-27 – 0,231 b; Fe-0 – 2,56 b; Zr-0 – 0,185 b [1]), it is obvious that the value of the capture cross-section for natural platinum is an order of magnitude higher than for structural materials. This leads to the fact that platinum is easily activated in radiation fields.

For measuring the temperature of the equipment installed in the containment, it may be useful in the future to use the method of re-mote transmission of the temperature signal in the form of infrared radiation from the controlled element using infrared optical fibers.

In this work, an assessment was made of the possibility of using flexible infrared optical fibers made of radiation-resistant crystal systems AgBr – TlI and AgBr – $TlBr_{0.46}I_{0.54}$, for transmission of the temperature signal of the bearings of own needs mechanisms from containment of the NPP.

2. Materials and Methods

The use of IR PIR fibers in fiber temperature sensors provides several advantages over existing methods and control devices, namely:

- in comparison with the thermoelectric method, it allows contactless temperature measurement in conditions of electromagnetic interference;
- in comparison with the thermoresistive method, it is inertialess, and also removes the problem of the influence of the temperature dependence of the resistance of communication lines on the accuracy of temperature measurement;
- compared to quartz fiber sensors, it directly measures from –20 to +300 °C due to direct transmission of infrared radiation from its sources to the receiver;
- allows to avoid the operation of activated sensitive elements in the conditions of gamma and neutron fields of the containment of the NPP.

In this paper, for the purpose of using in the containment of the NPP, the flexible infrared PIR fibers are considered, manufactured by scientific laboratory «Fiber Technologies and Photonics» at the Chemical-Technological Institute of the Ural Federal University named after the first President of Russia B. N. Yeltsin.

For the mid and far infrared range, there is a limited number of trans-parent materials with low optical loss. In particular, these are glasses based on chalcogenides, tellurides, fluorides, as well as crystals of silver halides and monovalent thallium.

The last class of materials presented is distinguished by a wide trans-mission range in the range from 0.4 to 60.0 μm for crystals and from 2.0 to 25.0 μm for PIR fibers without absorption windows, low optical losses (up to 0,1 dB/m), high flexibility. The transmission spectra of some silver and thallium halide PIR fibers are shown in figure 1.

Among halide materials stand out crystalline systems AgBr – TlI and AgBr – $TlBr_{0.46}I_{0.54}$, since crystals and fibers based on them are photo- and radiation-resistant, have tensile strength and flexibility up to 250 MPa, the critical bending radius of fibers depends on the PIR fiber diameter and reaches 1 mm

Photo resistance provides the use of these materials without a special shell. Radiation resistance makes it possible to use crystals and PIR fibers under conditions of strong ionizing radiation up to 500 kGy, which opens up wide possibilities for the introduction of such optical materials into nuclear energy, radiation physics, and space technologies [2].

Studies of radiation resistance were carried out with β-irradiation of fibers with doses up to 500 kGy. As a result of the study, a significant drop in transmission was not observed; at a high content of thallium and iodine in the crystals, an increase in their transmission up to 5% was found. Figure 2 shows the relative transmittance obtained by comparing the transparency of the AgBr – $TlBr_{0.46}I_{0.54}$ optical fibers before and after irradiation with a dose of 200 kGy. The relative transmittance is estimated as a percentage, while the transmittance before irradiation is taken as 100%. The area of increasing transmission (values over 100%) is highlighted in Figure 2 in green. The figure shows two groups of curves, the left group of curves is plotted for systems with a $TlBr_{0.46}I_{0.54}$ content in AgBr up to 27%, and the right group of curves for systems with a $TlBr_{0.46}I_{0.54}$ content of more than 70%. The transmission of the fibers before and after irradiation was estimated for fiber diameters of 3 μm, 5 μm, 10.6 μm, and 14 μm.

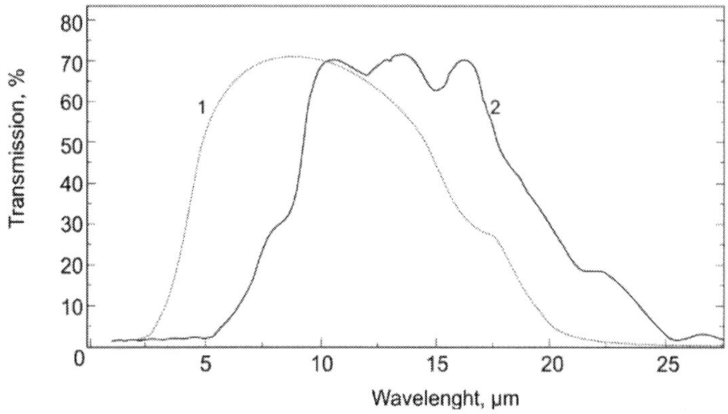

Figure 1. Wavelength dependence of the fiber transmission: 1 – comparison of the transmission spectra of the fibers based on the AgBr-TlI system with AgCl-AgBr; 2 – transmission spectre of the PIR fibers based on the AgBr – $TlBr_{0.46}I_{0.54}$.

A similar study was carried out for fibers based on crystals of the AgBr – TlI system under β- and γ-irridation, as a result of which the resistance of the fibers to radiation was also observed without a de-crease in their optical transmission [3].

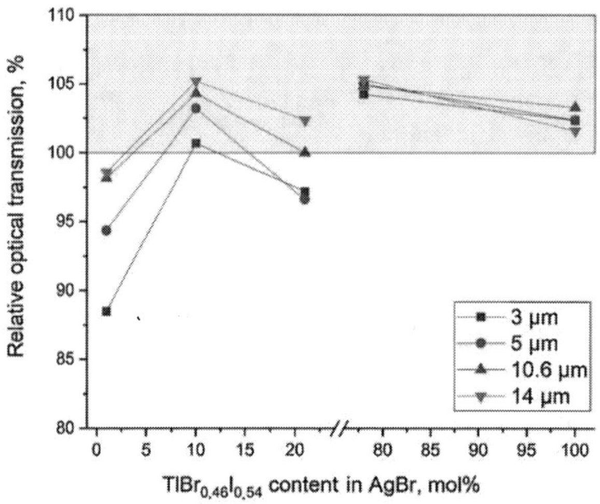

Figure 2. Relative optical transmission of the PIR fibers with different $TlBr_{0.46}I_{0.54}$ content in AgBr.

3. Results

The optical fibers proposed in this work are transparent in the mid-infrared range from 2 to 25 μm, which, according to Planck's and Wien's laws, corresponds to temperatures from -200 to +1100 °C, respectively (figure 1 and figure 3). This temperature range and the properties listed above make it possible to use infrared PIR fibers as fiber channels for delivering an optical signal from hard-to-reach places due to a wide spectral transmission range, flexibility and mechanical strength, in particular, for transmitting a bearings of own needs mechanisms temperature signal from the containment of a NPP.

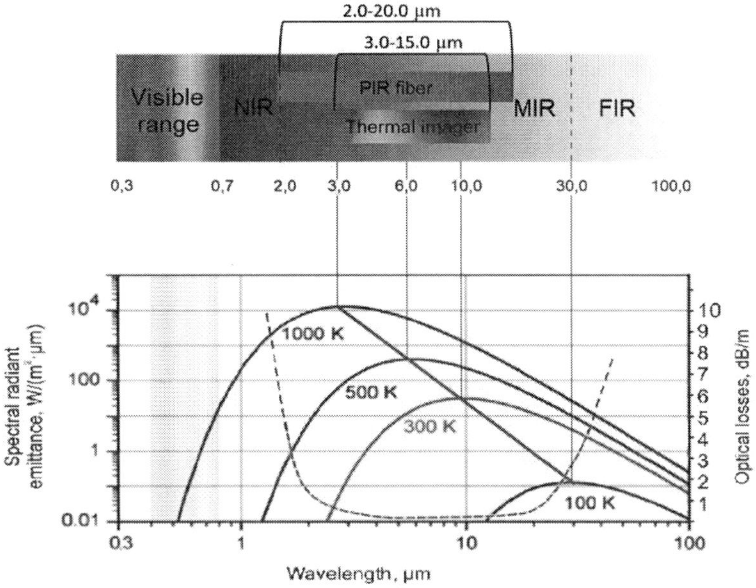

Figure 3. Planck's and Wein's laws for the optical range. Working areas of infrared fiber and thermal imagers are indicated (NIR – near infrared range; MIR – medium infrared range; FIR – far infrared range).

To study the functional properties of infrared PIR fibers, authors used fibers with a cross-sectional diameter of 1.1 mm, without a protective sheath and with a critical bend radius of 1.24 mm. Since the influence of the number and radii of fiber bends plays a key role in laying fiber channels, the dependence of the optical transmittance on the bending radius of the fiber was studied [4]. As can be seen from the results shown in figure 4, transmission decreases with decreasing radius according to a logarithmic dependence, which is in good agreement with the laws of wave optics.

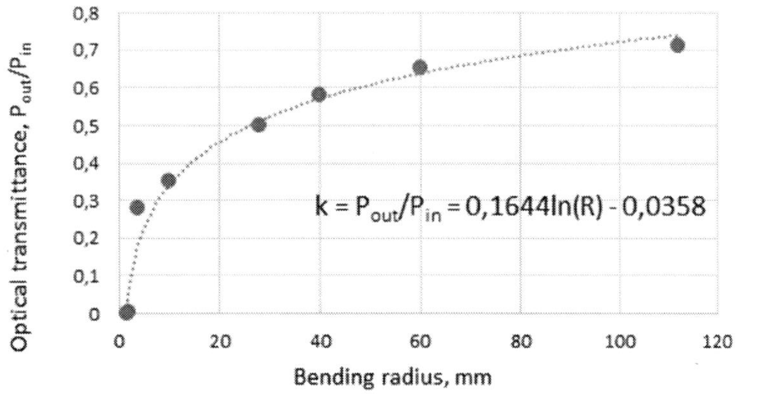

Figure 4. Dependence of the optical transmission coefficient from bend radius.

The second important parameter is the length of the fiber channel. Since the losses in halide fibers reach 0.1 dB/m, the maximum length of such channels is tens of meters with a limiting value of ≈ 50 m.

The properties of IR optical fibers open prospects for the use of IR optical fibers of systems AgBr – TlI and AgBr – $TlBr_{0.46}I_{0.54}$ in NPP devices. In this regard, the possibility of introducing optical fibers into the systems of external temperature measurement of equipment installed in the containment of a nuclear power plant with a VVER reactor was considered, namely, for monitoring the temperatures of bearings of own needs mechanisms (figure 5). The optical scheme for measuring temperature is shown in figure 6.

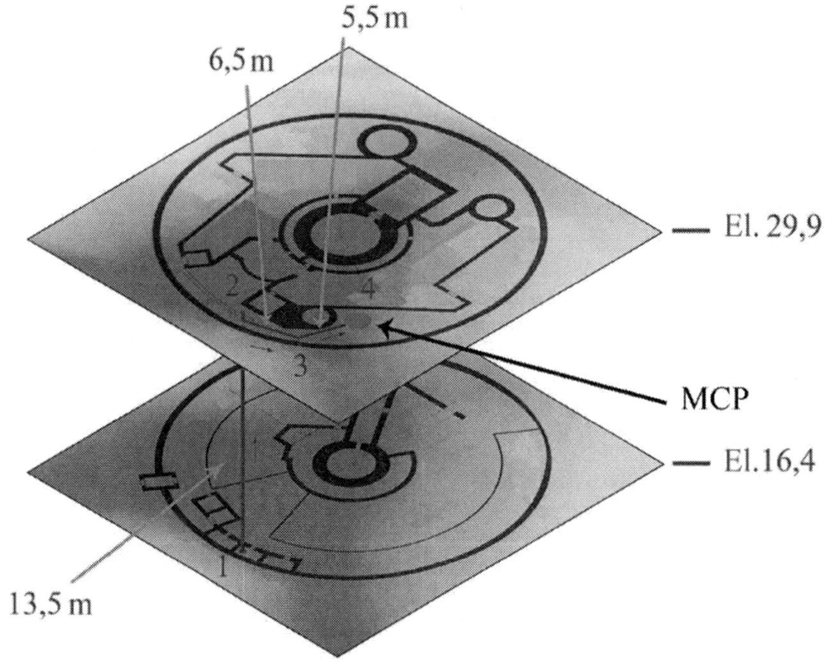

Figure 5. Route of PIR fiber to the Own Needs Mechanisms' (ONM) bearings.

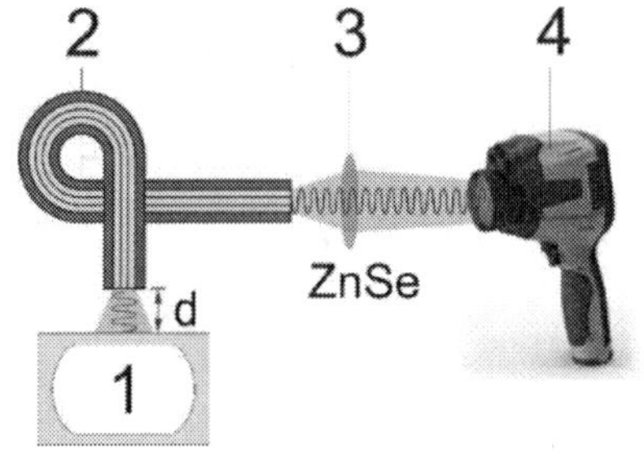

Figure 6. Overall scheme of object temperature measurements by thermographic method: 1 – an IR radiation source (bearing); 2 – IR-fiber; 3 – ZnSe-lens; 4 - thermal imager

The PIR fiber is laid into the inner part of the containment through the sealed lead-in (point 1). Further, taking into account the proportions of the containment and the height of the marks, the optimal path for laying the fiber to the measured surface of the ONM bearings is determined (point 4).

4. Discussion

Since the routing of cables of impulse lines and other instrumentation and automation devices from the containment to the serviced area depends on the layout of the equipment, it is an un-changeable factor and involves multiple bends. The equipment layout is determined by the compactness of the equipment arrangement.

The length of the fiber is determined by approximate calculations based on the available data. Thus, the path of laying the fiber, shown in figure 5, assumes the required length of the section of the fiber passing in the containment is about 24 m. With an allowable fiber length of up to 50 m, there is a margin for adding the fiber to a secondary device, which may be, for example, a thermal imaging camera. The number of fiber bends in a laying similar to that shown in figure 5 can reach 3-4, with only one bend at an angle of 90 degrees, the rest at angles less than 90 degrees. The bending radii were taken many times larger than the critical one. With such a number of bends, the optical loss is about 27%, which is an acceptable value providing high measurement accuracy.

5. Conclusion

Radiation resistance of flexible infrared PIR fibers made of AgBr – TlI and AgBr – $TlBr_{0.46}I_{0.54}$ crystal systems, their permissible maximum length, insignificance of optical losses with a small number of bends, allows positive to assess the possibility of their use for transmitting the signal of the surface temperature of the NPP equipment, in particular, the moving parts of the equipment installed in the pressurized volume of the NPP.

The advantages of using infrared PIR fibers for measuring the surface temperatures of equipment installed in a pressurized volume include: inertialessness, clarity, radiation resistance. It should be especially noted that the use of this method makes it possible to avoid the operation of activated sensitive elements in conditions of gamma and neutron fields of the NPP containment volume.

Currently, many issues of applying the method of measuring temperature and transmitting a temperature signal using infrared fibers made of AgBr – TlI and AgBr – $TlBr_{0.46}I_{0.54}$ crystal systems require in-depth research, in particular, the question of more simpler than a thermal imaging camera, a secondary device for receiving a temperature signal and transmitting it to the SHC (software and hardware complex). At present, these issues are being investigated in the research laboratory «Fiber Technologies and Photonics» at the Chemical-Technological Institute of the Ural Federal University named after the first President of Russia B. N. Yeltsin.

References

[1] Koscheev V N, Manturov G N and Nikolaev M N 2014 Biblioteka gruppovyh konstant BNAB-RF dlya raschetov reaktorov i zashchity [Group Constants Library for Nuclear Reactor and Shielding Calculations] *Izvestiya Vuzov. Yadernaya Energetika* **3** 93-103

[2] Yuzhakova A A, Zhukova L V, Akif'eva N N, Krasnov D A and Korsakov A S 2020 Application of Infrared Polycrystalline Fibers in Thermal Imaging Temperature Control Systems *Sensors and Actuators: Physical* **314** 1-8

[3] Salimgareev D D, Lvov A E, Korsakova E A, Korsakov A S and Zhukova L V 2019 Crystals of AgBr–$TlBr_{0.46}I_{0.54}$ System: Synthesis, Structure, Properties, and Application *Materials Today Communications 20* 1-10

[4] Korsakova E A, Lvov A E, Salimgareev D D, Korsakov A S, Markham S, Mani A, Silien C, Syed T and Zhukova L V 2018 Stability of MIR Transmittance of Silver and Thal-lium Halide Optical Fibres in Ionizating β- and γ-Radiation from Nuclear Reactors Infrared *Physics and Technology* **93** 171-7

IOP Conf. Series: Materials Science and Engineering 1089 (2021) 012048 doi:10.1088/1757-899X/1089/1/012048

[5] Zhukova, L V, Primerov N V, Korsakov A S and Chazov A I 2008 $AgCl_xBr_{1-x}$ and $AgCl_xBryI_{1-x-y}$ Crystals for IR Engineering and Optical Fiber Cables *Inorganic Materials* **12** 1372-77
[6] Korsakov A S 2017 *Struktura Fotonno-Kristallicheskih Svetovodov na Baze Modificirovannyh Galogenidserebryannyh Kristallov i Issledovanie ih Funkcional'nyh Svojstv* (Ekaterinburg: URFU)
[7] Zhukova L V, Lvov A E, Korsakov A S, Salimgareev D D and Korsakov V S 2018 Domestic Developments of IR Optical Materials Based on Solid Solutions of Silver Halogenides and Monovalent Thallium *Optics and Spectroscopy* **6** 933-43
[8] Korsakova E A, Korsakov A S, Korsakov S S and Zhukova L V 2018 IR Thermographic System Supplied with an Or-dered Fiber Bundle for Investigation of Power Engineering Equipment and Units *Sino-Russian ASRTU Conference Alternative Energy: Materials, Technologies, and Devices Ural Federal University 13–16 July 2018* (Dubai: Knowledge E) pp 209-16
[9] Korsakova E A, Zhukova L V, Korsakov A S, Shmygalev A S and Korsakov M S 2018 Thermal Imaging by Means of IR-Fiber Bundle for Medical Applications *International Conference on Laser Optics (ICLO), Saint-Petersburg, 04-08 June 2018* (New York: IEEE) p 529
[10] Gordov A N, Zhagullo O M and Ivanova A G 1992 *Osnovy Temperaturnyh Izmerenij* (Moscow: Energoatomizdat)
[11] Liu Y, Tian Y, Fan X, Bu Y, He P, Li H, Yn J and Zheng X 2018 Feasibility Study of Transformer Winding Temperature and Strain Detection Based on Distributed Optical Fibre Sensors *Sensors* **18** 1-14
[12] Artyushenko V, Wojciechowski C, Ingram J, Kononenko V, Lobachev V, Sakharova T, Ludczak J, Grzebieniak A and Wojciechowski Z 2005 Specialty Fibers for Broad Spectra of Wave-Length and Power *Optical Fibers: Technology* **5951** 1-7
[13] Marini D 2017 Micro-Opto-Mechanical Sensors for Tactile Width Meas-urements of Surface Opening Cracks in Concrete *IEEE Sensors* **5** 1-3
[14] Lavi Y, Millo A and Katzir A 2006 Flexible Ordered Bundles of Infrared Transmitting Silver-Halide Fibers: Design, Fabrication, and Optical Measurements *Applied Optics* **23** 8-14
[15] Dziedzic A, Golonka L, Kozlowski J, Licznerski B and Nitsch K 1997 Thick-Film Resistive Temperature Sensors Measurement *Science and Technology* **8** 78-85
[16] Barker D G and Jones M R 2003 Inversion of Spectral Emission Measurements to Re-construct the Temperature Profile Along a Blackbody Optical Fiber Thermometer Inverse *Problems in Engineering* **11** 495-513
[17] Prasad A 2013 Measurement of Temperature and Pressure on the Surface of a Blunt Cone Using FBG Sensor in Hypersonic Wind Tunnel Measurement *Science and Technology* **9** 1-10
[18] Artyushenko V G, Bocharnikov A, Colquhoun G, leach C A, Lobachov V, Pirogova L, Sakharova T, Satitskij D, Ezevskaya T and Bublikov A 2007 Mid IR-Fiber Spectroscopy in the 2-17 μm Range *SPIE – The Int. Society for Optical Eng., Florence, 7 November 2007* (Florence: Society of Photo-Optical Instrumentation Engineers) pp 1-12
[19] Planck M 1914 *The Theory of Heat Radiation* (Philadelphia: P. Blakiston's Son & Co.)
[20] Perić D, Livada B, Perić S and Vijić S 2019 Thermal Imager Range: Predictions, Expectations, and Reality *Sensors* **15** 1-8

Simulation of the hydrodynamics in polymerizers based on the CFD model for the synthesis of rubbers and plastics using the method of the solution polymerization

I O Antonova[1], R R Dmitricheva[1], G V Manuyko[1], G A Aminova[1], V V Bronskaya[2], O S Kharitonova[3], A T Gimaltynov[1], T V Ignashina[2] and D S Balsamov[4]

[1]Department of Technology of Engineering Materials, Kazan National Research Technological University, Karl Marx St., 68, Kazan, 420015, Russia
[2]Department of Chemical Process Engineering, Kazan National Research Technological University, Karl Marx St., 68, Kazan, 420015, Russia
[3]Department of Chemical Technology of Petroleum and Gas Processing, Kazan National Research Technological University, Karl Marx St., 68, Kazan, 420015, Russia
[4]Department of Power Supply of Enterprises and Energy Resource Saving Technologies, Kazan State Power Engineering University, Krasnoselskaya St., 51, Kazan 420066, Russia

E-mail: olga.220499@mail.ru

Abstract. Mixing in a viscous medium, such as in polymerization processes, is worth noting. To control the polymerization process for improving the product quality, it is important to understand the mixing process. Also, it is essential to understand the flow behaviour in the stirred-tank reactor for the equipment design and from the point of view of the large-scale transition of the process. Local and general characteristics of mixing in the reactor are related to the design and selected speed of the agitator. 3D-modeling of the flow in the cylindrical reactor with the paddle stirrer was carried out using the potential of a supercomputer. Numerical results showed good consistency with experimental results and correlation developed by other researchers. To determine the best mode of introduction into the paddle space of the stirrer, two efficiency criteria were used: with respect to power and mixing time.

1. Introduction

Earlier research on mixing is based on the development of mathematical models of mixing [1-11]. The problems that occur in these models for periodic or continuous polymerization in the stirred-tank reactor are related to the fact that the increase in viscosity during the polymerization quickly changes the initial assumptions about the homogeneity, which regulate the main aspects of the mixing process. Another problem for the stirred-tank reactors is the difficulty of combining mathematically complex polymerization kinetics with mixing models. These problems can be partially overcome using the computational fluid dynamics (CFD). Computational fluid dynamics is a design tool for developing new processes and optimizing existing ones. The development of computational fluid dynamics has made it possible to visualize mixing processes without conducting the real-time experiments, which are not feasible in some cases. Recent developments in the computer technology have made the computational fluid dynamics an attractive tool for designing, optimizing and visualizing various processes. Mixing models which are implemented using the computational fluid dynamics can be useful in the transition from laboratory installations to industrial production [12-22]. CFD provides the information about the turbulent zone and introducing reagents in areas with the intense turbulence can help improve the product yield.

Radial fluid flows are mainly created in the stirred-tank reactor with turbine agitators. When working with turbine agitators with a large number of revolutions, along with the radial flow, a tangential (circular) flow of the device contents and the formation of a funnel may occur. In this case, the infringement baffles are installed in the device.

2. Materials and Methods

In this paper, the velocity fields in the stirred-tank reactor were studied for a stationary process using ANSYS Fluent.

It is built the geometry of the device with a radius of 0.15 m and a height of 1 m, the agitator radius of 0.05 m, the mixing arm height of 0.022 m, the mixing arm length of 0.04 m, the mixing arm thickness of 0.004 m, the agitator shaft diameter of 0.02 m, the speed of rotation of the agitator varied within 1-20 rpm. Hexane was chosen as the working fluid so it has an absolute viscosity at 20°C of $0.3*10^{-3}$ (dyne·s/cm^2) and a density of 655 kg/m^3.

The stirred-tank reactor creates a three-dimensional flow, so the basis of the mathematical solution of the problem is a system of differential equations describing the movement in the reactor with the turbine agitator, which are based on the Navier-Stokes equations and the continuity equation of the three-dimensional problem statement. In this system of 4 equations, the 3 components of velocity and pressure are the independent required parameters.

The flow of liquid in the reactor due to the action of the agitator is represented by current lines. In the solution area, the velocity profiles, such as the average axial, tangential and radial velocities, were obtained.

3. Results

As a result of numerical experiments in the stirred-tank reactor with a turbine agitator (figure 1–8), the existence of an axial flow that provides secondary circulation is revealed.

Creating a model (hydrodynamic module) for the synthesis of synthetic rubber (SR) and software implementation was carried out on the basis of the FLUENT 16 software package using the supercomputer complex of the Joint Supercomputer Center of the Russian Academy of Sciences.

Computational fluid dynamics algorithms for an incompressible fluid are based on solving the Navier-Stokes or Reynolds equations (elliptical type) to determine the pressure that connects the entire flow region at each moment of time.

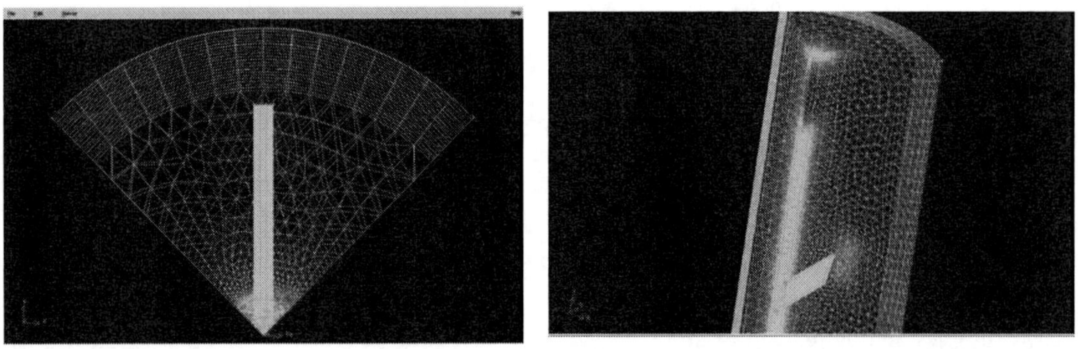

Figure 1. The network in the sector of the reactor.

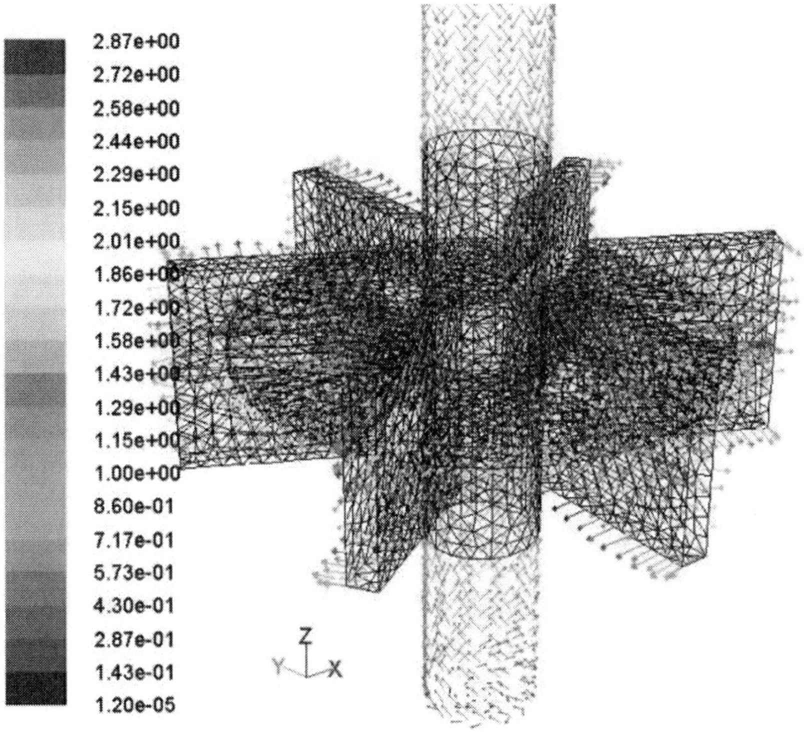

Figure 2. The field of total velocities in the stirred-tank reactor.

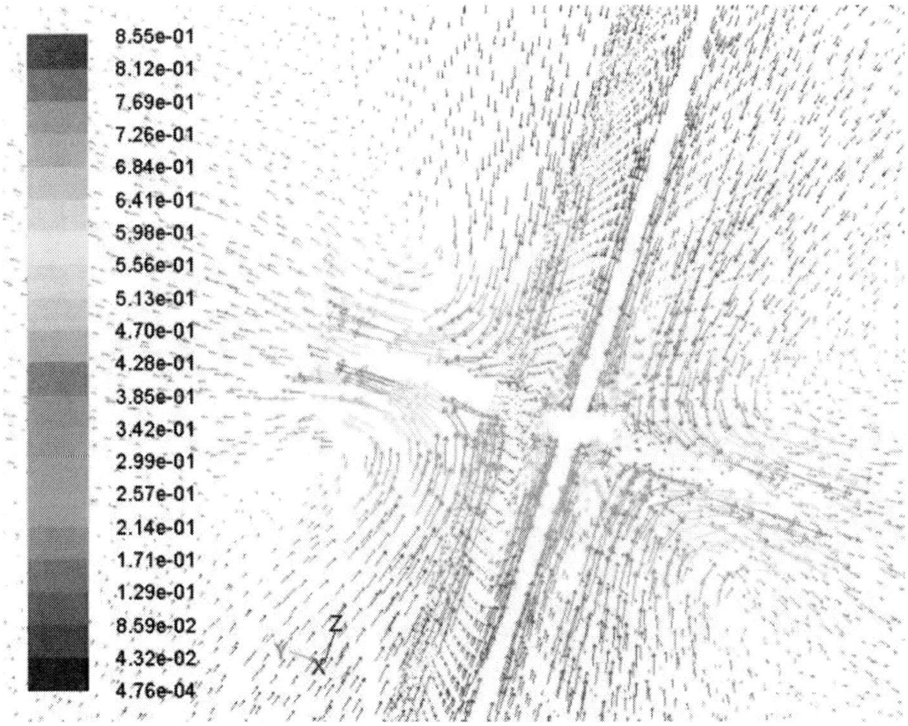

Figure 3. The profile of total velocities in the cross section of the mixing arms.

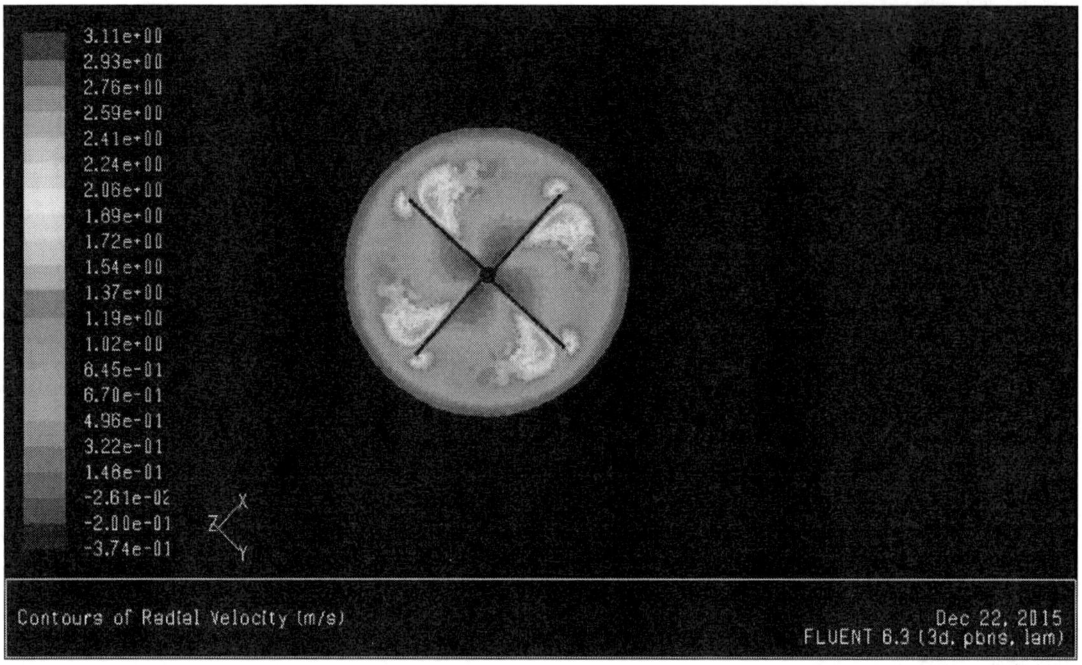

Figure 4. The radial velocity profile in the cross section of the mixing arms, laminar viscosity model.

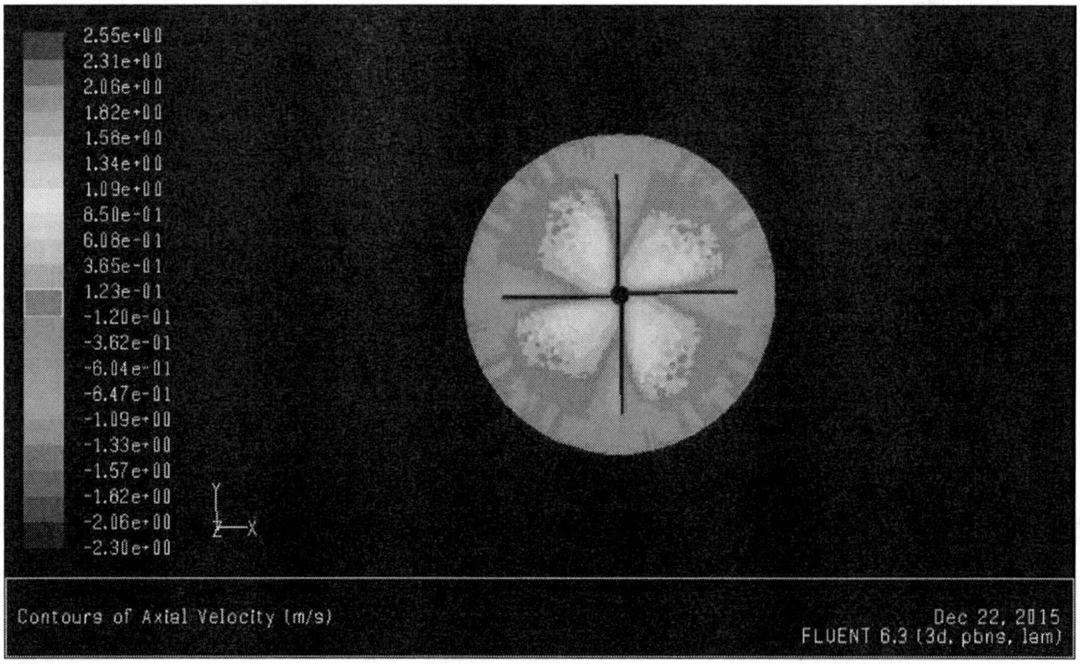

Figure 5. The axial velocity profile in the cross section of the mixing arms, viscous laminar model.

IOP Conf. Series: Materials Science and Engineering 1089 (2021) 012049 doi:10.1088/1757-899X/1089/1/012049

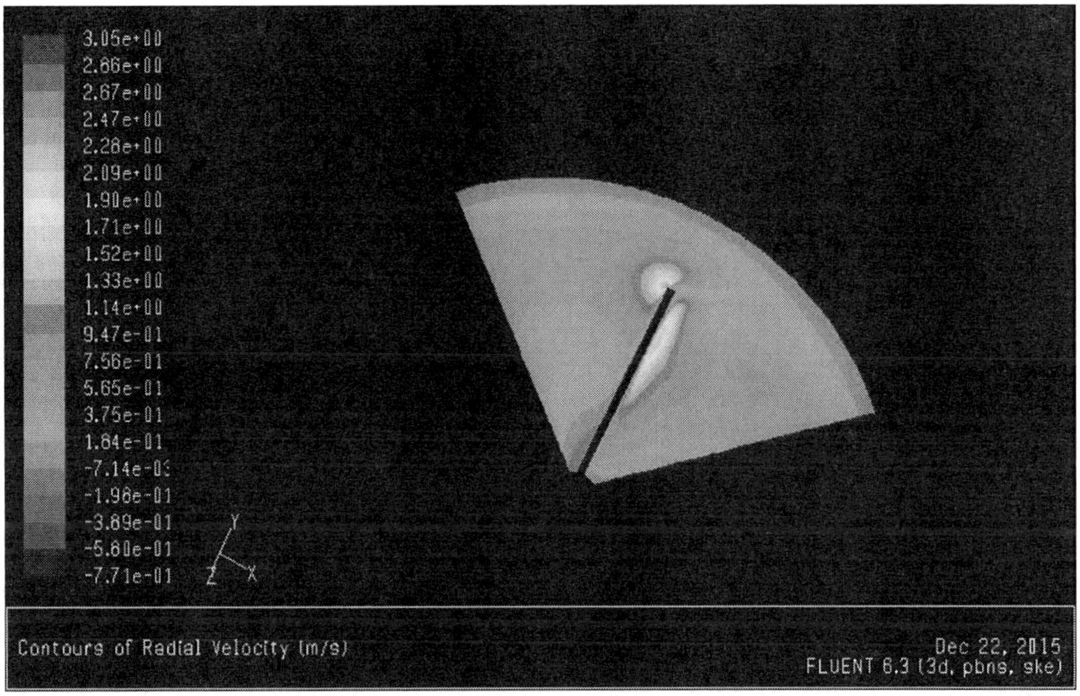

Figure 6. The radial velocity profile in the cross section of the agitator arms, k-ε model.

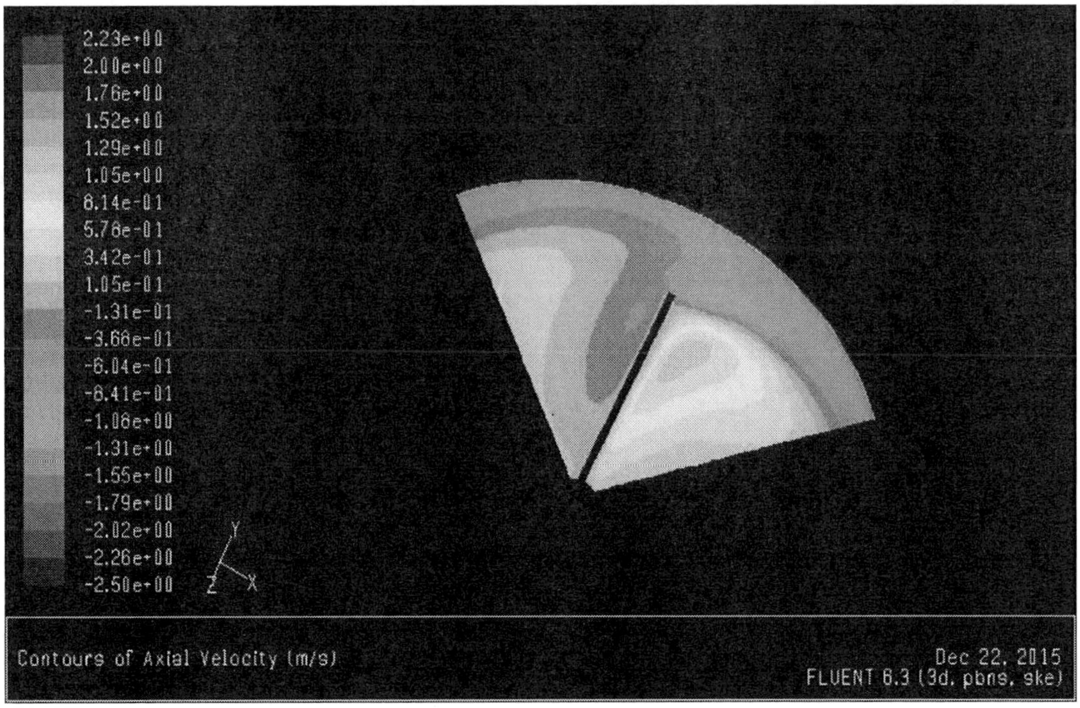

Figure 7. The axial velocity profile in the cross section of the agitator arms, k-ε model.

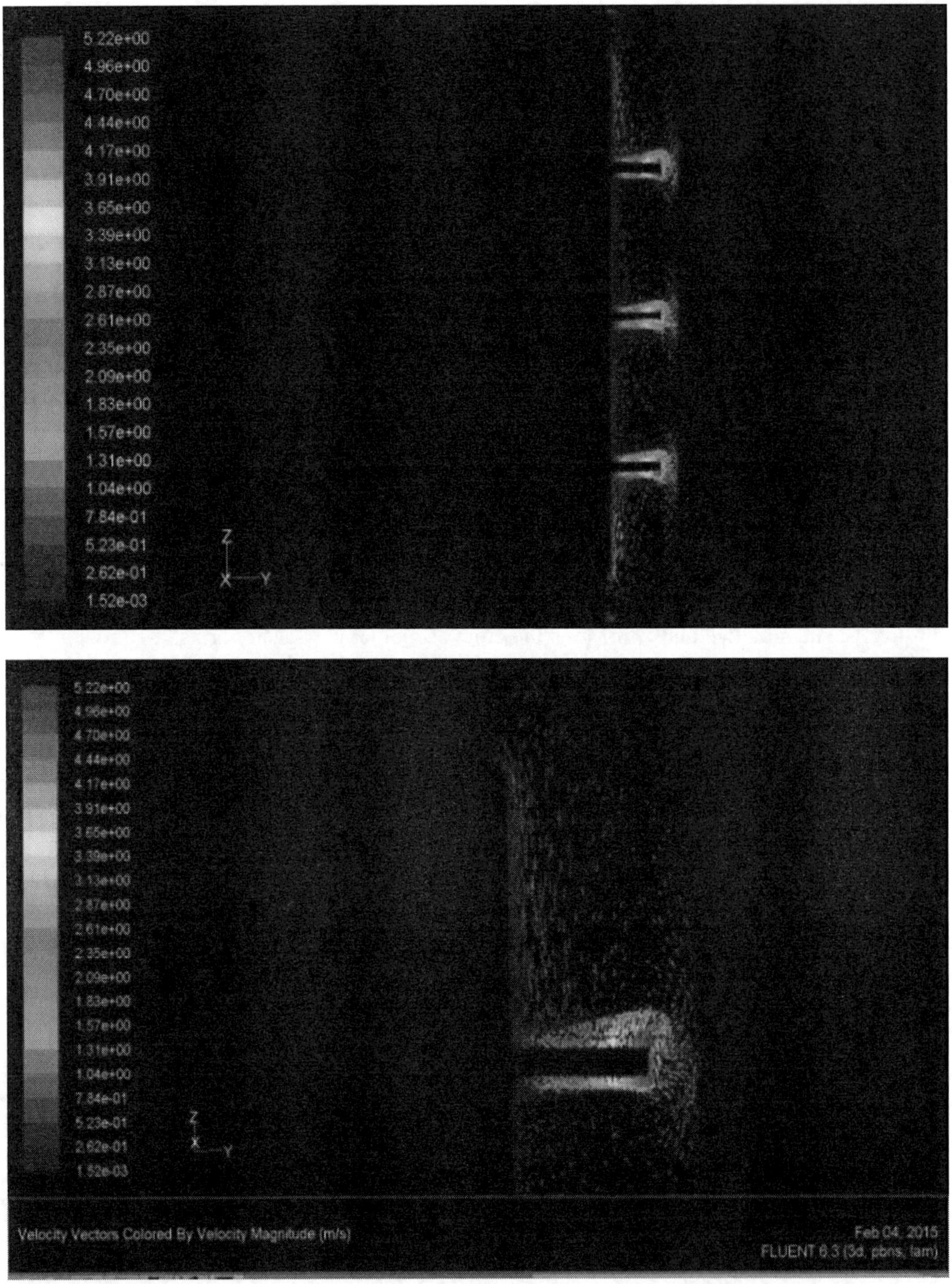

Figure 8. The total velocity profile in the cross section of the agitator arms, laminar viscosity model.

The implementation of computations of all the main resource-intensive operations leads to the need to parallelize the task on the supercomputer. An increase in the number of computational units for solving the problem leads to "saturation" in which there is no increase in performance (Figure 9). This is due to an increase in the consumption of exchanging information between computing units. Moreover, the larger the size of the computational stack is, the more the number of units can be used for computation on it. The exchange between computing units is limited by the speed of transfer of information. These phenomena are associated with the limitation of the growth of a computer network by Amdahl's law, while the network topology, memory access problems and its optimization difficulties, possible problems in optimizing the MPI parallelization process are important.

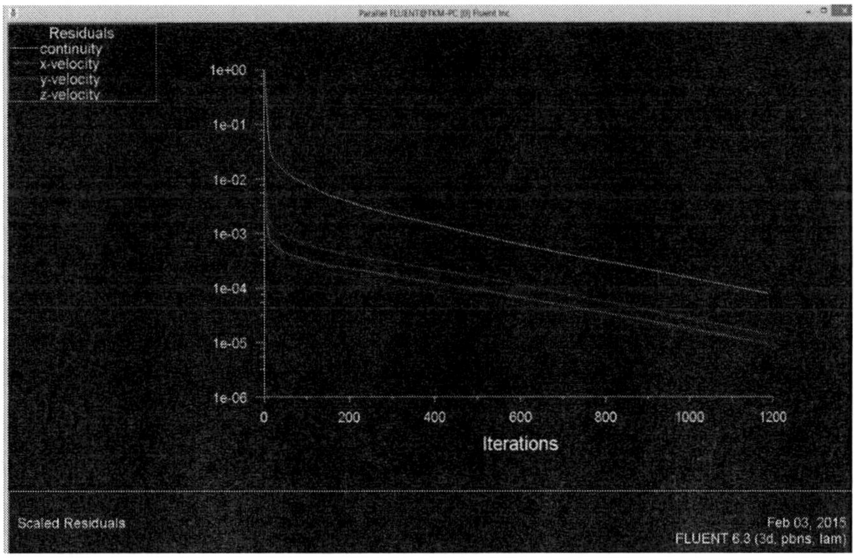

Figure 9. The dependence of the computational error on the number of iterations.

In the calculations, such components of the CFD-model as the geometry of the reactor with the paddle stirrer, the computational grid, the boundary conditions and physical properties of materials, the k-ε turbulence model and the parameters of the numerical solution of the system of transport equations were specified.

The equations are written in the cylindrical coordinates. In this system of 4 equations, the independent required parameters are 3 components of velocity and pressure.

The numerical algorithm is based on the final volume method for the unstructured grid. The complexity of numerical algorithms based on the grid methods for the Navier-Stokes and Reynolds equations is the determination of the relationship between the pressure field and the velocity field, which is provided by the continuity equation implemented using the SIMPLE algorithm on aligned grids. The values of the velocity and pressure fields are stored in the same units that is the centers of the control volumes (aligned grids). From the point of view of software implementation, aligned grids are the most economical.

The calculations were performed on different grids using single precision. Calculations were carried out on a grid containing 1 million cells using 50 processors (200 cores) of a supercomputer.

In addition, the calculation and analysis of the technological characteristics of the agitator was carried out. The power criterion K_N was determined by

$$K_N = z_0 \frac{\pi^2}{6} \lambda \frac{h_0}{d_m} \left[1 - \left(\frac{r_{in}}{r_0}\right)^3\right] Re_c^{-1} \tag{1}$$

where: z_0 is the number of agitator arms,
λ is the resistance coefficient of the central plates, h_0 is the height of projection of the arm on the meridional plane, m,
r_{in} is the radius of the inner edge of the arm, m,
$r_0 = 0,5d_m$,
$Re_c = \frac{\rho n d_m^2}{\mu}$ is the centrifugal Reynolds criterion, where ρ is the density of the medium, kg/m^3, n is the rotational speed of the agitator, rps,
μ is the dynamic viscosity coefficient of the medium, Pa·s,
d_m is the diameter of the agitator, m.

The power consumed for mixing N [W] was found as follows:

$$N = K_N \rho d_m^5 n^5 \tag{2}$$

where: n is the speed of the agitator, rps or rpm,
ρ is the density of the medium, kg/m^3,
d_m is the diameter of the agitator, m.

Also, one of the most important characteristics is the ram effect (or supply) V_c , i.e. the volume of liquid flowing to the agitator per unit of time and, thus, V_c determines the circulation of the liquid in the reactor:

$$V_c = \frac{Q}{n d_m^3} \tag{3}$$

where Q is the volumetric flow rate, m^3/s.

4. Discussion

The copolymerization of isobutylene with isoprene proceeds according to the cationic mechanism. Cationic polymerizations are the least studied which is explained by the very high reaction rate that consists of a large number of stages. This reaction is highly exothermic and is carried out at temperatures which are close to the freezing point of the reaction mass (nearly -95^{0}C). Maintaining such a low temperature in the reaction zone is necessary to ensure the quality of the produced butyl rubber, i.e. its molecular weight, composition, viscosity. It is known that when the temperature in the reaction zone rises above minus 50^{0}C, the rubber of low molecular weight is obtained, which makes it unsuitable for further use. Therefore, to provide the required low temperature in the polymerization zone, the reactor is equipped with 6 heat exchange elements, which are tube bundles with 205 tubes. The reaction medium is cooled with boiling liquid ethylene, which circulates through the tubes of heat exchangers. Also, to intensify the heat transfer process inside the reactor, it is equipped with a high-speed turbine agitator, which has a rotational speed of about 230 rpm.

Optimal for the mixing process is to reduce the consumed power and increase the ram effect (Figure 10).

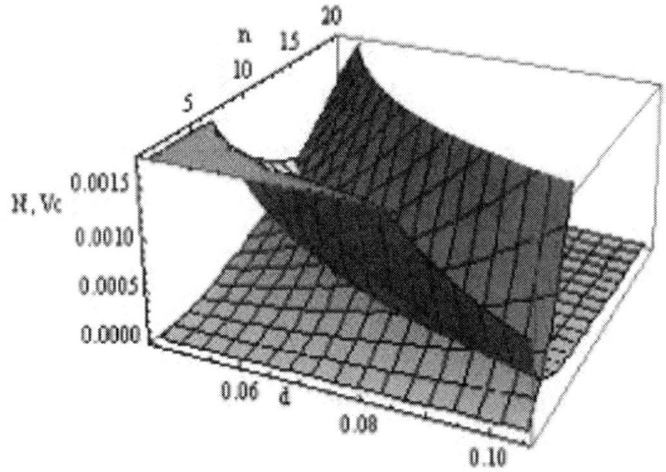

Figure 10. The dependence of the power N and ram effect V_c on the radius and speed of the agitator under $\lambda = 42$, $Q = 10^{-6}$ m^3/s, $h = 0.3$.

5. Conclusion
Thus, it is obtained the results of numerical simulation of the stirred-tank reactor with the turbine agitator in order to analyze the behavior of the power criterion and other characteristics of the process.

6. Acknowledgments
The study was carried out within the framework of a scientific project of The Russian Foundation for Basic Research № 13-03-01308-a.

References
[1] Markina E A 2010 *Sintez butilkauchuka s ispol'zovaniyem modifitsirovannoy kataliticheskoy sistemy na osnove khloristogo alyuminiya: dissertatsiya na soiskaniye uchenoy stepeni kandidata khimicheskikh nauk* [Synthesis of butyl rubber using a modified catalytic system based on the aluminium chloride: PhD] (Kazan: KGTU) [in Russian]

[2] Sangalov Yu A and Minsker K S 2001 *Polimery` i sopolimery` izobutilena: Fundamental`ny`e problemy` i prikladny`e aspekty`* [Polymers and copolymers of isobutylene: Fundamental problems and applied aspects] (Ufa: Gilem) [in Russian]

[3] Friebe L, Nuyken O, Windisch H and Obrecht W 2002 Polymerization of 1,3-butadiene initiated by neodymium versatate/diizobutilaluminium hydride/ethylaluminium sesquichloride: kinetics and conclusions about the reaction mechanism *Macromolecular Chem. and Phys.* **203(8)** 1055-64

[4] Balzamov D S, Balzamova E Yu, Ibatullin S R and Sabitov L S 2019 Efficiency increase of gas turbine work in the summer period *IOP Conf. Ser.: Mater. Sci. Eng.* **570** 012008

[5] Balzamov D S, Balzamova E Yu, Bronskaya VV and Valitov N V 2020 Possibility of using associated petroleum gas as a fuel for a production boiler house *IOP Conf. Ser.: Mater. Sci. Eng.* **791** 012006

[6] Dmitricheva R R, Antonova I O, Manuyko G V, Aminova G A, Bronskaya V V, Shaikhetdinova R S, Balzamov D S and Kharitonova O S 2020 Obtaining medium molecular characteristics of cation copolymerization of isobutylene and isoprene *IOP Conf. Ser.: Mater. Sci. Eng.* **862** 062045

[7] Manuiko G V, Salakhov I I, Aminova G A, Akhmetov I G, Dyakonov G S, Bronskaya V V and Demidova E V 2010 Mathematical modeling of 1,3-butadiene polymerization over a neodymium-based catalyst in a batch reactor with account taken of the multisite nature of the catalyst and chain transfer to the polymer *Theoretical Foundations of Chemical Eng.* **44** 139-49
[8] Quirk R P, Kells R P, Yunlu K nad Cuif J P 2000 Butadiene polymerization using neodymium versatate-based catalysts: catalyst optimization and effects of water and excess versatic acid *Polymer* **41(15)** 5903-8
[9] Aminova G A, Manuiko G V, Bronskaya V V, Ignashina T V, D'yakonov G S, Bashkirov D V and Demidova É V 2008 Influence of chain-transfer reactions on the molecular-weight-distribution function of diene rubber on a neodymium-containing catalyst system *J. of Eng. Physics and Thermophysics* **81** 1247-51
[10] Manuiko G V, Aminova G A, Bronskaya V V., Ignashina T V, D'Yakonov G S and Bashkirov D 2008 Calculation of the molecular weight distribution of the polymer produced in a cascade of reactors with allowance for chain transfer to the polymer *Theoretical Foundations of Chemical Engineering* **42** 336-9
[11] Aminova G A, Manuiko G V, Litvinenko G I, D'Yakonov G S, Zolotarev V L, Filipova A G, Ignashina T V, Davydova V V and Antonova O V 2002 Mechanism of butadiene polymerization on cobalt-containing catalyst *Russian J. of Applied Chem.* **75(7)** 1146-50
[12] Manuyko G V, Bashkirov D V, Bronskaya V V, Ignashina T V, Kharitonova O S and Aminova G A 2020 Creating molecular weight distribution of butadiene rubber on neodymium-based catalytic system *IOP Conf. Series: Mater. Sci. Eng.* **734** 012066
[13] Iovu H, Hubca G, Racoti D and Hurst J S 1999 Modelling of the butadiene and isoprene polymerization processes with a binary neodimium-based catalyst *European Polymer J.* **35(2)** 335-44
[14] Pros A, Marquardt P, Reichert K-H, Nentwig W and Knauf T 1993 Modelling the polymerization of 1,3-butadiene in solution with a neodymium catalyst *Die Angewandte Macromolekulare Chemie* **211(1)** 89-101
[15] Quiteria V R-S, Sierra C A, Gómez-Fatou J M, Galan C and Fraga L M 1999 Tin-coupled styrene-butadiene rubbers (SBRs) – Relationship between coupling type and properties *Macromolecular Materials and Eng.* **246(1)** 85-96
[16] Uraneck C A and Short J N 1970 Solution-polymerized rubbers with superior breakdown properties *J. Applied Polymer Sci.* **14(6)** 1421-32
[17] Kharitonova O S, Bronskaya V V, Ignashina T V, Al-Muntaser A A and Khairullina L E 201 Modeling of absorption process using neural networks *IOP Conf. Series: Earth Environ. Sci.* **315** 032025
[18] McIntosh W C, Hamann E, Oertel R and Bartke M 2007 Estimation of Kinetic Parameters for the Polar-Modified Anionic Solution Copolymerization of 1,3-Butadiene and Styrene *Macromolecular Symposia* **259(1)** 102-9
[19] Iovu M C, Buzdugan E, Teodorescu M, Britchi A G, Hubca G and Iovu H 1999 Copolymerization of styrene with butadiene using methyl tertbutyl ether as active center modifier *Macromolecular Materials and Engineering* **271(1)** 18-23
[20] Worsfold D J and Bywater S 1964 Anionic polymerization of isoprene *Canadian J. of Chemistry* **42(12)** 2884-92
[21] Chang C C, Halasa A F, MillerJr J W and Hsu W L 1994 Modelling studies of the controlled anionic copolymerization of butadiene and styrene *Polymer Int.* **33(2)** 151-9
[22] Hsieh H L and Yeh H C 1985 Polymerization of Butadiene and Isoprene with Lanthanide Catalysts; Characterization and Properties of Homopolymers and Copolymers *Rubber Chem. Technol.* **58(1)** 117-45 (https://doi.org/10.5254/1.3536054)

IOP Conf. Series: Materials Science and Engineering 1089 (2021) 012050 doi:10.1088/1757-899X/1089/1/012050

Stress-strain behavior monitoring of metal structures of NPP

A M Baldin, A V Kurteev, K E Orlov, M M Sevastyanov and O L Tashlykov

Department of Nuclear Plants and Renewable Energy Sources, Ural Federal University named after the first President of Russia B. N. Yeltsin, Mira St., 19, Yekaterinburg, 620002, Russia

E-mail: smanisty@mail.ru

Abstract. Conclusiveness of NPP safety assurance has been actualized. Methods of reliability and operability assurance of NPP elements during operation, life extension and putting out of service have been presented. The importance of diagnostics and monitoring of the condition of equipment, pipelines and structures of a NPP during operation, life extension and putting out of service have been pointed out. Distinctive features of working in radiation hazardous conditions and importance of remote diagnostic and monitoring methods as a part of the doze rates reduction have been presented. The developed and constructed system for monitoring the stress-strain behavior of a metal structures and the principles of its operation have been presented. The schematic diagram of the system and its components have been described. The description of the research methodology, its purpose, main stages and the results of carried out experiments have been presented. The operability of the system and the reliability of its readings have been confirmed. The dependence of the monitoring system's readings on the mechanical loading of the sample tested in the corresponding experiment has been obtained. The applicability of the developed system for monitoring the stress-strain behavior of a metal structures of NPP elements has been confirmed.

1. Introduction

The general purpose of safety assurance during NPP operation is reliable protection of the population and the environment from unacceptable radiation exposure. The main purpose of technical safety is high-reliable prevention of accidents and major malfunctions at the NPP [1].

Systems of diagnostics and monitoring of the equipment, pipelines and structures condition at all stages of the NPP life cycle are playing an important role in the safety assurance of nuclear power plants and well-timed procedures to prevent failures and accidents [2-4].Technical diagnostics should be carried out during operation and repair in order to maintain the established level of reliability, to ensure safety requirements and efficiency of equipment usage, and should be focused on following interrelated tasks:

- Evaluation of technical condition;
- Localization of failure or malfunction area;
- Forecasting of technical condition.

Early detection of signs of defects and deviations in the operation of the reactor plant is one of the main standards of the accident-prevention protection. As high priority actions to improve safety, VVER power plants are equipped with diagnostic systems (early detection of leaks, detection of foreign and weakly-fixed objects in the core coolant circulation circuit, etc.).

Maintaining the reliability and the technical conditions of NPP equipment and systems during operation include: regular monitoring of the condition of equipment and systems, scheduled periodic or unplanned (in case of malfunction or accident) impact on equipment and systems in order to eliminate detected unacceptable changes in their condition - restoration of their operability and renewal of its resources, analysis and evaluation of the effectiveness (quality) of equipment and systems condition management [5].

Safety of operation depends largely on the quality of maintenance and repair of NPP systems and equipment. Maintenance and repair of systems and equipment is carried out in order to keep up their technical condition and reliability at the required level, to ensure that the nuclear plant (NP) satisfy the safety requirements during the entire period of operation.

Special feature of the NPP is the presence of radioactive equipment and pipelines, which presumes specific requirements for work in the controlled access area. All over the world, the NPP staff exposure has been steadily decreasing in the last decades, which was conditioned by many factors (improvement of the water chemical conditions of the core coolant, optimization of the operation process, replacement of alloys containing cobalt, improvement of repairing work, etc.).However, due to the aging of the NPP equipment, there is an increase of the amount of works carried out to restore its operability. Among the ways of reducing the staff exposure [6, 7], the most effective one lies in the integration of remote and robotic devices for diagnostics of the state of radioactive equipment, systems, structures of nuclear power plants and their repair. It is important to take into account the principle of optimization, which is one of the basic principles of radiation safety [8-10]. At the moment, to determinate the technical condition of metal structures under the effect of dynamic and static stress, the stressed surfaces of the equipment are put to monitoring, according to the results of which, an expert statement about the stress-strain behavior (SSB) of the metal and the risk of its destruction is made.

In recent years, there is a high demand for magnetic inspection methods based on the relationship between the magnetic parameters of structures and the mechanical stress of its metal. Such methods for evaluating mechanical stress and SSB of a metal have a number of important advantages over other methods of metal inspection. These methods are a non-destructive and they provide the ability to control under almost any operating conditions of the inspected equipment and to monitor SSB of metal construction remotely during its operation [11].

The usage of such inspection method is highly important when diagnosing vessel equipment and metal structures with significant overall size and specific amount of metal (energy facilities, oil and gas production, pipeline networks for various purposes, etc.).For such equipment early diagnostics of metal damage significantly reduces the risks of its sudden destruction, the consequences of which cause significant damage to the economy and the environment [12].

These problems can be solved by using magnetic inspection of the SSB of a metal, based on the usage of the Forster effect, according to which the magnetic permeability of a section of a metal structure depends on the direction and intensity of mechanical stresses acting upon this section. which has a high magnetoelastic sensitivity and locality of control of the stress-strain state of the metal of a steel product, metal structure [11, 13].

For this reason, this study has been carried out, the device has been designed and the method of local magnetic inspection of the stress-strain behavior of metal, in particular, elements of vessel equipment and metal structures, has been developed.

2. Materials and Methods

2.1. The operating principle of the device for monitoring the stress-strain behaviour of metal structures

The device considered in this study (figure 1) is intended for non-destructive inspection of mechanical stresses in ferromagnetic materials and can be used to organize active monitoring of the stress-strain behavior of various construction and environmentally hazardous objects, such as nuclear power facilities, oil storage tanks and also for the development of intelligent building projects with controlled operational reliability.

.

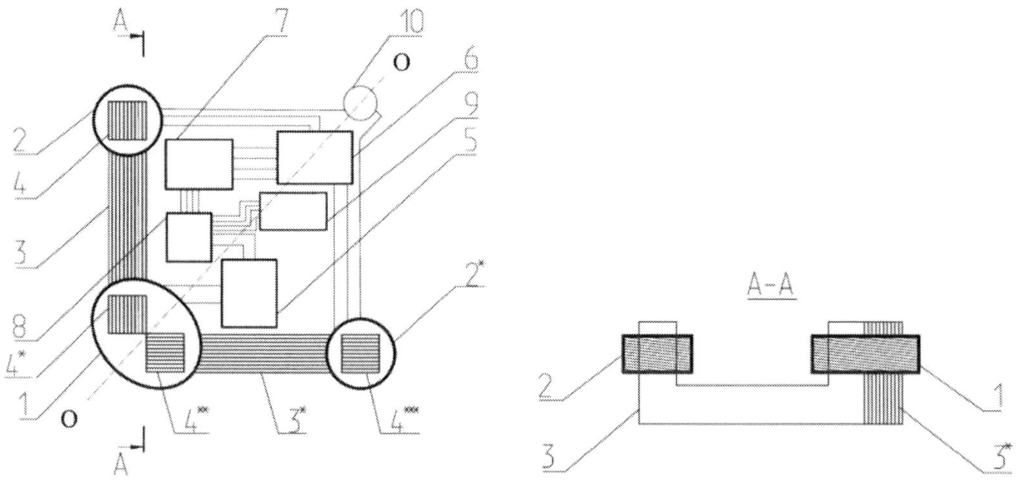

Figure 1. Schematic diagram of the device for monitoring the stress-strain behaviour of metal structures: 1 – driving coil; 2, 2* – two measuring coils symmetrically and perpendicularly located relative to the "O-O" axis; 3, 3* – magnet core with square cross-section, connecting the driving and measuring coils; 4, 4*, 4**, 4*** – poles of magnet cores, located symmetrically relative to the "O-O" axis; 5 – generator of signal, symmetrical about null-line; 6 – rectifiers with low voltage drop on operational amplifiers; 7 – analog-to-digital converter; 8 – processor; 9 – data transmission unit via wireless interface; 10 – support with height adjustment.

After switching on the device, the signal generator 5, controlled by the central processor 8, provides driving coil with the signal with the preassigned frequency 1. To except the processes of magnetization of the measured surface, a signal is maintained symmetrical about null-line. The device is installed on a flat area of the surface of the inspected section of the metal structure, while the magnetic lines passing through the U-shaped magnetic core 3, 3* are closed up through the surface area of the inspected section of the metal structure. Due to this EMF is induced in the measuring coils 2, 2*, the value of which depends on magnetic permeability of the sample, which, according to the Forster effect, depends on the direction and intensity of the mechanical stresses effecting on the structure [14].

To convert AC voltage into DC voltage, the electrical signals received from the measuring coils 2, 2* are forwarded to rectifiers with a low voltage drop, made using low-noise operational amplifiers 6. The rectified signal in the analog-to-digital converter 7 is converted into a digital signal, after which enters the central processor 8, where after processing is transmitted through the data transmission unit via the wireless interface 9 to the data output device.

Before the start of mechanical stress monitoring, the device is calibrated in accordance with the material and shape of the inspected structure to determine the coefficient for conversion the incoming electrical signal into the value of mechanical stress.

The proposed device for monitoring the stress-strain behavior of metal structures allows measurements under conditions of a biaxial stress behavior, which makes it possible to determine the direction and intensity of both static and dynamic stress of building structures, bridges, buildings, equipment, etc.

2.2. Experimental technique

The aim of the study is determination and development of a method of local magnetic inspection of the stress-strain behavior of metal, elements of vessel equipment and metal structures based on the Forster method.

During the research, the following objectives were accomplished:

- Construction of a prototype device for monitoring the stress-strain behaviour of metal structures;
- Carrying out the experiments;
- Analysis of obtained data.

When constructing a prototype, the driving and measuring coils were interconnected by magnetic cores with a square cross-section, made of separate U-shaped plate elements made of high-quality vacuum transformer steel.

Experimental studies were carried out at the educational and training facility (ETF) of the Department of Nuclear Plants and Renewable Energy Sources of the Ural Federal University. ETF is an important part of specialists' preparation for the maintenance and repair of NPP equipment at UrFU. The ETF have been founded with the support of Rosenergoatom Concern and equipped with training facilities, demonstration models and diagrams, samples of equipment for remote control and repair, laboratory facilities. ETF allows you to carry out the necessary practical and laboratory training [15].

To accomplish objectives following experiments have been carried out:

- Straining of the industrial I-beam 30Б1 (GOST 260020-83), operability confirmation of monitoring system in conditions similar to industrial facilities;
- Straining of the metal plate made of steel (GOST 103-2006), determination of the measurement sensitivity;
- Approbation of the experimental data, obtaining the dependence between the monitoring system readings and the mechanical stress in the metal.

2.2.1. Testing of the stress-strain behaviour monitoring system at industrial facilities. I-beam30Б1 has been chosen as the test sample.

The characteristics of the I-bean are presented in table 1. Its configuration is shown in figure 2.

The I-beam has been strained with load weighing 32 kg. The load has been suspended with a rafter in the middle of the inspected section of the I-beam (figure 3).

Table 1. Dimensions of I-bean 30Б1.

Dimension	Value (mm)
Height, h	298
Web height, h_ω	282
Flange width, b	149
Flange overhang, b_ω	71.75
Flange thickness, t	8
Web thickness, s	5.5
Radius of conjugation, r	13
Length, L	5500

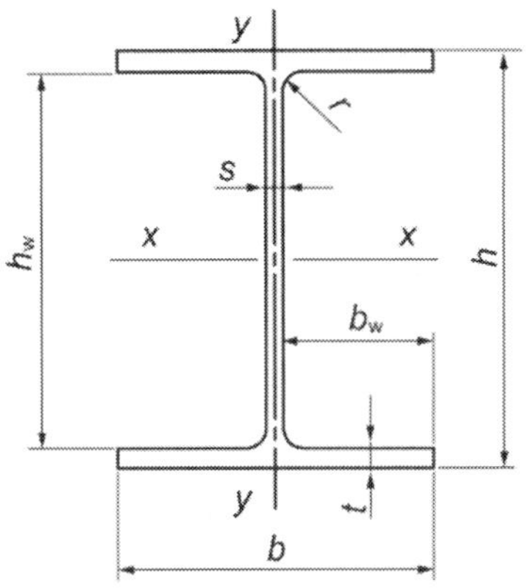

Figure 2. Cross-section of I-beam.

Figure 3. Picture of experiment "Straining of the industrial I-beam".

Before straining the I-beam, the object was at quiescent state without external actions applied to it. In 20 seconds after the start of the experiment, a load weighing 32 kg was placed on the rafter. After 25 seconds, the load was removed, the experiment lasted until 68 second.

2.2.2. Straining of the metal plate, determination of the measurement sensitivity of SSB monitoring system. Device for monitoring stress-strain behaviour of metal structures has been installed on a plate made of sheet metal and fixed as an anchorage.

The characteristics of the plate are presented in table 2.

Table 2. Characteristics of plate.

Characteristic	Value
Thickness, S, (mm)	6
Width, B, (mm)	150
Length, L, (mm)	600
Material	Steel CT-3

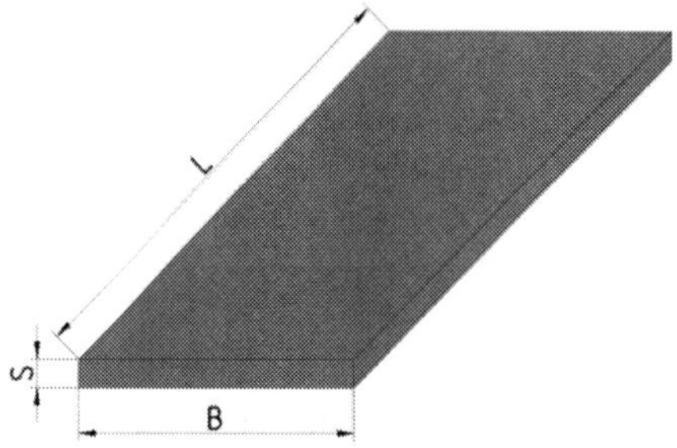

Figure 4. Plate.

The straining has been carried out in the form of successive opposite in direction pointed loadings on the loose end of the plate. The values of the loadings were equal to 0.2; 0.4 and 0.5 N. At the initial moment of time, there was no loading on the plate, after 25 seconds a loading was applied in the form of a force equal to 0.2 N, after 15 seconds the force was increased to 0.4 N, after another 15 seconds, the force was increased to 0.5 N. After that same procedures have been repeated but in opposite direction.

2.2.3. Obtaining the dependence between the monitoring system readings and the mechanical stress in the metal. To obtain the relationship between the recorded parameter (mV) and the actual values of stresses (MPa) in the metal, the following experiment has been performed.

Constant loads (0.111; 0.131; 0.301 and 0.32 MPa) were applied to the loose end of the fixed with anchorage steel plate. In the similar manner to the previous experiments, basing on obtained data using MS Excel a graph of the strain of the material has been constructed, and using the "trend line" tool, the dependence between the measured value and the actual stresses in the metal has been determined and the R-squared was calculated. Then, at the second phase, constant loads (0.242; 0.51; 1.334 and 8.175 MPa) were applied to the same point at the same conditions. Obtained in the second phase results have been compared to calculated with determined equation's values.

3. Results

Basing on the results of the experiment on loading the I-beam, the dependance diagram showing temporal variation of the SSB monitoring system readings has been obtained (figure 5).

As a result of straining the metal plate, the material was tested in tension and compression with the application of equal force in both directions.

The experiment results are shown in figure 6.

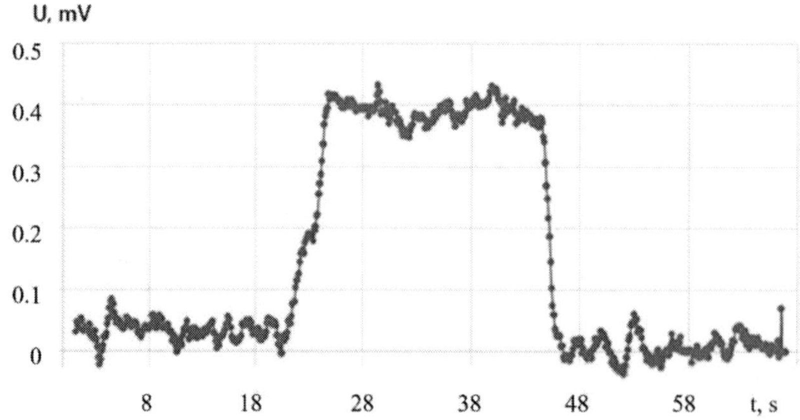

Figure 5. Dependence of the readings of the monitoring system on the time during which the loading was performed in case of straining I-beam.

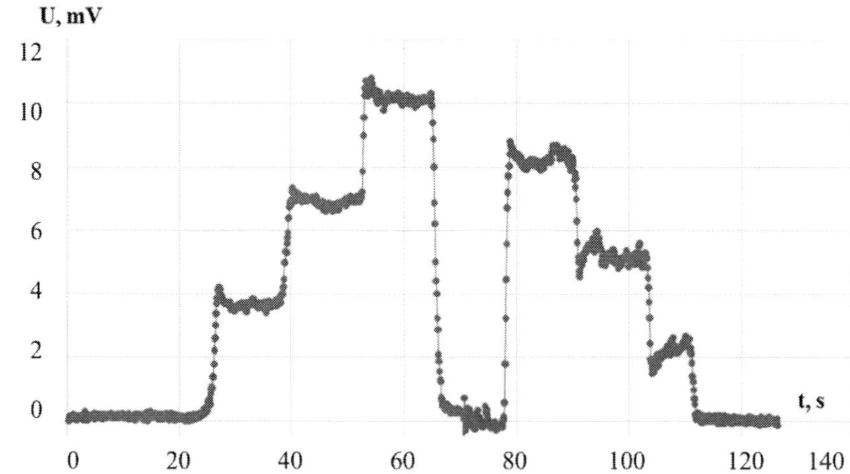

Figure 6. Dependence of the readings of the monitoring system on the time during which the loading was performed in case of straining plate.

The results of construction of graph showing dependence of the readings of the monitoring system on the actual values of straining in the metal are shown in figure 7.

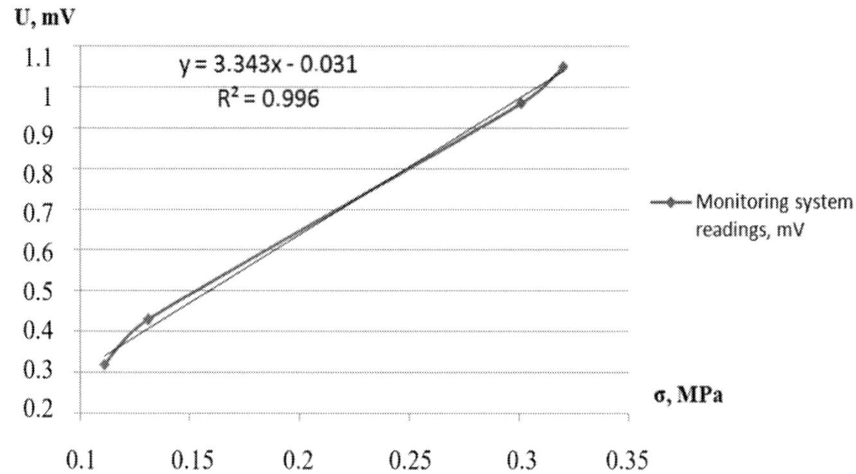

Figure 7. Dependence of the readings of the monitoring system on the mechanical stress under tension of the metal.

Based on the results of plotting the graph, the following results were obtained. The equation of the dependence of the readings of the monitoring system on the mechanical stress under tension in the metal is:

$$y = 3.343 \cdot x - 0.031 \quad (1)$$

where:
y – recorded readings of the monitoring system, mV;
x – values of metal stress, MPa.

The obtained dependence (1) was used to extrapolate and interpolate to calculate values, which have been compared to experimental readings of the second phase. The experimental and calculated values are given in table 3.

Table 3. Comparison of SSB monitoring system readings and calculated with equation (1) values.

Mechanical stress in metal, MPa	Reading, mV	Calculated value, mV
0.242	0.78	0.77801
0.51	1.67	1.67393
1.334	4.42	4.42856
8.175	27.2	27.298

4. Discussion

The graph of dependence of the readings of the monitoring system on the time during which the loading was performed in case of straining I-beam (figure 5) clearly shows the jump of the strain in metal at the moment of application the force, as well as the constancy of the system readings during loading. After removing the loading, the recorded value reduces back to zero, which confirms that the system if operational and the sample was not magnetized.

In case of plate straining, from the graph (figure 6) the dependence of the readings jumps and their heights on value of applied to sample loading is clearly visible. The constancy of the readings of the monitoring system at the moment when the loading does not change again indicates the absence of magnetization of the sample. The smooth, stepped structure of the graph indicates the inertial response of the device to a change in the loading in the material. Difference in the recorded parameters during the experiment on sample compression is explained by the action of gravity on the loading in metal plate. The experiment was carried out by bending the plate with equal force in the vertical upward direction against the vector of the force of gravity, thus the strain created in the sample appears to be smaller. This is can be observed on the graph (figure 6).

In the case of obtaining the dependence between the monitoring system readings and the mechanical stress in the metal, the readings of the monitoring system deviate from the theoretical straight line by 0.4%. This fact confirms that the proposed SSB monitoring system demonstrate the consistent and proportional readings under the conditions of time-dependent stress-strain behavior of metal constructions.

5. Conclusion

Magnetic inspection methods of metal elements and constructions are inalienable part of safety assurance during NPP and other industrial facilities operation and repair. SSB inspection methodbased on the usage of the Forster effect is indeed one of the most perspective methods of this type at the moment remaining unexpanded. Under this study such SSB monitoring system has been designed and constructed.

According to the results of maintained experiments, this monitoring system showed its operability, stability, accuracy and proportionality of its readings depending on straining of inspected structures. Due to this qualities, proposed SSB monitoring system it can be efficiently used at facilities that requiring highly accurate inspection of its element's mechanical stress-strain behavior.

References

[1] Novikov G A, Tashlykov O L and Shcheklein S E 2017 *Obespechenie bezopasnosti v oblasti ispolzovaniya atomnoj energii* [Ensuring safety in the field of nuclear energy use] (Ekaterinburg: Ural State University) [In Russian]

[2] Tashlykov O L, Shcheklein S E, Sesekin A N, Chentsov A G, Nosov Yu and Smyshlaeva O Y 2014 Ecological features of fast reactor nuclear power plants (NPPs) at all stages of their life cycle *WIT Transactions on Ecology and the Environment* **190(2)** 907-18
[3] Tashlykov O L and Shcheklein S E 2016 Ecological foresight in the nuclear power of XXI century *Int. J. of Energy Production and Management* **1(2)** 133-40
[4] Nosov Y V, Rovneiko A V, Tashlykov O L and Shcheklein S E 2019 Decommissioning Features of BN-350,-600 Fast Reactors *Atomic Energy* **125(4)** 219-23
[5] Tashlykov O L 2018 *Remont oborudovanija atomnyh stanzij* [Repair of equipment of nuclear power plants] (Ekaterinburg: Ural State University) [In Russian]
[6] Tashlykov O L, Shcheklein S E, Russkikh I M, Seleznev E N and Kozlov A V 2017 Composition Optimization of Homogeneous Radiation-Protective Materials for Planned Irradiation Conditions *Atomic Energy* **121(4)** 303-7
[7] Russkikh I M, Seleznev E N, Tashlykov O L and Shcheklein S E 2015 Experimental and Theoretical Study of Organometallic Radiation-Protective Materials Adapted to Radiation Sources with a Complex Isotopic Composition *Physics of Atomic Nuclei* **12(78)** 1451-6
[8] Tashlykov O L, Shcheklein S E, Lukyanenko V Yu, Mikhailova A F, Seleznev E N and Kozlov A V 2015 Optimizaziya sostava radiazionnoj zashchity [Optimization of radiation protection composition] *Izvestiya Vuzov Yadernaya Energetika* **4** 36-42 [In Russian]
[9] Mikhailova A. F and Tashlykov O L 2018 Puti realizazii prinzipa optimizazii v radiologicheskoj zashchite personala [The Ways of Implementation of the Optimization Principle in the Personnel Radiological Protection] *Yadernaya Fizika i Inzhiniring* **9(4)** 393-401 [In Russian]
[10] Kropachev Y A, Tashlykov O L and Shcheklein S E 2019 Optimizaziya radiazionnoj zachity na etape vyvoda energoblokov AES iz ekspluatazii [Optimization of radiation protection at the stage of nuclear power plant units decommissioning] *Izvestiya vuzov. Yadernaya Energetika* **1** 119-30 [In Russian]
[11] Wilson J, Liu J, Karimian N, Davis C L and Peyton A J 2014 Assessment of microstructural changes in Grade 91 power station tubes through permeability and magnetic Barkhausen noise measurements *ECNDT 2014* (Prague: NDT)
[12] Chopra H D, Hicho G E 2001 Review of the jumpsum based nondestructive testing method for evaluating mechanical properties of ferromagnetic materials *Materials Evaluation* **59(10)** 1215-22
[13] Turner M J, Clouhg R W, Martin H C and Topp LJ 1956 Stiffness and Deflection Analysis of Complex Structures *J. Aeronaut. Sci.* **23** 805-24
[14] Zhou J T, Zhao Y Y, He Q, Zhang P and Zhang H 2019 Experimental of corrosion detection of galvanized steel strands based on magnetic memory *J. of Chang'an University* **39(1)** 81-90
[15] Shcheklein S E, Tashlykov O L, Velkin V I, Shastin A G, Dementjev V N, Markelov N I and Sorokin Yu I 2004 Opy't sotrudnichestva kontserna "Rosenergoatom" i Uralskogo Gosudarstvennogo Tekhnicheskogo Universiteta v oblasti podgotovki spetsialistov dlya tekhnicheskogo obsluzhivaniya i remonta oborudovaniya AES [Experience of cooperation of Rosenergoatom Concern and the Ural State Technical University in the field of personnel training for the maintenance and repair of the NPP equipment] *Izvestiya vuzov. Yadernaya Energetika* **3** 121-9 [In Russian]

The outlook for multivibrator application in the present day electronic systems

N Yu Savvin[1], D D Garbuzov[2], R S Singatulin[1] and P V Roshchubkin[1]

[1]Department of Electric Power Engineering and Automation, Belgorod State Technological University named after V.G. Shukhov, Kostyukov St., 46, Belgorod, 308012, Russia
[2]Department of Energy Engineering of Heat Technologie, Belgorod State Technological University named after V.G. Shukhov, Kostyukov St., 46, Belgorod, 308012, Russia

E-mail: n-savvin@mail.ru

Abstract. The paper discusses challenges that the present-day electronic industry of the Russian Federation faces today, analyses reasons for lagging in this sector in volume and production scale of the finished products in comparison to the leading in this area countries, suggests ways for reaching a new level in this industry by implementing a nationwide digitalization, reviews a place of digital signals in modern-day electronic, their types, creation methods, main characteristics, devices for generation and processing. The method of digital signals forming using the multivibrator was suggested in this paper. Different multivibrator types were reviewed, including the multivibrator with AND-NOT gates, its advantages were considered in comparison with the transistor stage multivibrators, parameters of the electronic circuit elements in the AND-NOT gate multivibrator for generating a signal with the required characteristics were calculated, then the multivibrator was assembled and the output signal parameters were taken down. The output signal parameters and the calculated parameters were identical. The conclusion was made about the possibility of the suggested AND-NOT gate multivibrator application in the complex automatic control circuits and electronic systems indication circuits as a device that generates signals with the required parameters.

1. Introduction

Advancement and progress in the electronic industry are some of the key elements for economic growth in developed countries. Currently, this industry sector is a leading scientific sector [1] and its main purpose is to develop, produce, and implement electronic equipment in all areas of life. The level of electronics development in any country has a direct impact on people`s quality of life, volume, and complexity of product output, sovereignty, and integrity of the state. The electronic industry is critical for providing all aspects of sovereignty, security, and progress in any country, and also for national policy towards implementation of the economy`s digital transformation. Electronic industry development in Russia is the most crucial task at the moment and demands close attention.

The reason that keeps Russia behind in electronic production, in comparison with the USA and China, is the fact that this industry was practically built from nothing, which is the result of degradation during the '90s. In this context, the state plays an active role in supporting and developing the national electronic industry, implementing programs that allow significantly increase electronic devices sales at the world`s markets within a short period of time.

A notable example of such programs is the «Development strategy of the electronic industry of the Russian Federation until 2030», the framework of which outlines the main directions of the state policy for the national electronic industry development. The strategy specifies a complex solution for nine key areas: «Scientific and technological development, «Means of production», «Industry standards», «Staff», «Management», «Cooperation», «Industrial data environment», «Markets and products», «Economic efficiency». Implementation and progressive realization of this program require close attention to the area of training and qualification of professionals. This demands the promotion of a specific sector as well as science in general.

2. Materials and Methods

2.1. Materials

Trends in electronics development entail its full digitalization. The function of modern electronic equipment involves digital signal processing in real-time. Fuzzy logic controller development is one such example of fuzzy rules application for automatic control systems architecting. Electrical signals are commonly used for electronic equipment control.

An important issue in the development of circuits, which are based on the multivibrators, is the task of providing precise parameters of the circuit elements that give maximum reliability and precision of the device functioning. This issue was discussed in the paper «Choice of tolerance limits for multivibrator components» by A.A. Kazakov and E.G. Prugov. The examined in this paper program allows us to determine tolerance limits for the components of the circuit in development based on the required precision of the objective function. This is especially important for the circuits with components that have nonlinear parameters.

Insufficient front slope of the generated pulses is one of the shortcomings in traditional multivibrator design. This issue was resolved by using a solution suggested in the patent «Multivibrator» by I.M. Zhovtis. The patent describes a design when blocking oscillator and multivibrator are combined in one stage. This design uses an additional retroactive circuit.

2.2. Methods

The main purpose of the multivibrator is the creation of a square-wave pulse with predetermined parameters which allows us to use it in an indicating system of different electronic devices. The inventor's certificate «Device for output of information from electronic counters to the display» by Leshcinsky A.S., Zelenin A.N., Vasilyev V.M. analyzed a possibility of functional capability enhancement by way of information readout from several sources. This invention can be useful in devices where single display use creates difficulties of information output from several sources and allows the output of information in a certain sequence with predetermined time duration.

The development of the multivibrator circuits based on logic gates has been discussed in scientific publications many times. The inventor's certificate «One-shot multivibrator» by Bizyaev V.A., Zharkov L.K. Voronov V.I. reviewed the multivibrator circuit which allows us to increase the output signal stability substantially. For this purpose, an inverter was introduced into the device's circuit which is based on the fourth AND-NOT gate.

The use of programmable multivibrators can be helpful in circuits of the automatic control devices. This type of multivibrator, which was reviewed in the inventor's certificate «Programmable multivibrator» by Petrovsky A.K., Antonov V.V., allows us to control frequency, relative duration, and pulse number in series. The function of this multivibrator type is based on a comparison of the program and ramp voltage values, which form pulse width and pauses in series, that were generated by the integrated into the device trigger flip-flop.

Signals can be continuous or discrete. The continuous (analog) signal differs from discrete in a way that the information parameter, which such signal carries, can take any instantaneous value within the preset limits. Discrete signal, on the contrary, has only two states: high and low level of the information parameter. Discrete square-wave signals and meander are of special interest for use in the control circuits. [2] Meanders and square-wave signals are the basic signals, which comprise the backbone of the entire electronics. Meander is a voltage that switches between fixed levels at equal intervals. Usually, these signals are used for amplifier testing which should process rapid transitions between two levels of voltage. Meander is an ideal clock frequency signal for digital systems – computers, wireless communication devices, high-definition TV, and many others. [3] Square-wave signal has characteristics that are similar to meander with the exception that intervals of high and low levels are not equal against each other.

To generate square-wave electrical pulses with preset parameters a device called a multivibrator is used. The purpose of the multivibrator is to generate signals that are close to a square shape. This type of signal is essentially a sum of a large number of harmonics, and each of them is multiple of the main oscillation frequency. [4] In electronic devices multivibrators have a support function, therefore, there are no strict requirements for stability of time characteristic of the generated output signal.

3. Results

3.1. Pulse generation

There are several variants of how multivibrators can be constructed. One of them is transistor based (Figure 1). In this case, the multivibrator is a two-stage amplifier with a retroactive circuit (RC) [5]. At the same time, the output of the second transistor stage is connected to the input of the first transistor stage through the capacitor, which through charging and discharging takes part in the pulse generation. In this way, the function of pulse generation is implemented using the amplifier.

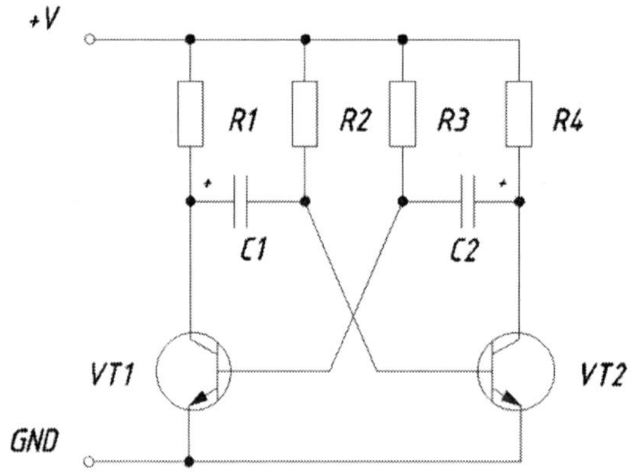

Figure 1. Multivibrator circuit with transistor stages.

In this circuit resistances of each resistor are equal, i.e. R1 = R2 = R3 = R4, also capacitor`s capacitances are equal C1 = C2, and the oscillogram of the output signal (Figure 2) is a square-wave signal with equal pause time and pulse duration [6].

According to equation (1) pulse frequency of the symmetrical multivibrator is:

$$f = \frac{700}{(\mathrm{C1} * R2)'} \tag{1}$$

where: C1= C2 – capacitance of the capacitors in stages,
R1= R2 – resistance of the resistors in the multivibrator`s arms.

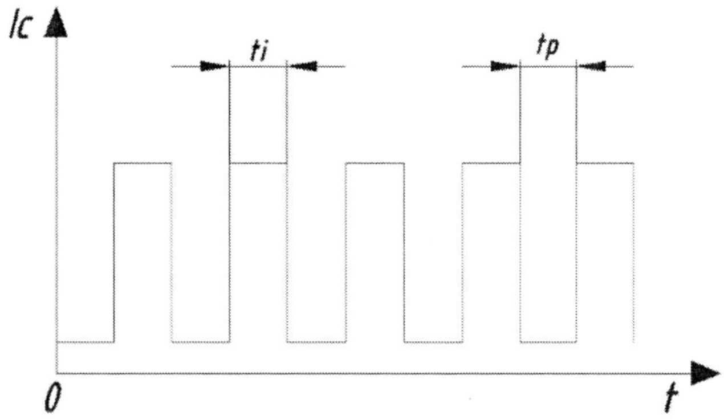

Figure 2. Output signal oscillogram of the symmetrical multivibrator.

The multivibrator is characterized by parameters such as frequency f, Hz, pulse duration t_i , ms, output signal amplitude U, V, relative duration (T/t_i).

The multivibrator, which is based on logic gates, presents some features of interest [7]. This circuit is free from shortcomings of the transistor circuit because of the low current drain and it is not locked to the power supply of +5V.

3.2. Circuits

Presented in Figire 3 circuit is considered a classic circuit of multivibrator which is based on logic gates. In this circuit, three logic gates are generating pulses, and the fourth is improving an output signal waveform [8]. The circuit is characterized by two quasi-stable states. In the first state LG1 (logic gate) is open, LG2 and LG3 are closed. While the capacitor is charging the voltage at the LG3 input becomes higher than the threshold which makes it open [9]. The regenerative process initiates circuit switching. In the second state LG2 and LG3 are open, LG1 is closed.

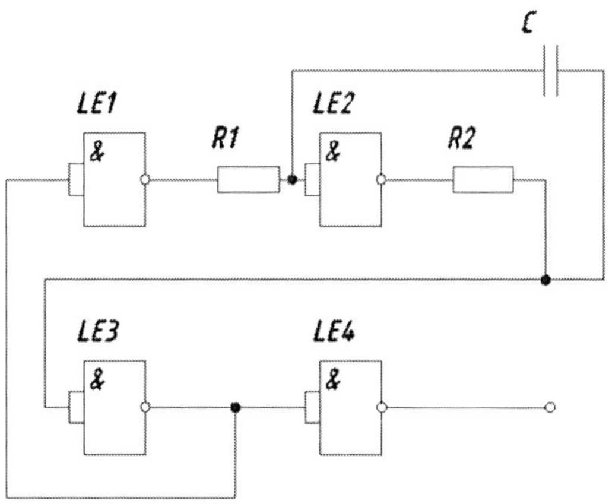

Figure 3. Multivibrator circuit based on AND-NOT gates.

To assemble the multivibrator on logic gates it is suggested to use two SN54LS22 microchips [10]. The microchip of this type consists of two AND-NOT gates and has 10 input legs, two output legs, and power supply legs.

It is necessary to test the logic gates of the microchip for defects before the assembly. In order to do that the inputs of every logic gate should be connected by energizing it with a high/low level signal and then take down the signal from the output. If a signal of a certain level doesn`t disrupt AND-NOT gate logic functioning then it means that the logic gate is working properly.

4. Discussion

Microchip based multivibrator generates square-wave signal parameters of which can be changed by way of resistor`s resistance and capacitor`s capacitance change in the multivibrator circuit [11]. Pause duration time of the pulse is calculated by the equation (2):

$$t_{u1} \approx (R_1 + R_2) * C * \ln\left(\frac{2 * R_2}{R_1 + R_2} + \frac{1.5 * R_2}{4 + R_2}\right), \tag{2}$$

where: R1 – resistance of the resistor between LG1 and LG2,
R2 – resistance of the resistor between LG2 and LG3,
C – capacitance of the capacitor in the feedback circuit [12].

Pulse working time is given by equation (3):

$$t_{u2} \approx (R_1 + R_2) * C * \ln(2.5 + \frac{3 * R_1}{R_1 + R_2}) \tag{3}$$

The period of oscillation is given by equation (4):

$$T = t_{u1} + t_{u2} = 10{,}1 \text{ мс } \text{ ms} \tag{4}$$

The oscillation frequency is calculated according to equation (5):

$$f = \frac{1}{T} \tag{5}$$

Relative pulse duration is calculated using the equation (6):

$$Q = \frac{T}{t_{u2}} = 1 + \frac{t_{u1}}{t_{u2}} \tag{6}$$

When the necessary values of resistance and capacitance in the multivibrator circuit are determined, it is time to wire the elements according to Figure 4.

When the device is assembled it is necessary to test it with the oscillograph by taking down parameters of the output signal from the fourth logic gate`s output [13].

The output signal parameters testing requires voltage supply to the inputs +5V and GND, the fourth logic gate output should be connected to the input of the oscillograph [14]. The received oscillogram (Figure 5) allows us to estimate the output signal parameters and observe their change while resistances R1, R2, and capacitances C are changing [15].

The received output signal parameters should be compared with the calculated to make sure that the multivibrator works properly [16]. The presented AND-NOT gate based multivibrator can be used in circuits of devices that automatically control load for operation mode indication control, and also to generate controlling pulses for the operation of other devices which use the square-wave signal as a controlling signal.

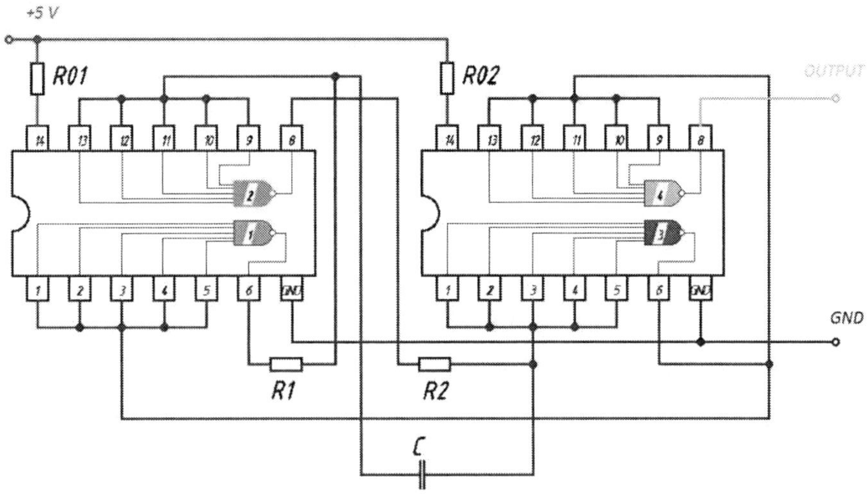

Figure 4. Wiring diagram of the SN54LS22 based multivibrator.

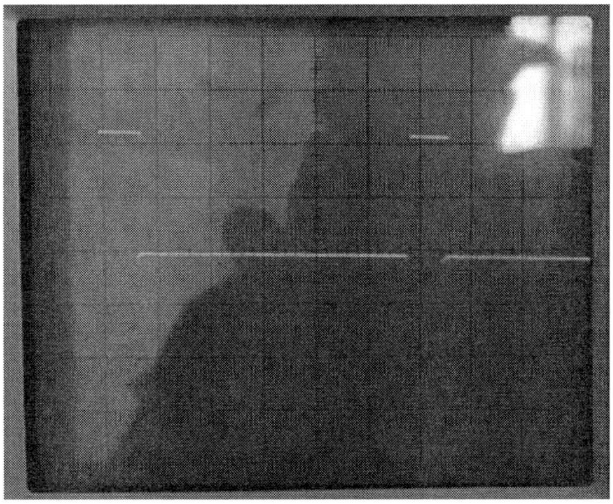

Figure 5. Oscillogram of the output signal.

5. Conclusion

The results of the conducted experiment allow us to conclude that an AND-NOT gate based multivibrator has good prospects in industrial automation as an indication instruments control device and in devices that are controlled by way of square-wave signal sequence with preset parameters. The circuit, which is realized during the experiment, provides output signal parameters that are high in accuracy when compared to the calculated parameters, and that allows us to use it as an integral part of complex control devices and automation.

6. Acknowledgments

The work is realized in the framework of the Program of flagship university development on the base of the Belgorod State Technological University named after V.G. Shukhov, using equipment of High Technology Center at BSTU named after V.G. Shukhov.

References

[1] Krokhin O N and Lebedev P N 2007 Quantum electronics 50th jubilee *Proceedings of SPIE* **6595** 659502

[2] He R and Yun H 2020 Electronic Differential Control of Rear-Wheel Independent-Drive Electric Vehicle *SAE Int. J. Veh. Dyn., Stab., and NVH* **4(1)** 49-65

[3] Haseda Y, Bonefacino J, Tam H-Y, Chino S, Koyama S and Ishizawa H 2019 Measurement of pulse wave signals and blood pressure by a plastic optical fiber FBG sensor *Sensors* **19(23)** 5088

[4] Nanda M A, Seminar K B, Nandika D and Maddu A 2019 Development of termite detection system based on acoustic and temperature signals *Measurement* **147** 106902

[5] Hardiyanto D, Priyambodo S, Kristiyana S and Enrico C A R 2020 Automatic Broadcast Transmission Control with 8 Relay Channels Based on Arduino *IOP Conf. Ser.: Mater. Sci. Eng.* **846** 012047

[6] Getting I A 1938 Multivibrator geiger counter circuit *Physical Review J. Archive* **53** 103

[7] Watanabe Y 1930 Some remarks on the multivibrator *Proceedings of the Institute of Radio Engineers* **18(2)** 327-35

[8] Ma P, Gao L, Ginzburg P and Noskov R E 2020 Nonlinear Nanophotonic Circuitry: Tristable and Astable Multivibrators and Chaos Generator *Laser and Photonics Reviews* **14(3)** 1900304

[9] Vlasov A I, Prisyazhnuk S P and Zhalnin V P 2020 Analysis of memristor modules as an element base of microprocessor control systems: Contradictions and prospects *Int. Conf. on Industrial Eng. Applications and Manufacturing* (Sochi: IEEE) 9111917

[10] Rajput D S and Singh R 2015 Timer circuit using OTRA and its application as astable and Monostable multivibrator *2nd Int. Conf. on Electronics and Communication Systems* (Coimbatore: IEEE) pp 1047-50

[11] Hu W T, Zhang Y X, Sun Y, Liu W S and Zhang L Z 2014 Research on small-capacity capacitor type sensor detection method *Advanced Materials Research* **981** 594-7

[12] Filanovsky I M, Jarvenhaara J and Tchamov N T 2013 Push-pull RC-oscillator/multivibrator *56th Int. Midwest Symp. on Circuits and Systems* (Columbus: IEEE) pp 141-4

[13] Shaa P, Sun H J, Zhou W L, Xu X H, Zhang J J, Wang Q, Hu C Y, Yan P and Miao X S 2012 Programmable frequency Multivibrator based on memristor *Information Optoelectronics, Nanofabrication and Testing* (Wuhan: OSA Publishing) IF5A.4

[14] Abrams Z R and Zhang X 2011 Signals and circuits in the Purkinje neuro *Front. Neural Circuits* **5** 11

[15] Mitra N and Mukhopadhyay S 2011 A method of developing all-optical mono-stable multivibrator system exploiting the Kerr non-linearity of medium *Optik* **122(2)** 92-4

[16] Khentov V Ya and Semchenko V V 2018 On the connection of the physical properties of metal elements with the temperature of debay *Chemical Bulletin* **1(2)** 19-23

The usage of water conditioning sludge from heat power industry enterprises in wall ceramics production

I V Starostina[1], O V Peganova[2], R G Shevtsova[3] and N V Shakurova[4]

[1]Department of Industrial Ecology, Belgorod State Technological University named after V.G. Shukhov, Kostyukov St., 46, Belgorod, 308012, Russia
[2]Veropharm JSC (Abbott group), Rabochaya St., 14, Belgorod, 308009, Russia
[3]Department of Theoretical and Applied Chemistry, Belgorod State Technological University named after V.G. Shukhov, Kostyukov St., 46, Belgorod, 308012, Russia
[4]Department of Glass and Ceramics Technology, Belgorod State Technological University named after V.G. Shukhov, Kostyukov St., 46, Belgorod, 308012, Russia

E-mail: starostinairinav@yandex.ru

Abstract. The paper considers the ecological problems of water conditioning systems at heat power industry facilities. It is noted that the stable operation of water-heating equipment is provided with the quality of feed water. Various water-soluble salts, which can incrust the inner walls of boilers or pipelines, are removed from the natural water. For desalination of natural water the physical-chemical methods are mostly used. The by-product of desalination or chemical water-conditioning systems is the sludge waste. This work presents the results of using chemical water conditioning sludge from JSC Verofarm in the raw charge for wall ceramics production. The sludge is a finely-divided powder, the mineral composition of which is presented with calcium and magnesium carbonates, brucite and iron compounds, which make the sludge pink. It has been determined that the baking of sludge-containing ceramic samples at temperature 1000°C results in dissociation of calcium and magnesium carbonates with the formation of gaseous products H_2O and CO_2, which make the structure of ceramic materials porous. The presence of iron compounds in the raw charge promotes the formation of low-temperature silicate melt, which intensifies the baking process of ceramic samples. The optimal sludge content in the charge that doesn't reduce the strength characteristics of wall ceramics is 10%. Such sludge content forms the liquid phase on the surface of the raw mix's particles, which solidifies the forming pores. The increase of sludge content up to 20% intensifies gas generation; the amount of the formed liquid phase is not sufficient, which increases porosity and reduces the density and strength of ceramic materials. At that, the engobe layer is formed on the surface of the obtained wall ceramics samples. The wall ceramics products, obtained with the use of chemical water conditioning sludge, can be used as structural heat-insulation materials in making building envelopes or interior partitions.

1. Introduction

The power industry enterprises belong to water-using branches of industry. Near 50% of the water, taken by the industrial enterprises of the Russian Federation from natural water sources, is used for the needs of heat power industry. The stable operation of water-heating equipment depends on the quality of the feed water. For example, for gas-turbine steam boilers the total water hardness shouldn't exceed 35 μmol /dm^3, the feed water for water-tube boilers with working steam pressure up to 4 MPa should have hardness no more than 30 μmol /dm^3. So, the natural feed water is subject to the compulsory conditioning and purification. This process consists in removing from natural water various water-soluble salts, which can incrust the inner walls of boilers and pipelines. Desalination of natural water is performed mostly by physical and chemical methods – like filtering it through special ionite filters, which selectively extract certain microelements. The by-product of desalination or chemical water conditioning systems is the sludge waste, which contains trace elements, extracted from the natural water, metals and reagents of flushing fluids – silicon, aluminum, iron, calcium and magnesium compounds, some organic impurities etc.

In amounts of its generation, the chemical water conditioning sludge can be classified among large-scale waste products. The heat power industry facilities, depending on the amount of the produced heat or electric energy, generate up to 7 thousand tons of sludge annually [1]. Sludge has a high moisture content, with the water existing as free water, pore water, surface adhesion water and internal combined water [2]. Sludge in the form of pulp with moisture content up to 90% is transported to interim storage sites for dehydrating or to sludge depositories for permanent storage.

The arrangement of sludge reservoirs or sludge depositories requires considerable capital investments and inevitably results in environmental pollution. The sludge suspensions can leak during pumping them through pipelines, filter into soil or ground waters, get into the air with evaporation of gaseous products from the surface or the secondary blowing of dusty components, can cause condemnation or pollution of lands, or irreparable changes of landscapes.

Besides, the flooding or erosion of certain elements of sludge depositories can disrupt the operation of drainage facilities, and cause the filtrates' penetration into ground waters and their pollution. The accidents at sludge depositories are accompanied with spillage of millions of cubic meters of loose sludge sediments to the neighboring territories and water bodies.

So, developing methods of the sustainable utilization of such waste, which would allow reducing the amounts of their accumulation and storage, is a vital task.

At present, chemical water conditioning sludge doesn't find any large-scale application. Though there are some research works in using sludge waste in building materials production or in road construction [3, 4].

The work [5] considers the usage of sludge wastes, formed in the technological cycle of vanadium pentoxide and ferrovanadium alloys production, in producing ceramic wall materials. It is shown that the optimal amount of sludge in the raw charge of ceramic bricks, not reducing their physical and mechanical properties, amounts from 3 to 5% by weight.

In the works [1, 6, 7] the possibility of using chemical water conditioning sludge, generated at the stage of raw water preliminary lime, coagulation and electrochemical treatment, for obtaining a sorption material, is demonstrated. The obtained adsorbents showed the high efficiency in purifying sewage waters from heavy metal ions and petroleum products, and exhaust gases from hydrogen sulfide [8, 9].

According to data in the work [10] the carbonate sludge from heat power stations' chemical water conditioning systems can be used as an additive to high-sulfur furnace oils. The findings have shown that the sludge compounds are adsorbed on the surface of furnace oil's paraffin components, preventing their crystallization and the thickening of oil. This reduces the viscosity and chilling temperature of furnace oil, which in its turn reduces the energy consumption for heating furnace oil and pumping it through pipelines.

The sludge, generated at CHPP in the process of water conditioning, was utilized as filler for gypsum binders [11]. The adding of sludge doesn't influence the water demand and water resistance of binders, but improves the setting time of a composite binder. The optimal content of sludge is 15%.

On the basis of sludge waste, thermally modified in conditions of lack of oxygen, the hydrophobic oleophilic carbon-bearing sorption materials were obtained for removing heavy metal ions from sewage waters, and petroleum or petroleum products spillages from water surface [12, 13].

The work [14] suggests the technology of recultivation materials production on the basis of sludge. It includes the three-stage dehydration of sludge with the use of geotextile containers system and mixing it with hardening additives – ashes from CHPP, milled boiler refractory waste and sooty waste. The obtained composite material corresponds in its properties to natural soils and can be used for construction and maintenance activities at recultivated lands – for filling pits and ditches, for technical shielding of grounds and vertical land leveling.

The purpose of this research is to study the properties of chemical water conditioning sludge wastes from boiler plants of a pharmaceutical enterprise JSC Verofarm, Belgorod, Russia, and to consider the possibility of its utilization in ceramic bricks production.

2. Materials and Methods

In a sludge depository the sludge appears as waterlogged pink mass with moisture content 83%. After drying at temperature 105-110°C it acquires the properties of a powder, i.e. has no binding properties. The dry sludge is a polydisperse system, with high dispersion; particle size is from 0.1 μm to 50 μm (figure 1). The chemical composition of sludge is presented in table 1.

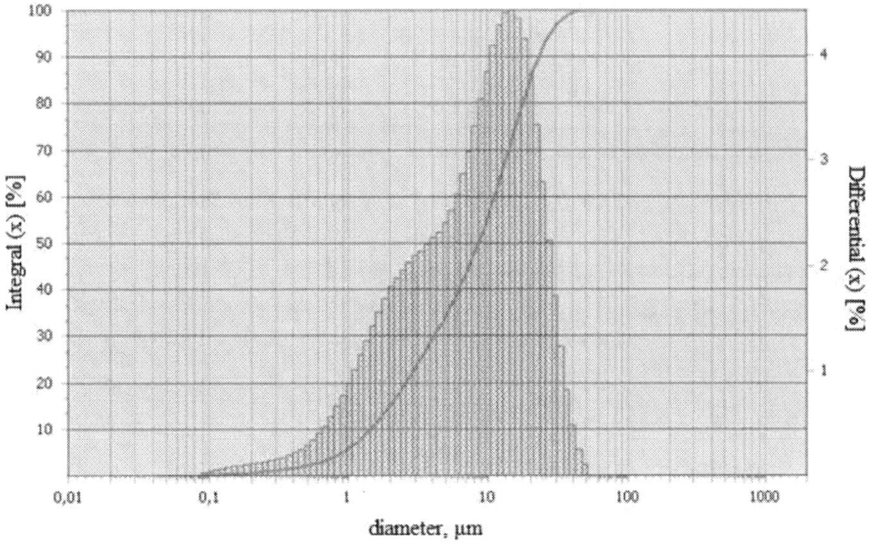

Figure 1. Results of granulometric composition analysis of chemical water conditioning sludge from JSC Verofarm.

Table 1. The oxide composition of chemical water conditioning sludge from JSC Verofarm (wt.%).

CaO	MgO	SiO_2	R_2O	Fe_2O_3	Al_2O_3	SO_3	chlorides
46.6 – 61.8	16.4 – 20.7	3.1 – 14.7	2.4 – 7.0	1.7 – 2.7	0.1 – 0.4	0.4 – 0.7	03- 19.7

According to the X-ray phase analysis results (figure 2-a), the sludge contains the following minerals: calcium and magnesium carbonates, brucite ($Mg(OH)_2$), iron compounds FeOOH and Fe_2O_3, which makes the sludge pink. By the intensity and area of characteristic peaks we can conclude that the prevailing mineral is calcium carbonate, and the sludge can be characterized as carbonaceous.

In the paper the utilization of chemical water conditioning sludge in the raw charge for ceramic bricks has been considered. For this purpose the thorough selection and optimization of the raw mix composition and its burning conditions are required, which provides the controlled phase formation in the process of ceramic charge material sintering. As an argillous raw material, the clay from Bykovskoye deposit, Belgorod region, Russia, was used.

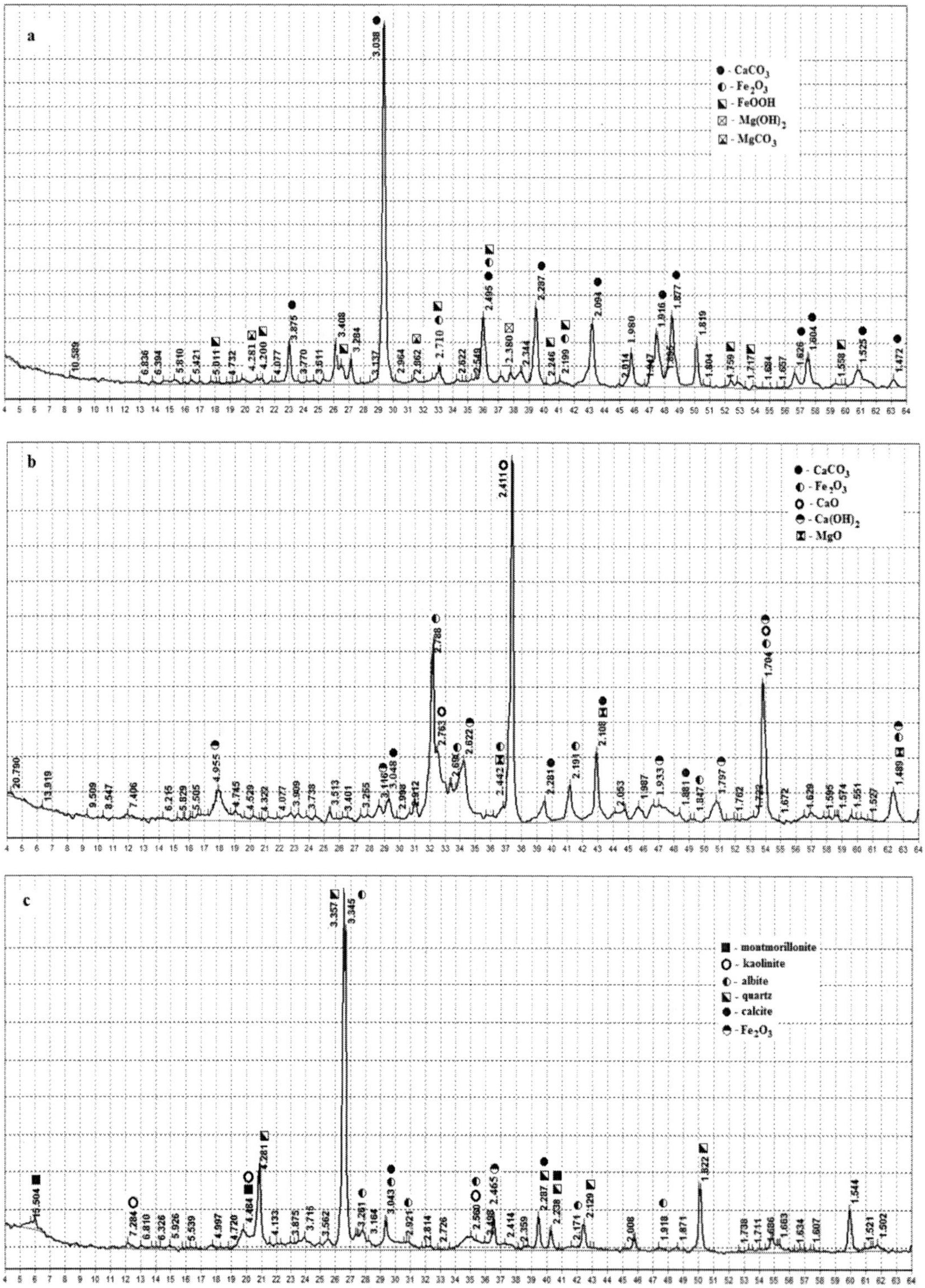

Figure 2. X-ray diffractograms: a – dry initial chemical water conditioning sludge; b – chemical water conditioning sludge, heat-treated at temperature 1000°C; c - clay from Bykovskoye deposit.

The mineral composition of the clay was evaluated according to the X-ray phase analysis results (figure 2-c). In the X-ray diffraction pattern the strong lines are registered, characteristic for the following minerals:

- clay minerals: montmorillonite d(Å)= 15.504; 4.484; 2.238, albite (anorthite) d(Å)= 3.345; 3.261; 2.560; 2.171; 1.918, kaolinite d(Å)= 7.284; 4.484; 2.560;
- accessory minerals: quartz SiO_2 d(Å)=4.281; 3.357; 2.287; 2.238; 2.129; 1.822, calcite d(Å)= 3.043; 2.287, hematite (rutile) Fe_2O_3 d(Å)= 2.465. The clay from Bykovskoye deposit is characterized as polymineral with the prevailing montmorillonite-hydromicaceous components, and contains impurities of quartz, calcite and hematite.

Dry chemical water conditioning sludge was added to 100% of raw charge in amount from 0.2 to 20% of solid components mass. The prepared charge was mixed to a homogeneous condition and moistened up to 20%. Then the cylindrical ceramic samples 3 cm high and 3 cm in diameter were made by means of plastic molding. The samples were dried at temperature 105-110°C to the constant weight, and then baked at temperature 1000 °C.

3. Results and discussion

The adding of chemical water conditioning sludge to the raw charge promotes the formation of porous structure of the obtained ceramic materials (figure 3). Such tendency is explained by the ability of the sludge's components – calcium and magnesium carbonates – to decompose under the action of high temperatures by the following reaction:

$$CaCO_3 \rightarrow CaO + CO_2$$

$$MgCO_3 \rightarrow MgO + CO_2$$

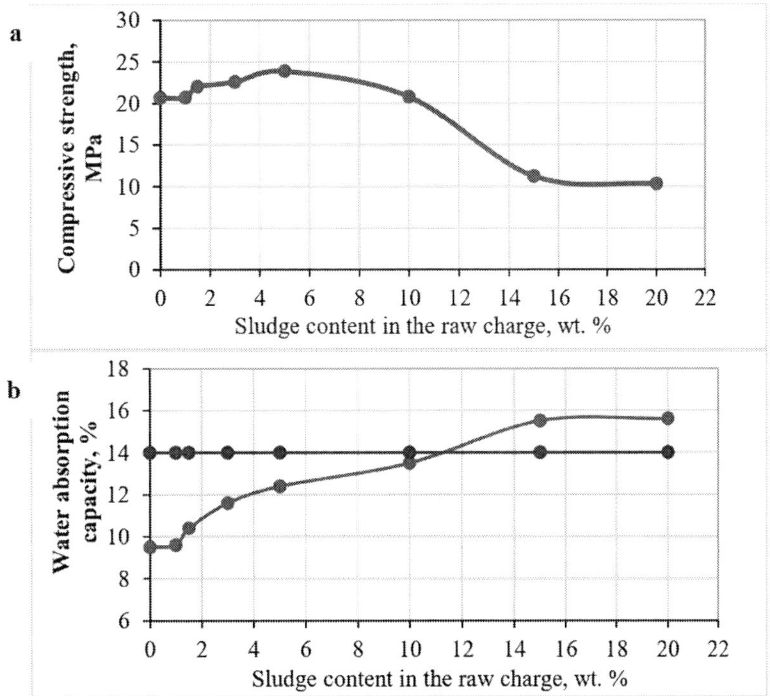

Figure 3. The influence of sludge content in the raw charge on physical and chemical properties of ceramic materials:
a – compressive strength; b – water absorption capacity.

The heat treatment of sludge, apart from intensive gas release, is accompanied with the generation of Fe_2O_3 and CaO (figure 2-b), which in clay-containing masses act as strong fluxing agents due to forming relatively low-melt compounds with Al_2O_3 and SiO_2. The mineralizing role of iron oxides in ceramic masses has been studied in a number of works [15] It is noted that protoxidic iron compounds, having high reactivity, intensively form ferrous silicate melts or glasses [15].

Thus, the sludge content in the raw charge up to 10% doesn't reduce strength properties of the obtained materials. This allows the use of water treatment sludge from heat power enterprises in the production of building materials [16-18]. As a result of samples' baking, the forming liquid phase on the surface of raw mix's particles penetrates into their capillaries, formed as a result of slight gas development, and acts as a cementing binder, forming thin films at phase contacts. The gases are also entrapped by the melt and the pores are fixed, which provides stability of the obtained samples' compressive strength – 21.0 MPa (in the control sample) and 24.0 MPa (5% of sludge) (figure 3). As a result, the more porous structure, than in the control sample, is formed – the water absorption value is in-creased from 9.5% (in the control sample) to 12.4% (5% of sludge) and up to 13.5 % (10% of sludge). The formed porosity in its water absorption capacity is close to the upper acceptance limit - 14% (figure 3).

With increase of the sludge content in the raw charge up to 20%, the process of CaO, generated at decomposition of sludge's carbonate components, absorption with the formed liquid phase is intensified. But at the same time, the volume of the formed liquid phase is not enough for entrap-ping gaseous products or solidifying pores, which eventually results in the increase of porosity up to 15.6%, and decrease of ceramic materials' density, which adversely affects the strength proper-ties of samples.

It should be noted that the increased sludge content (10-20%) in the raw charge results in the formation of engobe layer on the surface of ceramic materials. This is due to the excess amount of CaO, formed in the process of thermal dissociation of calcium carbonate, incompletely bonded with clay minerals. The excess of CaO interacts with gaseous components with the formation of light (engobe) superficial layer on ceramic materials samples.

4. Conclusion

So, the usage of chemical water conditioning sludge in the raw charge for wall ceramics without reducing their physical-mechanical characteristics is possible in amount up to 10% by weight over 100%. Ceramic products with the increased sludge content – from 10 to 20% – are characterized with the increased porosity and lowered physical-mechanical properties, so they can be used only as structural heat-insulation materials.

5. Acknowledgments

The work is realized in the framework of the Program of flagship university development on the base of the Belgorod State Technological University named after V.G. Shukhov, using equipment of High Technology Center at BSTU named after V.G. Shukhov.

References

[1] Nikolayeva L A and Kotlyar M N 2019 Ochistka stochny`x vod gal`vanicheskix cexov ot ionov medi modificirovanny`m karbonatny`m shlamom [Purification of electroplating shops' sewage waters from copper ions with modified carbonate sludge] *Vodnoe Xozyajstvo Rossii: Problemy`, Texnologii, Upravlenie* **5** 124-32

[2] Shen K, Xu H, Ding M and Cui J 2016 Dewatering of drinking water treatment sludge by vacuum electro-osmosis *Separation Science and Technology* **51(13)** 2255-66

[3] Zamora R M R, Ayala F E, Garcia L C, Moreno A D and Schouwenaars R 2008 Optimization of the preparation conditions of ceramic products using drinking water treatment sludges *J. of Environmental Sci. and Health, Part A* **43(13)** 1562-8

[4] Alzoubi A E, Ghunimat D M, Al-Rawashdeh M and Hanandeh S 2020 The potential of using water purification wastes as fine aggregates in concrete mixes: an initial study *Australian J.*

of Civil Eng.

[5] Starostina I, Simonov M, Volodchenko M, Starostina Yu, Fomin A and Fokina E 2018 The usage of iron-containing sludge wastes in ceramic bricks production *IOP Conf. Ser.: Mater. Sci. Eng.* **365** 032066

[6] Nikolaeva L A and Khanzina D A 2019 Purification of Water Sources from Oil Contamination by Hydrophobic Carbonate Sludge *IOP Conf. Ser.: Earth Environ. Sci.* **288** 012018

[7] Nikolaeva L A and Iskhakova R Ya 2019 Adsorption of industrial wastewater from oil products with application of mathematical modeling *IOP Conf. Ser.: Earth Environ. Sci.* **288** 012017

[8] Orescanin V, Kollar R, Halkijevic I, Kuspilic M and Flegarl V 2014 Neutralization/purification of the wastewaters from printed circuit boards production using waste by-products *J. of Environmental Sci. and Health, Part A* **49(5)** 540-4

[9] Nikolaeva L A, Zainullina R and Al-Okbi A K 2019 Adsorption Drying of Natural Gas by Carbonate Sludge *Chemical and Petroleum Engineering* **54** 919-25

[10] Zvereva E R, Burganova F I, Khabibullina R V, Zverev L O and Sheshukov E G 2019 The scheme of dosing additives to fuel oil and evaluation of the effectiveness of its implementation at the enterprises of the fuel and energy complex *E3S Web Conf.* **124** 01033

[11] Valeev R Sh and Shaykhiev I G 2010 Utilizaciya shlamovy`x otxodov teploe`nergeticheskix centralej pri proizvodstve stroitel`ny`x materialov [Recycling of sludge waste from combined heat and power plants in building materials production] *E`kologiya i Promy`shlennost` Rossii* **2** 28-9

[12] Starostina I V, Stolyarov D V and Kosukhin M M 2017 Carbonic and Mineral Adsorbent, Received out of a Slurry Withdrawal of Oil-Extracting *Solid State Phenomena* **265** 319-24

[13] Starostina I, Nikitina A, Kosukhin M and Starostina Yu 2018 Efficient carbonaceous sorbent based on the waste sludge of oil extraction industry for purifying wastewater from petroleum products *Int. J. of Eng. and Technology (UAE)* **7(2.23)** 266-9

[14] Kondratiev O I, Chertes K L, Tupitsyna O V, Pystin V N and Petrenko E N 2019 Perspektivy` utilizacii shlamov na osnove sero- i formal`degidsoderzhashhix otlozhenij [Prospects for the Recovery of Sludge on the Basis of Grayand Formaldehyde-Containing Sediments] *E`kologiya i Promy`shlennost` Rossii* **23(5)** 32-6

[15] Kairakbaev A K, Abdrakhimova E S and Abdrakhimov V Z 2019 Ispol`zovanie otxodov obogashheniya czvetnoj i chernoj metallurgii kazaxstana v proizvodstve keramicheskix materialov [Use of Waste of Non-Ferrous and Ferrous Metallurgy of Kazakhstan in the Production of Ceramic Materials] *E`kologiya i Promy`shlennost` Rossii* **23(6)** 12-6

[16] Alfimova N I, Pirieva S Yu, Gudov D V, Shurakov I M and Korbut E E 2018 Optimization of receptural-technological parameters of manufacture of cellular concrete mixture *Construction Materials and Products* **1(2)** 30-36

[17] Elistratkin M Yu, Minakova A V, Jamil A N, Kukovitsky V V and Issa Jamal Issa Eleyan 2018 Composite binders for finishing compositions *Construction Materials and Products* **1(2)** 37-44

[18] Chernysheva N V, Shatalova S V and Evsyukova A S 2018 Fisher Hans-Bertram Features of the selection of the rational structure of the compositional gips binder *Construction Materials and Products* 1(2) 45-52

Use of a group of underwater hydraulic tarans for the water supply system

V Y Sokolov and S A Naumov

Department of Electrical and Heat Power Engineering ,Orenburg state University, Shevchenko St., 28, Orenburg, 460005, Russia

E-mail : naum090909@rambler.ru, teploosu@mail.ru

Abstract. When choosing water-lifting equipment, a number of issues related to the type of water intake structure, the energy for driving the plant, the adopted pressure and control structures, and the automation system are solved. Their structures should be simple, convenient to operate, have an affordable cost, provide the required water supply to the above areas. For agricultural water supply and irrigation of land plots located above the water source, the equipment for lifting water using the energy of the water flow - a hydraulic ram - fully meets the requirements of all known means of mechanizing the water lift. The cascade hydraulic unit operates due to dispersed energy of open water flows and hydraulic impact energy. The use of a cascade hydraulic ram plant on the river will allow the water supply of individual consumers without the cost of electricity, which is absolutely safe for the environment. The proposed topic of work is therefore very relevant.

1. Introduction

Water supply of power plant is performed using hydrotarane plant, in accordance with Figure 1.

If the pumps are replaced by a group of underwater hydraulic tarans, the need for a water intake device will run out.

A group of hydraulic tarans will need to be installed in the river channel, where the flow of water will reach the highest speed, near the farm.

If it is necessary to artificially increase the flow rate of water in the river, it is possible to locally narrow the bed by creating a fill [1, 2].

In addition, it will be necessary to build a base platform for a taran group connected by a generalizing pipeline. The base is necessary to ensure rigid fixation of each element of the underwater structure, as well as to create a single zero mark for all hydrotarans, ensuring equal initial parameters of the devices operation.

However, for all its positive qualities, the hydraulic ram also has drawbacks, which, as relatively inexpensive and convenient electricity and motorized equipment spread, ultimately led to the almost complete displacement of these motor-free pumps by conventional pumps with electric or gasoline drive. Some of these disadvantages can be compensated quite easily, but others cannot be eliminated because, as is often the case, they are a direct extension of the advantages[3-5].

Firstly, to ensure the acceleration of the flow after the next opening of the baffle valve, there should no longer be water passed there in the previous cycle. If for any reason it does not leave during the hydraulic impact, then it will prevent the acceleration of a new portion of water in the delivery pipe, which will not gain a speed sufficient to close the baffle valve. In the best case, the flow will pick up the desired speed much longer than would have happened in the absence of water behind the baffle valve - this is the unproductive loss of water through the baffle valve and a decrease in the efficiency of the plant. Naturally, water can escape only if there is a drain, so the discharge of the delivery pipeline (more precisely, the location of the bump valve) cannot be below the level of the drain reservoir, otherwise the passed water will not be able to release the bump valve.

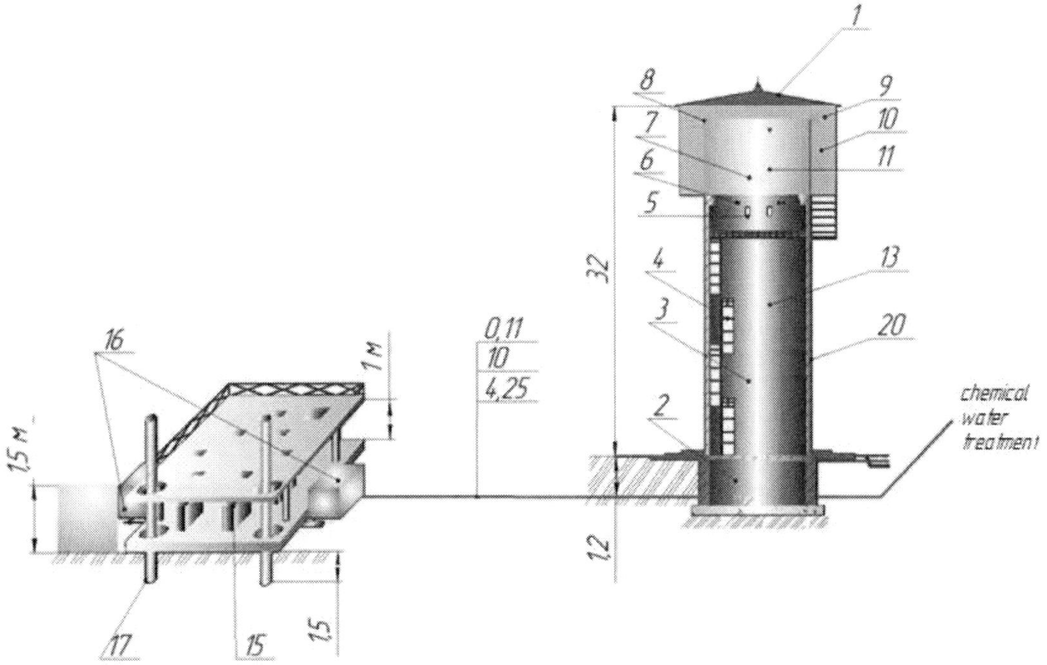

Figure 1 Power plant water supply: 1 - water tower; 2- foundation and basement; 3 - supply - discharge pipeline; 4 - ladder; 5 - gland compensators; 6 - pipe for water removal for fire extinguishing; 7 - pipe for water withdrawal for household needs; 8 - tank; 9 - tent; 10 - ladder; 11 - overflow pipe; 12 - mud pipe; 13 - discharge pipe; 14 - barrel; 15 - hydrotarane unit; 16 - ballast tank; 17 - pile

Secondly, in order to accelerate the flow in the delivery pipeline to a good speed (at least a meter per second), it is necessary to ensure an altitude difference of at least several centimeters in a section several meters long.

For these reasons, hydraulic rams cannot work in reservoirs with a constant surface level, such as ponds and lakes, as well as in flat sections of rivers, where hundreds of meters, or even kilometers of flow, there is a difference in levels in centimeter or other[6,7].

Third, a substantial portion of the water is "lost" through the discharge pipe drain. Moreover, the "lost" volume is usually many times larger than the raised volume. Of course, this water is "lost" not in vain, but does its job - its energy goes to the rise of another part of the stream. However, when the total amount of available water is low, this "wastefulness" may be unacceptable. In general, the efficiency of such pumps is determined by the correct choice of the length and volume of the delivery pipe, the ratio of the sections of the baffle and pressure valves and the forces necessary for their opening and closing, depending on the required lifting height and flow rate in the delivery pipeline, that is, in the end, the working difference in levels and water flow. Therefore, ideally, each instance of such a pump should be configured individually for specific installation conditions[8].

Fourth, when using a "classic" storage cap with air, air can gradually dissolve in the injected water, which is facilitated by increased pressure. Therefore, air must be periodically replenished. To solve this problem, the use of a membrane hydraulic accumulator tank as such a cap, which in recent years has become an integral part of autonomous water supply systems in cottages and cottages, will help. Another way to solve this problem is that with the close location of the baffle and pressure valves and strong working hydraulic impacts with the separation of liquid from the baffle valve, it is possible to try to organize automatic pumping of air through these valves, although a number of technical problems will have to be overcome [9,10].

Finally, fifthly, the hydraulic ram has considerable dimensions. Thus, it is generally believed that the optimum length of the injection pipe is in the range of up to ten meters or more. This is due to the fact that the mass of moving and then stopping water should be large enough to provide good energy for the working hydraulic shock. Since the mass of water is directly proportional to its volume, this imposes inevitable restrictions on the minimum dimensions of more or less productive structures. And the duration of the hydraulic impact should also be sufficient so that the pressure valve has time to open and pass a noticeable volume of water, and this time is also directly proportional to the distance from the baffle valve to the supply reservoir or reservoir. However, by turning the injection pipe into a spiral, it is possible to reduce the linear dimensions of the installation several times. But the weight, determined by the necessary strength and rigidity of the structure, is unlikely to be significantly reduced [11,12].

On the other hand, the capacity of the hydraulic tank is limited by its size. Too large dimensions of the hydraulic ram will also cause problems, since all elements of the structure in the area of the working hydraulic shock must have not only sufficient strength, but also maximum rigidity. As linear dimensions grow, providing the necessary rigidity may require too thick walls and, as a result, too massive parts.

2. Materialsand Methods

Nevertheless, the classic hydraulic ram remains an extremely simple, unpretentious and very unusual device, which has been completely undeservedly almost forgotten recently.

The maximum (steady) speed can be defined as [13]:

$$v^c = \varphi\sqrt{2gH} = \sqrt{\frac{2gH}{l} + \varepsilon^c}\ ; \tag{1}$$

$$\varepsilon^C = \sum \varepsilon + \mu l / d\ ; \tag{2}$$

$$h_p = \frac{\varepsilon v^2}{2g}\ ; \tag{3}$$

$$h_l = \frac{\mu l v_c^2}{d 2g}\ , \tag{4}$$

Where:

u_c is the speed of steady-state water movement;
φ – the resistance factor constant for the given system;
ε_c – is the coefficient of local resistance;
ε – coefficient of hydraulic friction;
h_p – pressure loss on local resistances;
h_l – head losses along the length of the feed line;
l, d – pipeline length and diameter, respectively.

It should be noted that the operation of the hydraulic ram is carried out at an unknown water speed. Studies have established that its value is defined as:

$$V = 0,99u_c \tag{5}$$

Upon reaching the speed value and, by force from the water velocity head, the impact valve rises rapidly upwards, stopping the flow of liquid. Hydraulic shock occurs (injection stroke). The force of the velocity (dynamic) head can be determined by the formula:

$$P_{кd} = \frac{p \cdot v^2 \cdot w \cdot t}{2}, \tag{6}$$

where:
$w{\cdot}t$ is the midelian section of the valve plate;
p – water density.

Maximum pressure of hydraulic impact is determined by formula (6). The discharge period begins from the start of the shock valve closing (opening of the pressure valve) and ends with its closing. The duration of the injection period is defined by:

$$t_н = \frac{2nt}{c} \tag{7}$$

where n – is the wave run from the working chamber to the feed tank repeated n times.

3. Results

At the same time, the hydraulic ram unit is a simple and unpretentious device, so the use of such units is relevant for energy saving.

The technologies used today in agriculture are very simplified. The reason for this is both the poor financial condition of farms and the low technological equipment of production. Thus, the development of energy-saving technologies in the agro-industrial complex is relevant, and the use of energy-saving plants using renewable energy sources is a priority, especially for remote areas with poorly developed infrastructure.

The platform will be a frame of steel profile pipes of rectangular section 80x60x4 (GOST 13663 Steel profile pipes). For the platform (frame 16 m and 4 ribs of 4 m) 32 m of steel profile pipe will be needed. As the 80x60x4mm pipe size calculations have shown, the framework will require 32 m of pipe weighing 258.24 kg[13,14,15].

The frame will need to be clad with a 5 mm thick steel corrugated sheet.

To secure the platform, piles will need to be installed using a special mechanical device to clog piles into the ground called a coper. The coper can function by manual labor or use different kinds of hammers to lift the cargo. Current species of hammers include diesel and hydraulic hammers as well as vibromolots. The principle of functioning of hammers is based on conversion of mechanical energy into kinetic energy: heavy weight rises to maximum height using the force of air, steam or diesel engine, and then descends on the pile, hitting it into the ground.

Depending on the type of soil and piles, different types of copra are used. A coper with a hydraulic hammer is most effective and is good at scoring steel, reinforced concrete and wooden piles. In addition to specialized copier installations for pile scoring, universal machines - excavators are used, for which purpose they are equipped with a suspended mast. Due to the installation of the guide on a standard crane boom, the excavator performs the functions of a pile machine in a short period of time. Wooden, reinforced concrete, steel piles and tongue fences are clogged. In our case, the piles will be electric welded seamless pipes with a diameter of 273 mm and a wall thickness of 12 mm. Metal piles and tongue fences immersed with clogging are coated with anticorrosive coating.

IOP Conf. Series: Materials Science and Engineering 1089 (2021) 012053 doi:10.1088/1757-899X/1089/1/012053

The cost of pile works (the price of pile scoring) depends on a number of factors: the method of immersion, the depth of pile immersion, the section of pile, the category of soil, the scope of works, the complexity of the project.

Piles will need to be submerged into the ground to a depth of 1.5 m. In summer the water level in the line is 1.5 m, and in the flood in spring rises by a maximum of 0.8 m. When raising the hydrotaran assembly to the surface, the structure should be held on piles, as well as not damaged against the upper control platform, so it is necessary to provide another 1 m of pipes above the water surface. Thus, the length of each pipe will be 4.8 m. We need 2 pipes. Total length - 9.6 m. Weight of one running meter of electric welded seamless pipe 273x8 mm is 52.28 kg, then total weight - 502 kg.

The diameter of the pile hole in the platform shall be greater than twice the diameter of the pile. In our case, the diameter will be 0.45 m. This is necessary so that when the platform of the hydrotarane assembly is lifted to the surface of the water, the plastic pipeline (extending to the water tower and laid in the ground) can bend behind the platform to the surface[16,17].

The fact is that this pipe will be taken with a margin of length, which will be compensated by the inclination of the pipeline underwater. And when the platform is lifted, the gap between the hole in the platform and the pile will ensure free straightening of the pipeline. The path of the platform relative to the pile is shown in Figure 2.

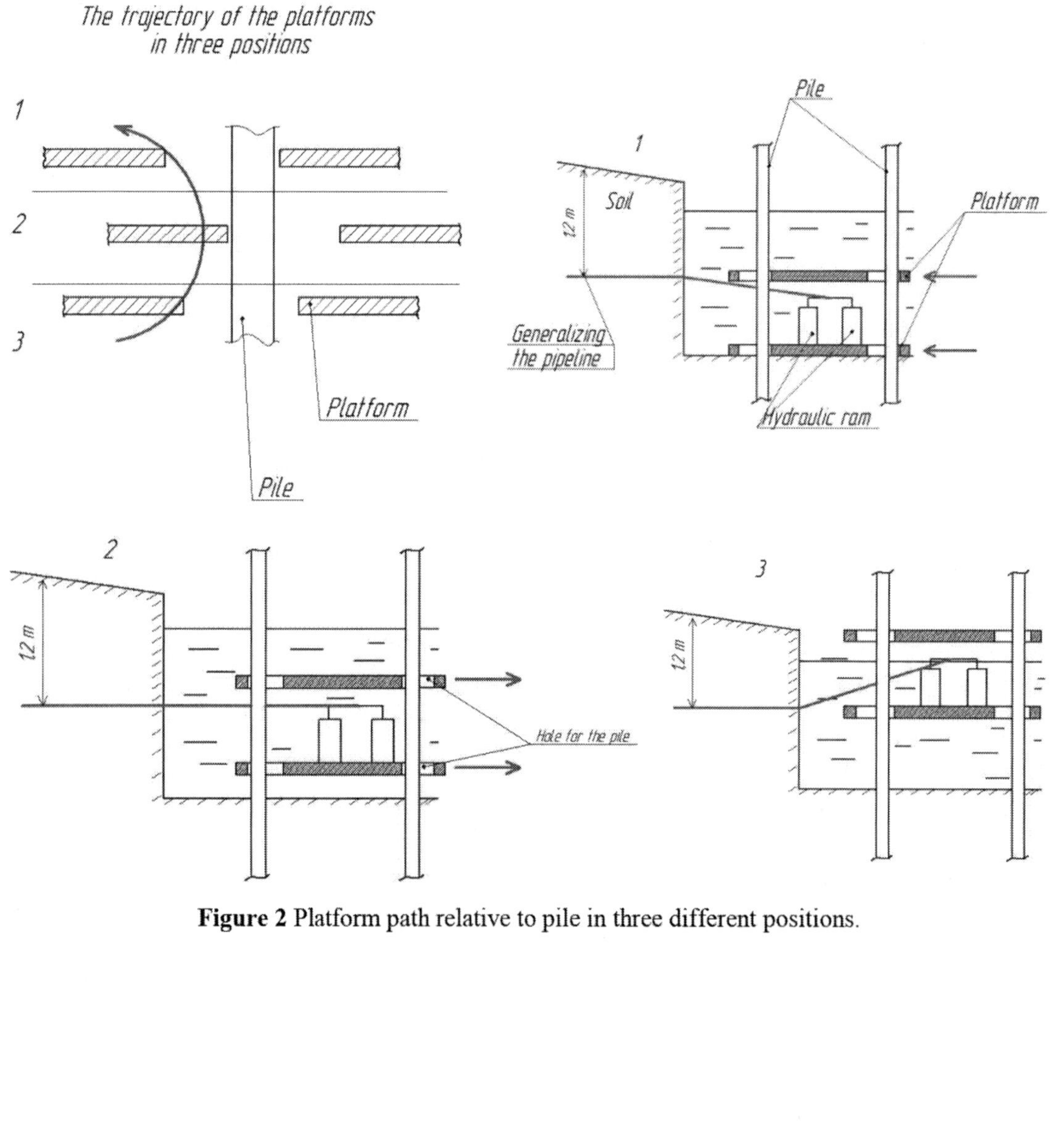

Figure 2 Platform path relative to pile in three different positions.

All plastic pipe connections will be provided with compression fittings. They are used for small diameter pipes (110 mm or less), i.e. where flange joints, butt welding joints are not economically feasible or practical, compression fittings (crimping fittings, collet crimping joint) are used. Collet crimping joints are designed for pressure up to 16 atm. In these joints resistance to mechanical loads is provided due to cutting of teeth of split plastic bushing into the pipe, and tightness - rubber sealing ring.

The main advantages of compression fittings are: easy installation; Repairs that do not require special skills and tools; Multiple assembly-disassembly (up to 10 times); Absence of metal elements in the joint and, consequently, high corrosion resistance, hygienicity and durability; Possibility of connection with elements of other materials (cranes, pipes, hoses).

Coarse filters in the form of mesh with a cell of 0.005 m to catch garbage and prevent ingress of fish and other hydrotaran into the pipeline should be installed directly on the ram in the form of a removable nozzle for periodic mechanical cleaning. Almost general (full) surface area of the mesh filter choose such that it exceeded the area of an entrance opening in the case (or pipe cross-sectional area) for 40%. In such a case, we will increase the diameter of the inlet section of the filter nozzle by 40%, and it will become 0.4 x 0.228 = 0.319 m.

So-called filters will need to be periodically maintained (cleaned), turning off one group of hydrotarans at the time, with the second group switched on. The filters must be cleaned in turn.

In order to clean the filter, as well as to carry out repairs (if absolutely necessary), each hydraulic ram must be disconnected from the system by means of a cover that will cover the inlet section of the supply pipe. At that, at disconnection the device will automatically turn off locally. At the same time, however, the pressure in the common (generalizing) pipeline will decrease by an amount (at the output of one ram) of about 21500, which is 0.215 at - this is very small and does not significantly affect the supply of water to the water tower [19,20].

Hydrotaran C 1 2465488 RU F04F 7/02 (2006.01) is offered for water supply to the thermal power plant. Underwater hydraulic ram Zagumennikova N Y, Sokolov VY , Naumov S A, Sadchikov A V, Gorichev SV, Lavrentiev A V, Korobkov A I - № 201112658/069; Zayav l. 16.06.2011. The hydraulic ram according to Figure 3 comprises a housing, a supply pipe including a hydraulic pipe with an impact valve on the rod, connected to the delivery pipe by means of a delivery valve and an air cap, and to the discharge pipes according to the invention, the supply pipe is made in the form of a confuser, according to the example of an inlet pipe. A (tapered) portion of the venturi flowmeter, the diameter of the inlet section D being four times the diameter of the section d, Which increases the flow rate of the liquid by 16 times, and the shape of the drain nozzles is conoidal; Having the highest capacity and flow rate, the speed required to operate the impact valve is achieved faster, Which increases the frequency and force of the hydraulic shock, thereby increasing the frequency of water supply to the consumer; In the housing on the hydraulic pipe at the point of maximum flow velocity there installed is a catchment ring communicating with it by means of holes uniformly located along the pipe section, the total area of which should be equal to the area of flow section of the hydraulic pipe, which reduces friction.

Fluid molecules against the walls of the hydraulic tube and provides a more complete use of the hydraulic impact energy, hence a greater supply of water to the consumer; Impact valve consists of a shut-off element made in the form of a drop on a rod and a drive mechanism, which includes two cylinders combined by a coupling, in one of which a permanent magnet of cone shape is built, and the rod contains rigidly fixed thrust and steel ring, which simplifies the valve mechanism and ensures its reliable and uninterrupted operation. Figure 3 shows an underwater hydraulic ram - general view.

The body 1 of the underwater hydraulic ram is immersed in the river so that the supply pipe 2 is at a depth from the surface of 150-200 mm free end to the water flow. Water through supply pipe 2, due to difference in diameters of its inlet D and outlet d sections, Accelerating in direction of arrow B goes to hydraulic pipe 3, With the impact valve 4 open, consisting of a drop-shaped locking element 5; Providing lower resistance to liquid flow, and drive mechanism 6, the main mass falling into drain pipes 7, Conoidal shapes for providing greater fluid flow and accelerating fluid flow; And partially into the catchment ring 8 through the drainage holes 9, Note here that total area of said holes should be equal to that of hydraulic tube 3.

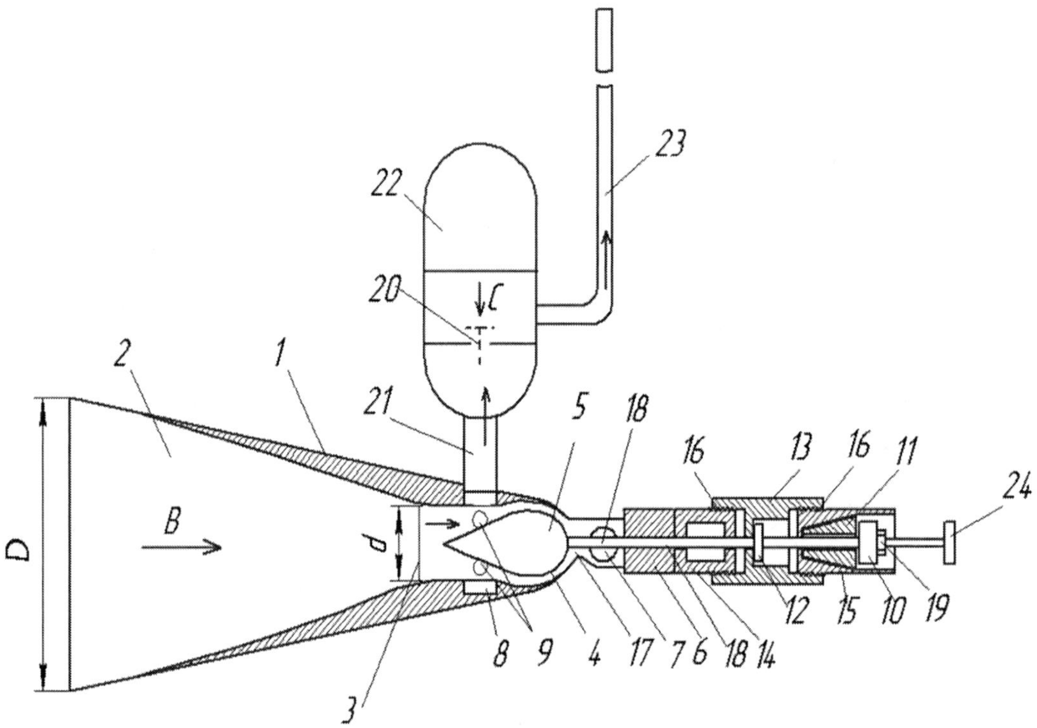

Figure 3. Underwater hydraulic ram – general view:1 – housing; 2 – supply pipe; 3 – hydraulic pipe; 4 – impact valve; 5 –shows a drop-shapedlocking member; 6 – drive mechanism; 7 – drain pipes; 8 – catchment ring; 9 – water passage holes; 10 – steel ring; 11 is a permanent magnet; 12 – thrust ring; 13 – coupling; 14, 15 – cylinders; 16 – threaded connection; 17 is a passage hole; 18 – rod; 19 – nut; 20 – pressure valve; 21 – delivery pipe; 22 – air cap; 23 – delivery pipe.

In the extreme open position of the impact valve 4, the steel ring 10 is as close as possible to the constant cone-shaped magnet 11, but their full connection is prevented by the thrust ring 12 which, in this position, abuts against the wall of the coupling 13 connecting the cylinders 14 and 15 by the threaded connection 16. This connection allows the frequency of operation of the impact valve to be adjusted.

Once the flow of water has gained the required speed, the shut-off member 5 will begin to move towards the passage hole 17 (blocking water access to the drain pipes 7) and with it the rod 18 with the rings 10 and 12 fixed on it. The steel ring 10 contained in the cylinder 15 and secured by the nut 19 will begin to break off from the magnet 11 by a force counteracting the magnetic attraction force to the extreme position at which the locking member 5 will completely cover the passage hole 17 - the closed position of the impact valve 4. At this point, a hydraulic shock will occur, accompanied by an increase in the pressure in the hydraulic pipe 3 due to the continuing inertia of the water flow. At this time, the pressure valve 20 opens, and water flows through the water-passing holes 9 from the water-collecting ring 8 through the pressure pipe 21 into the air cap 22, and from there, under air pressure in the air cap 22 through the pressure pipe 23 into the pressure tank to the consumer.

After the pressure in the hydraulic pipe 3 drops, when the hydraulic shock energy is consumed, the pressure valve 20 is closed in the direction of the arrow C from the pressure in the air cap 22. The impact valve 4 opens, the water starts to flow again through the drain pipes 7, quickly gaining speed in the supply pipe 2 to a value necessary for closing the passage hole 17 by the shut-off member 5 so that the cycle is repeated.

The offered work is carried out within the resolution of the government of the Orenburg region of 07.12.2006 No. 394-p "About the priority directions of development of science, technologies and the equipment and the list of critical technologies in the Orenburg region", is considered in the section critical technologies in the Paragraph 49 "Technologies New and Renewables".

4. Conclusion

At present, the Faculty of Electric Power Engineering at the Department of Electric and Thermal Power Engineering of "Orenburg State University" is developing design and technical documentation on the basis of the proposed patents, determining and calculating design-mode parameters of the presented structures.

The proposed developments will reduce the costs of heat and electric power generation for remote consumers, as well as save the consumer from the high costs of tracing networks both thermal and electric. The use of hydrotarane plants will eliminate electric pumps, which in turn will significantly reduce the cost of electricity and maintenance personnel.

As a result, the proposed equipment allows to solve problems of energy supply to remote consumers with minimal consumption of energy resources.

References

[1] Young B W 1996 Simplified Analysis and Design of the Hydraulic Ram Pump *Proc. of the Institution of Mechanical Engineers, Part A: J. of Power and Energy* **210(4)** 295-303

[2] Nambiar P, Shetty A, Thatte A, Lonkar S and Jokhi V 2015 Hydraulic ram pump: Maximizing efficiency *Int. Conf. on Technologies for Sustainable Development* (Mumbai: IEEE) 7095840

[3] Schiller E J and Kahangire P 1984 Analysis and computerized model of the automatic hydraulic ram pump *Canadian J. of Civil Eng.* **11(4)** 743-50

[4] Hussin S M, Gamil S A, Amin N A M, Safar M J A, Majid M S A, Kazim M M and Nasir N F M 2017 Design and analysis of hydraulic ram water pumping system *J. Phys.: Conf. Ser.* **908** 012052

[5] Jafri M 2019 Evaluation of field eligibility and efficiency of a single compression 3hydraulic-ramp-pump *SAINSTEK* **4(1)** 306-11

[6] Rennie LC and Bunt E A 1990 The Automatic Hydraulic Ram Experimental results *Proc. of the Institution of Mechanical Engineers, Part A: J. of Power and Energy* **204(1)** 23-31

[7] Sarma D, Das M, Brahma B, Pandwar D, Rongphar S and Rahman M 2016 Investigation and parameter optimization of a hydraulic ram pump using taguchi method *J. of The Institution of Engineers (India):Series C* **97** 551-9

[8] Sampath S S, Sawan S, Allan M P, Waleed J and Chithirai P S M 2015 Estimation of Power and Efficiency of Hydraulic Ram Pump with Re-circulation System *Int. J. of Computer-aided Mechanical Design and Implementation* **1(1)** 7-18

[9] Shuaibu M 2007 Design and Construction of a Hydraulic Ram Pump*Leonardo Electronic J. of Practices and Technologies* **6(11)** 59-70

[10] Matthias I, Suchard S, Johannes F J M, Johannes M and Wolfram S 2015 Hydraulic Ram Pump s for Irrigation in Northern Thailand *Agriculture and Agricultural Science Procedia* **5** 107-14

[11] Rohan DB, Swapnil PR, Pradeep SB and Manoday RR 2015 Designing of Hydraulic Ram Pump *Int. J. of Eng. And Computer Sci.* **4** 11966-71

[12] Zambrano J C I, Michavila J, Pinilla E A, Diehl J C and Ertsen M W 2019 Water Lifting Water: A Comprehensive Spatiotemporal Review on the Hydro-Powered Water Pumping Technologies *Water* **11(8)** 1677

[13] Watt SBA 1975 *Manual on the Hydraulic Ram for Pumping Water* (London: Intermedate Technology Publication Ltd)
[14] Glover P B M 1994 Computer Simulation and Analysis Method in the Development of the Hydraulic Ram Pump: PhD Thesis (Coventry: University of Warwick)
[15] Diwan P, Patel A and Sahu L 2016 Design and Fabrication of Hydraulic Ram with Methods of Improving Efficiency *Int. J. of Current Eng. and Sci. Research* **3(4)** 5-13
[16] Deo A, Pathak A, Khune S and Pawar MM 2016 Design Methodology for Hydraulic Ram Pump International *J. for Innovative Research in Science & Technology* **5(4)** 4737-45
[17] Tomboulides AG and Orszag SA 2000 Numerical investigation of transitional and weak turbulent flow past a sphere *J. of Fluid Mechanics* **416** 45-73
[18] Filipan V, Vireg Z and dan Bergant A 2003 Mathematical Modelling of a Hydraulic Ram Pump System *J. of Mechanical Eng.* **49(3)** 137-49
[19] Harith M N, Bakar R A, Ramasamy D and Quanjin M A 2017 A significant effect on flow analysis & simulation study of improve design hydraulic pump *IOP Conf. Ser.: Mater. Sci. Eng* **257** 012076
[20] Shende P B, Choudhary S K and Ninawe A P 2015 Analysis and Enhancement of Hydraulic Ram Pump Using Computational Fluid Dynamics (CFD) *Int. J. for Innovative Research in Science & Technology* **2(03)** 109-33

AUTHOR INDEX

Institute of Physics
Dirac House, Temple Back
Bristol BS1 6BE UK

ISSN: 1757-8981
ISBN 978-1-7138-4590-4

9 781713 845904

CURRAN ASSOCIATES INC.
proceedings.com